Phenolic Compounds in Food

Characterization and Analysis

Food Analysis & Properties

Series Editor

Leo M. L. Nollet

University College Ghent, Belgium

Phenolic Compounds in Food: Characterization and Analysis (2018)
Edited by Leo M.L. Nollet, Janet Alejandra Gutierrez-Uribe

Spectroscopic Methods in Food Analysis (2018)
Edited by Adriana S. Franca and Leo M.L. Nollet

Multiresidue Methods for the Analysis of Pesticide Residues in Food (2017)
Edited by Horacio Heinzen, Leo M.L. Nollet, and Amadeo R. Fernandez-Alba

Marine Microorganisms: Extraction and Analysis of Bioactive Compounds (2016)
Edited by Leo M. L. Nollet

Flow Injection Analysis of Food Additives (2015)
Edited by Claudia Ruiz-Capillas and Leo M. L. Nollet

Phenolic Compounds in Food

Characterization and Analysis

Edited by

Leo M. L. Nollet
Janet Alejandra Gutierrez-Uribe

CRC Press
Taylor & Francis Group
Boca Raton London New York

CRC Press is an imprint of the
Taylor & Francis Group, an **informa** business

CRC Press
Taylor & Francis Group
6000 Broken Sound Parkway NW, Suite 300
Boca Raton, FL 33487-2742

Printed on acid-free paper

International Standard Book Number-13: 978-1-4987-2296-4 (Hardback)

Library of Congress Cataloging-in-Publication Data

Names: Nollet, Leo M. L., 1948- editor.
Title: Phenolic compounds in food : characterization & analysis / [edited by] Leo M.L. Nollet.
Description: Boca Raton : CRC Press, Taylor & Francis Group, 2018. | Includes bibliographical references.
Identifiers: LCCN 2017048106 | ISBN 9781498722964 (hardback)
Subjects: LCSH: Phenols. | Food--Analysis.
Classification: LCC TP453.P45 P45 2018 | DDC 661/.82--dc23
LC record available at https://lccn.loc.gov/2017048106

Visit the Taylor & Francis Web site at
http://www.taylorandfrancis.com

and the CRC Press Web site at
http://www.crcpress.com

Contents

SECTION III DIFFERENT GROUPS OF PHENOLIC COMPOUNDS RELATED TO FOODS

SECTION IV ANTIOXIDANT POWER

SECTION V PHENOLIC COMPOUNDS IN DIFFERENT FOODSTUFFS

Series Preface

There will always be a need for analyzing methods of food compounds and properties. Current trends in analyzing methods include automation, increasing the speed of analyses, and miniaturization. The unit of detection has evolved over the years from micrograms to picograms.

A classical pathway of analysis is sampling, sample preparation, cleanup, derivatization, separation, and detection. At every step, researchers are working and developing new methodologies. A large number of papers are published every year on all facets of analysis. So, there is a need for books that gather information on one kind of analysis technique or on analysis methods of a specific group of food components.

The scope of the CRC Series on Food Analysis & Properties aims to present a range of books edited by distinguished scientists and researchers who have significant experience in scientific pursuits and critical analysis. This series is designed to provide state-of-the-art coverage on topics such as

1. Recent analysis techniques on a range of food components
2. Developments and evolution in analysis techniques related to food
3. Recent trends in analysis techniques of specific food components and/or a group of related food components
4. The understanding of physical, chemical, and functional properties of foods

The book *Phenolic Coumpounds in Foods: Characterization and Analysis* is the fifth volume in this series.

I am happy to be a series editor of such books for the following reasons:

- I am able to pass on my experience in editing high-quality books related to food.
- I get to know colleagues from all over the world more personally.
- I continue to learn about interesting developments in food analysis.

A lot of work is involved in the preparation of a book. I have been assisted and supported by a number of people, all of whom I would like to thank. I would especially like to thank the team at CRC Press/Taylor & Francis, with a special word of thanks to Steve Zollo, Senior Editor.

Many, many thanks to all the editors and authors of this volume and future volumes. I very much appreciate all their effort, time, and willingness to do a great job.

I dedicate this series to

- My wife, for her patience with me (and all the time I spend on my computer)
- All patients suffering from prostate cancer; knowing what this means, I am hoping they will have some relief

Dr. Leo M. L. Nollet (Retired)
University College Ghent
Ghent, Belgium

Preface

Natural phenolic compounds have received a lot of attention in the last few years since a great amount of them can be found in plants and the consumption of vegetables and beverages with a high level of such compounds may reduce risks of the development of several diseases. This is partially due to their antioxidant power since other interactions with cell functions have been discovered.

Phenolic compounds are one of the biggest and most widely distributed groups of secondary metabolites in plants. They play a role of protection against insects and other plant stress elicitors. They are involved in many functions in plants, such as sensorial properties, structure, pollination, resistance to pests and predators, germination, processes of seed, development, and reproduction.

Phenolic compounds can be classified in different ways, ranging from simple molecules to highly polymerized compounds.

This book deals with all aspects of phenolic compounds in food. This book has five sections with 21 chapters:

Section I: Phenolic Compounds
Section II: Analysis Methods
Section III: Different Groups of Phenolic Compound Related to Foods
Section IV: Antioxidant Power
Section V: Phenolic Compounds in Different Foodstuffs

In the chapters of Section I, the classification and occurrence of phenolic compounds in nature and foodstuffs is addressed.

Section II discusses all major aspects of analysis of phenolic compounds in foods: extraction, clean-up, separation, and detection.

In Section III, the reader finds out more information about and facts on specific analysis methods of a number of classes of phenolic compounds, from simple molecules to complex compounds.

The antioxidant power of phenolic compounds is detailed in Section IV.

In Section V, specific analysis methods in different foodstuffs are discussed.

It is a great pleasure to thank all the contributors of each chapter. They did an excellent job, and spent a lot of time and effort to deliver outstanding manuscripts.

Leo M. L. Nollet
Janet Alejandra Gutiérrez-Uribe

Nothing great was ever achieved without enthusiasm.

Ralph Waldo Emerson

I congratulate my co-editor, Janet, for her superb and persistent work on this project.

Leo M. L. Nollet

About the Editors

Leo M. L. Nollet, PhD, earned an MS (1973) and PhD (1978) in biology from the Katholieke Universiteit Leuven, Belgium. He is an editor and associate editor of numerous books. He edited for M. Dekker, New York—now CRC Press of Taylor & Francis Publishing Group—the first, second, and third editions of *Food Analysis by HPLC* and *Handbook of Food Analysis*. The last edition is a two-volume book. Dr. Nollet also edited the *Handbook of Water Analysis* (first, second, and third editions) and *Chromatographic Analysis of the Environment*, third and fourth editions (CRC Press). With F. Toldrá, he coedited two books published in 2006, 2007, and 2017: *Advanced Technologies for Meat Processing* (CRC Press) and *Advances in Food Diagnostics* (Blackwell Publishing—now Wiley). With M. Poschl, he coedited the book *Radionuclide Concentrations in Foods and the Environment*, also published in 2006 (CRC Press). Dr. Nollet has also coedited with Y. H. Hui and other colleagues on several books: *Handbook of Food Product Manufacturing* (Wiley, 2007), *Handbook of Food Science, Technology, and Engineering* (CRC Press, 2005), *Food Biochemistry and Food Processing* (first and second editions; Blackwell Publishing—now Wiley—2006 and 2012), and the *Handbook of Fruits and Vegetable Flavors* (Wiley, 2010). In addition, he edited the *Handbook of Meat, Poultry, and Seafood Quality*, first and second editions (Blackwell Publishing—now Wiley—2007 and 2012). From 2008 to 2011, he published with F. Toldrá five volumes on animal product–related books: *Handbook of Muscle Foods Analysis*, *Handbook of Processed Meats and Poultry Analysis*, *Handbook of Seafood and Seafood Products Analysis*, *Handbook of Dairy Foods Analysis*, and *Handbook of Analysis of Edible Animal By-Products*. Also in 2011, with F. Toldrá, he coedited two volumes for CRC Press: *Safety Analysis of Foods of Animal Origin* and *Sensory Analysis of Foods of Animal Origin*. In 2012, they published the *Handbook of Analysis of Active Compounds in Functional Foods*. In a coedition with Hamir Rathore, *Handbook of Pesticides: Methods of Pesticides Residues Analysis* was marketed in 2009; *Pesticides: Evaluation of Environmental Pollution* in 2012; *Biopesticides Handbook* in 2015; and *Green Pesticides Handbook: Essential Oils for Pest Control* in 2017. Other finished book projects include *Food Allergens: Analysis, Instrumentation, and Methods* (with A. van Hengel; CRC Press, 2011) and *Analysis of Endocrine Compounds in Food* (Wiley-Blackwell, 2011). Dr. Nollet's recent projects include *Proteomics in Foods* with F. Toldrá (Springer, 2013) and *Transformation Products of Emerging Contaminants in the Environment: Analysis, Processes, Occurrence, Effects, and Risks* with D. Lambropoulou (Wiley, 2014). In the series Food Analysis &

Properties, he edited (with C. Ruiz-Capillas) *Flow Injection Analysis of Food Additives* (CRC Press, 2015) and *Marine Microorganisms: Extraction and Analysis of Bioactive Compounds* (CRC Press, 2016). With A.S. Franca, he edited *Spectroscopic Methods in Food Analysis* (CRC Press, 2017), and with Horacio Heinzen and Amadeo R. Fernandez-Alba he edited *Multiresidue Methods for the Analysis of Pesticide Residues in Food* (CRC Press, 2017).

Janet Alejandra Gutiérrez-Uribe, PhD, is an associate professor in the NutriOmics research group at the School of Engineering and Science from Tecnológico de Monterrey. She is the director of the Bioengineering Department in the South Region of Tecnológico de Monterrey. Dr. Gutiérrez-Uribe is a food engineer with graduate studies in biotechnology. For more than 10 years, she has worked on phytochemistry and in the nutritional biochemistry of phenolic compounds and other nutraceuticals. Her research is focused on Mexican foods such as black beans, cacti, agave, and maize. She has published more than 60 papers in different prestigious journals and is the inventor of more than 10 patents and applications in Mexico and abroad. She has graduated more than 25 graduate students and her teaching skills go beyond lectures. Work with industry and social service are her main drivers in the development of challenges related to biochemistry, molecular biology, cell culture, and nutraceutical discovery.

List of Contributors

Emilio Alvarez-Parrilla
Universidad Autónoma de Ciudad Juárez
Departamento de Ciencias Químico Biológicas
Ciudad Juárez, Chihuahua, México

Begoña de Ancos
Institute of Food Science, Technology and Nutrition—ICTAN
Spanish National Research Council—CSIC
Madrid, Spain

Marilena Antunes-Ricardo
Tecnológico de Monterrey
Centro de Biotecnología-FEMSA
Escuela de Ingeniería y Ciencias
Monterrey, Nuevo León, México

M. Pilar Cano
Tecnológico de Monterrey
Centro de Biotecnología-FEMSA
Escuela de Ingeniería y Ciencias
Monterrey, Nuevo León, México

and

Instituto de Investigación en Ciencias de la Alimentación (CIAL)
Spanish National Research Council (CSIC)
Madrid, Spain

Rocio Campos-Vega
Programa de Posgrado en Alimentos del Centro de la República (PROPAC)
Research and Graduate Studies in Food
Science School of Chemistry
Universidad Autónoma de Querétaro
Qro, México

Anaberta Cardador-Martínez
Tecnológico de Monterrey
Escuela de Ingeniería y Ciencias
Campus Querétaro
San Pablo, Querétaro, México

Yanis Caro
Université de La Réunion
Ecole Supérieure d'Ingénieurs Réunion Océan Indien Dpt. *Agroalimentaire*
Ile de La Réunion, France

Janu Chandran
Agro Processing and Technology Division
CSIR–National Institute for Interdisciplinary Science and Technology
Trivandrum, Kerala, India

J. Abraham Domínguez–Avila
Coordinación de Tecnología de Alimentos de Origen Vegetal
Centro de Investigación en Alimentación y Desarrollo, A.C. (CIAD, AC)
Hermosillo, Sonora, México

and

Universidad Autónoma de Ciudad Juárez
Departamento de Ciencias Químico Biológicas
Ciudad Juárez, Chihuahua, México

Laura A. de la Rosa
Universidad Autónoma de Ciudad Juárez
Departamento de Ciencias Químico Biológicas
Ciudad Juárez, Chihuahua, México

Laurent Dufossé
Université de La Réunion
Ecole Supérieure d'Ingénieurs Réunion Océan Indien Dpt. *Agroalimentaire*
Ile de La Réunion, France

Alessandra Durazzo
Consiglio per la ricerca in agricoltura e l'analisi dell'economia agraria
Centro di ricerca CREA - Alimenti e Nutrizione
Rome, Italy

Zamantha Escobedo-Avellaneda
Tecnológico de Monterrey
Centro de Biotecnología-FEMSA
Escuela de Ingeniería y Ciencias
Monterrey, Nuevo León, México

María Teresa Espino Sevilla
Universidad de Guadalajara
Centro Universitario de la Ciénega
Ocotlán, Jal., México

Mireille Fouillaud
Université de La Réunion
Ecole Supérieure d'Ingénieurs Réunion Océan Indien Dpt. *Agroalimentaire*
Ile de La Réunion, France

José Alberto Gallegos-Infante
Instituto Tecnológico de Durango
Felipe Pescador
Durango, Dgo., México

Claudia Ivette Gamboa-Gómez
Unidad de Investigación Biomédica
Instituto Mexicano del Seguro Social
Durango, Durango, México

Angelita Gambuti
University of Naples Federico II
Department of Agricultural Sciences
Division of Grape and Wine Sciences
Avellino, Italy

Andrea Gómez-Maqueo
Tecnológico de Monterrey
Centro de Biotecnología-FEMSA
Escuela de Ingeniería y Ciencias
Monterrey, Nuevo León, México

Gustavo A. González Aguilar
Coordinación de Tecnología de Alimentos de Origen Vegetal
Centro de Investigación en Alimentación y Desarrollo, A.C. (CIAD, AC)
Hermosillo, Sonora, México

Rubén Francisco González-Laredo
Instituto Tecnológico de Durango
Felipe Pescador
Durango, Dgo., México

Isabelle Grondin
Université de la Réunion
Laboratoire de Chimie des Substances Naturelles et des Sciences des Aliments
Ile de La Réunion, France

Janet Alejandra Gutiérrez-Uribe
Tecnológico de Monterrey
Centro de Biotecnología-FEMSA
Escuela de Ingénieria y Ciencias
Monterrey, Nuevo León, México

Erick Heredia Olea
Tecnológico de Monterrey
Centro de Biotecnología-FEMSA
Escuela de Ingeniería y Ciencias
Monterrey, Nuevo León, México

Angélica María Hernández-Arriaga
Programa de Posgrado en Alimentos del Centro de la República (PROPAC)
Research and Graduate Studies in Food Science
School of Chemistry
Universidad Autónoma de Querétaro
Qro, México

Daniel A. Jacobo-Velázquez
Tecnológico de Monterrey
Centro de Biotecnología-FEMSA
Escuela de Ingeniería y Ciencias
Zapopan, Jalisco, México

Gerold Jerz
Technische Universität Braunschweig
Institute of Food Chemistry
Braunschweig, Germany

Cristian Jiménez Martínez
Departamento de Graduados e Investigación en Alimentos
Escuela Nacional de Ciencias Biológicas
Instituto Politécnico Nacional
México

Eugenia Lugo-Cervantes
Food Technology Departament
Centro de Investigación y Asistencia en Tecnología y Diseño del Estado De Jalisco
Jalisco, México

Luigi Moio
Department of Agricultural Sciences
University of Naples "Federico II"
Avellino, Italy

Martha Rocío Moreno-Jiménez
Instituto Tecnológico de Durango
Felipe Pescador
Durango, Dgo., México

Nayana N
Agro Processing and Technology Division
CSIR–National Institute for Interdisciplinary Science and Technology
Trivandrum, Kerala, India

P Nisha
Agro Processing and Technology Division
CSIR–National Institute for
Interdisciplinary Science and
Technology
Trivandrum, Kerala, India

Ramón Pacheco-Ordaz
Coordinación de Tecnología de Alimentos de Origen Vegetal
Centro de Investigación en Alimentación y Desarrollo, A.C. (CIAD, AC)
Hermosillo, Sonora, México

B. Dave Oomah, Retired
Formerly with the National Bioproducts and Bioprocesses Program
Pacific Agri-Food Research Centre
Agriculture and Agri-Food Canada
Summerland, BC, Canada

Maribel Ovando-Martínez
Coordinación de Tecnología de Alimentos de Origen Vegetal
Centro de Investigación en Alimentación y Desarrollo, A.C. (CIAD, AC)
and
Departamento de Investigaciones Científicas y Tecnológicas de la Universidad de Sonora
División de Ciencias Biológicas
Universidad de Sonora
Hermosillo, Sonora, México

Esther Pérez-Carrillo
Tecnológico de Monterrey
Centro de Biotecnología-FEMSA
Escuela de Ingeniería y Ciencias
Monterrey, Nuevo León, México

Ana Elena Quirós-Sauceda
Coordinación de Tecnología de Alimentos de Origen Vegetal
Centro de Investigación en Alimentación y Desarrollo, A. C. (CIAD, AC)
Hermosillo, Sonora, México

Griselda Rabadan-Chavez
Food Technology Departament
Centro de Investigación y Asistencia en Tecnología y Diseño del Estado De Jalisco
Jalisco, México

Maria del Refugio Ramos-Jerz
Technische Universität Braunschweig
Institute of Food Chemistry
Braunschweig, Germany

Nina del Rocío Martínez-Ruíz
Universidad Autónoma de Ciudad Juárez
Departamento de Ciencias Químico Biológicas
Ciudad Juárez, Chihuahua, México

Nuria Elizabeth Rocha-Guzmán
Instituto Tecnológico de Durango
Felipe Pescador
Durango, Dgo., México

Jacqueline Ruiz-Canizales
Coordinación de Tecnología de Alimentos de Origen Vegetal
Centro de Investigación en Alimentación y Desarrollo, A. C. (CIAD, AC)
Hermosillo, Sonora, México

Norma Julieta Salazar-López
Departamento de Investigación y Posgrado en Alimentos
Universidad de Sonora
Sonora, Mexico

Concepción Sánchez-Moreno
Institute of Food Science, Technology and Nutrition–ICTAN
Spanish National Research Council–CSIC
Madrid, Spain

Georgina Sandoval
Centro de Investigación y Asistencia en Tecnología y Diseño del Estado de Jalisco (CIATEJ)
Guadalajara, Jalisco, México

Jesús Santana-Gálvez
Tecnológico de Monterrey
Centro de Biotecnología-FEMSA
Escuela de Ingeniería y Ciencias
Monterrey, Nuevo León, México

Liliana Santos-Zea
Tecnológico de Monterrey
Centro de Biotecnología-FEMSA
Escuela de Ingeniería y Ciencias
Monterrey, Nuevo León, México

Kenia Vázquez-Sánchez
Programa de Posgrado en Alimentos del Centro de la República (PROPAC)
Research and Graduate Studies in Food Science
School of Chemistry
Universidad Autónoma de Querétaro
Qro, México

Mekala Venkatachalam
Université de La Réunion
Ecole Supérieure d'Ingénieurs Réunion Océan Indien Dpt. *Agroalimentaire*
Ile de La Réunion, France

Gustavo R. Velderrain-Rodríguez
Coordinación de Tecnología de Alimentos de Origen Vegetal
Centro de Investigación en Alimentación y Desarrollo, A. C. (CIAD, AC)
Socorro, México

Socorro Josefina Villanueva-Rodríguez
Centro de Investigación y Asistencia en Tecnología y Diseño del Estado de Jalisco (CIATEJ)
Jalisco, Mexico

Mónica A. Villegas-Ochoa
Coordinación de Tecnología de Alimentos de Origen Vegetal
Centro de Investigación en Alimentación y Desarrollo, A. C. (CIAD, AC)
Hermosillo, Sonora, México

Jorge Welti-Chanes
Tecnológico de Monterrey
Centro de Biotecnología-FEMSA
Escuela de Ingeniería y Ciencias
Monterrey, Nuevo León, México

Peter Winterhalter
Technische Universität Braunschweig
Institute of Food Chemistry
Braunschweig, Germany

SECTION I

Phenolic Compounds

CHAPTER 1

Classification of Phenolic Compounds

Jesús Santana-Gálvez and Daniel A. Jacobo-Velázquez

CONTENTS

1.1 INTRODUCTION

Phenolic compounds or phenolics are substances that possess an aromatic ring with at least one hydroxyl group. They are the most widely distributed secondary metabolites, and universally present in the plant kingdom (Cheynier et al. 2013). More than 8,000 different phenolics have been identified (Kabera et al. 2014). Therefore, there is the need to classify them.

There is no consensus regarding how phenolic compounds should be classified. Most classifications of phenolics are based on their chemical structure. In this sense, phenolic compounds can be classified in four different ways, starting from the most general to the most specific: (1) Flavonoids and non-flavonoids (Kabera et al. 2014); (2) number of aromatic rings (Kabera et al. 2014); (3) carbon skeleton, which depicts in a very basic way how the carbon atoms of the molecule are organized (e.g., C_6, C_6–C_1, C_6–C_3–C_6, etc.); and (4) basic chemical structure, which is a description or image that specifies common atoms (e.g., C, H, O), functional groups

TABLE 1.1 Classification of Phenolic Compounds by Number of Aromatic Rings, Carbon Skeleton, and Basic Chemical Structure

Number of Aromatic Rings	Carbon Skeleton	Group Name	Basic Chemical Structure
One	C_6	Simple phenols	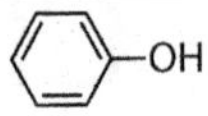
		Benzoquinones	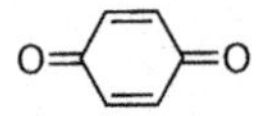
	C_6–C_1	Phenolic acids	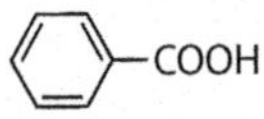
		Phenolic aldehydes	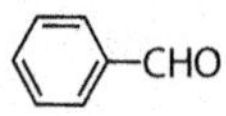
	C_6–C_2	Acetophenones	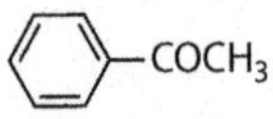
		Phenylacetic acids	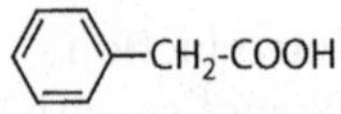
	C_6–C_3	Hydroxycinnamic acids	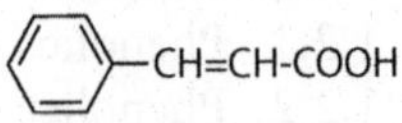
		Coumarins	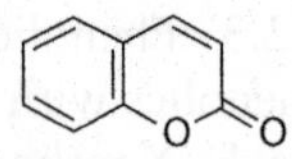
		Phenylpropenes	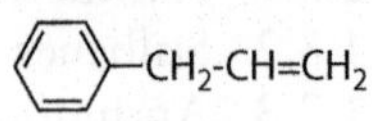
		Chromones	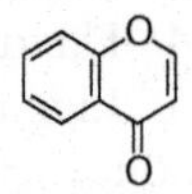
	C_6–C_4	Naphthoquinones	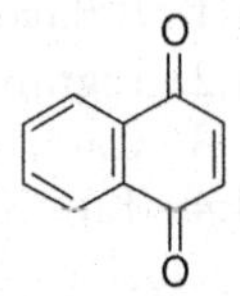
Two	C_6–C_1–C_6	Xanthones	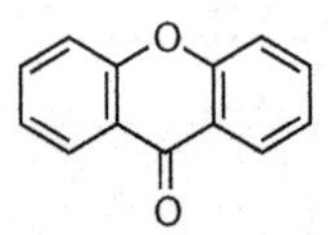
	C_6–C_2–C_6	Stilbenes	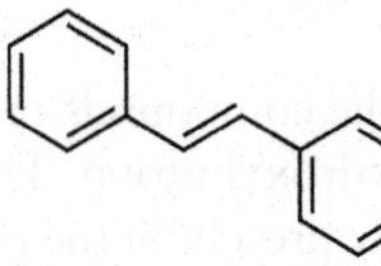
		Anthraquinones	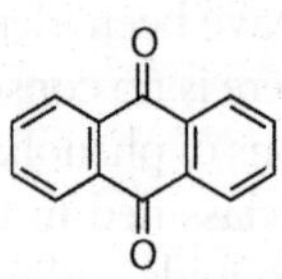

(Continued)

TABLE 1.1 (CONTINUED) Classification of Phenolic Compounds by Number of Aromatic Rings, Carbon Skeleton, and Basic Chemical Structure

Number of Aromatic Rings	Carbon Skeleton	Group Name	Basic Chemical Structure
	C_6–C_3–C_6	Flavonoids	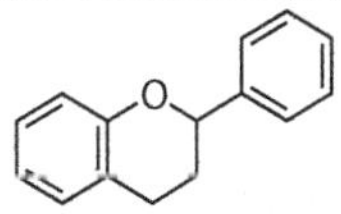
		Isoflavonoids	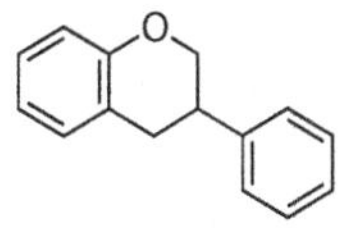
		Neoflavonoids	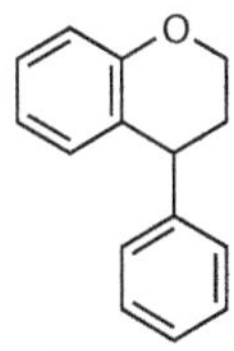
	$(C_6$–$C_3)_2$	Lignans	Multiple
		Neolignans	Multiple
Three or more	$(C_6)_n$	Catechol melanins	Multiple
	$(C_6$–$C_3)_n$	Lignins	Multiple
	$(C_6$–C_3–$C_6)_2$	Biflavonoids	Multiple
	$(C_6$–C_3–$C_6)_n$	Condensed tannins (Proanthocyanidins)	Multiple

(e.g., aromatic rings, hydroxyl groups, keto groups), saturations (e.g., single bonds, double bonds), types of bonds (e.g., covalent, ionic), and how are they all linked to one another.

From the methods of phenolic compounds classification described above, the last method (based on basic chemical structure), is the most widely used. In this regard, at least 20 different classes of phenolic compounds have been identified: Simple phenols, benzoquinones, phenolic acids, phenolic aldehydes, acetophenones, phenylacetic acids, hydroxycinnamic acids, coumarins, phenylpropenes, chromones, naphthoquinones, xanthones, stilbenes, anthraquinones, flavonoids, biflavonoids, lignans, lignins, catechol melanins, and tannins. However, method (4) is typically combined with method (3) (carbon skeleton) for a more organized classification (Bravo 1998; Cheynier et al. 2013; Pereira et al. 2009).

Other classification methods are based on the number of hydroxl groups (Kabera et al. 2014), distribution (shortly or widely distributed; Giada 2013), or solubility (soluble [free or soluble conjugated] or insoluble) Acosta-Estrada et al. 2014; Giada 2013; Sun et al. 2012).

The type of classification to be used will depend on what kind of information is desired. For instance, classifying phenolics as *soluble* or *insoluble* is useful from a nutritional point of view, since the metabolic fate of the compounds in the gastrointestinal tract and their physiological effects depend largely on their solubility characteristics (Giada 2013). For educational purposes, however, it is more illustrative to classify phenolics first by number of aromatic rings, followed by carbon skeleton, and finally by basic chemical structure, so that the student or researcher can become familiar with them more quickly than when compared with the other classification methods. Table 1.1 shows the classification of phenolic compounds using this method. Furthermore, a brief description of each phenolic group is included in this chapter.

1.2 PHENOLICS WITH ONE AROMATIC RING

1.2.1 Phenolics with C_6, C_6–C_1 and C_6–C_2 Carbon Skeletons

Simple phenols (C_6) are widely distributed in plants and include compounds such as thymol, resorcinol, and orcinol (Figure 1.1a). Phenolic acids (C_6–C_1) are derived from benzoic acid. The general formula and names of the main phenolic acids are shown in Figure 1.1b. Likewise, phenolic aldehydes (C_6–C_1) are derived from phenolic acids and include compounds such as vanillin, syringaldehyde, and *p*-hydroxybenzaldehyde (Figure 1.1c). Phenolic with C_6–C_2 carbon skeletons (acetophenones and phenylacetic acids) are less commonly described in literature.

1.2.2 Phenolics with C_6–C_3 Carbon Skeletons

Phenolic compounds with C_6–C_3 carbon skeletons, also known as phenylpropanoids, includes the hydroxycinnamic acids, coumarins, phenylpropenes, and chromones (Table 1.1). Among this group of phenolics, hydroxycinnamic acids and coumarins are the most studied, whereas little is known about phenylpropenes and chromones. The most important phenylpropanoids are hydroxycinnamic acids. These compounds are produced via the

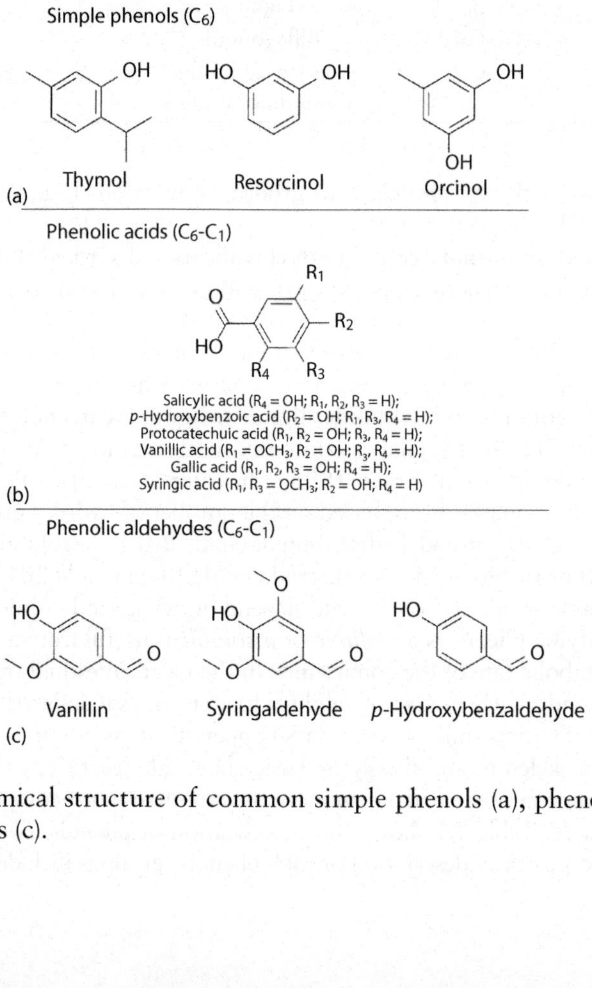

FIGURE 1.1 Chemical structure of common simple phenols (a), phenolic acids (b), and phenolic aldehydes (c).

phenylpropanoid metabolism, which starts by the conversion of L-phenylalanine into trans-cinnamic acid by the action of phenylanaine ammonia–lyase (PAL). The general chemical structure of hydroxycinnamic acids and the main compounds from this group are shown in Figure 1.2.

Hydroxycinnamic acids are commonly found in the form of esters, along with a cyclic alcohol–acid ester such as the quinic acid. The most abundant and studied hydroxicinnamic acid is chlorogenic acid and its derivatives (Giada 2013). Chlorogenic acid is an ester of caffeic acid with quinic acid (Figure 1.3). Important sources of chlorogenic acid include apples, artichoke, betel, burdock, carrots, coffee beans, eggplants, eucommia, grapes, honeysuckle, kiwi fruit, pears, plums, and potatoes, among others (Santana-Gálvez et al. 2017).

On the other hand, coumarins are considered phytoalexins, since they are over-accumulated in plant tissues as a result of pathogen attack or abiotic stresses. An example of a phytoalexin is the isocoumarin, which is accumulated in carrots due to wounding stress and is responsible for bitter flavor in fresh-cut produce such as carrot (Lafuente et al. 1996). The general structure and examples of some coumarins are shown in Figure 1.4.

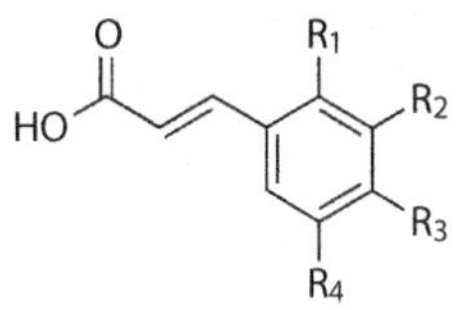

Cinnamic acid (R_1, R_2, R_3, R_4 = H);
p-Coumaric acid (R_3 = OH; R_1, R_2, R_4 = H);
Caffeic acid (R_2, R_3 = OH; R_1, R_4 = H);
Ferulic acid (R_2 = OCH_3; R_3 = OH; R_1, R_4 = H);
Sinapic acid (R_2, R_4 = OCH_3; R_3 = OH; R_1 = H).

FIGURE 1.2 General chemical structure of hydroxycinnamic acids and main compounds from this group.

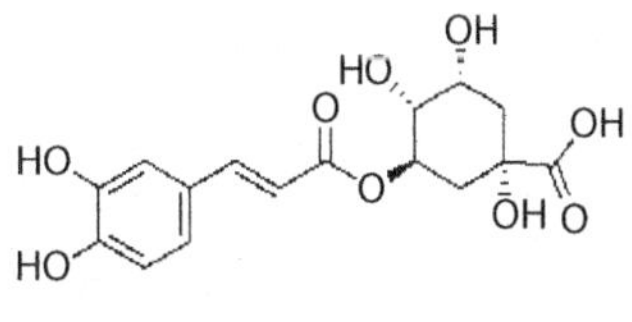

FIGURE 1.3 Chemical structure of chlorogenic acid.

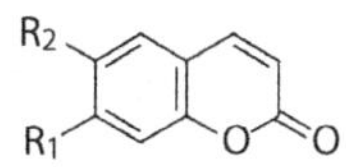

Coumarin (R_1, R_2 = H);
Umbelliferone (R_1 = OH; R_2 = H);
Aesculetin (R_1, R_2 = OH);
Scopoletin (R_1 = OH; R_2 = OCH_3)

FIGURE 1.4 General chemical structure of coumarins.

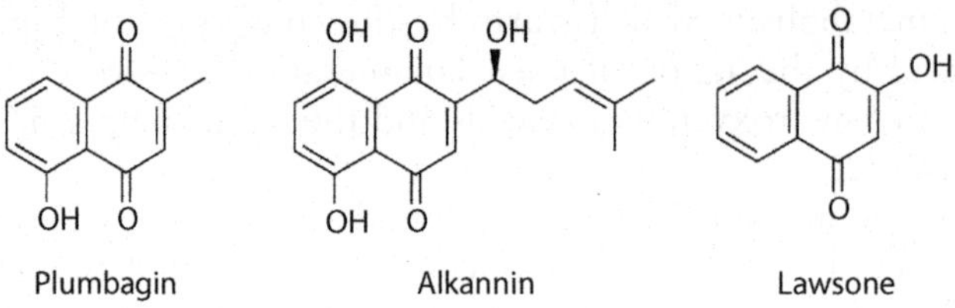

FIGURE 1.5 Chemical structure of common naphtoquinones.

1.2.3 Phenolics with C_6–C_4 Carbon Skeletons (Naphtoquinones)

Naphtoquinones are phenolic compounds widely distributed in higher plants, fungi, and bacteria. Examples of naphtoquinones include the plumbagin, lawsone, and alkannin (Figure 1.5). These naphtoquinones are biosynthesized via different pathways, including the acetate and mevalonate pathway (plumbagin), shikimate/succinyl CoA combined pathway (plumbagin), and shikimate/mevalonate pathway (alkannin), respectively. The pharmacological activities of naphtoquinones have been attributed to their redox potential (Goulart et al. 1997), and they have shown activity against malaria (Likhitwitayawuid et al. 1998), cancer (Itoigawa et al. 2001), and inflammation (Pinho et al. 2011).

1.3 PHENOLICS WITH TWO AROMATIC RINGS

1.3.1 Xanthones

Xanthones are secondary metabolites present in higher plants, fungi, and lichens (Peres et al. 2000). They have a C_6–C_1–C_6 carbon skeleton, which consists of two aromatic rings bonded to an O-heterocycle. The aromatic ring at the right side of the chemical structure is called the *A ring*, while the one at the left is called the *B ring* (Figure 1.6).

The A ring is derived from an acetic acid pathway, while the B ring is from the shikimic acid pathway (El-Seedi et al. 2010). The chemical structure of xanthones is considered "privileged," as it can interact with diverse drug targets (Lesch and Bräse 2003). Isoprene, methoxyl, and hydroxyl groups at various locations on the A and B rings results in a diverse array of xanthone compounds (Gutierrez-Orozco and Failla 2013). Therefore, xanthones have been classified into five major groups: Simple oxygenated xanthones, xanthone glycosides, prenylated and related xanthones, xanthonolignoids, and miscellaneous xanthones (Peres et al. 2000). The prenylated derivatives are the most abundant group of xanthones (Paiva et al. 2012). Xanthones can be found in fruits such as mangosteen (Gutierrez-Orozco and Failla 2013). An example of a xanthone is depicted in Figure 1.7.

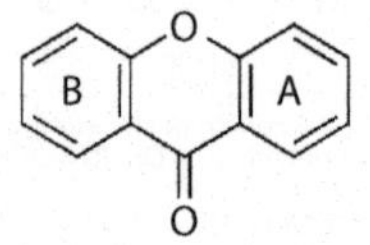

FIGURE 1.6 Basic chemical structure of xanthones.

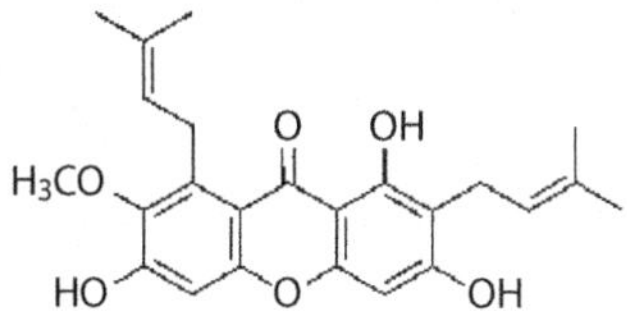

FIGURE 1.7 Chemical structure of the xanthone α-mangostin.

1.3.2 Stilbenes

Stilbenes are compounds with a C_6–C_2–C_6 carbon skeleton that comprises two aromatic rings joined by an ethylene bridge. A large diversity of stilbenes results from this structure: Monomers that vary in the number and position of hydroxyl groups; substitution with sugars, methyl, methoxy, and other groups; steric configuration (*cis-trans* isomerism); and oligomers (dimers, trimers, tetramers, etc.) that result from the condensation of stilbene monomers (Roat and Saraf 2015). Stilbenes can be found in a wide range of dietary sources, including red wine, grapes, peanuts, legumes, berries (blueberries, bilberries, cranberries, mulberries), plum fruits, and other sources (Kasiotis et al. 2013). Their biosynthesis results from the activity of stilbene synthase, which catalyzes the condensation of one molecule of *p*-coumaroyl-CoA obtained from the shikimic acid pathway, and three molecules of malonyl-CoA obtained from acetyl-CoA (Roat and Saraf 2015). The most studied stilbene is *trans*-resveratrol, whose structure is shown in Figure 1.8.

1.3.3 Anthraquinones

Anthraquinones constitute the largest group of natural pigments, with about 700 compounds described so far (Duval et al. 2016). About 200 of these compounds come from flowering plants, while the rest are produced by fungi and lichens (Seigler 1998). They can be found in all parts of plants: Roots, rhizomes, fruits, flowers, and leaves (Duval et al. 2016). Similarly to stilbenes, the chemical structure of anthraquinones consists of a C_6–C_2–C_6 carbon skeleton, but instead of the two aromatic rings connected by an ethylene bridge, they are connected by a ring formed by two keto groups. Anthraquinones are especially common in the families of Fabaceae (*Cassia*), Liliaceae (*Aloe*), Polygonaceae (*Rheum*, *Rumex*), Rhamnaceae (*Rhamnus*), Rubiaceae (*Asperula*, *Coelospermum*, *Coprosma*, *Galium*, *Morinda*, and *Rubia*), and Scrophulariaceae (*Digitalis*) (Seigler 1998). The two main biosynthetic pathways leading to anthraquinones in higher plants are the polyketide pathway and the chorismate/*o*-succinylbenzoic acid pathway (Han et al. 2002). The anthraquinones most frequently reported are emodin, physcion, catenarin, and rhein (Figure 1.9; Duval et al. 2016).

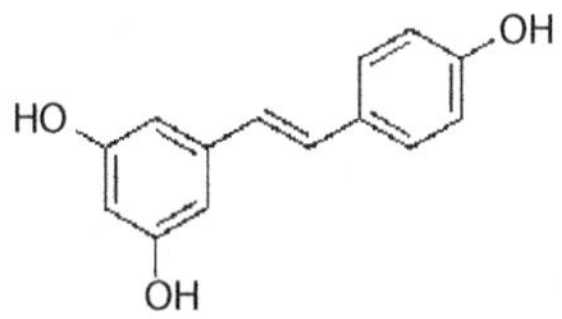

FIGURE 1.8 Chemical structure of the stilbene *trans*-resveratrol.

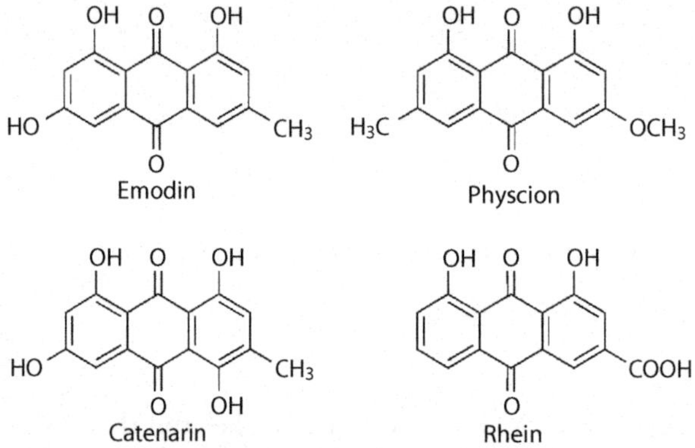

FIGURE 1.9 Chemical structures of common anthraquinones.

1.3.4 Flavonoids

Flavonoids are the largest group of phenolic compounds in nature. Over 8,000 plant phenolics have been identified, and more than 5,000 of them are flavonoids (Harborne 1993; Kabera et al. 2014). Flavonoids are universal within the plant kingdom; they are the most common pigments, next to chlorophyll and carotenoids (Stalikas 2007). The term *flavonoid* is generally used to describe phenolics with a chemical structure consisting of a C_6–C_3–C_6 carbon skeleton; more specifically, an aromatic ring attached to a benzopyran moiety (Marais et al. 2006). The aromatic ring in the benzopyran moiety is called the *A ring*, the aromatic ring attached to the benzopyran moiety is called the *B ring*, and the O-heterocyclic ring is called the *C ring* (Figure 1.10). Depending on the position of the B ring, flavonoids are classified into three major groups: Flavonoids (2-phenylbenzopyrans), isoflavonoids (3-phenylbenzopyrans), and neoflavonoids (4-phenylbenzopyrans) (Figure 1.10; Marais et al. 2006).

The flavonoids (2-phenylbenzopyrans) are further divided into eight groups based on the degree of oxidation and saturation of the C ring: Flavanones, flavones, flavonols, dihydroflavonols, flavan-3-ols, flavan-4-ols, flavan-3,4-diols (leucoanthocyanidins), and anthocyanidins (Figure 1.11). Furthermore, the isoflavonoids are divided into 11 groups: Isoflavans, isoflavones, isoflavanones, isoflav-3-enes, isoflavanols, rotenoids, coumestanes, 3-arylcoumarins, coumaronochromenes, coumaronochromones, and pterocarpans (Figure 1.12). Moreover, the neoflavonoids are divided into three groups: 4-arylcoumarins, 3,4-dihydro-4-arylcoumarins, and neoflavenes (Figure 1.13). Finally, other natural compounds that contain a C_6–C_3–C_6 backbone are considered to be minor flavonoids, and

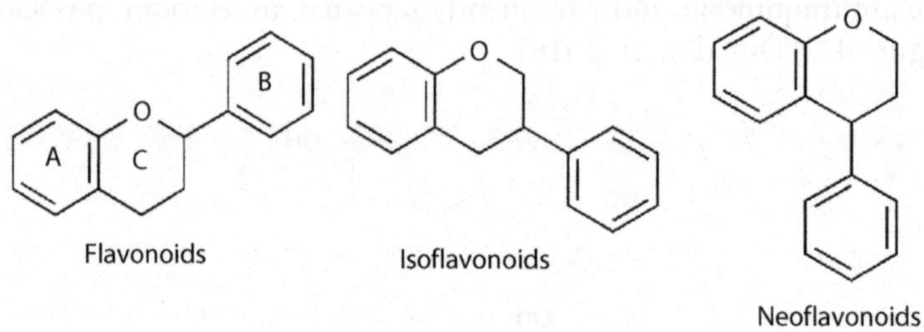

FIGURE 1.10 Basic chemical structures of flavonoids, isoflavonoids, and neoflavonoids.

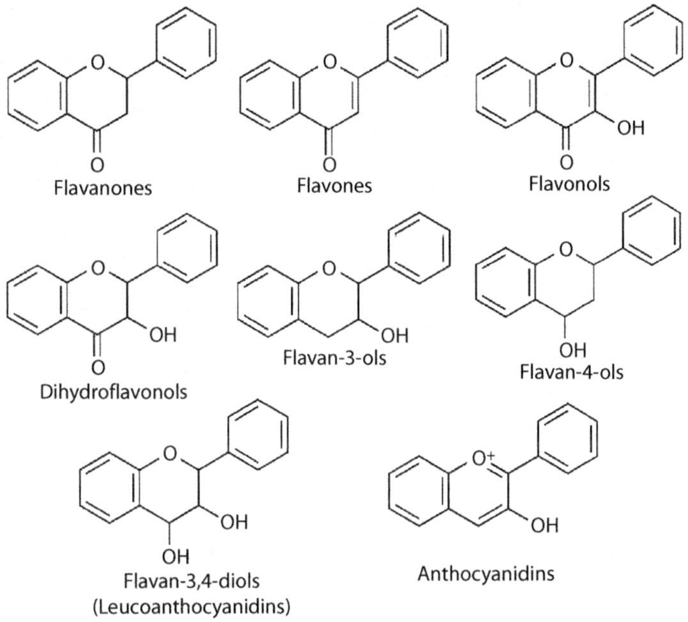

FIGURE 1.11 Basic chemical structures of the main classes of flavonoids.

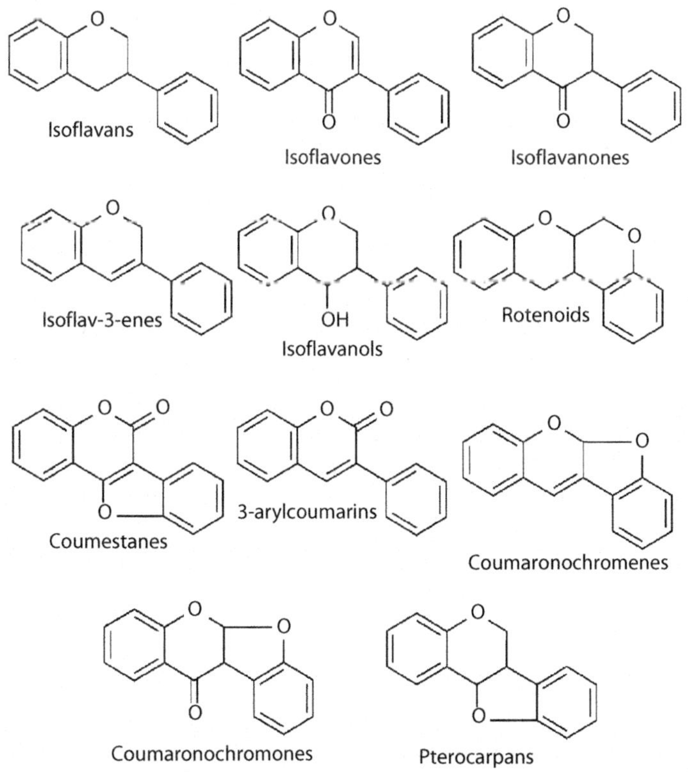

FIGURE 1.12 Basic chemical structures of the main classes of isoflavonoids.

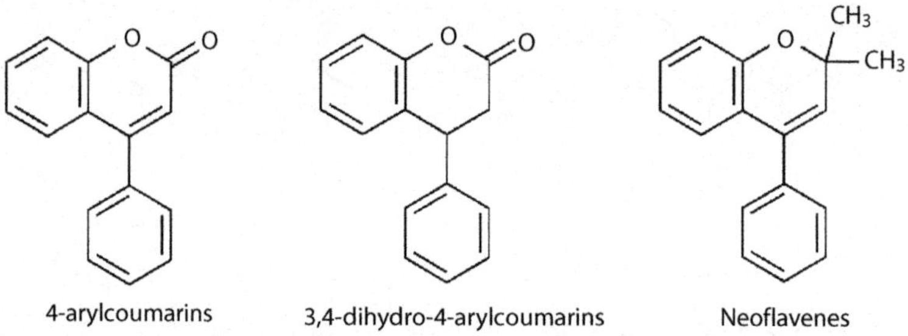

FIGURE 1.13 Basic chemical structures of the main classes of neoflavonoids.

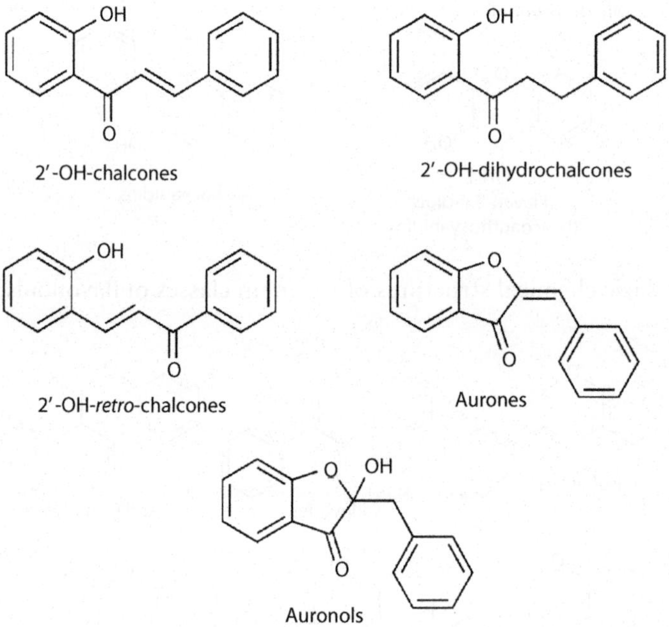

FIGURE 1.14 Basic chemical structures of the main classes of minor flavonoids.

are divided into five groups: 2'-hydroxychalcones, 2'-OH-dihydrochalcones, 2'-OH-retro-chalcone, aurones, and auronols (Figure 1.14; De Rijke et al. 2006; Marais et al. 2006).

Similarly to stilbenes, flavonoids are the result of the condensation between one molecule of *p*-coumaroyl-CoA and three molecules of malonyl-CoA. However, the condensation is catalyzed by a different enzyme, known as chalcone synthase, which gives rise to chalcones. Chalcones are then processed by many different enzymes to produce the great variety of flavonoids (Davies et al. 2006).

1.3.5 Lignans and Neolignans

Lignans and neolignans are two major groups of compounds with a carbon skeleton of $(C_6–C_3)_2$, which is the result of the chemical union of two phenylpropanes. For nomenclature purposes, the carbons of each $C_6–C_3$ unit are numbered from 1 to 6 in the aromatic ring, starting with the propyl group, and from 7 to 9 in the propyl group, starting with the benzene ring. To distinguish between the C_6–C_3 units, the numbers are primed in one of them. When both C_6–C_3 units are covalently linked between the carbons at positions 8 and 8′, the compound is known as *lignan*; in the absence of such a bond, the compound is termed *neolignan*. Furthermore, compounds in which the C_6–C_3 units are joined by an oxygen atom are classified under neolignans and are called *oxyneolignans*.

Lignans are further classified into eight groups: Arylnaphthalenes, aryltetralins, dibenzocyclooctadienes, dibenzylbutanes, dibenzylbutyrolactols, dibenzylbutyrolactones, furans, and furofurans (Figure 1.15; Teponno et al. 2016). On the other hand, at least 15 subtypes of neolignans have been identified, which include benzofurans, dihydrobenzofurans, diarylethanes, benzodioxins, and alkylarylethers, among others (Ríos et al. 2002; Teponno et al. 2016). Other kinds of lignans include three minor groups: Oligomeric lignans and neolignans, hybrid lignans, and norlignans. The oligomeric lignans and neolignans are oligomeric phenylpropanoids. Hybrid lignans refer to a single C_6–C_3 unit bound to other kinds of phenolics or non-phenolic compounds, yielding flavanolignans, isoflavanolignans, xanthonolignans, coumarinolignans, stilbenolignans, sesqui-, di-, and triterpenelignans. Finally, norlignans have a C_{16} to C_{17} core structure and are usually found to co-occur with lignans or neolignans (Ríos et al. 2002; Teponno et al. 2016).

Lignans and neolignans are widely spread within the plant kingdom and are derived from the shikimic acid pathway (Teponno et al. 2016). They are present in nuts, oilseeds, cereals, breads, legumes, fruits, vegetables, soy products, meat products, coffee, tea, red and white wines, and lager beer (Landete 2012). Together with isoflavonoids, lignans constitute one of the main groups of phytoestrogens (López-Biedma et al. 2016).

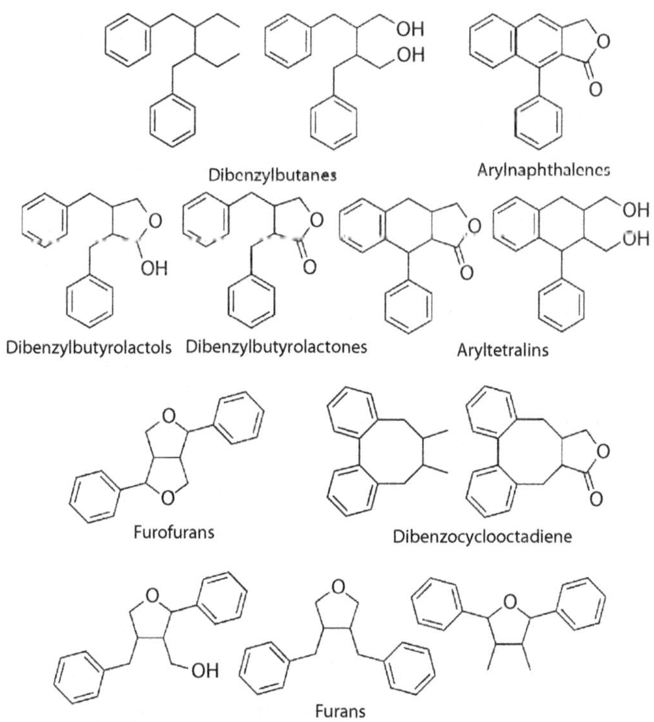

FIGURE 1.15 Basic chemical structures of the main classes of lignans.

1.4 PHENOLICS WITH THREE OR MORE AROMATIC RINGS

1.4.1 Biflavonoids

As the name suggests, biflavonoids are the result of the covalent union between two flavonoids. Common biflavonoids include dimers of flavone–flavone, flavanone–flavone, flavanone–flavanone, or flavone–flavonol subunits (Bagla et al. 2014; Jamila et al. 2014). The flavonoid dimers are connected either by C–C or C–O–C bonds (Sasaki et al. 2015). Unlike their monomeric constituents, the distribution of biflavonoids is restricted to some species, highlighting the *Ginkgo biloba* and species from the genera Garcinia and Selaginella (Carrillo-Hormaza et al. 2016). Examples of biflavonoids joined by C-C or C-O-C bonds are shown in Figure 1.16.

1.4.2 Lignins

Lignins are described as natural aromatic polymers of 4-hydroxyphenylpropanoids [$(C_6$—$C_3)_n$] with their units connected by C–O–C or C–C linkages (Bunzel and Ralph 2006). They are a component of the plant cell wall, and second to cellulose as the most abundant organic substance in plants (Moura et al. 2010). Together with polysaccharides and oligosaccharides, lignins form part of dietary fiber (Bunzel and Ralph 2006). The primary monomers that are used for the biosynthesis of lignins are three *p*-hydroxycinnamyl alcohols, also known as monolignols: *p*-coumaryl, coniferyl and sinapyl alcohols (Figure 1.17; Ralph et al. 2004). Other monomers that may be incorporated into lignins are 5-hydroxyconiferyl

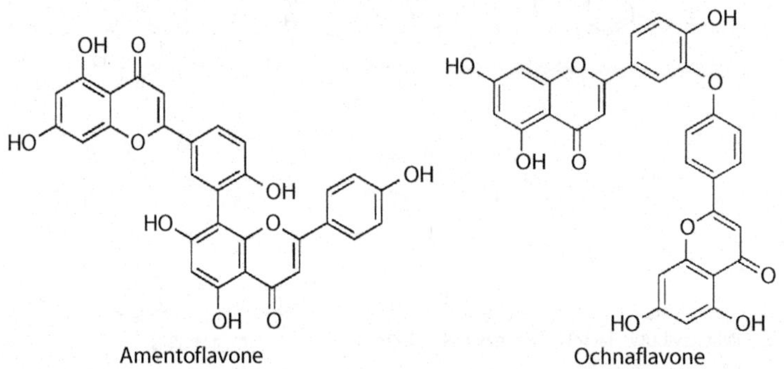

FIGURE 1.16 Chemical structures of the biflavonoids amentoflavone and ochnaflavone.

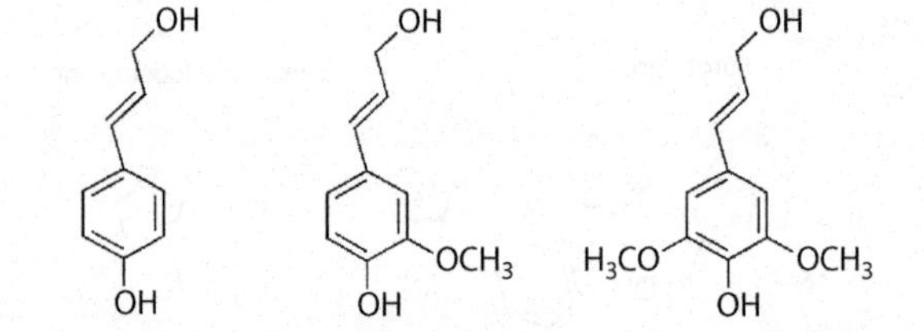

FIGURE 1.17 Primary monomers used for the biosynthesis of lignins.

alcohol, hydroxycinnamaldehydes, hydroxycinnamyl *p*-hydroxybenzoates, and hydroxycinnamyl *p*-coumarates (Bunzel et al. 2005). Food sources of lignins include cereal grains, brans, fruits, and vegetables (Bunzel and Ralph 2006).

1.4.3 Catechol Melanins

From the Greek word *melanos* meaning "dark," melanins are a group of complex pigments of high molecular weight that are usually dark (Lattanzio 2013; Solano et al. 2014). Structurally speaking, though, there is no consensus about what melanins exactly are, since they are widely diverse in composition, size, functions, occurrence, and color (black, brown, yellow, or red) (d'Ischia et al. 2013; Lattanzio 2013; Solano et al. 2014). However, melanins can be defined as heterogeneous polymers derived by the oxidation and polymerization of phenolic compounds and their resulting quinones (d'Ischia et al. 2013; Lattanzio 2013; Solano et al. 2014).

Melanins are present in all life kingdoms and therefore, some authors classify them into five groups according to their origin: Animal melanin, plant melanin, fungal melanin, bacterial melanin, and synthetic melanin (Solano et al. 2014). Other authors classify them into three groups based on their chemistry: Eumelanins, pheomelanins, and allomelanins. Eumelanins and pheomelanins are found mainly in animals, whereas allomelanins are found in plants (Lattanzio 2013). Since most food sources covered in this book are from plant origin, our discussion will focus on allomelanins.

Most allomelanins are synthesized using the phenolic compound catechol as a precursor; therefore, they are also known as catechol melanins [$(C_6)_n$] (Lattanzio 2013; Solano et al. 2014). Other precursors include caffeic, chlorogenic, protocatechuic, or gallic acids (Solano et al. 2014). Catechol melanins are synthesized by the activity of catechol oxidase, the most frequent type of polyphenol oxidase in plants (Solano et al. 2014). Examples of dietary sources of catechol melanins include black tea, black soybeans, black sesame seeds, and grapes (Hsieh and Lien 2012; Kamei et al. 1997; Novikov et al. 2001).

1.4.4 Tannins

Tannins are phenolic compounds of high molecular weight (500–20,000 Da) synthesized via the shikimic acid pathway (Giada 2013; Kabera et al. 2014; Khanbabaee and van Ree 2001). They are classified into two major groups: (1) Hydrolyzable tannins (HTs), and (2) non-hydrolyzable tannins, also known as condensed tannins or proanthocyanidins (PAs). HTs have a center of glucose or a polyhydric alcohol partially or completely esterified with phenolic groups, such as gallic acid (gallotannins) or ellagic acid (ellagitannins) (Giada 2013; Kumari and Jain 2012). However, there are exceptions; for instance, vescalagin is a non-hydrolyzable ellagitannin (Khanbabaee and van Ree 2001). On the other hand, PAs are polymers of flavonoids [$(C_6–C_3–C_6)_n$], specifically flavan-3-ols like (+)-catechin and (−)-epicatechin, or flavan-3,4-diols (leucoanthocyanidins), or a mixture of the two (Giada 2013; Kumari and Jain 2012). Unlike PAs, HTs are readily hydrolyzed with hot water, acids, bases, or enzymes such as tannases, yielding carbohydrates and phenolic acids (Giada 2013; Khanbabaee and van Ree 2001; Kumari and Jain 2012). Examples of HTs and PAs are depicted in Figure 1.18.

Tannins are largely found in the plant kingdom, where PAs are more widespread than HTs (de Jesus et al. 2012; Kumari and Jain 2012). Some dietary sources of HTs include

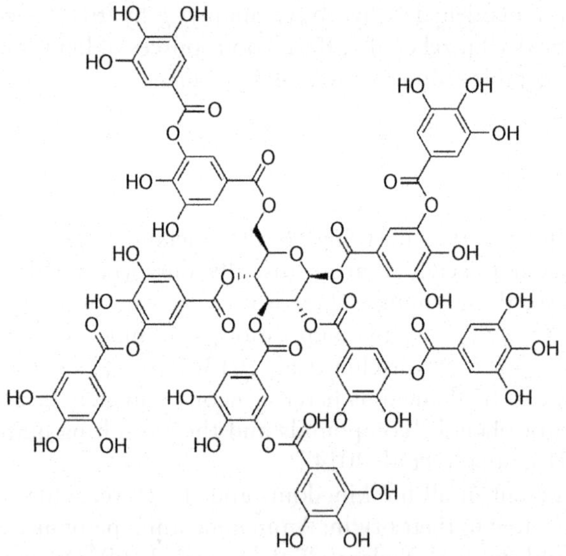

Tannic acid

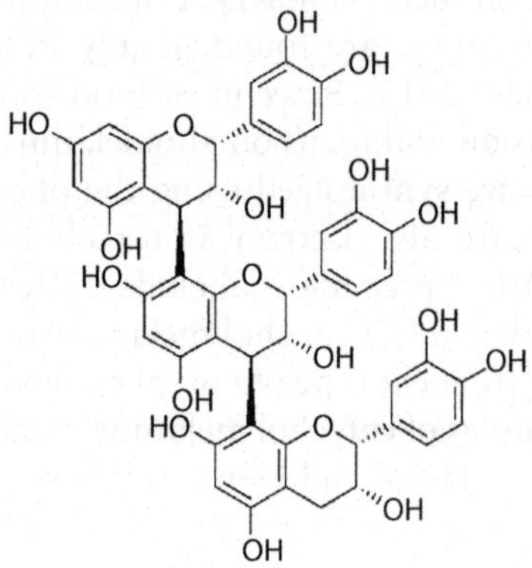

Procyanidin C1

FIGURE 1.18 Chemical structures of the hydrolysable tannin tannic acid and the proanthocyanidin procyanidin C1.

pomegranate, strawberries, raspberries, barley, rice, and oat, while sources containing PAs include coffee, tea, wine, grapes, cranberries, strawberries, blueberries, apples, apricots, peaches, dry fruits, mint, basil, and barley, among others (Kumari and Jain 2012). Tannins are responsible for the astringent taste of many fruits and vegetables (de Jesus et al. 2012).

Other kinds of tannins can be found in nature, such as complex tannins and phlorotannins. Complex tannins have a catechin unit bound glycosidically to a gallotannin or an ellagitannin unit, and, unlike HTs, they are only partially hydrolyzable due to the C–C bond of their catechin unit with the glycosidic moiety (Khanbabaee and van Ree 2001). Phlorotannins are a special kind of tannins that have been found only in brown seaweeds and are not commonly consumed by humans (Giada 2013).

1.5 CONCLUSIONS

Herein we presented a method to classify phenolic compounds, which included the number of aromatic rings, followed by carbon skeletons, and finally by basic chemical structures. Since the scientific relevance of phenolic compounds has increased in the last two decades, it is mandatory to establish a standard procedure for their classification. The updated classification method should include other aspects of phenolics such as their bioavailability and the metabolites produced in the gastrointestinal track after human consumption.

ACKNOWLEDGMENT

This study is based upon research supported by research funds from Tecnológico de Monterrey (Bioprocess and Synthetic Biology Research Group). Author J.S.-G. also acknowledges the scholarship (291137) from CONACYT. The authors would like to thank Mariana Elizondo for technical support on the drawings of the chemical structures shown in the chapter.

REFERENCES

Acosta-Estrada, B.A., Gutiérrez-Uribe, J.A., and Serna-Saldívar, S.O. 2014. Bound phenolics in foods, a review. *Food Chemistry*, 152:46–55.

Bagla, V.P., McGaw, L.J., Elgorashi, E.E., and Eloff, J.N. 2014. Antimicrobial activity, toxicity and selectivity index of two biflavonoids and a flavone isolated from *Podocarpus henkelii* (Podocarpaceae) leaves. *BMC Complementary and Alternative Medicine*, 14:383–388.

Bravo, L. 1998. Polyphenols: Chemistry, dietary sources, metabolism, and nutritional significance. *Nutrition Reviews*, 56:317–333.

Bunzel, M., and Ralph, J. 2006. NMR characterization of lignins isolated from fruit and vegetable insoluble dietary fiber. *Journal of Agricultural and Food Chemistry*, 54:8352–8361.

Bunzel, M., Seiler, A., and Steinhart, H. 2005. Characterization of dietary fiber lignins from fruits and vegetables using the DFRC method. *Journal of Agricultural and Food Chemistry*, 53:9553–9559.

Carrillo-Hormaza, L., Ramírez, A.M., Quintero-Ortiz, C. et al. 2016. Comprehensive characterization and antioxidant activities of the main biflavonoids of *Garcinia madruno*: A novel tropical species for developing functional products. *Journal of Functional Foods*, 27:503–516.

Cheynier, V., Comte, G., Davies, K.M., Lattanzio, V., and Martens, S. 2013. Plant phenolics: Recent advances on their biosynthesis, genetics, and ecophysiology. *Plant Physiology and Biochemistry*, 72:1–20.

Davies, K., Schwinn, K., Gould, K., and Lister, C. 2006. Molecular biology and biotechnology of flavonoid biosynthesis, flavonoid functions in plants. In Ø.M. Andersen, and K.R. Markham (eds.), *Flavonoids: Chemistry, Biochemistry and Applications* (pp. 143–180, 397–425). USA: CRC Press.

De Jesus, N.Z.T., Falcão, H.S., Gomes, I.F. et al. 2012. Tannins, peptic ulcers and related mechanisms. *International Journal of Molecular Sciences*, 13:3203–3228.
De Rijke, E., Out, P., Niessen, W.M.A., Ariese, F., Gooijer, C., and Brinkman, U.A.T. 2006. Analytical separation and detection methods for flavonoids. *Journal of Chromatography A.*, 1112:31–63.
d'Ischia, M., Wakamatsu, K., Napolitano, A. et al. 2013. Melanins and melanogenesis: Methods, standards, protocols. *Pigment Cell & Melanoma Research*, 26:616–633.
Duval, J., Pecher, V., Poujol, M., and Lesellier, E. 2016. Research advances for the extraction, analysis and uses of anthraquinones: A review. *Industrial Crops and Products*, 94:812–833.
El-Seedi, H.R., El-Barbary, M.A., El-Ghorab, D.M.H. et al. 2010. Recent insights into the biosynthesis and biological activities of natural xanthones. *Current Medicinal Chemistry*, 17:854–901.
Giada, M.L.R. 2013. Food phenolic compounds: Main classes, sources and their antioxidant power. In J.A. Morales-González (ed.), *Oxidative Stress and Chronic Degenerative Diseases: A Role for Antioxidants* (pp. 87–106). INTECH.
Goulart, M.O.F., Zani, C.L., Tonholo, J., Freitas et al. 1997. Trypanocidal activity and redox potential of heterocyclic- and 2-hydroxy-naphthoquinones. *Bioorganic and Medicinal Chemistry Letters*, 7:2043–2048.
Gutierrez-Orozco, F., and Failla, M.L. 2013. Biological activities and bioavailability of mangosteen xanthones: A critical review of the current evidence. *Nutrients*, 5: 3163–3183.
Han, Y.-S., van der Heijden, R., Lefeber, A.W.M., Erkelens, C., and Verpoorte, R. 2002. Biosynthesis of anthraquinones in cell cultures of *Cinchona* 'Robusta' proceeds via the methylerythritol 4-phosphate pathway. *Phytochemistry*, 59:45–55.
Harborne, J.B. 1993. *The Flavonoids: Advances in Research since 1986*. London: Chapman and Hall.
Hsieh, P.-H., and Lien, T.-F. 2012. Study of the physico-chemical properties and antioxidant activity of extracted melanins. *Journal of Agricultural Science*, 4:217–229.
Itoigawa, M., Ito, C., Tan, H.T.-W., Okuda, M. et al. 2001. Cancer chemopreventive activity of naphthoquinones and their analogs from *Avicennia* plants. *Cancer Letters*, 174:135–139.
Jamila, N., Khairuddean, M., Khan, S.N., and Khan, N. 2014. Complete NMR assignments of bioactive rotameric (3→8) biflavonoids from the bark of *Garcinia hombroniana*. *Magnetic Resonance in Chemistry*, 52:345–352.
Kabera, J.N., Semana, E., Mussa, A.R., and He, X. 2014. Plant secondary metabolites: Biosynthesis, classification, function and pharmacological properties. *Journal of Pharmacy and Pharmacology*, 2:377–392.
Kamei, H., Koide, T., Kojima, T., Hasegawa, M., and Umeda, T. 1997. Suppression of growth of cultured malignant cells by allomelanins, plant-produced melanins. *Cancer Biotherapy and Radiopharmaceuticals*, 12:47–49.
Kasiotis, K.M., Pratsinis, H., Kletsas, D., and Haroutounian, S.A. 2013. Resveratrol and related stilbenes: Their anti-aging and anti-angiogenic properties. *Food and Chemical Toxicology*, 61:112–120.
Khanbabaee, K., and van Ree, T. 2001. Tannins: Classification and definition. *Natural Product Reports*, 18:641–649.
Kumari, M., and Jain, S. 2012. Tannins: An antinutrient with positive effect to manage diabetes. *Research Journal of Recent Sciences*, 1:1–8.

Lafuente, M.T., López-Gálvez, G., Cantwell, M., and Yang, S.F. 1996. Factors influencing ethylene-induced isocoumarin formation and increased respiration in carrots. *Journal of the American Society for Horticultural Science*, 121:537–542.

Landete, J.M. 2012. Plant and mammalian lignans: A review of source, intake, metabolism, intestinal bacteria and health. *Food Research International*, 46:410–424.

Lattanzio, V. 2013. Phenolic compounds: Introduction. In K.G. Ramawat, J.-M. Mérillon (Eds.), *Natural products: Phytochemistry, Botany and Metabolism of Alkaloids, Phenolics and Terpenes* (pp. 1543–1580). Berlin Heidelberg: Springer.

Lesch, B., and Bräse, S. 2003. A short, atom-economical entry to tetrahydroxanthenones. *Angewandte Chemie International Edition*, 43:115–118.

Likhitwitayawuid, K., Kaewamatawong, R., Ruangrungsi, N., and Krungkrai, J. 1998. Antimalarial Naphthoquinones from *Nepenthes thorelii*. *Planta Medica*, 64: 237–241.

López-Biedma, A., Sánchez-Quesada, C., Delgado-Rodríguez, M., and Gaforio, J.J. 2016. The biological activities of natural lignans from olives and virgin olive oils: A review. *Journal of Functional Foods*, 26:36–47.

Marais, J.P.J., Deavours, B., Dixon, R.A., and Ferreira, D. 2006. The stereochemistry of flavonoids. In E. Grotewold (ed.), *The Science of Flavonoids* (pp. 1–4). New York: Springer Science + Business Media, Inc.

Moura, J.C.M.S., Bonine, C.A.V., Viana, J.O.F., Dornelas, M.C., and Mazzafera, P. 2010. Abiotic and biotic stresses and changes in the lignin content and composition in plants. *Journal of Integrative Plant Biology*, 52:360–376.

Novikov, D.A., Kurchenko, V.P., and Azarko, I.I. 2001. Photoprotective properties of melanins from grape (*Vitis vinifera*) and black tea (*Thea sinensis*). *Radiatsionnaia Biologiia, Radioecologiia*, 41:664–670.

Paiva, A.M., Sousa, M.E., Camões, A., Nascimento, M.S.J., and Pinto, M.M.M. 2012. Prenylated xanthones: Antiproliferative effects and enhancement of the growth inhibitory action of 4-hydroxytamoxifen in estrogen receptor-positive breast cancer cell line. *Medicinal Chemistry Research*, 21:552–558.

Pereira, D.M., Valentão, P., Pereira, J.A., and Andrade, P.B. 2009. Phenolics: From chemistry to biology. *Molecules*, 14:2202–2211.

Peres, V., Nagem, T.J., and de Oliveira, F.F. 2000. Tetraoxygenated naturally occurring xanthones. *Phytochemistry*, 55:683–710.

Pinho, B.R., Sousa, C., Valentao, P., and Andrade, P.B. 2011. Is nitric oxide decrease observed with naphthoquinones in LPS stimulated RAW 264.7 macrophages a beneficial property? *PLoS ONE*, 6(8):e24098.

Ralph, J., Lundquist, K., Brunow, G. et al. 2004. Lignins: Natural polymers from oxidative coupling of 4-hydroxyphenylpropanoids. *Phytochemistry Reviews*, 3: 29–60.

Ríos, J.L., Giner, R.M., and Prieto, J.M. 2002. New findings on the bioactivity of lignans. In A.U. Rahman (ed.), *Studies in Natural Products Chemistry, volume 26, Bioactive Natural Products (Part G)* (pp. 189). Amsterdam: Elsevier Science B. V.

Roat, C., and Saraf, M. 2015. Stilbenes: Chemistry and pharmacological properties. *Journal of Applied Pharmaceutical Research*, 3:1–7.

Santana-Gálvez, J., Cisneros-Zevallos, L., and Jacobo-Velázquez, D. 2017. Chlorogenic acid: Recent advances on its dual role as a food additive and a nutraceutical against metabolic syndrome. *Molecules*, 22(3):358.

Sasaki, H., Kitoh, Y., Tsukada, M. et al. 2015. Inhibitory activities of biflavonoids against amyloid-β peptide 42 cytotoxicity in PC-12 cells. *Bioorganic and Medicinal Chemistry Letters*, 25:2831–2833.

Seigler, D.S. 1998. Benzoquinones, naphthoquinones, and anthraquinones. In *Plant secondary metabolism* (pp. 85). Norwell, MA: Kluwer Academic Publishers.

Solano, F. 2014. Melanins: Skin pigments and much more—Types, structural models, biological functions, and formation routes. *New Journal of Science*, 2014:498276.

Stalikas, C.D. 2007. Extraction, separation, and detection methods for phenolic acids and flavonoids. *Journal of Separation Science*, 30:3268–3295.

Sun, L., Zhang, H., and Zhuang, Y. 2012. Preparation of free, soluble conjugate, and insoluble-bound phenolic compounds from peels of rambutan (*nephelium lappaceum*) and evaluation of antioxidant activities *in vitro*. *Journal of Food Science*, 77, C198–C204.

Teponno, R.B., Kusari, S., and Spiteller, M. 2016. Recent advances in research on lignans and neolignans. *Natural Product Reports*, 33:1044–1092.

CHAPTER 2

Phenolic Compounds in Nature

Rubén Francisco González-Laredo, Nuria Elizabeth Rocha-Guzmán, José Alberto Gallegos-Infante, Martha Rocío Moreno-Jiménez, and Claudia Ivette Gamboa-Gómez

CONTENTS

2.1 INTRODUCTION

In this chapter, a historical and conceptual review on natural polyphenols is given. The general and classical concepts as well as recent contributions to the fascinating field of chemistry of natural products are presented concisely. Fifty years ago, phenolics were more known or perceived for their effects rather than for the science behind them and their potential biological activities that would be revealed later. The development of modern analytical tools would be a key factor in allowing the emergence of the world of antioxidant and polyphenol science.

Phenolic compounds are secondary metabolites ubiquitously present in nature, particularly in the plant kingdom. Nominally these compounds have one or more hydroxyl groups (OH) bonded to an aromatic ring, being phenol the simpler example. Phenolics are the most investigated group of secondary metabolites so far. They are a large class of chemical compounds that show a diversity of structures, from rather simple ones like phenolic acids to polyphenols such as flavonoids that comprise several groups, including polymeric compounds based on these different classes. Unlike other phytochemicals such as alkaloids, the investigation of phenolics is more structured and predictable independently if the field of study is ecology, food science, or agronomy. Tea, coffee, and chocolate (cocoa) are the more common natural edible sources of beneficial polyphenols.

As hydroxyl groups holders, phenolics have the capacity to form hydrogen bonds and at the same time show slight acidity, forming the corresponding negative and nucleophilic phenoxide ions. The implication of this ionization property enhances its solubility in water and its chemical stability. Additionally, the presence of aromatic rings assists the delocalization of electrons or moving the negative charge along the ring system. The phenoxy ions may further release another electron to form the corresponding radical, which may undergo oxidative coupling reactions. These properties allow the biosynthesis of phenolic compounds and the interactions among plants and their environments. Most phenolic compounds have their origin in the shikimic acid pathways and from the acetyl coenzyme A and its activated form (Waterman and Mole 1994).

2.2 MILESTONES IN POLYPHENOL HISTORY AND RESEARCH

Phenol was discovered in 1834 by Friedlieb Ferdinand Runge, who extracted it in an impure form from coal tars. Later, in 1836, Auguste Laurent used the name *phène* for benzene; from there, the terms *phenol* and *phenyl* emerged. In 1843, French chemist Charles Gerhardt created the name *phenol*, referring to this compound. However, it wasn't until 1894 that the first known use of the word *polyphenol* appeared, according to the International Scientific Vocabulary.

Sir Joseph Lister (1827–1912) reported his pioneering surgery techniques and use of the antiseptic properties of phenol in surgery in 1867.

In the early 1900s, Folin and colleagues carried out intensive work on protein determinations (Folin and Ciocalteu 1927). The major publication dealt with variants of the Folin–Denis reagent applied to the determination of tyrosine and tryptophane in proteins. Eventually this would be known widely as the Folin–Ciocalteu phenol reagent.

Lowry et al. (1951) proclaimed that using the Folin–Ciocalteu phenol reagent for the measurement of proteins, that is, for the reaction with copper in alkali conditions and then reading the change in color was simpler and more sensitive than other methods. The Lowry paper became the most cited publication ever. Although more precise methods have been introduced since then, it still holds a rank in Thomson Reuters Web of Science, with more than 300,000 citations (Van Noorden et al. 2014).

Meanwhile, Bate-Smith (1953) developed chromatographic methods for isolation and the identification of leuco-anthocyanins as precursors to the anthocyanidins formed when boiling with mineral acid. This method would be improved and applied to the leaves and other tissues of numerous species of plants.

Bate-Smith (1954) proposed the concept of astringency to explain the conversion of plant tissues that became unpalatable by precipitating the salivary proteins when reacting with tannins. This biological property was important, as tannins were effective in deterring animal or microbial predators by immobilizing enzymes and impeding the invasion of the host plant tissues by the parasites.

In 1957, at the University of Cambridge, E.C. Bate-Smith and Tony Swain cofounded the Plant Phenolics Group, a precursor to the Phytochemical Society of Europe, which they founded in 1977 along with the renowned flavonoid specialist Jeffrey B. Harborne.

Singleton and Rossi (1965) improved the method using the Folin–Ciocalteu reagent instead of the Folin–Denis. They proposed gallic acid as a phenolic equivalent standard, obtaining more reproducibility in the time–temperature color development, which was similar to the tannin values obtained from previous methods. Their results were less variable and avoided interference from nonphenolic components. Eventually this procedure would be adapted and used for the determination of phenolic content in a plant or food sample (Singleton et al. 1999).

2.3 POLYPHENOL DEFINITIONS

Polyphenols are intrinsically related to tannins; consequently, the concepts of *vegetable tannins* and *plant polyphenols* have sometimes resulted in confusion. It can be said that all vegetable tannins are polyphenolics, but cannot be affirmed that the contrary is necessarily true. Tannins refer strictly to plant polyphenolic materials, having molecular masses between 500 Da and 3000 Da and large numbers of phenolic groups to form hydrogen-bonded cross-linked structures with collagen molecules, but also to precipitate alkaloids, gelatin, and other proteins from solution. In this traditional definition it was remarked the collagen-specific tanning action of the plant polyphenols. Particularly, the reaction of tannins with collagen was known as *tanning*. For these properties the history of polyphenols is strongly related to the tanning process and particularly to the development of the leather industry. Typical natural sources of *vegetable tannins* for industrial use included oak galls, quebracho and acacia heartwood extracts, and oak barks.

Many common low molecular weight *plant polyphenols* such as catechins and gallic acids are blue- or black-colored complexes when reacting with iron salts and also when oxidized with permanganate (Swain and Bate-Smith 1962). However, they are not able to cross-link with collagen, which would be distinctive of a tannic action, although they may be adsorbed by animal skin and also precipitate gelatin, a hydrolytically and thermally denatured form of collagen. Many of these polyphenols have promoted historical research attention not just because they are abundant plant secondary metabolites, but also as compounds that present many properties with diverse potential applications in varied human interest fields such as agriculture, ecology, medicine, food science, and nutrition. Polyphenols' capacity to form complexes with other biomolecules explains the roles they can play as secondary metabolites in plants chemical defense, in their distinguishing effects in herbal medicines, in plant beverages and foodstuffs, in floral copigmentation, and in the continued making of leather.

Later, Haslam and Cai (1994) proposed his definition of polyphenols as descriptors for water-soluble plant phenolic compounds having molecular masses from 500 Da to 4000 Da and holding 12–16 phenolic hydroxy groups on 5–7 aromatic rings per 1000 Da of relative molecular mass. Additionally, the compounds should experience typical phenolic reactions and have the ability to precipitate some alkaloids, gelatin, and other proteins from solution.

The basic meaning of a *phenolic compound* would include both the arene ring and its hydroxyl substituents. Consequently, the term *polyphenol* should be restricted to structures holding at least two phenolic moieties, regardless of the number of hydroxyl groups they bear, even considering in the definition polyphenolic compounds with no tanning action. However, according to Harborne (1989), such a rigorous polyphenol definition would need additional constraints, since many natural products from diverse biosynthetic origins contain more than one phenolic unit. This may be the case of some alkaloids derived from tyrosine amino acid.

There are simpler phenols such as catechol, resorcinol, pyrogallol, and phloroglucinol (di- and trihydroxylated substituted benzenes) that hold more than one hydroxy group on a benzene ring or in another joint arene ring, but this does not convert them in polyphenols, as the IUPAC (1997) still recognizes them as phenols.

A reconsidered definition by Quideau et al. (2011) about "true" plant polyphenols states that the term *polyphenol* should be used to define "plant secondary metabolites derived exclusively from the shikimate derived phenylpropanoid and/or the polyketide pathways, featuring more than one phenolic ring and being devoid of any nitrogen-based functional group in their most basic structural expression."

A modern distinction for polyphenols should consider their assumed antioxidative property. This capacity enhances their scavenging or stabilizing action against free radical species, as those derived from lipids and nucleic acids, which has been underlined as the essential factors behind degenerative and age-related diseases. However, it is known that being an antioxidant is not an exclusive property of polyphenols, as many simple plant phenols are also strong antioxidants. That is why another topic of confusion may be using the term *plant phenols* by industry instead of *polyphenols* as preferred for commercial and scientific communications.

There are three classes of "true" polyphenols:

1. Proanthocyanidins or condensed tannins, including procyanidins, prodelphinidins, and profisetinidins. They are formed from the oligomerization of flavan-3-ols units such as catechins, epigallocatechins, and fisetinidol.
2. Gallotannins and ellagitannins, known as hydrolyzable tannins, result from the metabolism of gallic acid with glucose.
3. Phlorotannins, found in red-brown algae, are essentially derived from the coupling of phloroglucinol monomers.

However, there are other groups of more or less complex plant phenolics, also called *tannins* without any firm evidence of their tanning action. Theatannins include the flavanols from green tea (epicatechin gallate and epigallocatechin gallate), which give place by oxidative couplings to the tropolone containing dimeric theaflavins and complex oligo and polymeric thearubigins in black tea (Tanaka et al. 2005). Caffetannins include hydroxycinnamic acids of varied hydroxy and methoxylation pattern. Usually these monophenolic carboxylic acids are esterified to polyols such as caffeic and caffeoylquinic acids from coffee beans, and known generically also as chlorogenic acids (Iwai et al. 2004). Labiataetannins are oligomeric structures of the dimeric rosmarinic acid formed from phenolic oxidative coupling reactions of caffeic acid, which are almost exclusive of the Lamiaceae family (Okuda et al. 1992). Hamamelitannins are composed of two galloyl units coupled to the unusual sugar hamamelose and are reported in significant amounts in the bark of *Hamamelis virginiana* (Hartisch and Kolodziej 1996).

2.4 FROM ANTI-NUTRITIOUS TO ANTIOXIDANT ROLES

The common chemical and biosynthetic background of plants makes them recognizable after key evolution patterns that are present today. This is the case of some phenolics or vegetal polyphenols that may show up and act distinctively as allelochemicals and phytoalexins in nature. Both phytochemical types are produced by plants as defense mechanisms. Allelochemicals are usually compounds produced and targeted toward other plants to avoid competition, while phytoalexins are chemicals of general response toward insect or microbial attacks or external threats such as drought and fire. Special attention on the allelochemicals' mechanisms of action, the energy cost of xenobiotic metabolism, and the adaptation process to the defense traits have centered in many phytochemicals and most phenolics classes. In all cases, the point is trying to respond to the question of why plants chose to depend on the production of metabolites with diverse phenolic functionality (Harborne 1988).

Plant phenolics have been relegated to the selected diets of many animal species, from elephants to ants, from placental mammals to ruminants. When humans consume plant-derived foods, they may respond in different ways to phenolics such as eugenol from banana, vanillin from vanilla, naringin from citrus, or dihydrochalcone from *Lindera lucida*. Reactions to closely related compounds could be repellent or attractive, bitter or sweeter. The sensorial response is complemented with nutritional and physiological consequences. Since Feeny (1976) claimed that tannins were allelochemicals acting in defense by reducing digestion of nutrients in herbivores, many models were set and proven. These studies have been conducted also to evaluate potential synergistic interactions of polyphenols with other phytochemicals from the diet.

While digestibility of complex high molecular weight tannins is very low, smaller, simpler phenolics may be absorbed into the body and disrupt other physiological processes. However, this absorption is still limited, though hydrolysis and lower polarity may make them more compatible when interacting across lipid membranes. Toxicity of free phenolics is another issue to consider, although few phenolics are toxic up to acute levels. This is the case of the cycled polyketide hypericin and phenolic furocoumarins (Waterman and Mole 1994).

2.5 POLYPHENOL–PROTEIN INTERACTIONS

It is well known that polyphenols have a high affinity for binding proteins. Many polyphenol properties could be explained by mechanisms of interaction with proteins. This complex phenomenon leads to changes in the structural, functional, and nutritional properties of both compounds, particularly in the digestibility, solubility, and molecular weight of proteins. The main parameters that affect protein–phenolic interactions are pH, temperature, the type and structure of phenolic compounds, and the protein type and concentration. Although the precise mechanism of how polyphenols impact proteins is still not well known, studies on these structural and functional changes are under investigation (Ozdal et al. 2013). Protein secondary and tertiary structures may change, while their solubility, amino acid concentration, and protein digestibility might be reduced, showing also undesirable changes in color and taste when they interact with phenolic compounds. However, proteins' thermal stability may be enhanced as well.

In general, the interaction with proteins could decrease the antioxidant capacity of polyphenols, although results are not conclusive due to differences in outcomes from the antioxidant capacity methods used, which are based in different principles. Likewise, in bioavailability experiments, contradictory results were obtained. Research on individual phenolic compounds is still missing both in antioxidant activity and bioavailability studies. Selecting the appropriate antioxidant methods and refined high-performance chromatographic techniques coupled to mass spectrometry are needed to better understand the mechanisms underlying protein–phenolic interactions and the factors affecting the degree of this interaction. Only after that can process conditions and products be designed to warrantee the maximum beneficial health effects for consumers, supplying optimum nutritional and functional properties to phenolic–protein products.

In particular, the inhibitory effects of dietary polyphenols against α-amylase have attracted great interest (Xiao et al. 2013). The molecular and structural features that influence the inhibition are the degree of hydroxylation of flavonoids; an unsaturated 2,3-bond in conjugation with a 4-carbonyl group; otherwise the glycosylation, methylation, and methoxylation of flavonoids decreased the inhibitory effect on α-amylase. Comparatively, galloylated catechins showed higher inhibition than the nongalloylated; the catechol-type catechins were stronger than the pyrogallol-type; the inhibition activities of catechins (2,3-trans structures) were higher than the epicatechins (2,3-cis structures). Cyanidin-3-glucoside showed higher inhibition than its cyanidin aglycone, galactoside, and diglycoside counterparts. Ellagitannins with β-galloyl groups at glucose C-1 positions show higher inhibitory activity than the α-galloyl and the nongalloylated versions in spite of the molecular weight of ellagitannins.

2.6 BIOACTIVE POLYPHENOLS

The antioxidant activity of phenolic compounds may be characterized by their ability to scavenge free radicals, donate electron or hydrogen atoms, or chelate metal ions. The phenolic compound structure is a significant factor for their radical scavenging and metal chelating activities (Balasundram et al. 2006).

The structure related to antioxidant activity depends on the number and position of the hydroxyl groups in relation to the carboxyl or keto functional groups in the case of phenolic acids and flavonoids, respectively. The structural skeletons and substitution patterns in B and C rings determine the antioxidant activity on flavonoids. For example, a catechol group (e.g., ortho-dihydroxylation of B-ring) or a pyrogallol group (e.g., hydroxylation at 3,4,5 in B-ring) has shown better antioxidant activity than mono B ring hydroxylated flavonoids. Besides a double bond between C-2 and C-3, in conjugation with a keto group (e.g., 4-oxo in C-ring) enhances their radical scavenging capacity (Pietta 2000). Also, an insaturation (e.g., a double bond between C-2 and C-3) joined to a hydroxyl at C-3 as in the case of kaempferol improves the active radical scavenging capacity of flavonoids (van Acker et al. 1996). Glycosylation, methoxylation, or substitution at hydroxylated carbons increases torsion angles, generates loss of coplanarity (Seeram and Nair 2002), and alters redox potential of the molecule, which finally affect its radical scavenging response (Pietta 2000; Seeram and Nair 2002). The contrary may be said from the galloylation, which enhances their activity.

Having two adjacent hydroxy groups on a phenyl ring enables the compound to undergo metal chelation, which is also a recognized advantage of some plant phenolics in their process to catalyze plant pigmentation or cationic nutrient promotion in the case of plant–litter–soil interactions (Scalbert et al. 1999).

In particular, astringency is ascribed to precipitation of salivary proteins by polyphenols, a mechanism possibly involved in their anti-nutritional effects. Plant phenolics and products formed from them during processing and storage are important contributors to the quality of plant-derived foods. Recent studies have "rediscovered" the quantitative importance of tannins in the diet and the difficulties of extracting them from plant material and analyzing them (Cheynier 2012).

Reaction mechanisms appear universal, involving all members of a given molecular family in similar ways. The resulting products share some similar properties. For instance, all pyranoanthocyanins are pigments and more resistant to hydration and sulfite bleaching than anthocyanins. Nevertheless, differences in molecular sizes or conformation may induce different behaviors, and in particular different abilities to establish molecular associations, with potential impact on organoleptic (e.g., color, taste, haze), technological, and health properties. Addressing the molecular weight and conformation distribution together with characteristic structural features is thus essential to establish relationships between polyphenol composition and food product quality.

It is important to highlight that polyphenols' bioactivity is not just centered in their capacity to exert antioxidant actions or their tendency to chelate metals or precipitate proteins in a rather nonspecific manner. There is convincing evidence that suggests mechanisms by which plant polyphenols display their protective actions against degenerative diseases is not only by their redox properties, but also by their ability to bind target peptides. This action mode would induce the inhibition of key enzymes and the modulation of transcription factors or cell receptors, as well the perturbation of protein aggregates. Consequently, this may regulate cell functions related to growth, proliferation, inflammation, angiogenesis, metastasis, apoptosis, and immune responses, affecting the signal transduction pathways (Spencer 2009).

2.7 POLYPHENOL BIOAVAILABILITY

As polyphenols may be key actors in interacting with proteins and enzymes as molecular targets, their potential effect is limited by their low bioavailability, according to Lipinski's rule (Lipinski et al. 2001). In general, the Lipinski's rule of five states that an orally available drug should agree with at least three of the following criteria:

1. Have no more than five hydrogen bond donors (i.e., the total number of nitrogen–hydrogen and oxygen–hydrogen bonds).
2. Have no more than ten hydrogen bond acceptors (i.e., all nitrogen or oxygen atoms).
3. Have a molecular mass lower than 500 daltons.
4. Have an octanol–water partition coefficient (Log P) not greater than 5.

Coincidentally, all numbers are multiples of five, which is the origin of the rule's name. As happens with other rules of thumb, there are many exceptions to Lipinski's rule.

Although experimental evidences of polyphenol bioavailability are discouraging in regards to the actual potential benefit in human health, still they are weakly absorbed and rapidly metabolized, exhibiting at long-term range valuable preventing effects. Thus, the synthesis of phenolic analogs or the modification of natural ones with optimum bioavailability response becomes a window of possibilities for research.

2.8 FUTURE POLYPHENOLS APPLICATIONS

2.8.1 Food Packaging

Phenolic compounds are among the most promising and abundant bioactive compounds that can be obtained feasibly from different plant materials, agricultural wastes, and industrial by-products at relatively low costs (Bonilla et al. 1999). Particularly, the use of natural phenolic compounds for food packaging is encouraging since they may improve food oxidative status and antimicrobial condition, displaying multiple benefits to human health (Coma 2008).

As mentioned before, the diversity of polyphenols' structure, presence of hydroxyl groups, and varied molecular weights enhance them to form H-bonding with peptide carbonyl groups of proteins. As an example, zein, a water insoluble hydrophobic storage protein from corn, presents special interest as a biopolymer since it may form good quality biodegradable and biocompatibility films. This biopolymer is likely to be produced in large amounts in the near future as a by-product of the rapidly developing bioethanol industry (Zhang et al. 2011). Flavanols such as catechins and phenolic acids including gallic acid, p-hydroxy benzoic acid, and ferulic acid can reduce the traditional brittleness problem of zein films and increase their flexibility significantly (Arcan and Yemenicioğlu 2011). These films containing phenolic compounds preserve the antioxidant and antimicrobial activities of their ingredients, mainly because most of the phenolic fraction is present in soluble form. Physical evidence has displayed the viability of applying polyphenols in the formulation of flexible bioactive packaging materials made of zein. To successfully test the active potential of these films in real packaging applications with functionalized materials, optimizing their mechanical properties is a current research possibility.

2.8.2 Sulfated Polyphenols

Sulfation in the plant kingdom may be related to the inactivation of toxic products by increasing their hydrosolubility in a similar way to animals. Particularly, more than 150 flavonols and flavones as smaller phenolic molecules have been reported so far (Sousa et al. 2008) as the more sulfated derivatives described from plants and sea grasses. However, there are references of natural sulfate esters of other phytochemicals classes such as anthraquinones, cyanogenic glycosides, and coumarins.

From the wide distribution of sulfated flavonoids in plants growing under aquatic environments, flavonoids sulfation looks like an ecological adaptation. Also, due to their increased water solubility, sulfated flavonoids appear to have an important role in copigmentation. They can give a place to stable molecular complexes with phenolic pigments, such as anthocyanins, producing natural intense colorations. These photochemical properties can be applied in bioimaging techniques as now performed in fluorometric assays, where sulfated coumarins are utilized as substrates (Aslam et al. 1997).

In recent years there have been multiple biological activities reported for sulfated flavonoids from natural sources. The list includes properties as antiviral, antitumor, anti-inflammatory, anticoagulant, antiplatelet, and immunomodulatory activities. The latter has increased the interest in research on sulfated bioactive polyphenols and mainly in the synthesis of sulfated small molecules. Besides, the need for knowledge on the physiological role of the sulfation processes highlights the potential of small phytochemicals in the already described sulfoproteomics exploration (Correia-da-Silva et al. 2014).

Hydrophobic and hydrophilic components are combined in a single sulfated small molecule. Moreover, the high water solubility and the nontoxic nature of this class of compounds must be underlined. This hydrophilic–hydrophobic balance looks like Mothers Nature's work. Thus, doing mimetic synthesis of non-natural sulfated small molecules or transforming natural ones could be an innovative way of generating new bioactive agents (Sousa et al. 2008).

As sulfated analogues, chemists still can formulate or design novel polyphenol nutraceuticals to act on specific protein targets, which represent another encouraging course for future research on polyphenols. Plant polyphenols are inspiring the quest for novel drugs.

2.9 CONCLUSIONS

A review on phenolic definitions was presented. However, the historical confusion on vegetable tannins and plant polyphenols is possibly still there, but it may be understood in terms of their anecdotic and historic reasons. Analytical techniques are revealing more insights on the chemistry and properties of polyphenols. A research window in the polyphenols' ability to bind important molecular targets is open. At firsthand, new applications for polyphenols as bioactive ingredients in food packaging are waiting, as well new sulfated active molecules that might be synthesized or transformed from natural polyphenols and benefit from their optimum solubility and bioavailability properties.

REFERENCES

Arcan, I. and Yemenicioğlu, A. 2011. Incorporating phenolic compounds opens a new perspective to use zein films as flexible bioactive packaging materials. *Food Research International* 44:550–556.

Aslam, M., Sheehan, M.T., Kvakoyszky, G., Davenport, K.G., and Gordon, D.J. 1997. *Photoactive coumarin sulfonate compounds* 5739295 USP.

Balasundram, N., Sundram, K., and Samman, S. 2006. Phenolic compounds in plants and agri-industrial by-products: Antioxidant activity, occurrence, and potential uses. *Food Chemistry* 99:191–203.

Bate-Smith, E.C. 1953. Colour reactions of flowers attributed to (a) Flavanols and (b) Carotenoid Oxides. *Journal of Experimental Botany* 4:1–9.

Bate-Smith, E.C. 1954. Astringency in foods. *Food* 23:124.

Bonilla, F., Mayen, M., Merida, J., and Medina, M. 1999. Extraction of phenolic compounds from red grape marc for use as food lipid antioxidants. *Food Chemistry* 66:209–215.

Cheynier, V. 2012. Phenolic compounds: From plants to foods. *Phytochemistry Reviews* 11:153–177.

Coma, V. 2008. Bioactive packaging technologies for extended shelf life of meat-based products. *Meat Science* 78:90–103.

Correia-da-Silva, M., Sousa, E., and Pinto, M.M.M. 2014. Emerging sulfated flavonoids and other polyphenols as drugs: Nature as an inspiration. *Medicinal Research Reviews* 34:223–279.

Feeny, P. 1976. Plant apparency and chemical defense. In J.W. Wallace and R.L. Manzell (eds.), *Biochemical interactions between plants and insects*. New York: Plenum Press.

Folin, O., and Ciocalteu, V. 1927. On tyrosine and tryptophane determinations in proteins. *The Journal of Biological Chemistry* 73:627–650.

Gerhardt, C. 1843. Recherches sur la salicine. *Annales de Chimie et de Physique*, series 3, 7:215–229.

Harborne, J.B. 1988. *Introduction to Ecological Biochemistry*, 3rd ed. London: Academic Press.

Harborne, J.B. 1989. Plant phenolics. In J.B. Harborne (ed.), *Methods in Plant Biochemistry*, vol. 1 (pp. 1–28). London: Academic Press.

Hartisch, C., and Kolodziej, H. 1996. Galloylhamameloses and proanthocyanidins from *Hamamelis virginiana*. *Phytochemistry* 42:191–198.

Haslam, E., and Cai, Y. 1994. Plant polyphenols (vegetable tannins): Gallic acid metabolism. *Natural Product Reports* 11:41–66.
IUPAC. 1997. *Compendium of Chemical Terminology*, 2nd ed. Compiled by A.D. McNaught and A. Wilkinson. Oxford: Blackwell Science, p. 299.
Iwai, K., Kishimoto, N., Kakino, Y., Mochida, K., and Fujita, T. 2004. *In vitro* antioxidative effects and tyrosinase inhibitory activities of seven hydroxycinnamoyl derivatives in green coffee beans. *Journal of Agricultural and Food Chemistry* 52:4893–4898.
Laurent, A. 1836. Sur la chlorophénise et les acides chlorophénisique et et chlorophénèsique. *Annales de Chemie et de Physique* 63:27–45.
Lipinski, C.A., Lombardo, F., Dominy, B.W., and Feeney, P.J. 2001. Experimental and computational approaches to estimate solubility and permeability in drug discovery and development settings. *Advanced Drug Delivery Reviews* 46:3–26.
Lister, J. 1867. Antiseptic principle of the practice of surgery. Retrieved from http://legacy.fordham.edu/halsall/mod/1867lister.asp
Lowry, O.H., Rosebrough, N.J., Farr, A.L., and Randall, R.J. 1951. Protein measurement with the Folin phenol reagent. *The Journal of Biological Chemistry* 193:265–275.
Okuda, T., Yoshida, T., and Hatano, T. 1992. In R.W. Hemingway and P.E. Laks (eds.), *Plant Polyphenols: Synthesis Properties, Significance*, pp. 539–569. New York: Plenum.
Ozdal, T., Capanoglu, E., and Altay, F. 2013. A review on protein–phenolic interactions and associated changes. *Food Research International* 51:954–970.
Pietta, P.G. 2000. Flavonoids as antioxidants. *Journal of Natural Products* 63:1035–1042.
Quideau, S., Deffieux, D., Douat-Casassus, C., and Pouységu, L. 2011. Plant polyphenols: Chemical properties, biological activities, and synthesis. *Angewandte Chemie International Edition* 50:586–621.
Runge, F.F. 1834. Ueber einige Produkte der Steinkohlendestillation (On some products of coal distillation). *Annalen der Physik und Chemie* 31:65–78.
Scalbert, A., Mila, I., Expert, D., Marmolle, F., Albrecht, A.M., Hurrell, R., Huneau, J.F., and Tom, D. 1999. In G.G. Gross, R.W. Hemingway, and T. Yoshida (eds.), *Plant Polyphenols 2: Chemistry, Biology, Pharmacology, Ecology*, pp. 545–554. New York: Kluwer Academic/Plenum.
Seeram, N.P., and Nair, M.G. 2002. Inhibition of lipid peroxidation and structure–activity related studies of the dietary constituents anthocyanins, anthocyanidins, and catechins. *Journal of Agricultural and Food Chemistry* 50:5308–5312.
Singleton, V.L. and Rossi, J.A. 1965. Colorimetry of total phenolics with phosphomolybdic-phosphotungstic acid reagents. *American Journal of Enology and Viticulture* 16:144–158.
Singleton, V.L., Orthofer, R., and Lamuela-Raventos, R.M. 1999. Analysis of total phenols and other oxidation substances by means of Folin-Ciocalteu reagent. *Methods in Enzymology* 299:152–178.
Spencer J.P.E. 2009. The impact of flavonoids on memory: Physiological and molecular considerations. *Chemical Society Reviews* 38:1152–1161.
Sousa, M.E., Correia-da-Silva, M., and Pinto, M.M.M. 2008. Sulfated flavonoids: Nature playing with the hydrophilic-hydrophobic balance. In G. Brahmachari (ed.), *Natural Products: Chemistry, Biochemistry and Pharmacology*. New Delhi: Narosa Publishing House.
Swain, T., and Bate-Smith, E.C. 1962. In H.S. Mason and A.M. Florkin (eds.), *Comparative Biochemistry*, vol. 3, pp. 755–809. New York: Academic Press.

Tanaka, T., Matsuo Y., and Kouno, I. 2005. A novel black tea pigment and two new oxidation products of epigallocatechin-3-O-gallate. *Journal of Agricultural and Food Chemistry* 53(19):7571–7578.

Van Acker, S.A.B.E., van den Berg, D.J., Tromp, M.N.J.L. et al. 1996. Structural aspects of antioxidant activity of flavonoids. *Free Radical Biology and Medicine* 20:331–342.

Van Noorden, R., Maher, B., and Nuzzo, R. 2014. The top 100 papers. Nature explores the most-cited research of all time. *Nature* 514:550–553.

Waterman, P.G., and Mole, S. 1994. *Analysis of phenolic plant metabolites.* Oxford: Blackwell Scientific Publications.

Xiao, J., Ni, X., Kai, G., and Chen, X. 2013. A review on structure–activity relationship of dietary polyphenols inhibiting α-Amylase. *Critical Reviews in Food Science and Nutrition* 53:497–506.

Zhang, B., Luo, Y., and Wang, Q. 2011. Effect of acid and base treatments on structural, rheological, and antioxidant properties of [alpha]-zein. *Food Chemistry* 124:210–220.

CHAPTER 3

Phenolic Compounds in Food

Andrea Gómez-Maqueo, Zamantha Escobedo-Avellaneda, M. Pilar Cano, and Jorge Welti-Chanes

CONTENTS

3.1 INTRODUCTION

Phenolic compounds are one of the most numerous, complex, and widely distributed groups of plant secondary metabolites. These compounds form an integral part of the human diet and are present in fruits, vegetables, legumes, spices, and herbs. They may also be found in animal tissues generally due to the ingestion of plants. In the past, they were considered as non-nutrients because of the adverse effects of compounds, such as tannins, on protein digestibility (Martin and Appel, 2010). Although phenolic compounds are not required for vital body functions in humans such as growth, reproduction, wound repair, and development, they have become of great interest in recent years due to their potential health benefits and reduction of risk of development of chronic diseases. Some of these functions include anti-allergenic, anti-artherogenic, anti-inflammatory, antimicrobial, antioxidant, antithrombotic, cardioprotective, and vasodilatory effects (Balasundram et al., 2006). There are numerous in vitro and in vivo studies that have proved beneficial physiological effects of phenolic compounds such as the inhibition of cell cancer proliferation, protection to neurons, improvement of insulin secretion, and reduction of vascularization and stimulation of vasodilatation (Haminiuk et al., 2012).

This chapter presents a general overview of the phenolic compounds in foods, beginning with a classification based on their relevant dietary sources and chemical structure. The different phenolic species will be discussed in terms of their health-promoting attributes and presence in food sources. Then the role and synthesis of phenolic compounds in plants as a protective response against adverse conditions or external factors will be addressed. Finally, the distribution of phenolic in different fruits and vegetable tissues will be discussed.

3.2 TOTAL PHENOLIC COMPOUNDS IN FOOD

Phenolic compounds may be found in a wide variety of dietary sources. They are found in high concentrations in red fruits such as blueberry, sour cherry, and blackberry; in herbs such as mint, sage, and lemonbalm, and in nuts such as walnuts and pistachios (Table 3.1). From the relevant food sources presented in Table 3.1, nuts present the highest total phenolic content with concentrations from 441.0 to 1,404.0 milligrams gallic acid equivalents/100 grams dry weight, the representative fruits presented in the table range from 50.9 to 670.9 milligrams gallic acid equivalents/100 grams fresh weight (Marinova et al., 2005; Rosales-Martínez et al., 2014). Herbs present values between 188.0 to 335.4 milligrams gallic acid equivalents/100 grams dry weight (Rababah et al., 2015). Vegetables present a range from 27.7 to 246.7 milligrams gallic acid equivalents/100 grams fresh weight, and sources of legumes and cereals present the lowest total phenolic content with 35.3 to 55.7 milligrams gallic acid equivalents/100 grams fresh weight and 6.5 to 15.3 milligrams total phenols/100 grams dry weight, respectively (Marinova et al., 2005; Chlopicka et al., 2012).

Currently, the food industry is moving toward the formulation of products with antioxidant constituents for specific health benefits. It is common knowledge that the main role of phenolic compounds is their action as antioxidants due to redox properties that allow them to absorb and neutralize free radicals, quench singlet and triplet oxygen, and decompose peroxides. In recent years there has been a large number of publications reporting the antioxidant activity of diverse foods; however, information regarding the total phenolic content in food vary from cultivar to cultivar (Anttonen and Karjalainen, 2005; Pantelidis et al., 2007; Mousavinejad et al., 2009).

Phenolic compounds should be classified into their corresponding species in order to establish specific health benefits associated with each class. In the following section, the proposed classification will be discussed with the objective of providing a general overview of the different phenolic compounds present in foods. The most abundant dietary phenolic compounds will be identified and subdivided into classes and species while highlighting their corresponding dietary sources.

3.3 CLASSIFICATION AND HEALTH BENEFITS

Due to the fact that phenolic compounds constitute a large number of heterogeneous structures, various classifications have been proposed. Phenolic compounds consist of an aromatic ring which bears one or more hydroxyl groups and their structure may vary from a simple phenolic molecule to a high molecular mass polymer (El Gharras, 2009). Harborne (1989) proposed a classification according to their carbon chain by categorizing these compounds in 16 major classes: Simple phenols (C_6), benzoquinones (C_6), phenolic acids (C_6–C_1), acetophenones (C_6–C_2), phenylacetic acids (C_6–C_2), hydroxycinnamic acids (C_6–C_3), phenylpropenes (C_6–C_3), coumarins (C_6–C_3), chromones (C_6–C_3), naphthoquinones (C_6–C_4), xanthones (C_6–C_1–C_6), stilbenes (C_6–C_2–C_6), anthraquinones (C_6–C_2–C_6), flavonoids (C_6–C_3–C_6), lignans and neolignans $(C_6–C_3)_2$, and lignins $(C_6–C_3)_n$. This classification, although useful from a chemical point of view, may not be the

TABLE 3.1 Important Food Sources of Phenolic Compounds and Their Total Phenolic Content

Group	Food Source	Total Phenolics[a]
Fruits	Blueberry	670.9
	Sour cherry	429.5
	Blackberry	355.3
	Plum	303.6
	Strawberry	244.1
	Black grape	213.3
	White grape	184.1
	Raspberry	178.6
	Red apple	125.4
	Pear	124.7
	Green apple	118.1
	Yellow apple	99.7
	Sweet cherry	78.8
	Fig	59.0
	Peach	50.9
Nuts[b]	Walnut	1404.0
	Mexican Pecan	1363.0
	Western nut	1225.0
	Iranian pistachio	710.0
	Mexican pistachio	566.0
	Virginia peanut	457.0
	Spanish peanut	441.0
Legumes	Yellow bean	55.7
	Green bean	35.3
Vegetables	Green pepper	246.7
	Red pepper	173.2
	Brussel sprout	161.5
	Radish	160.0
	Red onion	154.1
	Red cabbage	139.3
	Lettuce	124.5
	Salad	116.2
	Celery leaf	113.0
	Broccoli	101.7
	Carrot	96.0
	Tomato	76.9
	Kohlrabi	44.9
	Spring onion	36.5
	Leek	27.7

(*Continued*)

TABLE 3.1 (CONTINUED) Important Food Sources of Phenolic Compounds and Their Total Phenolic Content

Group	Food Source	Total Phenolics[a]
Herbs	Mint	335.4
	Sage	316.4
	Lemonbalm	303.2
	Thyme	299.2
	Parsley	188.0
Cereals and pseudocereals[c]	Buckwheat flour	15.3
	Quinoa flour	9.2
	Wheat flour	7.0
	Amaranth flour	6.5

Sources: Marinova, D. et al., *Journal of the University of Chemical Technology and Metallurgy*. 40(3): 255–260, 2005; Chlopicka, J. et al. *LWT–Food Science and Technology*. 46(2): 548–55, 2012; Rosales-Martínez, P. et al. *Journal of the Mexican Chemical Society*. 58(2): 185–193, 2014; Rababah, T.M. et al. *International Journal of Agricultural and Biological Engineering*. 8(2): 145, 2015.

[a] Expressed in milligrams gallic acid equivalents/100 grams fresh weight unless indicated otherwise.

[b] Milligrams gallic acid equivalents/100 grams dry weight.

[c] Milligrams total phenolics/100 grams dry weight.

most appropriate in classifying phenolics present in foods that include only some of those previously mentioned. Sánchez-Moreno (2002) classified phenolic compounds according to their solubility: Soluble (simple phenols, flavonoids, tannins of low and medium molecular weight not bound to membranes) and insoluble (condensed tannins, phenolic acids, and other low molecular weight compounds bound to cell wall polysaccharides or proteins). This classification is useful for the evaluation of the bioaccessibility of phenolic compounds and their metabolism in the gastrointestinal tract which may prove important for the further study of their health benefits.

Tapiero et al. (2002) classified phenolic compounds according to their abundance as dietary sources by defining two groups: The most abundant (flavonoids) and the least abundant (phenolic acids). The first group represents two-thirds of the total dietary intake, while the second group represents approximately one-third. In this chapter, the proposed classification divides phenolic compounds into three main groups according to their abundance as dietary sources: Most abundant (flavonoids), least abundant (phenolic acids), and others (lignans, tannins, and stilbenes). Furthermore, this classification maintains a structural organization by number of phenol rings and their bonds (Figure 3.1).

3.3.1 Flavonoids

Flavonoids are the most widely distributed phenolic compounds in foods and the main bioactive compounds found in fruits. They are characterized by a phenylbenzopyran chemical structure that includes a C_{15} (C_6–C_3 C_6) skeleton joined to a chroman ring (Pereira et al., 2009). Flavonoids may be divided into six classes as a function of the type

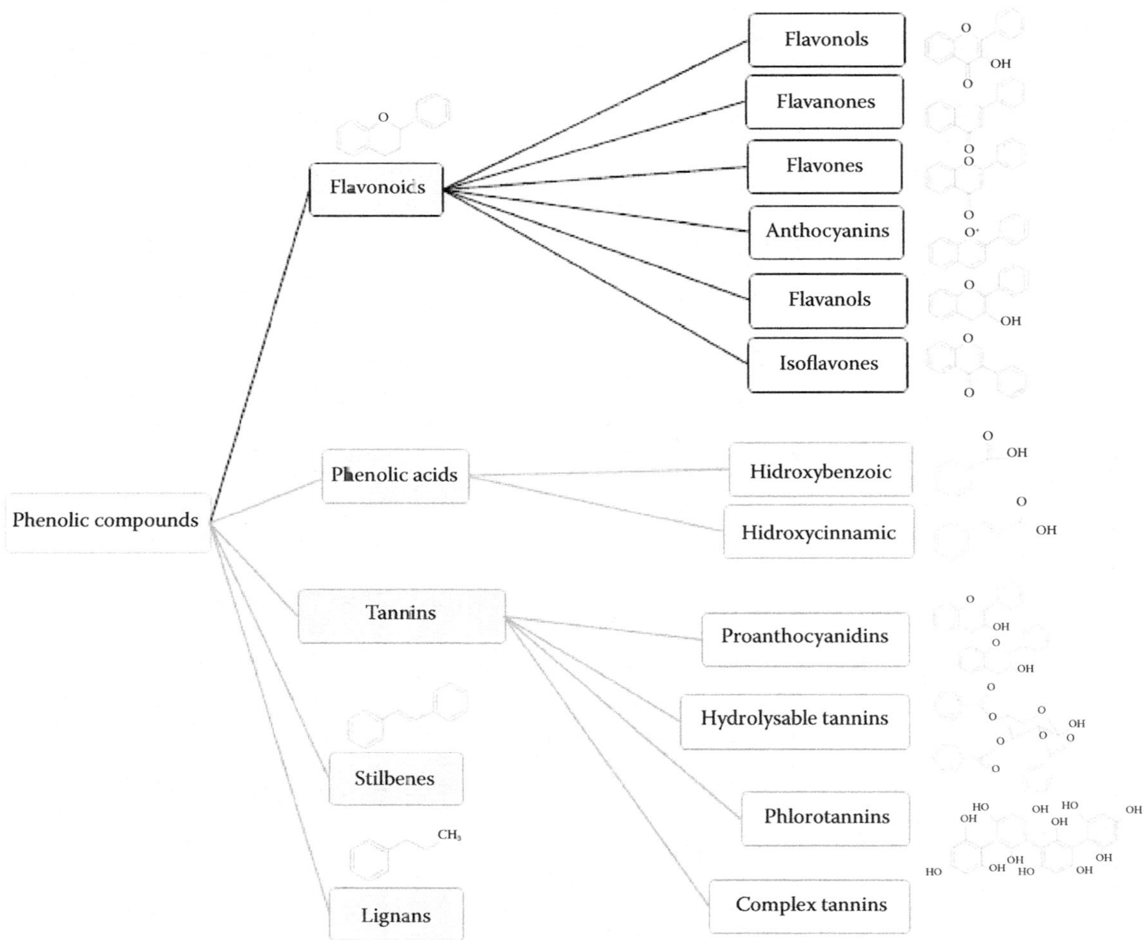

FIGURE 3.1 Classification of dietary phenolic compounds in foods and their general chemical structures.

of heterocycle: Flavonols, flavanones, flavones, anthocyanins, flavanols, and isoflavones (Figure 3.1). These compounds are present in nearly all plant species as a result of the plant's UV screening properties as a mean of protection. They also play an important role in fruits to attract pollinators and are responsible for most of the yellow, red, and blue colors present in flowers, fruits, and leaves (Lampila et al., 2009). Flavonoids are the most studied phenolic compounds in foods regarding their potential health-promoting effects. Epidemiological studies have shown that the daily intake of plant-derived foods may prevent some types of cancer, particularly in the gastrointestinal tract; cardiovascular diseases; and may also lower the incidence of diabetes (Liu, 2004; Bazzano et al., 2008; Wang et al., 2007).

Flavonoids are present in significant quantities in diverse food groups. Herbs and spices contain the highest flavonoid content, followed by fruits (Table 3.2). Mint, sage, thyme and lemonbalm contain in average 298.5, 273.5, 260.3, and 252.9 milligrams catechin equivalents/100 grams fresh weight, respectively. It has been widely reported that berries possess a higher flavonoid content than most fruits. Red berries are good sources of anthocyanins which are the predominating group of flavonoids in foods (Puupponen-Pimiä et al., 2001).

Flavanols are present in foods as monomers (catechins) and polymers (proanthocyanidins). Catechins are found in important quantities in tea as well as in many types of fruits, such as grapes and blackberries and apples (Table 3.3). Tea catechins have been found to be more efficient than α-tocopherol in inhibiting minced muscle lipid oxidation in fresh meats, poultry, and fish (Tang et al., 2001).

Flavanones are found in high concentrations in citrus fruits, their main aglycones are naringenin in grapefruit, hesperetin in orange, and eriodictyol in lemons. They are generally glycosylated by a disaccharide at position seven, either a neohesperidoside, which imparts a bitter taste (naringin in grapefruit), or a rutinoside, which is flavorless (El Gharras, 2009).

Flavonols are the most common type of flavonoids in foods, and their main representatives are kaempferol and quercetin (Manach et al., 2005). Rich sources of flavonols include dried parsley, saffron, kale, and onion (Table 3.3). These compounds are present in glycosylated forms in most food, commonly associated to glucose or rhamnose.

Although less abundant in fruits and vegetables, the main flavones found in foods, luteolin, and apigenin, are mainly present in their glycoside forms. Parsley and thyme represent the most important sources (Table 3.3). In celery, another important source, apigenin and luteolin contents in the leaves are approximately 40 times higher than the levels in the stalks (Justesen et al., 1998).

Anthocyanins are greatly found in berry-type fruits and possess colorant properties as well as biological, pharmacological, anti-inflammatory, antioxidant, and chemoprotective properties (Pascual-Theresa and Sánchez-Ballesta, 2008). Anthocyanins are water-soluble pigments that correspond to the glycoside or acyl-glycoside of anthocyanidins and are stored in the plant cell vacuole and impart color to ensure pollination, fertilization, and seed dispersal by animals (Harborne, 1998; Clifford, 2000; Routray and Orsat, 2011). Anthocyanins contribute to the color of most fruits and vegetables; some are responsible for yellow hues found mainly in the exterior of fruits, while others present colors that range from pink to deep purple or dark red and are found in high quantities in grapes, elderberry, red cabbage, and roselle, among others (Bridle and Timberlake, 1997; Gil et al., 2000). Cyanidin is the most common anthocyanin, and the 3-glucoside is the most active antioxidant anthocyanin (Einbond et al., 2004). It is present in high

TABLE 3.2 Total Flavonoid Content in Relevant Dietary Sources

Group	Food Source	Content
Fruits[a]	Blueberry	190.3
	Sour cherry	138.6
	Plum	136.2
	Black grape	77.1
	Pear	69.9
	Strawberry	69.7
	Blackberry	55.5
	Red apple	48.6
	White grape	36.5
	Yellow apple	34.9
	Raspberry	26.6
	Fig	20.2
	Sweet cherry	19.6
	Green apple	17.3
	Peach	15.0
	Orange[c]	6.1
	Banana[c]	0.7
	Kiwifruit[c]	0.4
Herbs and spices[a]	Mint	298.5
	Sage	273.5
	Thyme	260.3
	Lemonbalm	252.9
	Parsley[b]	1440.0
	Basil[b]	1230.0
	Chard[b]	1110.0
Vegetables[a]	Lettuce	97.5
	Salad	76.5
	Radish	48.5
	Celery	46.4
	Brussel sprout	33.1
	Green pepper	27.4
	Onion[d]	27.1
	Carrot	26.7
	Red cabbage	23.7
	Broccoli	18.8
	Red onion	18.7
	Spring onion	16.0
	Red pepper	13.7
	Tomato	12.8
	Kohlrabi	8.9
	Leek	2.6

(Continued)

TABLE 3.2 (CONTINUED) Total Flavonoid Content in Relevant Dietary Sources

Group	Food Source	Content
	Chinese cabbage[d]	1.9
	White cabbage[d]	1.1
	Potato[d]	0.1
Grains[e]	Buckwheat flour	15.3
	Quinoa flour	9.2
	Wheat flour	7.0
	Amaranth flour	6.5
	Wheat bread	2.0

Sources: Chu, Y.H. et al., *Journal of the Science of Food and Agriculture.* 80(5): 561–566, 2000; Marinova, D. et al., *Journal of the University of Chemical Technology and Metallurgy.* 40(3): 255–260, 2005; Tabart, J. et al. *Food Chemistry.* 105(3): 1268–1275, 2007; Chlopicka, J. et al. *LWT–Food Science and Technology.* 46(2): 548–55, 2012; Chandra, S. et al. *Evidence-Based Complementary and Alternative Medicine.* V 2014; and Rababah, T.M. et al. *International Journal of Agricultural and Biological Engineering.* 8(2): 145, 2015.

[a] Milligrams catechin equivalents/100 grams fresh weight.
[b] Milligrams quercetin equivalents/100 grams dry weight.
[c] Milligrams quercetin equivalents/100 grams fresh weight.
[d] Milligrams total flavonoids/kilograms fresh weight.
[e] Milligrams total flavanoids/100 grams dry weight.

quantities in berries such as black raspberries, elderberries, chokeberry, and blackberries. Other relevant dietary anthocyanins include delphinidin, malvidin, and peonidin (Table 3.3).

Isoflavones are found almost exclusively in leguminous plants, especially in soybean and soybean-derived products. Isoflavones are known to be estrogen analogues and bind to estrogen receptors; they may influence several biological processes including lipid and bone metabolism (Klein, 1998; Potter et al., 1998).

3.3.2 Phenolic Acids

Phenolic acids represent the second most widely distributed group of phenolic compounds in plant-derived foods. They may be found in high quantities in fruits such as blackcurrant, strawberry, raspberry, and gooseberry. They are also present in vegetables such as red cabbage, cucumber, carrot, and red beet (Table 3.4). Phenolic acids contain two distinguishing constitutive carbon frameworks and are classified as hydroxycinnamic and hydroxybenzoic acids differentiated by the position of the hydroxyl groups on the aromatic ring (Robbins, 2003). Hydroxycinnamic acids are more common than hydroxybenzoic acids and are found in most parts of plant fruits although higher concentrations are present in fruit peels and in plant leaves. The most widely distributed hydroxycinnamic

TABLE 3.3 Flavonoid Classes and Content in Relevant Dietary Sources

Flavonoid	Class	Food Source and Flavonoid Content (mg/100 g fresh weight)
Flavanoles	(–)-epicatechin	Grape seed (93.3), apple Red Delicious (9.8), apple *Malus domestica* (7.5), apple Granny Smith (7.1), blackberry (4.7)
	(+)-catechin	Grape seed (74.6), blackberry (37.1), apple *Malus domestica* (1.3)
	Thearubigins	Black tea prepared (49.0)
	(–)-epicallocatechin 3-gallate	Green tea brewed (70.2)
Flavanones	Erodictyol	Peppermint (30.9)
	Naringenin	Grape seed (53.0), grapefruit (32.6), rosemary (24.8), orange (15.3), artichoke (12.5)
	Hesperetin	Lime (43.0), orange (27.3)
Flavanols	Isorhamnetin	Dried parsley (331.2)
	Kaempferol	Saffron (205.5), kale (46.8), arugula (34.9), ginger (33.6), Chinese cabbage (22.5)
	Quercetin	Red onion (39.2), elderberry (26.8), kale (22.6), chia (18.4)
Flavones	Luteolin	Thyme (45.3), celery (24.8), oregano (25.0)
	Apigenin	Dried parsley (4503.50)
Anthocyanins	Cyanidin	Black raspberry (669.0), plum *Podocarpus elatus* (555.7), elderberry (485.3), chokeberry (433.1), red cabbage (209.8), blackberry (100.0), Molucca raspberry (90.2), black currant (62.5), cranberry (46.4), cherry (31.4)
	Malvidin	Blueberry (67.6), red grape (39.0)
	Peonidin	Cranberry (49.2)
	Delphinidin	Black currant (89.6), eggplant (85.7), grape *Vitis vinifera* (70.6)
	Pelargonidin	Radish (63.1), strawberry (24.9)
Isoflavones	Daidzein	Green soybean (61.7), sprouted soybean (12.9), red clover (11.0), pistachio nut (1.9)
	Genistein	Green soybean (60.1), sprouted soybean (18.8), red clover (10.0), pistachio nut (1.8)

Sources: Bhagwat, S.A. et al., USDA database for the isoflavone content of selected foods. Release 2.0, 2008; Bhagwat, S. et al., USDA. Database for the flavonoid content of selected foods, release 3.1. U.S. Department of Agriculture, Agricultural Research Service. Nutrient Data, 2014.

acids in foods are p-coumaric, caffeic, ferulic, and sinapic acid, while the most common hydroxybenzoic acids are p-hydroxybenzoic, vanillic, syringic and protocatechuic acids (Rice-Evans et al., 1996; Proestos et al., 2013).

Phenolic acids in foods are mainly found linked through ester, ether, or acetal bonds to structural components of the plant (Mattila and Helström, 2007). Hydroxybenzoic acids may act as components of complex structures such as tannins (gallotannins in mangoes and ellagitannins in red fruit; Clifford, 2000). These organic acids are considered

TABLE 3.4 Total Phenolic Acid Content in Relevant Dietary Sources

Group	Food Source	Total Phenolic Acid Content
Fruits[a]	Blackcurrant	68.2
	Strawberry	59.2
	Raspberry	54.6
	Gooseberry	23.1
	Grape	4.3
	Orange	2.4
	Pear	2.9
	Apple	1.7
	Banana	1.5
Legumes[b]	Soybean[a]	73.0
	Navy bean	48.3
	Black bean	38
	Pink bean	34.4
	Pinto bean	30.1
	Alubia bean	19.8
	Green bean[a]	3.5
	Pea[a]	0.54
Vegetables[a]	Red cabbage	41.0
	Cucumber	34.0
	Carrot	29.0
	Red beet	27.0
	Broccoli	15.0
	Spinach	11.0
	Chinese cabbage	7.7
	Iceberg lettuce	5.1
	Cauliflower	4.6
	White cabbage	3.8
	Tomato	3.5
	Garlic	1.7
	Red onion	1.4
	Celery root	1.3
	Onion	0.79
Other[c]	Red wine	49.4

Sources: Ghiselli, A. et al. *Journal of Agricultural and Food Chemistry* 46(2): 361–67, 1998; Luthria, D.L. and Pastor-Corrales, M.A., *Journal of Food Composition and Analysis*. 19(2): 205–211, 2006; Mattila, P. and Hellström, J., *Journal of Food Composition and Analysis*. 20(3–4): 152–60, 2007; Russell, W.R. et al. *Food Chemistry*. 115(1): 100–4, 2009.

[a] Milligrams total phenolic acids/100 grams fresh weight.

[b] Milligrams total phenolic acids/100 grams dry weight.

[c] Milligrams gallic acid equivalents/100 milliliters.

intermediates of lignin biosynthesis and are associated to diverse functions in plants. They strongly influence the cell membrane potentials and therefore possibly have an effect concerning ion uptake (Glass and Dunlop, 1974; Stevens et al., 2006).

Phenolic acids have reported diverse biological activities such as bile secretion increment, blood cholesterol, and lipid level reduction and antimicrobial activity (Ghasemzadeh and Ghasemzadeh, 2011). Studies using cell and animal models show the effects of phenolic acids on the expression and the activity of enzymes involved in the production of inflammatory mediators (Duthie et al., 2003). It has also been suggested that the regular consumption of phenolic compounds directly from plants may be more effective in combating oxidative damage in the body than in the form of dietary supplement due to the possible synergistic interactions with other food phenolic compounds (Martin and Appel, 2010). Although when present in complex structures phenolic acids are not absorbed by the mucosa, they have been defined as insoluble antioxidants in the gastrointestinal tract because of their ability to protect proteins, lipids, and carbohydrates from oxidative damage during the digestive process.

Dietary sources of phenolic acids and their classes are presented in Table 3.4. Caffeic acid is the most abundant phenolic acid and represents between 75 percent and 100 percent of the total hydroxycinnamic acid in most fruits (Proestos et al., 2013). Caffeic and quinic acid combine and form chlorogenic acid that is commonly found in many types of fruit, vegetables, and coffee (El Gharras, 2009). The p-coumaric acid content in wine is more dependent on genetic factors of grape than on exposure of light and climate (Clifford, 2000). Ferulic acid is the most abundant hydroxycinnamic acid in cereals. On the other hand, the content of hydroxybenzoic acids in edible plants is very low, except in certain red fruits, black radish, and onion. Red fruits are relevant sources of gallic acid, p-hydroxybenzoic acid, and procatechuic acid.

3.3.3 Silbenes, Lignans, and Tannins

Compared to flavonoids and phenolic acids, stilbenes and lignans contribute very little to the average dietary intake of food phenolics. Meanwhile tannins, although present in somewhat higher quantities, are difficult to assimilate as a consequence of their poor solubility. Despite this, these three groups are important in the analysis of phenolic compounds due to their potential biological effects. It has been reported that stilbenes possess cancer chemopreventive activity and that they protect lipoproteins from oxidative damage. Lignans are transformed by the intestinal microflora to produce enterolignans which can potentially reduce the risk of certain cancers and cardiovascular diseases. A small ingestion of the right kind of tannins have proved to possess antimutagenic, anticarcinogenic, and immunomodulation activity (Chung et al., 1998; Ribeiro de Lima et al., 1999; Milder et al., 2005). The main sources of these compounds are presented in Table 3.5.

Resveratrol is the most widely studied stilbene. It is present in red wine and has been associated with anticarcinogenic effects, although any protective effect of this molecule is highly unlikely at regular nutritional intakes (Vitrac et al., 2002; El Gharras, 2009). Besides being present in grapes and grape-derived products, this antioxidant is present in peanuts, pistachio, strawberries, currants, cranberries, and cranberry juice (Tosun and Inkaya, 2009). Resveratrol is toxic to plant pathogens and is produced as a self-defense agent in plants. It has garnered great interest in recent years due to its antioxidant properties and health benefits, which include anti-inflammatory, estrogenic, cardioprotective, anti-tumor, and anti-viral action (Heath, 2000).

TABLE 3.5 Phenolic Acid Classes and Content in Relevant Dietary Sources

Phenolic Acids	Class	Food Source and Flavonoid Content (mg/100 g fresh weight)
Hydroxycinnamic	Caffeic acid	Black chokeberry (141.1), cinnamon (24.2), nutmeg (16.30), thyme (11.7), blackcurrant (11.3), black bean (1.1[a]), apple (0.3), pear (0.1), coffee (0.03[b])
	p-coumaric acid	Strawberry (13.4), blackcurrant (12.4), navy bean (12.4[a]), black bean (11.6[a]), clove (8.5), green olive (5.9), raspberry (3.5), wine (2.2[b]) grape (0.5), orange (0.2)
	Cinnamic acid	Chinese cinnamon (20.1), green olive (14.3), lingonberry (4.1), black olive (0.8), strawberry (0.2)
	Ferulic acid	Hard wheat whole grain flour (72.2), navy bean (26.6[a]), black bean (25.5[a]), pinto bean (22.9[a]), hard wheat refined flour (14.1), bread rye whole grain flour (3.9), Blackcurrant (2.5), strawberry (1.5), banana (1.1)
	Chlorogenic acid	Lettuce (23.0), aubergine (31.0), carrot (17.0), tomato (0.9)
Hydroxybenzoic	Gallic acid	Wine (32.0[b]), raspberry (31.2), blackcurrant (21.0), strawberry (17.1)
	p-hydroxybenzoic acid	Raspberry (10.6), strawberry (5.2)
	Protocatechuic acid	Blackcurrant (7.5), gooseberry (5.3)
	Vanillic acid	Raspberry (1.5), blackcurrant (1.8)
	Salicyclic acid	blackcurrant (1.2)
	Syringic acid	Black olive (33.1), thyme (11.7[a]), oregano (3.8[a]), sage (3.4[a]), pear (0.5)
	Sinapic acid	Pinto bean (8.5[a])

Sources: Ghiselli, A. et al. *Journal of Agricultural and Food Chemistry* 46(2): 361–67, 1998; Luthria, D.L. and Pastor-Corrales, M.A., *Journal of Food Composition and Analysis*. 19(2): 205–211, 2006; Mattila, P. and Hellström, J., *Journal of Food Composition and Analysis*. 20(3–4): 152–60, 2007; Russell, W.R. et al. *Food Chemistry*. 115(1): 100–4, 2009; Neveu, V., Perez-Jiménez, J., Vos, F. et al. Phenol-Explorer: An online comprehensive database on polyphenol contents in foods. Database, 2010.

[a] Milligrams/100 grams dry weight.

[b] Milligrams/100 milliliters.

Lignans are one of the major classes of phytoestrogens in plants. They are conformed of two phenylpropane units and are present in small quantities in the human diet. They can be found in oleaginous seeds (linseed), leguminous plants (lentil), cereals (triticale and wheat), vegetables (garlic, asparagus, carrot), and fruits (pear, prune) in lesser quantities (El Gharras, 2009). The richest source of these compounds is flaxseed. In the gastrointestinal tract, lignans are converted into enterodiol and anterolactone and have been reported to possess estrogenic and anti-estrogenic properties (Meagher and Beecher, 2000).

Tannins are a group of polyhydroxy-flavan-3-ol oligomers and polymer with carbon–carbon linkages between flavonol subunits (Schofield et al., 2001; Haminiuk et al., 2012).

In the past, tannins have been considered undesirable because they precipitate proteins, inhibit digestive enzymes, and affect the absorption of vitamins and minerals. However, many of these molecules have proved to reduce mutagenicity, possess anticarcinogenic activity, and inhibit the growth of fungi, bacteria, and viruses (Chung et al., 1998). They may be classified according to their chemical structures and constitutive monomers into four groups: Proanthocyanidins, hydrolyzable tannins, phlorotannins, and complex tannins (Serrano et al., 2009). Proanthocyanidins may also be classified as condensed tannins and they are the polymerized product of flavan-3-ols (catechins) and flavan-3,4-diols or a mixture of both. Most fruits, incluing berries, are the major sources of proanthocyanidins in the human diet (Table 3.6). Wine, beer, and some fruit juices are also good sources, whereas most vegetables, legumes, nuts, and cereals contain only small amounts. On the other hand, hydrolyzable tannins refer to either gallotannins or ellagitannins which, when upon hydrolysis, respectively yield gallic acid or ellagic acid (Chung et al., 1998).

TABLE 3.6 Stilbens, Lignans, and Tannin Classes and Content in Relevant Dietary Sources

Compound	Class	Sources
Stilbenes	Resveratrol[a]	Mulberry (5.1), jamun seed (3.5), grape seed (0.6), grape skin (0.4), red wine (0.4[b])
	trans-astringin[b]	Red wine (1.1)
	trans-piceid[b]	Red wine (1.2)
Lignans[c]	Pinoresinol	Flaxseed (2.5), rye (0.4), oat (0.2), lemon (0.2)
	Lariciresinol	Flaxseed (3.7), buckwheat (0.4), rye (0.3)
	Medioresinol	Rye (0.1), lemon (0.6)
	Syringaresinol	Rye (1.0), wheat (0.4), oat (0.4), pineapple (0.1)
	Secoisolariciresinol	Flaxseed (323.7), buckwheat (0.1), asparagus (0.2), Kiwi (0.1)
	Matairesinol	Flaxseed (5.2), oat (0.1)
Tannins	*Proanthocyanidins*[c]	Cacao beans (9481.8), cranberry (418.8), Granny Smith apple (136.0), blueberry (176.5), barley (99.2), blackberry (23.3), sweet cherry (19.1)
	Hydrolyzable tannins	
	Gallotannin[d]	Chickpea, mango, persimmon, rhubarb, pistachio
	Ellagitannin[c]	Pecan (5358), walnut (1604), raspberry (487), blackberry (175), strawberry (106)
	Phlorotannins[a]	*Ascophyllum* seaweed (50.0)

Sources: Ribeiro de Lima, M.T. et al. *Journal of Agricultural and Food Chemistry*. 47(7): 2666–670, 1999; Bhagwat, S.A., Haytowitz, D.B., Prior, R.L. et al. USDA database for proanthocyanidin content of selected foods, 2004; Peñalvo, J.L. et al. *Journal of Agricultural and Food Chemistry*. 53(24): 9342–9347, 2005; Serrano, J. et al. *Molecular Nutrition & Food Research*. 53: S2, 2009; Holdt, S.L. and Kraan, S. *Journal of Applied Phycology*. 23(3): 543–597, 2011; Lipińska, L. et al., *Acta Scientiarum Polonorum Technologia Alimentaria*. 13(3): 289–299, 2014; Shrikanta, A. et al., *Journal of Food Science and Technology*. 52(1): 383–390, 2015.

[a] Milligrams/100 grams dry weight.
[b] Milligrams/100 milliliters.
[c] Milligrams/100 grams fresh weight.
[d] quantifiable data not found.

The third group of tannins, phlorotannins, are oligomeric or polymeric phloroglucinol (1,3,5-tirhydrozybenzene) derivatives found in marine brown algae (Serrano et al., 2009).

Given that mixtures of proanthocyanidins, hydrolyzable tannins, and phlorotannins are also possible in edible plants, the fourth group of tannins, complex tannins (Figure 3.1), refer to complex structures that contain structural elements of different tannin groups as well as other macromolecules.

3.4 THE ROLE OF PHENOLIC COMPOUNDS IN FOOD-GRADE PLANTS

Phenolic compounds are mainly found within the cell wall structures of different plant tissues and are present in seeds, roots, leaves, stems, flowers, and fruits. Phenolic compounds may also be classified according to the role they play in each stage of the plant's development: seed germination and dormancy, plant growth and development, response to stress factors and fruit set (Figure 3.2).

The first intervention of phenolic compounds in the life cycle of plants appears during seed germination. Some phenolic compounds are found in seed coats and embryos and affect seed germination and dormancy. Phenolic compounds such as hydroxycinnamic acids, coumarins, tannins, and ferulic acid are considered common seed germination inhibitors that act synergistically with other compounds (Sulusoglu, 2014). Unripe fruits may also possess high tannin contents which deter feeding on the fruits until their seeds are mature enough for dispersal (Lattanzio et al., 2006).

The second main role of phenolic compounds is directly related to plant growth and development. Phenolic compounds are physiological regulators or chemical messengers mainly found in the cell wall fraction. Their specific roles during plant growth and development are presented in Table 3.7.

Phenolic compounds play an important protective role against threatening external factors. They help the plant protect itself against pathogens, pollutants, tissue wounding, diseases, and insects. On one hand, phenylpropanoids and benzoic acid derivatives have been widely studied and reported to have allelopathic activity given that they reduce the growth of nearby plants in order to increase their own access to light, water, and nutrients. Phenylpropanoid compounds such as flavonoids, isoflavonoids, anthocyanins, and polyphenols are synthesized as a response to tissue wounding, nutritional stress, cold stress, and high-visible light (Shetty and McCue, 2003). On the other hand, certain classes of flavonoids protect cells from excessive UV radiation by absorbing the light strongly in the UV region while allowing the visible (photosynthetically active) wavelengths pass through without interruption (Özeker, 1999). This process involves peroxidases and the biosynthesis of polymeric phenolic compounds that lead to protective lignification (Shetty et al., 2002). During tissue wounding, their levels increase to act as precursors for the synthesis of lignin and suberin to prevent water loss (Boerjan et al., 2003). Lignin also plays a protective role in plants by blocking the growth of pathogens (Harborne, 1980). During pathogen attack, antimicrobial phenolics (phytoalexins) are synthesized around the site of infection while others participate in a signaling process that results in systemic acquired resistance (Shetty, 2004). Meanwhile, tannins are general toxins that reduce the growth of herbivores when added to their diets and act as feeding repellents and antimicrobials, and prevent against fungal and bacterial decay (Özeker, 1999).

Phenolic compounds during fruit set are presented from an evolutionary point of view since they serve as visual and aromatic signals for attracting animals to help pollination and fruit set by dispersing seeds (Lattanzio et al., 2006). The factors that influence

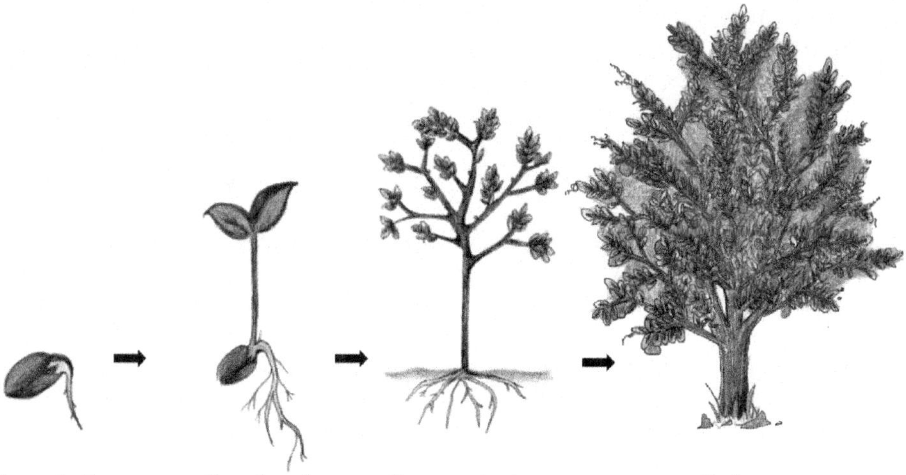

FIGURE 3.2 Main stages of a plant's life cycle where phenolic compounds play a significant role.

TABLE 3.7 Specific Roles of Phenolic Compounds during Plant Growth and Development

Phenolic Compound	Specific Role during Plant Growth
Ferulic acid	Cell wall development and lignin biosynthesis.
Flavonoids (quercetin, apigenin, kaempferol)	Natural auxin transport regulators in plants. Inhibitors of indole-3-acetic acid (IAA) transport (mono- and dihdroxy flavonoids).
	Promotion of leaf expansion.
	Callus growth stimulation.
	Increased rooting of cuttings.
Benzoic acids (including salicylic acid) and cinnamic acids	Strongly influence cell membrane potential therefore influencing ion uptake.
p-coumaric acid, ferulic acid and isomeric hydroxymethoxycinnamic acid	Bud dormancy.
Flavones and flavonols	Mediate the interaction of legumes and rhizobacteria (regulatory role).

Sources: Rolfe, B.G. and Gresshoff, P.M., *Annual Review of Plant Physiology and Plant Molecular Biology*. 39(1): 297–319, 1988; Sulusoglu, M., *Türk Tarım ve Doğa Bilimleri*. 6(6): 947–956, 2014.

the fruit's color are mainly established genetically; however, nutrients, temperature, and light may affect the flavonoid composition and the final color (Sulusoglu, 2014). Due to their role as visual and aromatic signaling metabolites, phenolic compounds contribute greatly to the sensory quality of fruit and other foods. Some of these compounds are responsible for the bitter taste in some fruits, such as oleuropein in olives and naringin in grapefruit, as well as in the caffeic, ferulic, and sinapic acid that is present in several fruits (Sulusoglu, 2014).

3.5 METABOLISM OF PHENOLIC COMPOUNDS IN FOOD-GRADE PLANTS

Phenolic metabolism is a dynamic system that involves ready-state concentrations of various phenolic compounds, which during certain phases of growth and development are subject to substantial qualitative and quantitative changes (Lattanzio et al., 2012). The accumulation of phenolic compounds in plant tissues has been considered a common adaptive response of plants to adverse environmental conditions (Swain, 1975; Lowry et al., 1980). There are three main reactions that take place during the synthesis of phenolic compounds: (1) Interconversions involved in biosynthetic sequences, (2) catabolic reactions that convert products into primary metabolic constituents, and (3) oxidative polymerization reactions that lead to insoluble structures of high molecular weight (Barz and Hoesel, 1979). However, phenolic metabolism in plants is the result of the interaction of at least five different pathways (Figure 3.3): (1) The glycolytic pathway that produces phosphoenolpyruvate, (2) the pentose phosphate pathway that produces erythrose-4-phosphate, (3) the Shikimate pathway that synthesizes phenylalanine, (4) the general

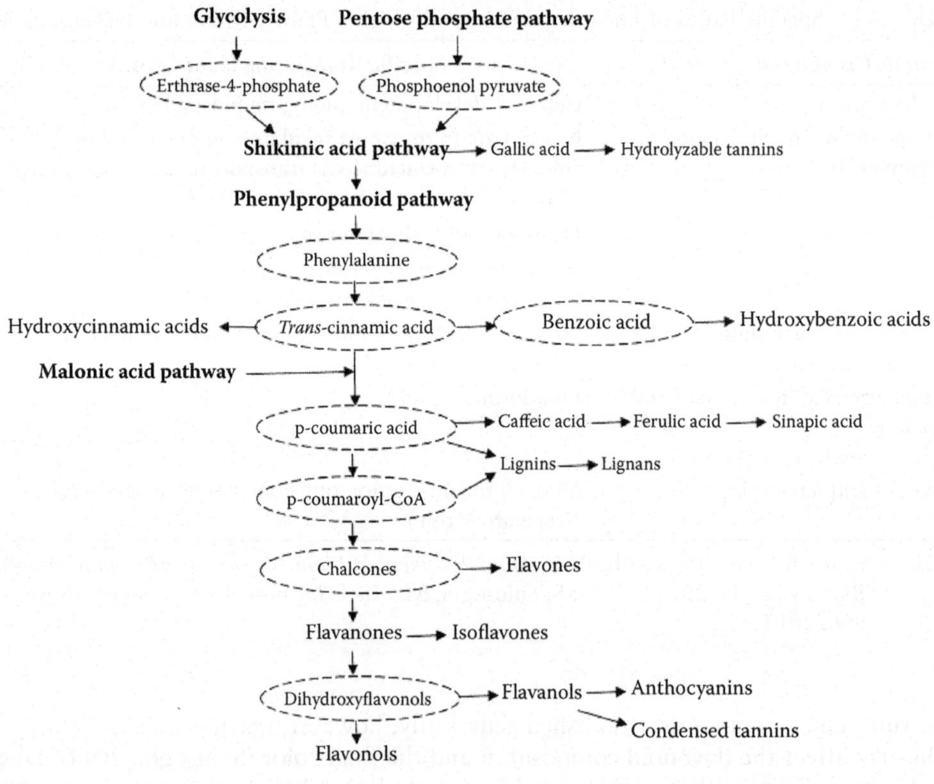

FIGURE 3.3 Principal metabolic routes and important intermediates in the synthesis of phenolic compounds in plants.

phenylpropanoid metabolism that produces the activated cinnamic acid derivatives, and (5) lignin and the diverse specific flavonoid pathways (Lattanzio et al., 2012).

The Shikimic acid pathway is the principal route for the synthesis of aromatic compounds in plants, such as amino acids including phenlalanine, tyrosine, and tryptophan (Ghasemzadeh and Ghasemzadeh, 2011). These amino acids are precursors for flavonoids, phenolic acids, coumarins, alkaloids, glucosinolates, and cyanogenic glycosides (Wink, 2010). As a result of this pathway, other precursors are also produced, such as lignin, growth hormone indole acetic acid, quinones of the electron transport chain, and storage compounds (Conn, 1986).

Most classes of phenolic compounds in plants are synthesized through the elimination of an ammonia molecule catalyzed by phenyl alanine ammonia lyase (PAL) in order to form cinnamic acid (Özeker, 1999; Ghasemzadeh and Ghasemzadeh, 2011). The activity of this important enzyme may increase due to environmental factors such as low nutrients, light, and fungal infection (Taiz et al., 2015). Subsequent reactions then lead to the addition of more hydroxyl groups and other substituents that form *trans*-cinnamic acid, *p*-coumaric acid, and their derivatives.

Plants may respond to elicitors (substances that induce physiological changes in the plant) by activating different mechanisms that affect the plant metabolism and enhance the synthesis of phenolic compounds (Baenas et al., 2014). Abiotic elicitation refers to metal ions, inorganic compounds, chilling, wounding, UV-light, and water stress; meanwhile, biotic elicitors refer to those provided by fungi, bacteria, viruses, cell wall components, and chemicals released at the attack site by plants under pathogen or herbivore attack (Matkowski, 2008; Becerra-Moreno et al., 2015; Świeca, 2016).

Elicitation may occur naturally during crop growth as a consequence of diverse conditions during plant development; however, specific treatments may be used to increase the production of phenolic compounds to produce functional foods or pharmaceutical products. The treatment of buckwheat sprouts with precursors of phenolics has shown to cause their accumulation by the stimulation of two key enzymes (tyrosine ammonia-lyase and phenylalanine ammonia-lyase) of the phenylpropanoid pathway (Świeca, 2016). Wounding stress has been reported to induce the activation of the primary and secondary metabolism of carrots, leading to an accumulation of phenolic compounds and lignin. It has also been found that the application of wounding in combination with additional stresses (UV-light, phytohormones, and hyperoxia) increases the wound-induced accumulation of phenolic compounds in carrots and manipulates their phenolic profiles (Becerra-Moreno et al., 2015). Low temperatures also affect and enhance phenolic metabolism under chill stress. Low critical temperatures below which an increase of phenylpropanoid and flavonoid metabolism is stimulated during the storage of plant tissues has been reported for various plant tissues, including artichoke, carrot, gherkin, maize, olive, pea, pear, potato, tomato, and watermelon (Lattanzio et al., 2012).

3.6 DISTRIBUTION OF PHENOLIC COMPOUNDS IN FOOD

Phenolic compounds are distributed unequally throughout different parts of foods. In most food groups, phenolics are present in higher concentrations in the peel rather than in the pulp, kernel, or grain. In some fruits, the peel contains from 1.6 to 32.5 times more phenolic compounds than those found in the pulp (Table 3.8). Apples, peaches, nectarines, oranges, and cherries contain from 1.6 to 4.5 times more phenolic compounds in

TABLE 3.8 Total Phenolic Compounds in the Pulp and Peel of Common Fruits

	Total Phenolic Content		
Food Source	**Pulp**	**Peel**	**Peel/Pulp Ratio[d]**
Blueberry (Northern highbush)[a]	274.2	3846.9	14.0
Apple[a]	1309.7	2193.9	1.7
Peach[a]	791.6	1288.4	1.6
Cherry (Capulí)[b]	331	1494	4.5
Mango (Keitt)[b]	28.5	927.2	32.5
Orange[b]	146.6	639.6	4.4
Plum (Wickson)[c]	22	163.1	7.4
Nectarine (Red Jim)[c]	41.5	140.3	3.4
Peach (Summersweet)[c]	22.8	67	2.9

Sources: Gil, M.I. et al., *Journal of Agricultural and Food Chemistry*. 50(17): 4976–4982, 2002; Vasco, C. et al., *Food Chemistry*. 111(4): 816–823, 2008; Manzoor, M. et al., *Molecules*. 17(6): 6491–6506, 2012a; Manzoor, M. et al., *Molecules*. 17(1): 390–407, 2012b; Wang, S.Y. et al., *Food Chemistry*. 132(4): 1759–1768, 2012; Escobedo-Avellaneda, Z. et al., *Journal of Functional Foods*. 6: 470–481, 2014; Abbasi, A.M. et al., *International Journal of Molecular Sciences*. 16(6): 13507–13527, 2015.

[a] Milligrams gallic acid equivalents/100 grams dry weight.
[b] Milligrams gallic acid equivalents/100 grams fresh.
[c] Milligrams total phenolics/100 grams fresh.
[d] Calculated by dividing total phenolics in peel by total phenolics in pulp on a weight-to-weight basis.

the peel than in the pulp, while mango and blueberry present even higher peel/pulp ratios such as 32.5 and 14.0 times more phenolics in peel than in pulp, respectively.

Total phenolic compounds present in lemon, orange, and grapefruit are usually 15 percent higher than those in the peeled fruit (Gorinstein et al., 2001). Inglett and Chen (2011) analyzed phenolic and flavonoid content in skin, pulp, and seed of miracle fruit and reported that the phenolic content in the peel was almost three times that of pulp and four times that of seed, the free flavonoid content in the peel was higher than that in the seed and pulp, and that the skin contributed about 52 percent of total flavonoids. In other fruits such as mango, longan, avocado, and jackfruit, the total phenolic content of the seed has been reported to be higher than that of the edible parts (Soong and Barlow, 2004). Chaovanalikit and Wrolstad (2004) reported that total phenolics and anthocyanins for one sour cherry cultivar (*Prunus cerasus L.*) and three sweet cherry cultivars (*P. avum L.*) were concentrated on the peel.

Phenolic compounds are also present in the non-edible fractions of vegetables, nuts, and oats. Carrot leaves, for example, have a higher phenolic content than the parts of the vegetable itself; peanut hulls and oat chaffs also present a higher total phenolic content compared to their kernel and grain, respectively (Table 3.9). Purple potato, red potato, and tomato contain from 1.9 to 2.3 times more phenolics in their peel, while peanuts contain about 100 times more phenolics in their peels than in the kernel.

In oats, phenolic acids are asymmetrically distributed within the grain, and some, such as ferulic acid and p-coumaric acid, occur mostly bound to the cell walls. Kähkönen et al. (1999) reported 30, 40, and 70 milligrams gallic acid equivalents/100 grams dry weight for oat grains, oat bran, and oat chaff, respectively. It has also been reported that

TABLE 3.9 Total Phenolics in the Edible and Non-Edible Parts of Diverse Food Sources (Vegetables, Nuts, and Oats)

Vegetables	Pulp	Peel	Non-Edible	Peel/Pulp Ratio[d]
Carrot[a]	60	660	740 (leaf)	11.0
Purple potato[b]	125	256	–	2.0
Red potato[b]	116	225	–	1.9
Tomato[c]	15	34.7	25.5 (seed)	2.3
Nuts	**Kernel**	**Peel**	**Non-edible**	**Peel/Kernel ratio[d]**
Brazilian nut[a]	406.8	1236.1	–	3.0
Peanut kernel[a]	92	9174	2759 (hull)	99.7
Oats	**Grain**	**Bran**	**Non-edible**	**Bran/Grain ratio[d]**
Oat grain[b]	30	40	70 (chaff)	1.3

Sources: Kähkönen, M.P. et al., *Journal of Agricultural and Food Chemistry*. 47(10): 3954–3962, 1999; Emmons, C.L. and David, M.P., *Cereal Chemistry* 76(6): 902–06, 1999; Reyes, L.F. et al., *American Journal of Potato Research*. 82(4): 271–277, 2005; Toor, R.K. and Savage, G.P., *Food Research International*. 38(5): 487–494, 2005; John, J.A. and Shahidi, F., *Journal of Functional Foods*. 2(3): 196–209, 2010; Win, M.M. et al., *Pakistan Journal of Botany*. 43(3): 1635–1642, 2011.

[a] Milligrams gallic acid equivalents/100 grams dry weight.

[b] Milligrams chlorogenic acid equivalents/100 grams fresh weight.

[c] Milligrams gallic acid equivalents/100 grams fresh weight.

[d] Calculated by dividing total phenolics in peel/bran by total phenolics in pulp/kernel/grain on a weight to weight basis.

the total cinnamic acid content of the hulls of Swedish oats (*Avena sativa L.*) resulted higher than that of the groats (23.6 compared to 3.6 milligrams/kilograms dry matter (Bryngelsson, 2002).

The study of the distribution of phenolic compounds in foods is relevant for the revaluation of waste products derived from the food processing industry. During the processing of fruits to produce juice and jams, for example, a substantial amount of phenol-rich residue (peels, seeds, pulp) is disposed. Such residue is a potential raw material to be incorporated in foods as such, or for the extraction of enriched phenolic fractions to be employed as a functional ingredient. Therefore, the study of the distribution of phenolics in different parts of foods is of great interest for the development of phenol-rich food products and nutraceuticals.

3.7 CONCLUSION

Foods such as fruits, vegetables, nuts, and herbs are important sources of phenolic compounds, and the consumption of these products has been associated with risk reduction of chronic degenerative diseases. The exact concentration of phenolics consumed daily in different regions of the world remains unknown and varies greatly between individuals. Phenolic compounds also influence the sensory characteristics of foods such as color and flavor, and their oxidation can cause detrimental sensory changes. The study of specific phenolic classes in food must be considered to argue specific health benefits. Flavonoids and their subgroups, flavonols, flavanones, flavones, flavanols, anthocyanidins, and isoflavones, represent two-thirds of the total dietary intake and are the main bioactive compounds in fruits. Meanwhile, phenolic acids are the second-most distributed group in edible plants. Phenolic compounds are synthesized by a combination of several pathways, which include glycolysis, pentose phosphate pathway, shikimic acid pathway, phenylpropanoid pathway, and malonic acid pathway. The main role of phenolic compounds in plants is to respond to physiological stress. Due to the high concentrations of phenolic compounds in process-derived by-products such as peels and seeds, these residues may serve as raw materials for extraction purposes to be further employed as antioxidants, colorants, nutraceuticals, and food additives.

REFERENCES

Abbasi, A.M., Guo, X., Fu, X. et al. 2015. Comparative assessment of phenolic content and in vitro antioxidant capacity in the pulp and peel of mango cultivars. *International Journal of Molecular Sciences*. 16(6): 13507–13527.

Anttonen, M.J. and Karjalainen, R.O. 2005. Environmental and genetic variation of phenolic compounds in red raspberry. *Journal of Food Composition and Analysis*. 18(8): 759–769.

Baenas, N., García-Viguera, C. and Moreno, D. 2014. Elicitation: A tool for enriching the bioactive composition of foods. *Molecules*. 19(9): 13541–3563.

Balasundram, N., Sundram, K. and Samman, S. 2006. Phenolic compounds in plants and agri-industrial by-products: Antioxidant activity, occurrence, and potential uses. *Food Chemistry*. 99(1): 191–203.

Barz, W. and Hoesel, W. 1979. Metabolism and degradation of phenolic compounds in plants. *Biochemistry of plant phenolics* (pp. 339–369). United States: Springer.

Bazzano, L.A., Li, T.Y., Joshipura, K.J. et al. 2008. Intake of fruit, vegetables, and fruit juices and risk of diabetes in women. *Diabetes Care*. 31(7): 1311–1317.

Becerra-Moreno, A., Redondo-Gil, M., Benavides, J. et al. 2015. Combined effect of water loss and wounding stress on gene activation of metabolic pathways associated with phenolic biosynthesis in carrot. *Frontiers in Plant Science*. 6: 837.

Bhagwat, S.A., Haytowitz, D.B. and Holden, J.M. 2008. USDA database for the isoflavone content of selected foods. Release 2.0.

Bhagwat, S., Haytowitz, D.B. and Holden, J.M. 2014. USDA. Database for the flavonoid content of selected foods, release 3.1. U.S. Department of Agriculture, Agricultural Research Service. Nutrient Data.

Bhagwat, S.A., Haytowitz, D.B., Prior, R.L. et al. 2004. USDA database for proanthocyanidin content of selected foods.

Boerjan, W., Ralph, J. and Baucher, M. 2003. Lignin biosynthesis. *Annual Review of Plant Biology*. 54(1): 519–546.

Bridle, P. and Timberlake, C.F. 1997. Anthocyanins as natural food colours-selected aspects. *Food Chemistry*. 58(1): 103–109.

Bryngelsson, S., Mannerstedt-Fogelfors, B., Kamal-Eldin, A. et al. 2002. Lipids and antioxidants in groats and hulls of Swedish oats (*Avena sativa* L.). *Journal of the Science of Food and Agriculture*. 82: 606–614.

Chandra, S., Khan, S., Avula, B. et al. 2014. Assessment of total phenolic and flavonoid content, antioxidant properties, and yield of aeroponically and conventionally grown leafy vegetables and fruit crops: A comparative study. *Evidence-Based Complementary and Alternative Medicine*. V 2014.

Chaovanalikit, A. and Wrolstad, R.E. 2004. Total anthocyanins and total phenolics of fresh and processed cherries and their antioxidant properties. *Journal of Food Science*. 69: FCT67–FCT72.

Chlopicka, J., Pasko, P., Gorinstein, A. et al. 2012. Total phenolic and total flavonoid content, antioxidant activity and sensory evaluation of pseudocereal breads. LWT–*Food Science and Technology*. 46(2): 548–55.

Chu, Y.H., Chang, C.L. and Hsu, H.F. 2000. Flavonoid content of several vegetables and their antioxidant activity. *Journal of the Science of Food and Agriculture*. 80(5): 561–566.

Chung, K.T., Wei, C.I., and Johnson, M.G. 1998. Are tannins a double-edged sword in biology and health? *Trends in Food Science & Technology*. 9(4): 168–75.

Clifford, M.N. 2000. Miscellaneous phenols in foods and beverages–nature, occurrence and dietary burden. *Journal of the Science of Food and Agriculture*. 80(7): 1126–1137.

Conn, E.E. 1986. *The Shikimic Acid Pathway (Recent Advance in Phytochemistry)*. New York: Springer, Plenum Press.

Duthie, G.G., Gardner, P.T., and Kyle, J.A.M. 2003. Plant polyphenols: Are they the new magic bullet? *Proceedings of the Nutrition Society*. 62(3): 599–603.

Einbond, L.S., Reynertson, K.A., Luo, X.D. et al. 2004. Anthocyanin antioxidants from edible fruits. *Food Chemistry*. 84(1): 23–28.

El Gharras, H. 2009. Polyphenols: Food sources, properties and applications. A review. *International Journal of Food Science & Technology*. 44(12): 2512–518.

Emmons, C.L. and David, M.P. 1999. Antioxidant activity and phenolic contents of oat groats and hulls. *Cereal Chemistry* 76(6): 902–06.

Escobedo-Avellaneda, Z., Gutiérrez-Uribe, J., Valdez-Fragoso, A. et al. 2014. Phytochemicals and antioxidant activity of juice, flavedo, albedo and comminuted orange. *Journal of Functional Foods*. 6: 470–481.

Ghasemzadeh, A. and Ghasemzadeh, N. 2011. Flavonoids and phenolic acids: Role and biochemical activity in plants and human. *Journal of Medicinal Plants Research*. 5(31): 6697–6703.

Ghiselli, A., Nardini, M., Baldi, A. et al. 1998. Antioxidant activity of different phenolic fractions separated from an Italian red wine. *Journal of Agricultural and Food Chemistry* 46(2): 361–67.

Gil, M.I., Tomás-Barberán, F.A., Hess-Pierce, B. et al. 2000. Antioxidant activity of pomegranate juice and its relationship with phenolic composition and processing. *Journal of Agricultural and Food Chemistry*. 48(10): 4581–4589.

Gil, M.I., Tomás-Barberán, F.A., Hess-Pierce, B. et al. 2002. Antioxidant capacities, phenolic compounds, carotenoids, and vitamin C contents of nectarine, peach, and plum cultivars from California. *Journal of Agricultural and Food Chemistry*. 50(17): 4976–4982.

Glass, A.D. and Dunlop, J. 1974. Influence of phenolic acids on ion uptake IV. Depolarization of membrane potentials. *Plant Physiology*. 54(6): 855–858.

Gorinstein, S., Martín-Belloso, O., Park, Y.-S. et al. 2001. Comparison of some biochemical characteristics of different citrus fruits. *Food Chemistry*. 74(3): 309–315.

Haminiuk, C.W.I., Maciel, G.M., Plata-Oviedo, M.S.V. et al. 2012. Phenolic compounds in fruits: An overview. *International Journal of Food Science & Technology*. 47(10): 2023–044.

Harborne, J.B. 1980. Plant phenolics. In E.A. Bell, B.V. Charlwood (eds.), *Encyclopedia of Plant Physiology, Volume 8 Secondary Plant Products*, pp. 329–395. New York: Springer-Verlag, Berlin Heidelberg.

Harborne, J.B. 1989. Methods in plant biochemistry. In P.M. Dey and J.B. Harborne (eds.), *Plant Phenolics*, p. 1. London: Academic Press.
Harborne, J.B. 1998. Phytochemical Methods: A guide to modern techniques of plant analysis. London: Chapman & Hall, 1973.
Heath, M.C. 2000. Nonhost resistance and nonspecific plant defenses. *Current Opinion in Plant Biology*. 3(4): 315–319.
Holdt, S.L. and Kraan, S. 2011. Bioactive compounds in seaweed: Functional food applications and legislation. *Journal of Applied Phycology*. 23(3): 543–597.
Inglett, G.E. and Chen, D. 2011. Contents of phenolics and flavonoids and antioxidant activities in skin, pulp, and seeds of miracle fruit. *Journal of Food Science*. 76(3): C479–C482.
John, J.A. and Shahidi, F. 2010. Phenolic compounds and antioxidant activity of Brazil nut (*Bertholletia excelsa*). *Journal of Functional Foods*. 2(3): 196–209.
Justesen, U., Knuthsen, P. and Leth, T. 1998. Quantitative analysis of flavonols, flavones, and flavanones in fruits, vegetables and beverages by high-performance liquid chromatography with photo-diode array and mass spectrometric detection. *Journal of Chromatography A*. 799(1): 101–110.
Kähkönen, M.P., Hopia, A.I., Vuorela, H.J. et al. 1999. Antioxidant activity of plant extracts containing phenolic compounds. *Journal of agricultural and food chemistry*. 47(10): 3954–3962.
Klein, K.O. 1998. Isoflavones, soy-based infant formulas, and relevance to endocrine function. *Nutrition Reviews*. 56(7): 193–204.
Lampila, P., Van Lieshout, M., Gremmen, B. et al. 2009. Consumer attitudes towards enhanced flavonoid content in fruit. *Food Research International*. 42(1): 122–29.
Lattanzio, V., Lattanzio, V.M., and Cardinali, A. 2006. Role of phenolics in the resistance mechanisms of plants against fungal pathogens and insects. *Phytochemistry: Advances in Research*. 661: 23–67.
Lattanzio, V., Cardinali, A. and Linsalata, V. 2012. Plant phenolics: A biochemical and physiological perspective. *Recent Advances in Polyphenol Research*. Vol 3.
Lipińska, L., Klewicka, E. and Sójka, M. 2014. The structure, occurrence and biological activity of ellagitannins: A general review. *Acta Scientiarum Polonorum Technologia Alimentaria*. 13(3): 289–299.
Liu, R.H. 2004. Potential synergy of phytochemicals in cancer prevention: Mechanism of action. *The Journal of nutrition*. 134(12): 3479S–3485S.
Lowry, B., Lee, D. and Hébant, C. 1980. The origin of land plants: A new look at an old problem. *Taxon*. 183–197.
Luthria, D.L. and Pastor-Corrales, M.A. 2006. Phenolic acids content of fifteen dry edible bean (*Phaseolus vulgaris* L.) varieties. *Journal of Food Composition and Analysis*. 19(2): 205–211.
Manach, C., Mazur, A. and Scalbert, A. 2005. Polyphenols and prevention of cardiovascular diseases. *Current Opinion in Lipidology*. 16(1): 77–84.
Manzoor, M., Anwar, F., Mahmood, Z. et al. 2012a. Variation in minerals, phenolics and antioxidant activity of peel and pulp of different varieties of peach (*Prunus persica* L.) fruit from Pakistan. *Molecules*. 17(6): 6491–6506.
Manzoor, M., Anwar, F., Saari, N. et al. 2012b. Variations of antioxidant characteristics and mineral contents in pulp and peel of different apple (*Malus domestica* Borkh.) cultivars from Pakistan. *Molecules*. 17(1): 390–407.
Marinova, D., Ribarova, F. and Atanassova, M. 2005. Total phenolics and total flavonoids in Bulgarian fruits and vegetables. *Journal of the University of Chemical Technology and Metallurgy*. 40(3): 255–260.

Martin, K.R. and Appel, C.L. 2010. Polyphenols as dietary supplements: A double-edged sword. *Journal of Nutrition and Dietary Supplements*. 2: 1–12.
Matkowski, A. 2008. Plant in vitro culture for the production of antioxidants—A review. *Biotechnology Advances*. 26(6): 548–560.
Mattila, P. and Hellström, J. 2007. Phenolic acids in potatoes, vegetables, and some of their products. *Journal of Food Composition and Analysis*. 20(3–4): 152–60.
Meagher, L.P. and Beecher, G.R. 2000. Assessment of data on the lignan content of foods. *Journal of food Composition and Analysis*. 13(6): 935–947.
Milder, I.E.J., Arts, I.C.W., Van De Putte, B. et al. 2005. Lignan contents of Dutch plant foods: A database including lariciresinol, pinoresinol, secoisolariciresinol and matairesinol. *British Journal of Nutrition*. 93(3): 393.
Mousavinejad, G., Emam-Djomeh, Z., Rezaei, K. et al. 2009. Identification and quantification of phenolic compounds and their effects on antioxidant activity in pomegranate juices of eight Iranian cultivars. *Food Chemistry*. 115(4): 1274–1278.
Neveu, V., Perez-Jiménez, J., Vos, F. et al. 2010. Phenol-Explorer: An online comprehensive database on polyphenol contents in foods. Database.
Özeker, E. 1999. Phenolic compounds and their importance. *Anadolu Ege Tarımsal Araştırma Enstitüsü Dergisi*. 9(2).
Pantelidis, G.E., Vasilakakis, M., Manganaris, G.A. et al. 2007. Antioxidant capacity, phenol, anthocyanin and ascorbic acid contents in raspberries, blackberries, red currants, gooseberries and Cornelian cherries. *Food Chemistry*. 102(3): 777–783.
Pascual-Theresa, S. and Sanchez-Ballesta, M.T. 2008. Anthocyanins: From plant to health. *Phytochemistry Reviews*. 7(2): 281–299.
Peñalvo, J.L., Haajanen, K.M., Botting, N. et al. 2005. Quantification of lignans in food using isotope dilution gas chromatography/mass spectrometry. *Journal of Agricultural and Food Chemistry*. 53(24): 9342–9347.
Pereira, D.M., Valentão, P., Pereira, J.A. et al. 2009. Phenolics: From chemistry to biology. *Molecules*. 14(6): 2202–2211.
Potter, S.M., Baum, J.A., Teng, H. et al. 1998. Soy protein and isoflavones: Their effects on blood lipids and bone density in postmenopausal women. *The American Journal of Clinical Nutrition*. 68(6): 1375S–1379S.
Proestos, C., Lytoudi, K., Mavromelanidou, O.K. et al. 2013. Antioxidant capacity of selected plant extracts and their essential oils. *Antioxidants*. 2(1): 11–22.
Puupponen-Pimiä, R., Nohynek, L., Meier, C. et al. 2001. Antimicrobial properties of phenolic compounds from berries. *Journal of Applied Microbiology*. 90(4): 494–507.
Rababah, T.M., Alhamad, M., Al-Mahasneh, M. et al. 2015. Effects of drying process on total phenolics, antioxidant activity and flavonoid contents of common Mediterranean herbs. *International Journal of Agricultural and Biological Engineering*. 8(2): 145.
Reyes, L.F., Miller, J.C. and Cisneros-Zevallos, L. 2005. Antioxidant capacity, anthocyanins and total phenolics in purple-and red-fleshed potato (*Solanum tuberosum* L.) genotypes. *American Journal of Potato Research*. 82(4): 271–277.
Ribeiro de Lima, M.T., Waffo-Téguo, P., Teissedre, P.L. et al. 1999. Determination of stilbenes (trans-astringin, cis- and trans-piceid, and cis- and trans-resveratrol) in Portuguese wines. *Journal of Agricultural and Food Chemistry*. 47(7): 2666–670.
Rice-Evans, C.A., Miller, N.J., and Paganga, G. 1996. Structure-antioxidant activity relationships of flavonoids and phenolic acids. *Free Radical Biology and Medicine*. 20(7): 933–956.

Robbins, R.J. 2003. Phenolic acids in foods: An overview of analytical methodology. *Journal of Agricultural and Food Chemistry*. 51(10): 2866–2887.
Rolfe, B.G. and Gresshoff, P.M. 1988. Genetic analysis of legume nodule initiation. *Annual Review of Plant Physiology and Plant Molecular Biology*. 39(1): 297–319.
Rosales-Martínez, P., Arellano-Cárdenas, S., Dorantes-Álvarez, L. et al. 2014. Comparison between antioxidant activities of phenolic extracts from mexican peanuts, peanuts skins, nuts and pistachios. *Journal of the Mexican Chemical Society*. 58(2): 185–193.
Routray, W. and Orsat, V. 2011. Blueberries and their anthocyanins: Factors affecting biosynthesis and properties. *Comprehensive Reviews in Food Science and Food Safety*. 10(6): 303–320.
Russell, W.R., Labat, A., Scobbie, L. et al. 2009. Phenolic acid content of fruits commonly consumed and locally produced in Scotland. *Food Chemistry*. 115(1): 100–04.
Sánchez-Moreno, C. 2002. Compuestos polifenólicos: Estructura y classificación: Presencia en alimentos y consumo: Biodisponibilidad y metabolismo. *Alimentaria*. 329: 19–28.
Schofield, P., Mbugua, D.M., and Pell, A.N. 2001. Analysis of condensed tannins: A review. *Animal Feed Science and Technology*. 91(1): 21–40.
Serrano, J., Puupponen-Pimiä, R., Dauer, A. et al. 2009. Tannins: Current knowledge of food sources, intake, bioavailability and biological effects. *Molecular Nutrition & Food Research*. 53: S2
Shetty, K. 2004. Role of proline-linked pentose phosphate pathway in biosynthesis of plant phenolics for functional food and environmental applications: A review. *Process Biochemistry*. 39(7): 789–804.
Shetty, K. and McCue, P. 2003. Phenolic antioxidant biosynthesis in plants for functional food application: Integration of systems biology and biotechnological approaches. *Food Biotechnology*. 17(2): 67–97.
Shetty, P., Atallah, M.T. and Shetty, K. 2002. Effects of UV treatment on the proline-linked pentose phosphate pathway for phenolics and L-DOPA synthesis in dark germinated *Vicia faba*. *Process Biochemistry*. 37(11): 1285–1295.
Shrikanta, A., Kumar, A. and Govindaswamy, V. 2015. Resveratrol content and antioxidant properties of underutilized fruits. *Journal of Food Science and Technology*. 52(1): 383–390.
Soong, Y.Y. and Barlow, P.J. 2004. Antioxidant activity and phenolic content of selected fruit seeds. *Food Chemistry*. 88(3): 411–17.
Stevens, J., Senaratna, T. and Sivasithamparam, K. 2006. Salicylic acid induces salinity tolerance in tomato (*Lycopersicon esculentum* cv. Roma): Associated changes in gas exchange, water relations and membrane stabilisation. *Plant Growth Regulation*. 49(1): 77–83.
Sulusoglu, M. 2014. Phenolic compounds and uses in fruit growing. *Türk Tarım ve Doğa Bilimleri*. 6(6): 947–956.
Swain, T. 1975. Evolution of flavonoid compounds. In *The Flavonoids*, pp. 1096–1129. United States: Springer.
Świeca, M. 2016. Potentially bioaccessible phenolics, antioxidant activity and nutritional quality of young buckwheat sprouts affected by elicitation and elicitation supported by phenylpropanoid pathway precursor feeding. *Food Chemistry*. 192: 625–32.
Tabart, J., Kevers, C., Sipel, A. et al. 2007. Optimisation of extraction of phenolics and antioxidants from black currant leaves and buds and of stability during storage. *Food Chemistry*. 105(3): 1268–1275.

Taiz, L., Zeiger, E., Møller, I.M. et al. 2015. Plant physiology and development. Sinauer Associates, Incorporated. ISBN 9781605352558.

Tang, S., Sheehan, D., Buckley, D.J. et al. 2001. Anti-oxidant activity of added tea catechins on lipid oxidation of raw minced red meat, poultry and fish muscle. *International Journal of Food Science & Technology*. 36: 685–692.

Tapiero, H., Tew, K.D., Ba, G.N. et al. 2002. Polyphenols: Do they play a role in the prevention of human pathologies? *Biomedicine & Pharmacotherapy*. 56(4): 200–207.

Toor, R.K. and Savage, G.P. 2005. Antioxidant activity in different fractions of tomatoes. *Food Research International*. 38(5): 487–494.

Tosun, I. and Inkaya, A.N. 2009. Resveratrol as a health and disease benefit agent. *Food Reviews International*. 26(1): 85–101.

Vasco, C., Ruales, J. and Kamal-Eldin, A. 2008. Total phenolic compounds and antioxidant capacities of major fruits from Ecuador. *Food Chemistry*. 111(4): 816–823.

Vitrac, X., Monti, J.P., Vercauteren, J. et al. 2002. Direct liquid chromatographic analysis of resveratrol derivatives and flavanonols in wines with absorbance and fluorescence detection. *Analytica Chimica Acta*. 458(1): 103–110.

Wang, S.Y. 2007. 16 fruits with high antioxidant activity as functional foods. *Functional Food Ingredients and Nutraceuticals: Processing Technologies*. 13: 371.

Wang, S.Y., Camp, M.J. and Ehlenfeldt, M.K. 2012. Antioxidant capacity and α-glucosidase inhibitory activity in peel and flesh of blueberry (*Vaccinium* spp.) cultivars. *Food Chemistry*. 132(4): 1759–1768.

Win, M.M., Abdul-Hamid, A., Baharin, B.S. et al. 2011. Phenolic compounds and antioxidant activity of peanut's skin, hull, raw kernel and roasted kernel flour. *Pakistan Journal of Botany*. 43(3): 1635–1642.

Wink, M. 2010. Introduction. In *Annual Plant Reviews Volume 39: Functions and Biotechnology of Plant Secondary Metabolites*, 2nd ed., 1–20. Oxford: Blackwell Publishing.

SECTION II

Analysis Methods

CHAPTER 4

Extraction Methods for Phenolic Compounds

Georgina Sandoval and Socorro Josefina Villanueva-Rodríguez

CONTENTS

4.1 INTRODUCTION

Extraction of phenolics is required for analytical quantification and use as antioxidants, bioactive ingredients, or supplements; for the production of pure ingredients and antioxidant extracts for application either as additive or nutraceutical products, and so on (Shi 2005). Before extraction, a pretreatment is usually performed in order to eliminate water and reduce particle size. These processes reduce the mass to be treated and increase the contact area of extraction solvents.

Table 4.1 summarizes the extraction methods for phenolics and their advantages and drawbacks. In general, pressurized systems and solid-phase extraction (SPE) are more expensive, but they produce higher purity extracts or compounds. Regarding energy savings and greener processes with lower solvent consumption, enzyme-assisted (EAE), microwave-assisted (MAE), and ultrasonic extraction (UE) are preferred.

Although solvent extraction is the most common method, more efficient or greener extraction methods are available. The selection of the best method depends on the nature of the matrix containing the phenolics and of the nature of the phenolic itself. For a particular compound, some solvents will be better than others, but once chosen, different extraction methods will have differences. Frequently, combining methods also increases extraction efficacy. In the following sections, extraction methods for phenolics are detailed.

TABLE 4.1 Extraction Methods for Phenolics

Extraction Method	Advantages	Drawbacks	Suitable for
Solvent (SE)	Simple. A lot of information available.	Solvent and energy costs. Auto-oxidation.	All types of phenolics.
Enzyme-Assisted (EAE)	Greener. Nontoxic. Increased yield and bioactivity. Energy savings.	Enzyme cost.	Thermolabile phenolics.
Microwave (MAE) and Ultrasonic (UAE)	Less heating cost. Lower solvent consumption. Accelerated extraction.	Uncontrolled heating. Polymerization (for microwave).	All types of phenolics.
Solid-Phase (SPE) & MIP[a]	Simple. Reduced extraction time. Increased purity.	Cost. Low batch-to-batch reproducibility.	All types of phenolics.
Pressurized Liquid (PLE)	Oxidation avoided.	Initial investment. Thermodegradation.	Non-thermolabile phenolics.
Supercritical fluids (SFE)	Facilitated purification.	Initial investment. Energy cost.	Polyphenols. Less polar phenolics.

[a] MIP: Molecularly Imprinted Polymers.

4.2 SOLVENT EXTRACTION (SE)

As previously stated, solvent extraction remains the most common extraction method. Other methods, such as MAE and UAE, still use solvents (but in lower quantity). The main parameters to consider in SE are solvent, temperature, solvent: solid ratio and extraction time. Besides cost, one must keep in mind the solubility of the phenolics to be extracted as well as the matrix that contains them (to avoid unwanted extractables) when choosing solvents. Table 4.2 presents a list of solvents used for extraction of some phenolic types.

Hot water, methanol, ethanol, and acetone (or their mixtures) are commonly used to extract polar phenolics, phenolic acids, and flavonoids. Higher molecular weight phenolics, such as tannins and anthocyanins, generally require a mixture of polar solvents with less polar solvents, alkalinized water, or diluted acids. Stilbenes (resveratrol for instance) have been successfully extracted with acetone, ethyl acetate, and alcohols.

Temperature is another important factor to consider in solvent extraction. Temperatures higher than 100°C could affect recovery yield of some phenolics

TABLE 4.2 Examples of Solvents Used for Extraction of Some Phenolic Types

Phenolic Type	Solvents	References
Phenolic acids Flavonoids	Water, methanol, ethanol, acetone & their mixtures	(Rodríguez-Pérez et al. 2015, Upadhyay et al. 2015, Vajić et al. 2015)
Tannins	Acetone, alkalinized water	(Chavan and Singhal 2013, Low et al. 2015)
Anthocyanins	Acetonitrile, alcohols + soft or diluted acids, chloroform	(Pompeu et al. 2009, Stalikas 2007, Yang et al. 2009)
Stilbenes	Acetone, ethyl acetate, alcohols	(Liu et al. 2013, Sun et al. 2006, Zhang et al. 2015)

(Palma et al. 2001) and have no effect on others (Amendola et al. 2010, Palma et al. 2001). A compromise between yield and stability must be chosen when higher temperatures are used. Arrhenius plots could be useful in such cases (Bucić-Kojić et al. 2015, Torun et al. 2015). For a chosen solvent, solvent: solid ratio is usually optimized together with other parameters such as temperature (Saikia et al. 2015) and extraction time (Yılmaz et al. 2015, Yingngam et al. 2015). Sometimes, particle size and bioactivity are also evaluated when optimizing solvent extraction (Aaby et al. 2013, Çam and İçyer 2015).

Emergent classes of solvents for phenolics (and natural products) extraction are ionic liquids (IL) and deep eutectic solvents (DES; Dai et al. 2013). IL are salts composed of organic cations and organic or inorganic anions, which are liquid at room temperature. IL exhibit interesting features besides their capacity to solubilize organic compounds: non-volatility (compared with volatile organic solvents), chemical and thermal stability, non-flammability, and high conductivity. However, IL could be toxic, depending on the anions. To overcome the potential toxicity of IL, "green-IL" or "Bio-IL" have been developed from biomaterial-derived products such as organic acids, amino acids, amines, and sugars. DES are composed of 2–3 pure compounds (not necessarily salts), which form hydrogen bonds. The resulting DES has a lower melting point than each of the individual compounds, and exhibits properties similar to those of ionic liquids. As for bio-IL, when DES come from natural compounds they are called *NADES*. Composition and melting points of several IL and DES can be found in Choi et al. (2011) and Dai et al. (2013). Examples of phenolics extracted with IL and DES are vanillin (Cláudio et al. 2010) and rutin (Choi et al. 2011).

4.3 ENZYME-ASSISTED EXTRACTION (EAE)

Solvent extraction has some drawbacks regarding safety, environmental risk, energy cost, and product quality. More environmentally friendly extraction methods are gaining interest, but in the case of phenolic compounds, cell walls can reduce aqueous in extraction efficiency. Enzyme-assisted extraction (EAE) takes advantage of enzyme abilities to perform selective disruption of cell matrices to release phenolics. EAE could also be used as pretreatment for other extraction methods (Mushtaq et al. 2014) or combined with other techniques (see Section 4.6). Additional advantages of EAE include lower extraction temperatures and thus protection of thermolabile compounds, shorter extraction times (Puri et al. 2012), higher extraction efficiency (Chen et al. 2011, Li et al. 2006), and increased bioactivity and stability (Dinkova et al. 2014, Puupponen-Pimiä et al. 2008). On the other hand, EAE could modify profiles of phenolics extracted, promote deglycosylation of flavonoids, and cause changes in sensory characteristics; but, in most cases, these modifications are favorable or could be controlled by extraction time or enzyme dose (Arnous and Meyer 2010, Coetzee et al. 2014, Kammerer et al. 2005, Puupponen-Pimiä et al. 2008, Sandoval and Villanueva 2013).

Table 4.3 presents examples of EAE of phenolics. The first step in EAE is determining which enzyme or enzymes are suitable for the matrix containing phenolics. For instance in seaweeds polysaccharides entail difficult phenolic extractions, therefore polysaccharidases are useful (Wijesinghe and Jeon 2012). Indeed, polysaccharide-degrading enzymes are the most used enzymes for EAE, but other accessory enzymes such as phenolic acid esterases are also useful for phenolic extraction (Coetzee et al. 2014, Ramirez et al. 2008). Once an enzyme or enzymes are chosen, parameters to optimize in EAE include enzyme concentration, extraction time, temperature, and pH (Puri et al. 2012).

TABLE 4.3 Examples of Enzymes Used for Extraction of Phenolics

Source	Enzymes	References
Seaweeds	Polysaccharydases and Proteases	(Wijesinghe and Jeon 2012)
Wine grapes	AEB	(Río Segade et al. 2015)
Cauliflower	Viscozyme, Rapidase	(Huynh et al. 2014)
Vanilla pods	β-glucosidase and pectinase	(Ruiz-Terán et al. 2001)
Grape skin and seeds	Cellulases, pectinases, β-glucosidase	(Arnous and Meyer 2010, Kammerer et al. 2005, Xu et al. 2014)
Citrus peel	Celluzyme MX	(Li et al. 2006)
Apple peel	Cellulases	(Kim et al. 2005)
Bilberry fruit peel & juice	Pectinex, PanzymPro Color, Panzym BE XXL	(Dinkova et al. 2014, Puupponen-Pimiä et al. 2008)
Watermelon rind	Kemzyme	(Mushtaq et al. 2014)
Ginko biloba leaves	Cellulases and pectinases	(Chen et al. 2011)
Olive[a]	Cellulases and pectinases	(De Faveri et al. 2008, Sharma et al. 2015)
Purple corn	Protease	(Jing and Giusti 2007)

[a] Enzymes increase olive oil yield and quantity of phenolics in olive oil.

However, for commercially available enzymes, producers could recommend optimal pH and temperature or a range of pH and temperature adequate for the enzyme. The main challenges in EAE are decreasing enzyme costs and developing specific enzymes or cocktails for phenolic extraction from a given matrix (Sandoval and Villanueva 2013).

4.4 ULTRASOUND (UAE) AND MICROWAVE (MAE) ASSISTED EXTRACTION

Ultrasound (UAE) and Microwave (MAE) assisted extraction are frequently used together with solvent extraction to improve extraction efficiency. An increasing number of phenolic extraction publications report the evaluation of these two technologies at the same time (separately or combined, in which case it is called UMAE) with varying results dependent on source (see Table 4.4).

TABLE 4.4 Examples of UAE and MAE of Phenolics

Source	Extraction Evaluated	References
Burdock leaves	Simultaneous IL-UMAE	(Lou et al. 2012)
Maize	UAE and MAE	(Biesaga 2011)
Blueberry	UAE and MAE	(Routray and Orsat 2014)
Pistacia lentiscus leaves	UAE and MAE	(Dahmoune et al. 2014)
Rosemary	UAE and MAE	(Švarc-Gajić et al. 2013)
Fig	IL-UAE	(Qin et al. 2014)
Flax seeds	UAE	(Corbin et al. 2015)
Rice	MAE	(Setyaningsih et al. 2015)
Eucalyptus	MAE	(Bhuyan et al. 2015)

Ultrasonic waves (sound waves with frequencies higher than 20 kHz) produce cavitation in the solvent forming gas bubbles that collapse near the surface of the solid sample increasing solvent penetration and further swelling. Both temperature and pressure are increased and therefore solubility, diffusivity, and penetration are also increased. A scheme of UAE apparatus is depicted in Figure 4.1.

Microwaves (nonionizing electromagnetic radiations with a frequency from 0.3 to 300) are absorbed by polar molecules, increasing energy and heating the whole sample simultaneously (provided that microwaves are homogeneously radiated over the sample). Figure 4.2 shows a scheme of MAE apparatus. Only dipolar materials are affected by microwaves and the higher the value of dielectric constant, the higher the level of microwave absorption, and the heating is more rapid for a given frequency. Dielectric constants of solvents used in MAE can be consulted in (Kaufmann and Christen 2002).

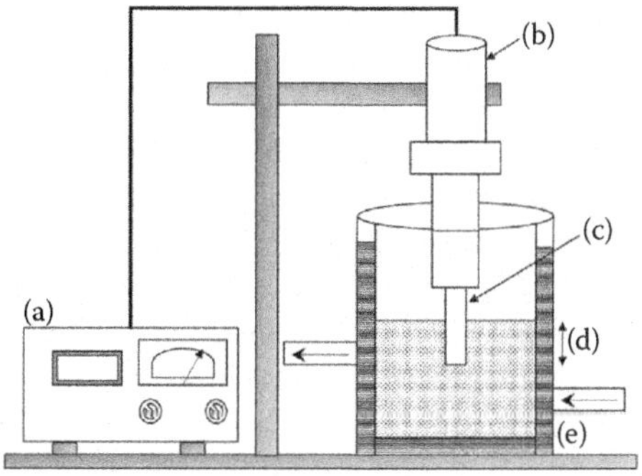

FIGURE 4.1 Scheme of UAE apparatus showing (a) ultrasound generator, (b) transducer, (c) ultrasonic probe, (d) depth of probe in liquid, and (e) jacketed beaker. (Reprinted with permission from Kadam, S.U., Tiwari, B.K., and O'Donnell, C.P. *Journal of Agricultural and Food Chemistry*, 61(20), 4667–4675, 2013. Copyright 2013 American Chemical Society.)

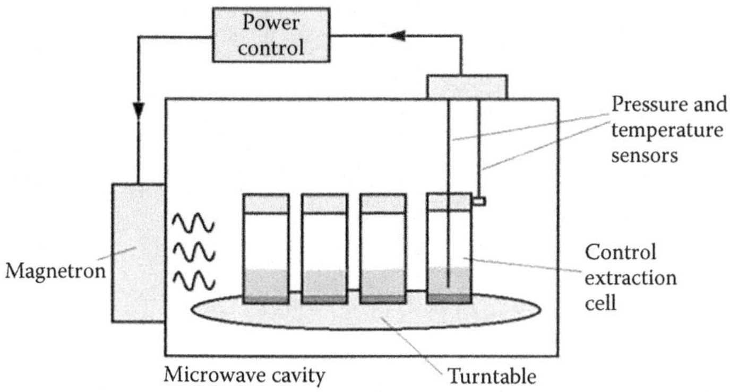

FIGURE 4.2 Scheme of MAE apparatus. Extraction vessels are placed on a turntable inside an oven and subjected to microwave irradiations generated by a magnetron. (Adapted from Portet-Koltalo, F., and Machour, N., in *Diesel Engine: Combustion, Emissions and Condition Monitoring*, 2013.)

Solvent: solid ratio in UAE and MAE is lower compared to traditional solvent extraction, a few milliliters per gram of sample compared with, for example, 100–150 milliliters. Parameters to optimize in UAE and MAE are selection of solvent, frequency (for UAE), or power (for MAE) and extraction time (which is also reduced compared with time required in traditional solvent extraction). UAE and MAE are usually operated at ambient temperature as both techniques tend to increase sample temperature. Depending on the equipment, temperature can be automatically controlled together with other parameters such as power and pressure. It is advisable to evaluate also stability of phenolics when UAE/MAE is used (Biesaga 2011). EAE could be also combined with UAE; for instance, the combination of these extraction techniques improved total phenolic content, antioxidant, and anti-tumor activities from *Trapa quadrispinosa* Roxb. residues (Li et al. 2017).

4.5 PRESSURIZED LIQUID (PLE) AND SUPERCRITICAL FLUID EXTRACTION (SFE)

Pressurized liquid extraction (PLE), also called accelerated solvent extraction, subcritical-fluid extraction, or high pressure and temperature extraction, uses liquid solvents at controlled elevated temperature and pressure below their critical point. Figure 4.3 illustrates the phase diagram for water, which in subcritical state is used as solvent for "subcritical water extraction" (SWE). Subcritical water is a good solvent even for non-polar flavonoids (Ko et al. 2014). PLE has other advantages over UAE/MAE: it requires smaller amounts of solvent and yields cleaner extracts. Automatization and coupling with separation techniques is also possible since the matrix residue is retained inside the sample extraction cell (see Figure 4.4) and no additional filtration step is required. As the sample is kept in an inert atmosphere, PLE improves the stability of light and oxygen-sensitive

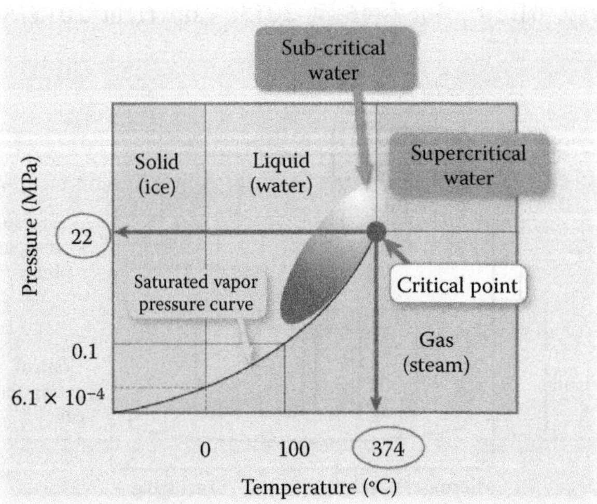

FIGURE 4.3 Phase diagram of water as a function of temperature and pressure. (From Mukherjee, N., Unusual properties of water: Phase behavior of Subcritical and Supercritical water, 2016. See https://www.linkedin.com/pulse/unusual-properties-water-phase-behavior-subcritical-mukherjee.)

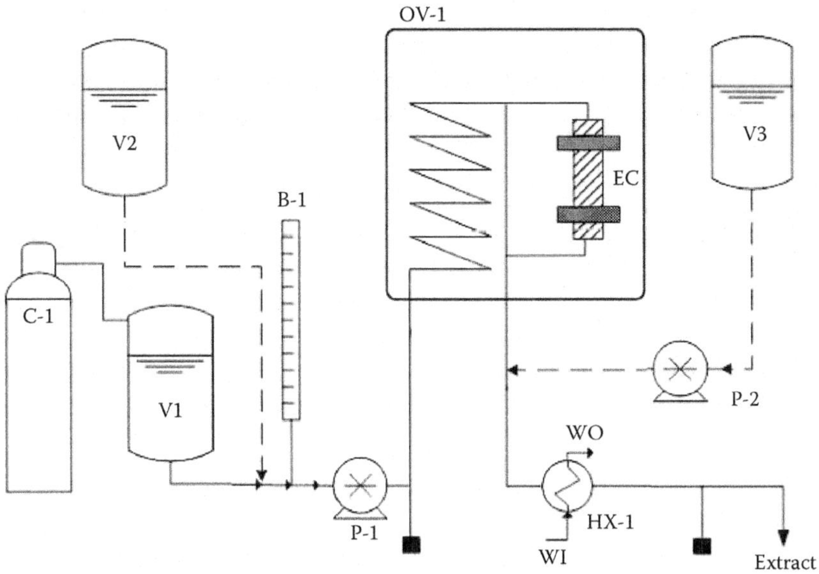

FIGURE 4.4 Schematic diagram of SWE system, B-1: Burette, C-1: Nitrogen cylinder, EC: Extraction cell, HX-1: Heat Exchanger, OV-1: Oven, P-1, 2: Pumps, V-1: Water tank, V-2: Solvent tank, V-3: Rinsing solvent tank, WI: Water inlet, WO: Water outlet. (Adapted from Haghighi Asl, A., and Khajenoori, M., in *Mass Transfer—Advances in Sustainable Energy and Environment Oriented Numerical Modeling*, pp. 459–487, 2013.)

phenolics. However, the high temperatures used in PLE can degrade thermolabile phenolics and browning, or polymerization can also occur (Wijngaard and Brunton 2009).

Another technique used to extract polyphenols is the supercritical fluids extraction (SFE). This is a very selective method for separating specific compounds from a whole plant extract. This extraction method is based on the properties acquired by the super critical fluids (SCF) when brought above its supercritical state, namely, in a state in which it behaves as a hybrid fluid, that is, can spread like a gas (effusion) and dissolve substances as a liquid (solvent; see also Figure 4.3). CO_2 is widely used in SFE because it has a low critical pressure and temperature (Pc ¼ 73.8 atm, Tc ¼ 31.1C); therefore, it is an ideal solvent for the treatment of natural products. SCF is characterized by the wide range of densities that can be taken. Above the critical conditions, small changes in pressure and temperature produce large changes in density. This property or behavior allows coverage of different polarity ranges and produces different relatively pure fractions, using only CO_2 that is nontoxic, nonflammable, and of a noncorrosive nature, making it the perfect solvent for natural products. The big disadvantage of the method is its high cost. The resolution of the method depends on controlling the extraction conditions. High pressure increases extraction yield. Increasing temperature also has a positive impact on the extraction performance as long as it does not damage the structure or quality of the substance of interest (Shi et al. 2005).

Another factor that may be involved in supercritical CO_2 extraction is the use of a cosolvent to obtain a specific polarity, as in the case of the work reported by Chafer et al. (Chafer et al. 2005) during the extraction of polyphenols from grape skin. In this case, extraction conditions were 20 percent v/v ethanol, 60°C, 250 bars, and flow rate two milliliters/minute. The extract was collected in water; the more polar polyphenols ((+)-catechin and (−)-epicatechin) remained in solution, while rutin, quercetin, and trans-resveratrol precipitated in this medium. The solution of the extracted polyphenols was filtered, and then dissolved with ethanol/H_2O (40:60). On the other hand, Pascual-Martí et al. (2001) selectively extracted resveratrol from grape skin using the following conditions: 40°C, 150 bar, 7.5 percent ethanol, and extraction time of 15 minutes. According to reports from Díaz-Reinoso et al. (2006), the extracts obtained by SFE showed a higher antioxidant capacity and performance than those obtained by other methods such as liquid solvent extraction simultaneously. Ramos-Jerz (2007) reported the extraction of polyphenolic compounds from the seed of avocado Hass through supercritical CO_2 using 207 bars at 50°C.

SFE and PLE were recently applied to extraction of phenolic compounds and anthocyanins from juçara (Garcia-Mendoza et al. 2017). SFE proved to be the most selective process for anthocyanin extraction, while PLE provided extracts rich in anthocyanins. The increase of temperature for PLE enhanced total phenolics.

As stated in Section 4.3, EAE could be combined with SFE. Sources of phenolics extracted using EAE/SFE are black pepper oleoresin (Dutta and Bhattacharjee 2015) and pomegranate peel (Mushtaq et al. 2015). Different methods were evaluated to extract phenolics from olive leaves (Taamalli et al. 2012), including PLE, SFE, and MAE. PLE resulted in higher extraction yields, but yield did not correlate with bioactivity evaluated as cytotoxicity against human breast cancer cells. Phenolics from *Cretan barberry* herb were extracted by PLE and SFE (Kukula-Koch et al. 2013). SFE enabled preliminary purification of the extract and subsequent extraction of the residue was performed using pressurized water: ethanol (50:50) mixture; leading to extracts with high total phenolic content and antioxidant properties. It was also observed (as for other techniques and sources) that the extraction method and solvents used affected both the bioactivity and the phenolics profile in the extract.

So, a number of substances and raw materials were obtained and processed by PLE and SFE. While these methods of extraction of polyphenols have advantages of great interest, the advantages must be weighed as the methodology is expensive and the use of these extraction processes depends on the added value that is obtained or desired given the substances' extraction.

4.6 CONCLUSIONS

As shown in previous sections of this chapter, a wide choice of extraction methodologies is available to extract phenolics. Equipment available and economic criteria will be considered to make a choice of the extraction strategy criteria that must be weighed against the needs of purity degree and quantity of substances to be obtained. Regarding optimization of parameters, there are well-known sources of phenolics (e.g., grapes, citrus, olive, and berries), for which reported extraction methods in scientific literature are good guides to optimize phenolics extraction from similar sources. However, on the other hand, there are also many new sources of polyphenols to explore that can be selected, adapting one or more of the methodologies available.

REFERENCES

Aaby, K., Grimmer, S., and Holtung, L. 2013. Extraction of phenolic compounds from bilberry (*Vaccinium myrtillus* L.) press residue: Effects on phenolic composition and cell proliferation. *LWT–Food Science and Technology*. 54 (1):257–264.

Amendola, D., De Faveri, D.M., and Spigno, G. 2010. Grape marc phenolics: Extraction kinetics, quality and stability of extracts. *Journal of Food Engineering*. 97 (3):384–392.

Arnous, A., and Meyer, A.S. 2010. Discriminated release of phenolic substances from red wine grape skins (*Vitis vinifera* L.) by multicomponent enzymes treatment. *Biochemical Engineering Journal*. 49 (1):68–77.

Bhuyan, D.J., Van Vuong, Q., Chalmers, A.C. et al. 2015. Microwave-assisted extraction of Eucalyptus robusta leaf for the optimal yield of total phenolic compounds. *Industrial Crops and Products*. 69:290–299.

Biesaga, M. 2011. Influence of extraction methods on stability of flavonoids. *Journal of Chromatography A*. 1218 (18):2505–2512.

Bucić-Kojić, A., Sovová, H., Planinić, M. et al. 2015. Temperature-dependent kinetics of grape seed phenolic compounds extraction: Experiment and model. *Food Chemistry*. 136 (3–4):1136–1140.

Çam, M., and İçyer, N. 2015. Phenolics of pomegranate peels: Extraction optimization by central composite design and alpha glucosidase inhibition potentials. *Journal of Food Science and Technology*. 52 (3):1489–1497.

Chafer, A., Pascual-Martí, M.C., Salvador, A. et al. 2005. Supercritical fluid extraction and HPLC determination of relevant polyphenolic compounds in grape skin. *Journal of Separation Science*. 28 (16):2050–2056.

Chavan, Y.V., and Singhal, R.S. 2013. Separation of polyphenols and arecoline from areca nut (*Areca catechu* L.) by solvent extraction, its antioxidant activity, and identification of polyphenols. *Journal of the Science of Food and Agriculture*. 93 (10):2580–2589.

Chen, S., Xing, X.H., Huang, J.J. et al. 2011. Enzyme-assisted extraction of flavonoids from Ginkgo biloba leaves: Improvement effect of flavonol transglycosylation catalyzed by Penicillium decumbens cellulase. *Enzyme and Microbial Technology*. 48 (1):100–105.

Choi, Y.H., van Spronsen, J., Dai, Y. et al. 2011. Are natural deep eutectic solvents the missing link in understanding cellular metabolism and physiology? *Plant Physiology*. 156 (4):1701–1705.

Cláudio, A.F.M., Freire, M.G., Freire, C.S. et al. 2010. Extraction of vanillin using ionic-liquid-based aqueous two-phase systems. *Separation and Purification Technology*. 75 (1):39–47.

Coetzee, G., Joubert, E., van Zyl, W.H. et al. 2014. Improved extraction of phytochemicals from rooibos with enzyme treatment. *Food and Bioproducts Processing*. 92 (4):393–401.

Corbin, C., Fidel, T., Leclerc, E.A. et al. 2015. Development and validation of an efficient ultrasound assisted extraction of phenolic compounds from flax (*Linum usitatissimum* L.) seeds. *Ultrasonics Sonochemistry*. 26:176–185.

Dahmoune, F., Spigno, G., Moussi, K. et al. 2014. *Pistacia lentiscus* leaves as a source of phenolic compounds: Microwave-assisted extraction optimized and compared with ultrasound-assisted and conventional solvent extraction. *Industrial Crops and Products.* 61:31–40.

Dai, Y., van Spronsen, J., Witkamp, G.-J. et al. 2013. Ionic liquids and deep eutectic solvents in natural products research: Mixtures of solids as extraction solvents. *Journal of Natural Products.* 76 (11):2162–2173.

De Faveri, D., Aliakbarian, B., Avogadro, M. et al. 2008. Improvement of olive oil phenolics content by means of enzyme formulations: Effect of different enzyme activities and levels. *Biochemical Engineering Journal.* 41 (2):149–156.

Díaz-Reinoso, B., Moure, A., Domínguez, H. et al. 2006. Supercritical CO_2 extraction and purification of compounds with antioxidant activity. *Journal of Agricultural and Food Chemistry.* 54 (7):2441–2469.

Dinkova, R., Heffels, P., Shikov, V. et al. 2014. Effect of enzyme-assisted extraction on the chilled storage stability of bilberry (*Vaccinium myrtillus* L.) anthocyanins in skin extracts and freshly pressed juices. *Food Research International.* 65, Part A:35–41.

Dutta, S., and Bhattacharjee, P. 2015. Enzyme-assisted supercritical carbon dioxide extraction of black pepper oleoresin for enhanced yield of piperine-rich extract. *Journal of Bioscience and Bioengineering.* 120 (1):17–23.

Garcia-Mendoza, M.P., Espinosa-Pardo, F.A., Baseggio, A.M. et al. 2017. Extraction of phenolic compounds and anthocyanins from juçara (*Euterpe edulis* Mart.) residues using pressurized liquids and supercritical fluids. *The Journal of Supercritical Fluids.* 119:9–16.

Haghighi Asl, A., and Khajenoori, M. 2013. Subcritical water extraction. In *Mass Transfer—Advances in Sustainable Energy and Environment Oriented Numerical Modeling*, ed. Hironori Nakajima, pp. 459–487.

Huynh, N.T., Smagghe, G., Gonzales, G.B. et al. 2014. Enzyme-assisted extraction enhancing the phenolic release from cauliflower (*Brassica oleracea* L. var. botrytis) outer leaves. *Journal of Agricultural and Food Chemistry.* 62 (30): 7468–7476.

Jing, P., and Giusti, M.M. 2007. Effects of extraction conditions on improving the yield and quality of an anthocyanin-rich purple corn (*Zea mays* L.) color extract. *Journal of Food Science.* 72 (7):C363–C368.

Kadam, S.U., Tiwari, B.K., and O'Donnell, C.P. 2013. Application of novel extraction technologies for bioactives from marine algae. *Journal of Agricultural and Food Chemistry.* 61 (20):4667–4675.

Kammerer, D., Claus, A., Schieber, A. et al. 2005. A novel process for the recovery of polyphenols from grape (*Vitis vinifera* L.) pomace. *Journal of Food Science* 70 (2):C157–C163.

Kaufmann, B., and Christen, P. 2002. Recent extraction techniques for natural products: Microwave-assisted extraction and pressurised solvent extraction. *Phytochemical Analysis.* 13 (2):105–113.

Kim, Y.J., Kim, D.O., Chun, O.K. et al. 2005. Phenolic extraction from apple peel by cellulases from *Thermobifida fusca. Journal of Agricultural and Food Chemistry.* 53 (24):9560–9565.

Ko, M.J., Cheigh, C.I., and Chung, M.S. 2014. Relationship analysis between flavonoids structure and subcritical water extraction (SWE). *Food Chemistry.* 143:147–155.

Kukula-Koch, W., Nektarios Aligiannis, Maria Halabalaki, Alexios-Leandros Skaltsounis, Kazimierz Glowniak, and Eleftherios Kalpoutzakis. 2013. Influence of extraction procedures on phenolic content and antioxidant activity of Cretan barberry herb. *Food Chemistry*. 138 (1):406-413. http://dx.doi.org/10.1016/j.foodchem.2012.10.045.

Li, F., Mao, Y.D., Wang, Y.F. et al. 2017. Optimization of ultrasonic-assisted enzymatic extraction conditions for improving total phenolic content, antioxidant and antitumor activities in vitro from *Trapa quadrispinosa* Roxb. residues. *Molecules*. 22 (3):396.

Li, B.B., Smith, B., and Hossain, Md. M. 2006. Extraction of phenolics from citrus peels: II. Enzyme-assisted extraction method. *Separation and Purification Technology*. 48 (2):189–196.

Liu, C., Wang, L., Wang, J. et al. 2013. Resveratrols in Vitis berry skins and leaves: Their extraction and analysis by HPLC. *Food Chemistry*. 136 (2):643–649.

Lou, Z., Wang, H., Zhu, S. et al. 2012. Ionic liquids based simultaneous ultrasonic and microwave assisted extraction of phenolic compounds from burdock leaves. *Analytica Chimica Acta*. 716:28–33.

Low, J.H., Rahman, W.A.W.A., and Jamaluddin, J. 2015. The influence of extraction parameters on spent coffee grounds as a renewable tannin resource. *Journal of Cleaner Production*. 101:222–228.

Mukherjee, N. 2016. Unusual properties of water: Phase behavior of Subcritical and Supercritical water https://www.linkedin.com/pulse/unusual-properties-water-phase-behavior-subcritical-mukherjee.

Mushtaq, M., Sultana, B., Anwar, F. et al. 2015. Enzyme-assisted supercritical fluid extraction of phenolic antioxidants from pomegranate peel. *The Journal of Supercritical Fluids*. Volume 104:122–131.

Mushtaq, M., Sultana, B., Bhatti, H.N. et al. 2014. RSM based optimized enzyme-assisted extraction of antioxidant phenolics from underutilized watermelon (*Citrullus lanatus* Thunb.) rind. *Journal of Food Science and Technology*. 1–9.

Palma, M., Piñeiro, Z., and Barroso, C.G. 2001. Stability of phenolic compounds during extraction with superheated solvents. *Journal of Chromatography A*. 921 (2):169–174.

Pascual-Martı́, M.C., Salvador, A., Chafer, A., and Berna, A. 2001. Supercritical fluid extraction of resveratrol from grape skin of *Vitis vinifera* and determination by HPLC. *Talanta*. 54 (4):735–740.

Pompeu, D.R., Silva, E.M., and Rogez, H. 2009. Optimisation of the solvent extraction of phenolic antioxidants from fruits of *Euterpe oleracea* using response surface methodology. *Bioresource Technology*. 100 (23):6076–6082.

Portet-Koltalo, F., and Machour, N. 2013. Analytical methodologies for the control of particle-phase polycyclic aromatic compounds from diesel engine exhaust. *Diesel Engine: Combustion, Emissions and Condition Monitoring*, ed. Saiful Bari, Chapter 4.

Puri, M., Sharma, D., and Barrow, C.J. 2012. Enzyme-assisted extraction of bioactives from plants. *Trends in Biotechnology*. 30 (1):37–44.

Puupponen-Pimiä, R., Nohynek, L., Ammann, S. et al. 2008. Enzyme-assisted processing increases antimicrobial and antioxidant activity of bilberry. *Journal of Agricultural and Food Chemistry*. 56 (3):681–688.

Qin, H., Zhou, G., Peng, G. et al. 2014. Application of ionic liquid-based ultrasound-assisted extraction of five phenolic compounds from fig (*Ficus carica* L.) for HPLC-UV. *Food Analytical Methods*. 1–9.

Ramirez, L., Arrizon, J., Sandoval, G. et al. 2008. A new microplate screening method for the simultaneous activity quantification of feruloyl esterases, tannases, and chlorogenate esterases. *Applied Biochemistry and Biotechnology*. 151 (2-3):711–723.

Ramos-Jerz, M. 2007. Phytochemical analysis of avocado Seeds (*Persea americana* Mill., c.v. Hass). Göttingen, Germany: Cuvillier.

Río Segade, S., Pace, C., Torchio, F. et al. 2015. Impact of maceration enzymes on skin softening and relationship with anthocyanin extraction in wine grapes with different anthocyanin profiles. *Food Research International*. 71:50–57.

Rodríguez-Pérez, C., Quirantes-Piné, R., Fernández-Gutiérrez, A. et al. 2015. Optimization of extraction method to obtain a phenolic compounds-rich extract from *Moringa oleifera* Lam leaves. *Industrial Crops and Products*. 66:246–254.

Routray, W., and Orsat, V. 2014. MAE of phenolic compounds from blueberry leaves and comparison with other extraction methods. *Industrial Crops and Products* 58:36–45. http://dx.doi.org/10.1016/j.indcrop.2014.03.038

Ruiz-Terán, F., Perez-Amador, I., and López-Munguia, A. 2001. Enzymatic extraction and transformation of glucovanillin to vanillin from vanilla green pods. *Journal of Agricultural and Food Chemistry*. 49 (11):5207–5209.

Saikia, S., Mahnot, N.K., and Mahanta, C.L. 2015. Optimisation of phenolic extraction from *Averrhoa carambola* pomace by response surface methodology and its microencapsulation by spray and freeze drying. *Food Chemistry*. 171:144–152.

Sandoval, G., and Villanueva, S. 2013. Extracción enzimática de antioxidantes. In *Antioxidantes en alimentos y salud*, ed. A. Parrilla et al. Mexico: Clave Editorial.

Setyaningsih, W., Saputro, I.E., Palma, M. et al. 2015. Optimisation and validation of the microwave-assisted extraction of phenolic compounds from rice grains. *Food Chemistry*. 169:141–149.

Sharma, R., Sharma, P.C., Rana, J.C. et al. 2015. Improving the olive oil yield and quality through enzyme-assisted mechanical extraction, antioxidants and packaging. *Journal of Food Processing and Preservation*. 39 (2):157–166.

Shi, J., Nawaz, H., Pohorly, J. et al. 2005. Extraction of polyphenolics from plant material for functional foods—Engineering and technology. *Food Reviews International*. 21 (1):139–166.

Stalikas, C.D. 2007. Extraction, separation, and detection methods for phenolic acids and flavonoids. *Journal of Separation Science*. 30 (18):3268–3295.

Sun, B., Ribes, A.M., Leandro, M.C. et al. 2006. Stilbenes: Quantitative extraction from grape skins, contribution of grape solids to wine and variation during wine maturation. *Analytica Chimica Acta*. 563 (1–2):382–390.

Švarc-Gajić, J., Stojanović, Z., Segura Carretero, A. et al. 2013. Development of a microwave-assisted extraction for the analysis of phenolic compounds from *Rosmarinus officinalis*. *Journal of Food Engineering* 119 (3):525–532.

Taamalli, A., Arráez-Román, D., Barrajón-Catalán, E. et al. 2012. Use of advanced techniques for the extraction of phenolic compounds from Tunisian olive leaves: Phenolic composition and cytotoxicity against human breast cancer cells. *Food and Chemical Toxicology*. 50 (6):1817–1825.

Torun, M., Dincer, C., Topuz, A. et al. 2015. Aqueous extraction kinetics of soluble solids, phenolics and flavonoids from sage (*Salvia fruticosa* Miller) leaves. *Journal of Food Science and Technology*. 52 (5):2797–2805.

Upadhyay, R., Jha, A., Singh, S.P. et al. 2015. Appropriate solvents for extracting total phenolics, flavonoids and ascorbic acid from different kinds of millets. *Journal of Food Science and Technology.* 52 (1):472–478.

Vajić, U.-J., Grujić-Milanović, J., Živković, J. et al. 2015. Optimization of extraction of stinging nettle leaf phenolic compounds using response surface methodology. *Industrial Crops and Products.* 74:912–917.

Wijngaard, H., and Brunton, N. 2009. The optimization of extraction of antioxidants from apple pomace by pressurized liquids. *Journal of Agricultural and Food Chemistry.* 57 (22):10625–10631.

Wijesinghe, W.A., and Jeon, Y.J. 2012. Enzyme-assistant extraction (EAE) of bioactive components: A useful approach for recovery of industrially important metabolites from seaweeds: A review. *Fitoterapia.* 83 (1):6–12.

Xu, C., Yagiz, Y., Borejsza-Wysocki, W. et al. 2014. Enzyme release of phenolics from muscadine grape (*Vitis rotundifolia* Michx.) skins and seeds. *Food Chemistry* 157:20–29.

Yang, Z., Chen, Z., Yuan, S. et al. 2009. Extraction and identification of anthocyanin from purple corn (*Zea mays* L.). *International Journal of Food Science & Technology.* 44 (12):2485–2492.

Yılmaz, F.M., Karaaslan, M., and Vardin H. 2015. Optimization of extraction parameters on the isolation of phenolic compounds from sour cherry (*Prunus cerasus* L.) pomace. *Journal of Food Science and Technology.* 52 (5):2851–2859.

Yingngam, B., Supaka, N., and Rungseevijitprapa, W. 2015. Optimization of process parameters for phenolics extraction of *Cratoxylum formosum* ssp. formosum leaves by response surface methodology. *Journal of Food Science and Technology.* 52 (1):129–140.

Zhang, J., Zhou, L., Zhang, P. et al. 2015. Extraction of polydatin and resveratrol from *Polygonum cuspidatum* root: Kinetics and modeling. *Food and Bioproducts Processing.* 94:518–524.

CHAPTER 5

Cleanup Methods

Liliana Santos-Zea, Janet Alejandra Gutiérrez-Uribe, and Georgina Sandoval

CONTENTS

5.1 INTRODUCTION

Crude extracts from plants, fungi, or algae obtained for analysis or isolation of phenolics usually contain large amounts of carbohydrates, proteins, amino acids, or lipids, while the concentration of the compounds of interest is usually low. Different methods, as reported in the literature, concentrate and recover polyphenol rich fractions before analysis, such as sequential extraction with different solvents, liquid–liquid partitioning (LLE), or solid phase extraction (SPE).

According to De Rijke et al. (2006), the choice of cleanup method following extraction depends on the matrix from which phenolics are to be extracted and the target compounds. When liquid samples are handled or the goal is the extraction of aglycones (usually following acid or enzymatic hydrolysis), LLE is considered a reasonable choice. On the other hand, in recent years SPE and similar techniques have been more widely used for the separation of glycosides from sugars and aglycones, particularly because these methods can ultimately be adapted as high-throughput techniques. In this chapter, several processes related to sample cleanup and fractionation of extracts are discussed.

5.2 SEQUENTIAL EXTRACTION AND LIQUID–LIQUID EXTRACTION (LLE) METHODS

Sequential extraction and liquid–liquid extraction (LLE) refer to the use of water, organic solvents, and mixtures of both to eliminate compounds that may cause interference during analysis of phenols. The main difference between both techniques is related to whether the cleanup is done on the dry extract (sequential extraction) or partitioning the extract between two immiscible phases (LLE).

In general terms, elimination of lipid and lipophilic pigments can be achieved by washing the crude extract with non-polar solvents such as hexane (Prieur et al. 1994; Ramírez-Coronel et al. 2004), dichloromethane (Neergjeen et al. 2006), chloroform

(Zhang et al. 2008), or petroleum ether (Foo et al. 2000). This step is particularly important when extracting bound phenolics by alkaline hydrolysis, which contributes in great measure to the correct quantitation of phenolic compounds in grains (Adom and Liu 2002) and other food matrices, such as the edible fungi *Pleurotus eryngii* (Li and Shah 2013). Fatty acids may interfere with phenolic determination (Krygier et al. 1982); therefore, due to their abundance in foods, it is important to remove them for a proper analysis. Extract defatting with hexane may be followed with subsequent partitioning steps, known as LLE, using different systems of two or more immiscible solvents (Table 5.1). By manipulating the combination of solvents used, extracts may be fractioned by collecting the different phases formed, according to their polarity. Ethyl acetate with water or aqueous solutions are commonly used in phenolic recovery, either in the organic phase, such as in bound phenolic extraction (Adom and Liu 2002; Li and Shah 2013), or to obtain phenolic compounds from olive oil production waste water (De Marco et al. 2007). In other cases, this LLE is used to reduce the amount of non-glycosylated flavonoids (Antunes-Ricardo et al. 2015) by collecting the aqueous phase instead. Changing the solvent polarity for partitioning an extract may yield fractions enriched in different types of phenolics, which in turn can have an effect on bioactivity, such as cytotoxic potential against cancer cells (García-Varela et al. 2016).

To reduce the use of organic solvents, new strategies for extract fractionation by LLE are being developed, such as the use of aqueous two-phase systems (ATPS). This process involves the use of certain additives, such as polymers and salts, above a critical concentration in aqueous solutions to allow the formation of two immiscible phases (Benavides et al. 2008). Although these systems are used mainly for the separation of proteins from crude extracts, the use of ethanol–salt systems allow the partition of phenolic compounds into the ethanolic phase, eliminating the need for other solvents, such as ethyl acetate (Table 5.1). Partial purification of chlorogenic acid from an ethanolic extract of carrot was achieved using an ethanol–potassium phosphate (pH 7) ATPS. Moreover, by modifying the pH of the salt solution, this system could be used directly for extraction to recover a fraction enriched in phenolic compounds (Sánchez-Rangel et al. 2016). ATPS can also be used for one-step extraction and partial purification of flavonoids from plants, such as apigenin and genistein from pigeon pea root. In this case, the flavonoids were preferentially partitioned into the upper phase of a 28 percent ethanol—22 percent potassium phosphate ATPS, allowing process intensification (Zhang et al. 2013).

Some LLE processes may be automatized and used as chromatographic techniques for partial or full purification of bioactive compounds, this is jointly known as (CCC) techniques. In general, these methods use two immiscible liquids as mobile and stationary phases and, depending on the particular kind of equipment used, partioning is facilitated by different methods. There are different technologies for continuous and automatized partitioning, such as centrifugal partition chromatography (CPC), high speed counter-current chromatography (HSCCC), droplet counter-current chromatography (DCCC), among others (Valls et al. 2009). By nature, this kind of LLE is highly versatile and small changes in the biphasic system composition allow the enrichment or purification of different kinds of phenolic compounds from diverse food matrices (Table 5.1). Zessner et al. (2008) used CPC as a first step in the purification of various types of phenolics from apples, yielding seven subfractions from most polar to least polar in an ethyl acetate–ethanol–water system (2:1:2) and each was tested for antioxidant potential. In general, the least polar fractions were the most effective, attributed to flavonols and procyanidins. Using a system with the same solvents but in different proportions, Delaunay et al. (2002) were able separate stilbenoids, including resveratrol, from flavonols in a single step. In the

TABLE 5.1 LLE Methods for Cleanup of Phenolic Extracts

Source	Recovered Compounds	LLE System	Reference
Whole grain flour	Bound phenolics fraction with high amounts of ferulic acid	Hexane-aqueous solution, followed by ethyl acetate-aqueous solution	Adom and Liu 2002
Fungus *Pleurotus eryngii* (bound phenolics)	Gallic acid, ferulic acid, naringenin	Hexane-aqueous solution, followed by ethyl acetate-aqueous solution	Li and Shah 2013
Opuntia ficus-indica cladode flour	Isorhamnetin glycosides	Hexane-aqueous HCl (pH 2) followed by ethyl acetate-aqueous HCl (pH 2)	Antunes-Ricardo et al. 2015
Olive oil mill wastewaters	Hydroxytyrosol, tyrosol, caffeic acid, vanillic acid, verbascoside, luteolin-7-glucoside, luteolin	Hexane-HCl (pH 2), followed by ethyl acetate-HCl (pH 2)	De Marco et al. 2007
Rhoeo discolor	Ferulic acid, vanillic acid, glycosylated vanillic acid, chlorogenic acid, p-coumaric acid, rhoeonin	Ethyl acetate-water (1:1), Hexane-ethyl acetate-methanol-water (1:1:1:1), hexane-methanol (1:1)	García-Varela et al. 2016
Carrot shreds	Hydroxycinnamic acid derivatives, p-coumaric acid derivatives, caffeoylquinic acid derivatives, ferulic acid, isocoumarin	Ethanol-potassium phosphate ATPS	Sánchez-Rangel et al. 2016
Pigeon pea (*Cajanus cajan*) root	Apigenin and genistein	Ethanol-potassium phosphate ATPS	Zhang et al. 2013
Apples	Phenol carboxylic acids, dihydrochalcones, flavan-3-ols, quercetin glycoside, procyanidins	Ethyl acetate-ethanol-water (2:1:2) in CPC	Zessner et al. 2008
Grape vine stalks	Stilbenoids and flavonoids	Ethyl acetate-ethanol-water (8:2:7) in CPC	Delaunay et al. 2002
Honey	Phenolic acids and flavonoids	$(NH_4)_2SO_4$ in aqueous solution with chloroform (extractant) and acetone (disperser)	Campone et al. 2014

case of CPC, it is not just important to consider the choice of solvent system to design an efficient purification process. Indeed, other important factors include the mobile phase flow rate, column capacity, and rotational speed, therefore providing the user with a diversity of parameters to obtain the desired purification degree for the compounds of interest (Valls et al. 2009).

Partitioning may result in the use of high amounts of solvents, but new techniques such as dispersive liquid–liquid microextraction (DLLME) reduce the amount of solvent used and therefore the cost and environmental impact (Fariña et al. 2007). In contrast to traditional partitioning systems, a complete dispersion of an extractant solvent into an aqueous sample is prepared with the help of a substance that is miscible within both components and causes the formation of a cloudy solution; this enlarges the contact area and facilitates mass transfer (Rezaee et al. 2006). DLLME has been used as a cleanup method to analyze phenolic compounds in honey that is faster and less expensive than SPE followed by LLE (Campone et al. 2014).

5.3 SOLID-PHASE EXTRACTION (SPE) AND MOLECULARLY IMPRINTED POLYMERS (MIP)

Solid-phase extraction (SPE) is a separation or concentration method based on the affinity of solutes dissolved or suspended in a liquid (known as the mobile phase) for a solid through which the sample is passed (known as the stationary phase). Compared to liquid–liquid extraction (LLE), SPE can reduce the time required, especially for automated methods, can handle small samples (50–100 μL), and consumes small amounts of solvent. There are several types of sorbents for SPE including normal-phase, reversed-phase, ionic, and other special sorbents.

SPE can be used, on one hand, for pre-concentration and cleanup of analytical samples to characterize polyphenols in food or in raw materials, and on the other hand, to concentrate polyphenols before their use as ingredients in functional foods or cosmetics. The analysis of phenolic compounds in honey is an example of the first case (Table 5.2); and resins such as Amberlite-XAD2 are used directly, mixing with a diluted sample or passing through a column to recover the flavonoids, since sugars and other polar compounds are not retained (Gómez-Caravaca et al. 2006). When using a 1 percent citric acid solution in water as an extraction agent to obtain anthocyanins from different berries, a recovery higher than 90 percent was observed in SPE while reducing the amount of sugars by 94 percent or more (Denev et al. 2010). Compared to other methods, SPE is a relatively new technique that has been developing over the last 30 years (Huck and Bonn 2000). When it is used to characterize polyphenols, it is usually coupled to a second separation method that allows verification of the chemical identity of individual polyphenols, including liquid chromatography (Christophoridou et al. 2005), HPLC (Dvořáková et al. 2007; Perez-Magariño et al. 2008), UPLC (Ferreiro-González et al. 2014), and gas chromatography (Ignat et al. 2011). These separation systems can be coupled to UV spectrophotometry, mass spectrometry (Lazarus et al. 1999), infrared spectroscopy, NMR (Chia-Chuan and Shoei-Sheng 2014; Christophoridou et al. 2005), or a combination of these methods (Christophoridou et al. 2005; Pérez-Trujillo et al. 2010).

Concentration can also be used to recover polyphenols from food industry effluents, such as the bioactive feruloyl putrescines found in maize lime cooking wastewater or "nejayote" (Acosta-Estrada et al. 2015), flavonoids from rose oil distillation (Rusanov et al. 2014), or hydroxytyrosol from olive mill waste water (Bertin et al. 2011).

Maize phenolics extract contains high concentrations of ferulic acid, making the purification of putrescine for bioactivity evaluation a difficult task. The use of SPE in a C18 cartridge produced an enriched fraction after washing with 20 percent aqueous methanol and eluting with 50 percent methanol (Acosta-Estrada et al. 2015). The use of resins as beads instead of columns make scaling up more feasible, as demonstrated

TABLE 5.2 Phases and Eluents for the Recovery of Phenolics from Various Sources

Solid Phase (Source)	Phenolics	Mobile Phase	Reference
Styerene-divinylbenzene (Honey)	Phenolic acids and flavonoids.	Methanol	Gómez-Caravaca et al. 2006
Divinilbenzene resine modified (Berries)	Anthocyanins.	Ethanol aqueous 90 percent	Denev et al. 2010
C18 (Maize processing waste water)	Ferulic acid, putrescines.	50 percent methanol	Acosta-Estrada et al. 2015
Brominated styrene-divinylbenzene (Rose oil distillation effluent)	Ellagic acid, 2-phenylethyl-O-β-glucopyranoside, kaempferol and quercetin glycosides.	90 percent ethanol.	Rusanov et al. 2014
Hydroxylated polystyrene-divinylbenzene (Olive mill waste water)	Hydroxytyrosol. Mixture of polyphenols.	Ethanol 0.5 percent HCl ethanol	Bertin et al. 2011
C18 (Alcohol-free beer)	(+)-Catechin Gentisic acid; Caffeic acid: (−)-Epicatechin; p-Coumaric acid; Ferulic acid; Salicylic acid; Quercetin.	Acetonitrile	Alonso Garcia et al. 2004
C18 (Apple musts and cider)	Chlorogenic acid; Epicatechin; Caffeic acid; p-Coumaric acid.	Methanol	Suárez et al. 1996
C18 (Black tea)	Theaflavins.	Ethanol (aq. Solut. 40 percent)	Nishimura et al. 2007
C18 (Red coffee cherries)	p-coumaroylquinic acid, hydroxycinnamic acids, flavonols, anthocyanidins, flavan-3-ols, caffeoylquinic acid, and p-coumaroylquinic acid.	Acetone/water	Ramírez-Coronel et al. 2004
Reversed-phase/strong anion-exchange (Beers)	Gallic, protocatechuic, caffeic, p-coumaric, ferulic and salicylic acids, of (+)-catechin, (−)-epicatechin, and quercetin.	Acetonitrile/ methanol	Dvořáková et al. 2007
Silica-gel and active matrix of Diol bonding (Crude extra virgin olive oil)	25 compounds, including simple phenols as hydroxytyrosol and tyrosol; lignanes as (+)-pinoresinol; secoiridoids as oleuropein aglycone.	Methanol	Pérez-Trujillo et al. 2010
m-Divinylbenzene & N-vinyl-pyrrolidone copolymer (Honey)	Gallic acid; p-HBA; Vanillic acid; Caffeic acid; Syringic acid; p-Coumaric acid.	Methanol	Michalkiewicz et al. 2008
m-Divinylbenzene & N-vinylpyrrolidone copolymer (Wine)	Hydroxybenzoic acids and derivatives including vanillic acid, syringic acid and ellagic acid. Hydroxycinnamic acids and derivatives including trans-caftaric acid and trans-fertaric acid. Flavanols as (+)-catechin and (-)-epicatechin.	Ethyl acetate	Pérez-Magariño et al. 2008

by Rusanov et al. (2014) who applied this method by mixing brominated styrene-divinylbenzene beads with rose oil distillation waste, allowing the particles to sediment after adsorption and desorption of a phenolic-rich fraction with 90 percent ethanol. The choice of elution solvent may affect which compounds are recovered. Bertin et al. were able to obtain a mixture of polyphenols using a hydroxylated polystyrene-divinylbenzene resin and eluting with 0.5 percent HCl in ethanol. However, they were able to obtain a hydroxytyrosol-enriched fraction with non-acidified ethanol using a smaller amount of the other phenols.

The application of SPE to increase selectivity of the chromatographic techniques for the identification of phenolics was reported by different authors. Diverse sorbent materials have been used for the fractionation between phenolic acids, flavonoids, and anthocyanins present in extracts (Table 5.2). The most commonly used sorbent is reverse-phase octadecylsilane (C18), which has a good affinity for phenolic compounds (Fontana et al. 2013). These are eluted from the cartridge with water or acidified water to remove polar non-phenolic compounds (sugars, organic acids, amino acids, proteins). Polyphenols are then eluted with methanol, acetonitrile, or ethyl acetate, as shown in Table 5.2.

Molecularly imprinted polymers (MIP) are synthetic polymers with highly specific recognition for targeted molecules. A MIP is synthesized using the targeted compound as template, followed by a polymerization and subsequent removal of the template with a solvent. This process leaves specific recognition sites for the targeted compound in the MIP. Some MIP have been developed for phenolic acids such as caffeic and *p*-hydroxybenzoic acids (Michailof et al. 2008) and for flavonoids like quercetin (Song et al. 2009). Regarding catechins, although the developed MIP have good molecular recognition, competitive adsorption was usually observed (Haginaka et al. 2007). In the case of resveratrol, MIP have helped to separate *cis* and *trans* isomers (Ma et al. 2007). Recent advances have demonstrated the intensification potential of this technique. In combination with magnetic separation, MIPs, or in this case MMIPs, have been used to extract and determine phenolic acids in aqueous systems (Shi et al. 2014). Polydopamine (PDA) coatings have been developed to reduce hydrophobicity of these materials and allow their dispersion into fruit juices to separate gallic acid within 10 seconds without a SPE column packing or a filtration operation (Hu et al. 2015). Another potential application of this technique is related to the extraction of phenolic compounds from biological fluids to facilitate the studies related to bioavailability and metabolism (Asfaram et al. 2017).

REFERENCES

Acosta-Estrada, B.A., Serna-Saldívar, S.O., and Gutiérrez-Uribe, J.A. 2015. Chemopreventive effects of feruloyl putrescines from wastewater (Nejayote) of lime-cooked white maize (*Zea mays*). *Journal of Cereal Science*, 64: 23–28. https://doi.org/10.1016/j.jcs.2015.04.012

Adom, K.K., and Liu, R.H. 2002. Antioxidant activity of grains. *Journal of Agricultural and Food Chemistry*, 50(21): 6182–6187. https://doi.org/10.1021/jf0205099

Alonso García, A., Cancho Grande, B., and Simal Gándara, J. 2004. Development of a rapid method based on solid-phase extraction and liquid chromatography with ultraviolet absorbance detection for the determination of polyphenols in alcohol-free beers. *Journal of Chromatography A*, 1054: 1287–1293.

Antunes-Ricardo, M., Gutiérrez-Uribe, J.A., López-Pacheco, F., Alvarez, M.M., and Serna-Saldívar, S.O. 2015. In vivo anti-inflammatory effects of isorhamnetin glycosides isolated from *Opuntia ficus-indica* (L.) Mill cladodes. *Industrial Crops and Products*, 76: 803–808. https://doi.org/10.1016/j.indcrop.2015.05.089

Asfaram, A., Ghaedi, M. and Dashtian, K. 2017. Rapid ultrasound-assisted magnetic microextraction of gallic acid from urine, plasma and water samples by HKUST-1-MOF-Fe 3 O 4-GA-MIP-NPs: UV–vis detection and optimization study. *Ultrasonics Sonochemistry*, 34: 561–570.

Benavides, J., Aguilar, O., Lapizco-Encinas, B.H., and Rito-Palomares, M. 2008. Extraction and purification of bioproducts and nanoparticles using aqueous two-phase systems strategies. *Chemical Engineering & Technology*, 31(6): 838–845. https://doi.org/10.1002/ceat.200800068

Bertin, L., Ferri, F., Scoma, A., Marchetti, L. and Fava, F. 2011 Recovery of high added value natural polyphenols from actual olive mill wastewater through solid phase extraction. *Chemical Engineering Journal*, 171: 1287–1293.

Campone, L., Piccinelli, A.L., Pagano, I., Carabetta, S., Di Sanzo, R., Russo, M., Rastrelli, L. 2014. Determination of phenolic compounds in honey using dispersive liquid–liquid microextraction. *Journal of Chromatography A*, 1334: 9–15. https://doi.org/10.1016/j.chroma.2014.01.081.

Chia-Chuan, C., and Shoei-Sheng, L. 2014. Application of HPLC–SPE–NMR in characterization of bioactive natural compounds. In G.K. Jayprakasha, B.S. Patil, and F. Pellati (Eds.), *Instrumental Methods for the Analysis and Identification of Bioactive Molecules* (Dec 2016, pp. 217–239). American Chemical Society.

Christophoridou, S., Photis, D., Tseng, L.H., and Manfred, S. 2005. Separation and identification of phenolic compounds in olive oil by coupling high-performance liquid chromatography with postcolumn solid-phase extraction to nuclear magnetic resonance spectroscopy (LC-SPE-NMR). *Journal of Agricultural and Food Chemistry*, 53: 4667–4679.

De Marco, E., Savarese, M., Paduano, A., and Sacchi, R. 2007. Characterization and fractionation of phenolic compounds extracted from olive oil mill wastewaters. *Food Chemistry*, 104(2): 858–867. https://doi.org/10.1016/j.foodchem.2006.10.005

De Rijke, E., Out, P., Niessen, W.M.A., Ariese, F., Gooijer, C., and Brinkman, U.A.T. 2006. Analytical separation and detection methods for flavonoids. *Journal of Chromatography A*, 1112: 31–63. https://doi.org/10.1016/j.chroma.2006.01.019

Delaunay, J., Castagnino, C., Cheze, C., and Vercauteren, J. 2002. Preparative isolation of polyphenolic compounds from *Vitis vinifera* by centrifugal partition chromatography. *Chromatographic Science*, 964: 123–128.

Denev, P., Ciz, M., Ambrozova, G., Lojek, A., Yanakieva, I., and Kratchanova, M. 2010. Solid-phase extraction of berries' anthocyanins and evaluation of their antioxidative properties. *Food Chemistry*, 123(4): 1055–1061.

Dvořáková, M., Hulín, P., Karabín, M., and Dostálek, P. 2007. Determination of polyphenols in beer by an effective method based on solid-phase extraction and high performance liquid chromatography with diode-array detection. *Czech Journal of Food Science*, 25: 182–188.

Fariña, L., Boido, E., Carrau, F., and Dellacassa, E. 2007. Determination of volatile phenols in red wines by dispersive liquid–liquid microextraction and gas chromatography–mass spectrometry detection. *Journal of Chromatography A*, 1157(1–2): 46–50 https://doi.org/10.1016/j.chroma.2007.05.006.

Ferreiro-González, M., Carrera, C., Ruiz-Rodríguez, A. et al. 2014. A new solid phase extraction for the determination of anthocyanins in grapes. *Molecules*, 19: 21398–21410.

Fontana, A.R., Antoniolli, A., and Bottini, R. 2013. Grape pomace as a sustainable source of bioactive compounds: Extraction, characterization, and biotechnological applications of phenolics. *Journal of Agricultural and Food Chemistry*, 61: 8987–9003.

Foo, L.Y., Lu, Y., Howell, A.B., and Vorsa, N. 2000. The structure of cranberry proanthocyanidins which inhibit adherence of uropathogenic P-fimbriated *Escherichia coli in vitro*. *Phytochemistry*, 54: 173–181. https://doi.org/10.1016/S0031-9422(99)00573-7

García-Varela, R., Fajardo Ramírez, O.R., Serna-Saldivar, S.O., Altamirano, J., and Cardineau, G.A. 2016. Cancer cell specific cytotoxic effect of *Rhoeo discolor* extracts and solvent fractions. *Journal of Ethnopharmacology*, 190: 46–58. https://doi.org/10.1016/j.jep.2016.05.051

Gómez-Caravaca, A.M., Gómez-Romero, M., Arráez-Román, D., Segura-Carretero, A., and Fernández-Gutiérrez, A. 2006. Advances in the analysis of phenolic compounds in products derived from bees. *J. Pharm. Biomed. Anal.* 41: 1220–1234. doi:10.1016/j.jpba.2006.03.002

Haginaka, J., Tabo, H., Ichitani, M., Takihara, T., Sugimoto, A., and Sambe, H. 2007. Uniformly-sized, molecularly imprinted polymers for (–)-epigallocatechin gallate, -epicatechin gallate and -gallocatechin gallate by multi-step swelling and polymerization method. *Journal of Chromatography A*, 1156(1–2): 45–50. doi: http://dx.doi.org/10.1016/j.chroma.2006.10.026.

Hu, X., Xie, L., Guo, J., Li, H., Jiang, X., Zhang, Y., and Shi, S. 2015. Hydrophilic gallic acid–imprinted polymers over magnetic mesoporous silica microspheres with excellent molecular recognition ability in aqueous fruit juices. *Food chemistry*, 179: 206–212.

Huck, C.W., and Bonn, C.K. 2000. Review: Recent developments in polymer-based sorbents for solid-phase extraction. *Journal of Chromatography A*, 885: 51–72.

Ignat, I., Volf, I., and Popa, V.I. 2011. A critical review of methods for characterisation of polyphenolic compounds in fruits and vegetables. *Food Chemistry*, 126: 1821–1835. https://doi.org/10.1016/j.foodchem.2010.12.026

Krygier, K., Sosulski, F., and Hogge, L. 1982. Free, esterified, and insoluble-bound phenolic acids. 1. Extraction and purification procedure. *Journal of Agricultural and Food Chemistry*, 30: 330–334. https://doi.org/10.1021/jf00110a028.

Lazarus, S.A., Adamson, G.E., Hammerstone, J.F., and Schmitz, H.H. 1999. High-performance liquid chromatography/mass spectrometry analysis of proanthocyanidins in foods and beverages. *Journal of Agricultural and Food Chemistry*, 47: 3693–3701.

Li, S., and Shah, N.P. 2013. Effects of various heat treatments on phenolic profiles and antioxidant activities of *Pleurotus eryngii* extracts. *Journal of Food Science*, 78: 1122–1129. https://doi.org/10.1111/1750-3841.12189

Ma, S., Zhuang, X., Wang, H., Liu, H., Li, J., and Dong, X. 2007. Preparation and characterization of trans-resveratrol imprinted polymers. *Analytical Letters*, 40(2): 321–333. doi: 10.1080/00032710600867689.

Michailof, C., Manesiotis, P., and Panayiotou, C. 2008. Synthesis of caffeic acid and p-hydroxybenzoic acid molecularly imprinted polymers and their application for the selective extraction of polyphenols from olive mill waste waters. *Journal of Chromatography A*, 1182(1): 25–33. doi: http://dx.doi.org/10.1016/j.chroma.2008.01.001.

Michalkiewicz, A., Biesaga, M., and Pyrzynska, K. 2008. Solid-phase extraction procedure for determination of phenolic acids and some flavonols in honey. *Journal of Chromatography A*, 1187: 18–24.

Neergjeen, V.S., Soobrattee, M.A., Bahorun, T., and Aruoma, O.I. 2006. Characterization of the phenolic constituents in Mauritian endemic plants as determinants of their antioxidant activities *in vitro*. *Journal of Plant Physiology*, 163: 787–799.

Nishimura, M., Ishiyama, K., Watanabe, A., Kawano, S., Miyase, T., and Sano, M. 2007. Determination of theaflavins including methylated theaflavins in black tea leaves by solid-phase extraction and HPLC analysis. *Journal of Agricultural and Food Chemistry*, 55: 7252–7257.

Perez-Magariño, S., Ortega-Heras, M., and Cano-Mozo, E. 2008. Optimization of a solid-phase extraction method using copolymer sorbents for isolation of phenolic compounds in red wines and quantification by HPLC. *Journal of Agricultural and Food Chemistry*, 56: 11560–11570.

Pérez-Trujillo, M., Gómez-Caravaca, A.M., Segura-Carretero, A., Fernández-Gutiérrez, A., and Parella, T. 2010. Separation and identification of phenolic compounds of extra virgin olive oil from *Olea europaea* L. by HPLC-DAD-SPE-NMR/MS. Identification of a new diastereoisomer of the aldehydic form of oleuropein aglycone. *Journal of Agricultural and Food Chemistry*, 58: 9129–9136.

Prieur, C., Rigaud, J., Cheynier, V., and Moutounet, M. 1994. Oligomeric and polymeric procyanidins from grape seeds. *Phytochemistry*, 36: 781–784.

Ramirez-Coronel, M.A., Marnet, N., Kolli, V. K., Roussos, S., Guyot, S., and Augur, C. 2004. Characterization and estimation of proanthocyanidins and other phenolics in coffee pulp (*Coffea arabica*) by thiolysis– high-performance liquid chromatography. *Journal of Agricultural and Food Chemistry*, 52(5): 1344–1349.

Rezaee, M., Assadi, Y., Hosseini, M.R.M., Aghaee, E., Ahmadi, F., and Berijani, S. 2006. Determination of organic compounds in water using dispersive liquid–liquid microextraction. *Journal of Chromatography A*, 1116(1): 1–9.

Rusanov, K., Garo, E., Rusanova, M. et al. 2014. Recovery of polyphenols from rose oil distillation wastewater using adsorption resins—A pilot study. *Planta Medica*, 80: 1657–1664. https://doi.org/10.1055/s-0034-1383145

Sánchez-Rangel, J.C., Jacobo-Velázquez, D.A., Cisneros Zevallos, L., and Benavides, J. 2016. Primary recovery of bioactive compounds from stressed carrot tissue using aqueous two-phase systems strategies. *Journal of Chemical Technology and Biotechnology*, 91: 144–154. https://doi.org/10.1002/jctb.4553

Shi, S., Guo, J., You, Q., Chen, X., and Zhang, Y. 2014. Selective and simultaneous extraction and determination of hydroxybenzoic acids in aqueous solution by magnetic molecularly imprinted polymers. *Chemical Engineering Journal*, 243: 485–493.

Song, X., Li, J., Wang, J., and Chen, L. 2009. Quercetin molecularly imprinted polymers: Preparation, recognition characteristics and properties as sorbent for solid-phase extraction. *Talanta*, 80(2): 694–702. doi: http://dx.doi.org/10.1016/j.talanta.2009.07.051.

Suárez, B., Picinelli, A., and Mangas, J.J. 1996. Solid-phase extraction and high-performance liquid chromatographic determination of polyphenols in apple musts and ciders. *Journal of Chromatography A*, 727: 203–209.

Valls, J., Millán, S., Martí, M.P., Borràs, E., and Arola, L. 2009. Advanced separation methods of food anthocyanins, isoflavones and flavanols. *Journal of Chromatography A*, 1216: 7143–7172. https://doi.org/10.1016/j.chroma.2009.07.030

Zessner, H., Pan, L., Will, F. et al. 2008. Fractionation of polyphenol-enriched apple juice extracts to identify constituents with cancer chemopreventive potential. *Molecular Nutrition & Food Research*, 52: S28–44. https://doi.org/10.1002/mnfr.200700317

Zhang, D.Y., Zu, Y.G., Fu, Y.J. et al. 2013. Aqueous two-phase extraction and enrichment of two main flavonoids from pigeon pea roots and the antioxidant activity. *Separation and Purification Technology*, 102: 26–33. https://doi.org/10.1016/j.seppur.2012.09.019

Zhang, Y., Seeram, N.P., Lee, R., Feng, L., and Heber, D. 2008. Isolation and identification of strawberry phenolics with antioxidant and human cancer cell antiproliferative properties. *Journal of Agricultural and Food Chemistry*, 56: 670–675.

CHAPTER 6

Separation and Detection Methods

Maria del Refugio Ramos-Jerz, Gerold Jerz,
Socorro Josefina Villanueva-Rodríguez, and Peter Winterhalter

CONTENTS

6.1 INTRODUCTION

The need to improve the extraction protocol for natural product isolation from raw materials as well as to speed up separation time has directed the focus of interest on countercurrent chromatography (CCC) as a very useful separation/extraction method. Ito performed the first applications of CCC in the late 1960s (Berthod 2002; Conway, 1990, 1995). During the last decades, CCC techniques were improved and many routine protocols were developed for natural product isolation (Ito and Conway 1996; Ito 2005; Winterhalter 2009; Jerz et al. 2014).

CCC is a liquid–liquid partition chromatographic technique (Weisz and Ito 2000) with several advantages. Some of them are the absence of solid adsorbents, inexpensive solvent mixtures, the fact that large sample loads can be applied, and, very importantly, the total recovery of the sample is assured. An immiscible solvent pair as well as different partition coefficients of the components in the mixture is required for a successful separation (Winterhalter 2009).

This liquid chromatographic technique has been implemented for the separation of polar as well as lipophilic compounds, and even mixtures of metal ions in inorganic mixtures have been separated.

The recovery of pure natural products from milligram to kilogram quantities (semi-industrial size) is feasible by the scale-up of the dimensions of the CCC device (Sutherland 2002). The complementary separation capacity of CCC is considered an advantage and gives another powerful tool compared to other more frequently used techniques, such as preparative HPLC.

In the case of CCC, both phases are liquids. The solute is introduced onto the stationary phase in the column to be separated according to Nernst's Partition law. Thus, the migration of the solutes through the coil is determined by the partition ratio (**K_D**) given by

$$K_D = Cs/Cm$$

Cs = concentration in the stationary phase
Cm = concentration in the mobile phase

A larger K value means a major affinity of the compound to the stationary phase with a consequently retarded elution time.

For the isolation of bioactive compounds from avocado seeds (*Persea americana* Mill. c.v. *Hass*), the CCC technique has been applied as an important tool in the separation of polyphenols.

In this context it is important to note that avocado seeds are not only rich in polyphenols but also in many other compound classes like olefins, hydroxy-acetogenins (Kashman et al. 1969; Werman et al. 1990; Adikaram et al. 1992), furanoic lipids (Rosenblat et al. 1995), and abscisic acid (ABA) derivatives (Ramos et al., 2004), for which CCC enabled excellent applications with regard to the initial fractionation of such compounds (Ramos-Jerz 2007).

Polyphenolic compounds such as tyrosol and hydroxytyrosol have been isolated before from Italian wines. They are also major substances in olive oil (Di Tommaso et al. 1998). Due to the antioxidant properties of such substances, there are numerous references about their origin, synthesis, and antioxidant activity (Amiot et al. 1989; Uccella 2003; Trujillo et al. 2006).

6.2 OBJECTIVE

The scavenging capacity of avocado cotyledon partitions (Ramos-Jerz 2007) as well as their activity in the HaCat assay (Ramos-Jerz et al. 2013) was tested in previous research with interesting results. The methanol–water partition (M:W) was, after ethyl acetate, the second-largest one, with 33 percent of scavenging activity related to ascorbic acid. Therefore, it was important to separate the compounds present in this fraction and purify them for further biological assays.

This chapter describes application of the HSCCC method to the separation of methanol–water partition from avocado seed (*Persea americana* Mill c.v. *Hass*) cotyledons on a triple coil equipment (Pharma Tech. Research Corp., U.S). In this way, it was possible to obtain the polar tyrosol derivatives **1** and **2** in pure form.

6.3 RESULTS

Dried avocado cotyledons yielded 5.4 percent of methanol–water partition (see Figure 6.1).

This partition was analyzed by LC–ESI–MS in negative mode. A mixture of polar compounds was present. Due to our previous research (Ramos-Jerz 2007), it was possible to identify proanthocyandins with *m/z* $[M\text{-}H]^-$ 865, dimers and trimers of proanthocyanidins with *m/z* $[M\text{-}H]^-$ 289, 577, and 863, as well as ABA derivatives with *m/z* $[M\text{-}H]^-$ 441

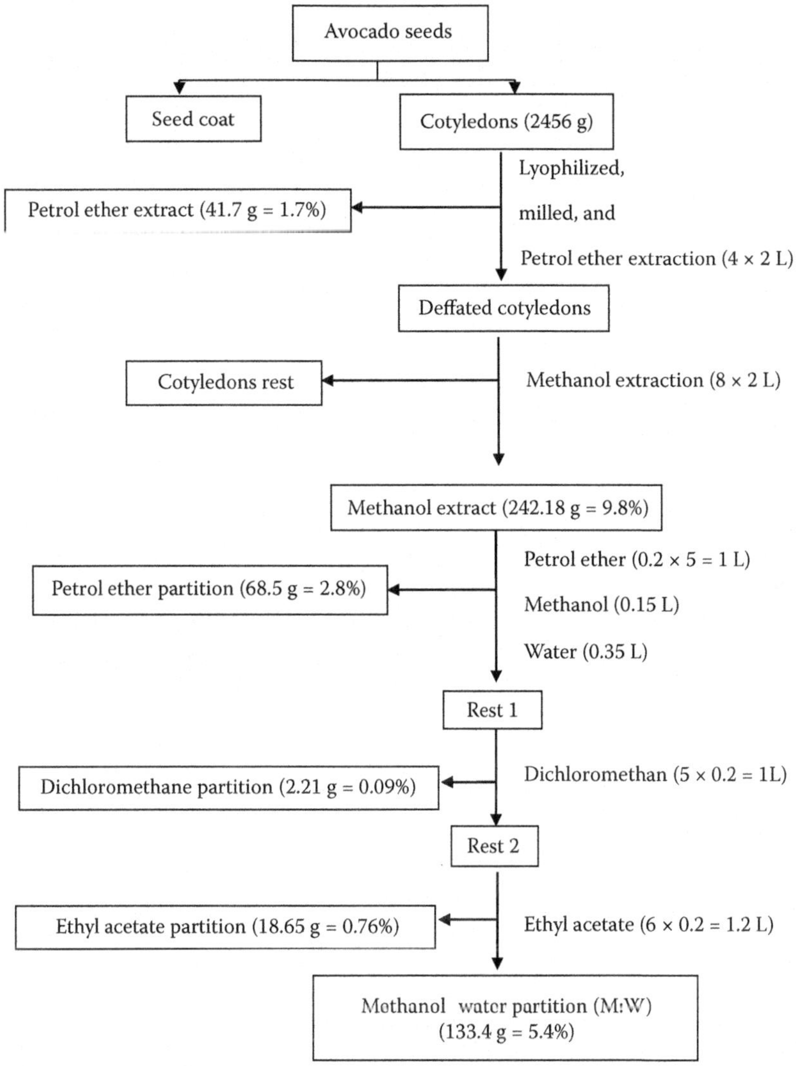

FIGURE 6.1 Avocado seeds methanol–water partition recovery processes.

and 443 (compounds **3** and **4**, respectively). The presence of quinic and chlorogenic acids with *m/z* of $[M\text{-}H]^-$ 191 and $[M\text{-}H]^-$ 353, respectively, was verified by using commercial standards. Two additional compounds with *m/z* of $[M\text{-}H]^-$ 299 and $[M\text{-}H]^-$ 315 were also detected (see Figure 6.2).

The fragmentation patterns of these compounds are shown in Table 6.1.

Biological assays were performed with the avocado seed partitions (not shown here). Moreover, the in vitro assays epidermal keratinocytes (NHEK) and normal human dermal fibroblasts (NHDF) also indicated bioactivity (Ramos-Jerz et al. 2013).

Fast evaluation of suitable solvent system for HSCCC separation was done by TLC analysis of the target compounds (**1** and **2**) in the biphasic solvent system. This assay is quite effective and quick in the practical approach of natural product recovery compared

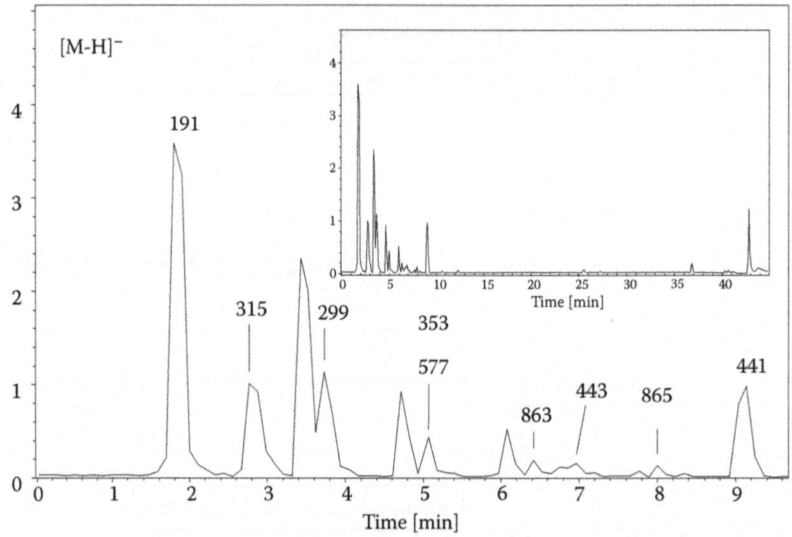

FIGURE 6.2 LC–ESI–MS of methanol–water partition from avocado cotyledons. [Column: Zorbax SB-C18 3.5 µm 150 × 4.6 mm. Solvents A: 2.5 percent CH_3COOH in H_2O, B: acetonitrile: A (80:20). Gradient: t (0,35,40,45 min), A (85,25,0,85), B (15,75,100,15). Flow rate: 0.8 $mLmin^{-1}$. Intensity × 10^6. ESI-neg.]

TABLE 6.1 Fragment Ions $[M–H]^-$ and Fragmentation Patterns (MS^2) of Identified Compounds in the Methanol–Water Partition of Avocado Cotyledons

Rt (min)	[M-H]-(MS^2)	Compound
1.8	191 (173, 127, 85)	Quinic acid
2.8	315 (135, 179, 119)	**2**
3.7	299	**1**
5.1	353 (191, 179, 173, 135)	Chlorogenic acid
5.1	577 (289)	Dimer of proanthocyanidins
6.4	863 (289)	Trimers of proanthocyanidins
7.0	443	**4**
8.0	865 (289)	Proanthocyanidins
9.1	441 (397, 330, 161)	**3**

to high sophisticated analytical devices such as HPLC–UV or mass spectrometry detection. Four different solvent systems were tested (Table 6.2) and the best partition ratio of compounds was obtained with system D.

With solvent system D, 5.4 grams of methanol–water partition were separated by preparative high-speed countercurrent chromatography (HSCCC). Six fractions were obtained (H1 to H6). The amount of each fraction in the avocado cotyledons is shown in Table 6.3. Fraction H5 appeared almost pure on the TLC analysis (see Figure 6.3.).

Each HSCCC fraction was further analyzed by LC–ESI–MS and the distribution of the compounds among the fractions are shown in Table 6.4.

TABLE 6.2 Tested Solvent Systems for HSCCC Separation of the Methanol–Water Partition from Avocado Cotyledon

Solvent System	*n*-Hexane	Ethyl Acetate	Methanol	Tert-Butylmethyl-Ether	*n*-Butanol	Acetonitrile	Water
A	4	5	4	–	–	–	5
B	0.5	5	0.5		–	–	5
C	–	4	–	–	1	–	5
D	–	–	–	1	3	1	5

TABLE 6.3 Yields of Whole Avocado Seed Extracts, Cotyledon Solvent Partitions, and HSCCC Fractions (H)

Extract, Partition, or HSCCC Fraction	Yield %
Methanolic extract	5.43[b]
Methanol–water partition (M:W)	5.4[a]
Fraction H1	1.7[a]
Fraction H2	1.4[a]
Fraction H3	0.10[a]
Fraction H4	0.09[a]
Fraction H5	0.11[a]
Fraction H6	0.22[a]

[a] Yield from dried cotyledons.
[b] Whole avocado seeds.

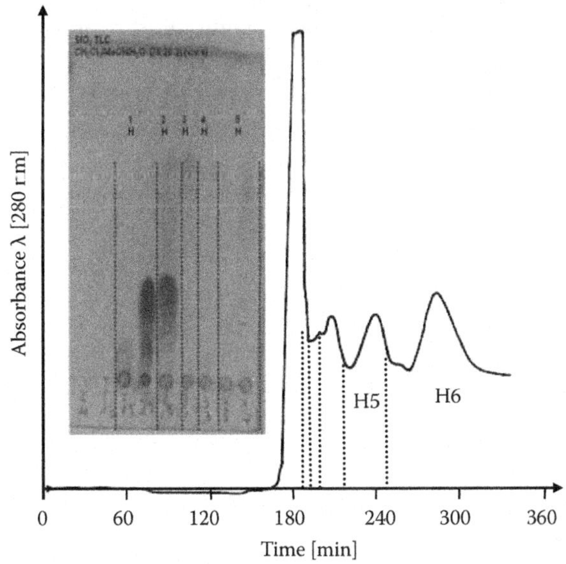

FIGURE 6.3 HSCCC separation of methanol–water partition of avocado seeds cotyledons. [Solvent system: *tert.*-butylmethylether: *n*-butanol: acetonitrile: water (1:3:1:5 v/v/v/v). Flow rate: 3.0 $mLmin^{-1}$. UV: $\lambda = 280$ nm, rotational velocity: 1000 rpm; mobile phase: lower phase, elution mode: head to tail.]

TABLE 6.4 Distribution of Identified Compounds in Methanol–Water Partition from Avocado Cotyledons and the Respective HSCCC Fractions (H)

Compound Present	M:W Partition	H1	H2	H3	H4	H5	H6
Quinic acid	+	+	+				
Compound **1**	+				+		+
Compound **2**	+			+		+	+
Chlorogenic acid	+		+			+	
Proanthocyanidins	+						
ABA derivatives (**3**, **4**)	+					+	+

The quinic and chlorogenic acids were mainly present in fractions H1 and H2. Compound **1** was found again in fractions H4 and H6 while compound **2** and the ABA derivatives (**3** and **4**) were present in fractions H5 and H6.

Nevertheless, fraction H5 was only present in an amount of 0.11 percent the avocado cotyledons. It was further separated by column chromatography (Ramos-Jerz 2007), resulting in 3.2 milligrams of compound **2** (31.7 milligram compound **2**/kilogram avocado cotyledon).

The LC–ESI–MS analysis (Figure 6.4) of **2** showed a quasi-molecular ion *m/z* $[M-H]^-$, 315 revealing a molecular weight of 316. By means of NMR data (Ramos-Jerz 2007), compound **2** was identified as 3-hydroxy-tyrosol-1'-β-D-glucoside.

Another part of the dried avocado seeds was extracted with methanol resulting in 54.3 grams of methanolic extract per kilogram of dried avocado seed. The separation of approximately 6 grams of methanolic extract by HSCCC (Figure 6.5) and further column chromatography (Ramos-Jerz 2007) resulted in 16.2 milligrams of compound **1** (144.2 milligram compound **1**/kilogram complete avocado seed).

The LC–ESI–MS analysis of compound **1** (Figure 6.6) resulted in *m/z* of $[M-H]^-$ 299, revealing a molecular weight of 300 uma and was identified by means of NMR-data as tyrosol 1'-β-D-glucoside. The structures of glucosides **1** and **2** as well as the abscicic acid derivatives **3** and **4** are onlined on Figure 6.7.

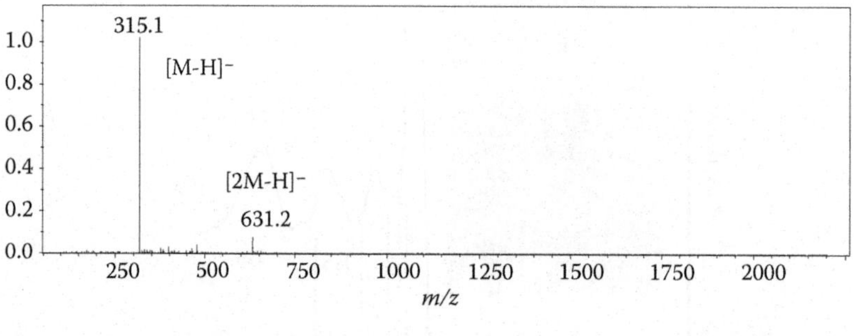

FIGURE 6.4 LC-ESI-MS of compound **2**. [Column: Zorbax SB-C18 3.5 µm 150 × 4.6 mm. Solvents A: 2.5 percent CH_3COOH in H_2O, B: acetonitrile: A (80:20). Gradient: t (0,35,40,45 min), A (85,25,0,85), B (15,75,100,15). Flow rate: 0.8 $mLmin^{-1}$. Intensity × 10^6. ESI-neg.]

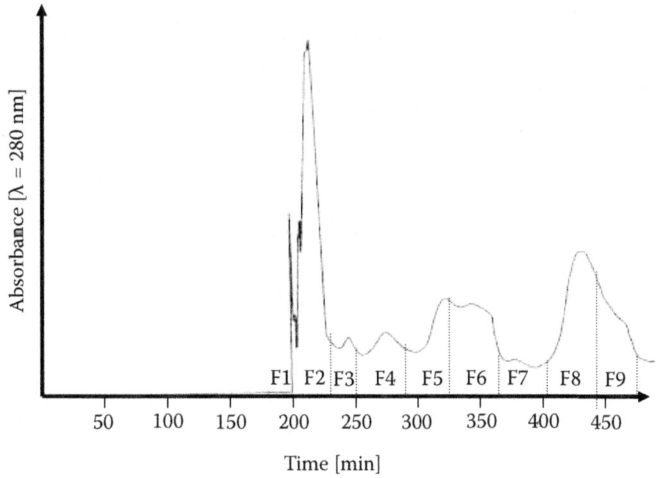

FIGURE 6.5 HSCCC separation of the crude methanolic extract of whole avocado seeds. (Flow rate: 2.5 mLmin^{-1}. Other conditions cf. Figure 6.3.)

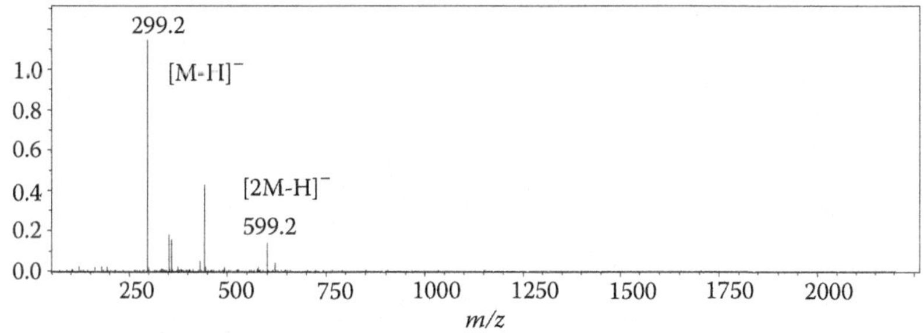

FIGURE 6.6 LC–ESI–MS of compound **1**. (Same conditions as Figure 6.4.)

Parallel to the chemical analysis, biological activities were also tested in the HSCCC fractions from the methanol–water partition, the commercial standard quinic and chlorogenic acids, as well as compounds **1** and **2**. The assay in vitro human epidermal keratinocytes (NHEK) and normal human dermal fibroblasts (NHDF) gave interesting results. Compounds **1** (also called salidroside) and **2** increased NHDF and the keratinocyte proliferation (Ramos-Jerz et al. 2013).

Many biological activities are reported for salidroside (Mao et al. 2010; Zhang et al. 2009; Qian et al. 2012), including antidiabetic activity (Li et al. 2008). Salidroside was isolated before from *Rhodiola sachalinensis* (Li and Chen 2001) and *Rhodiola crenulata* (Han et al. 2002) by HSCCC.

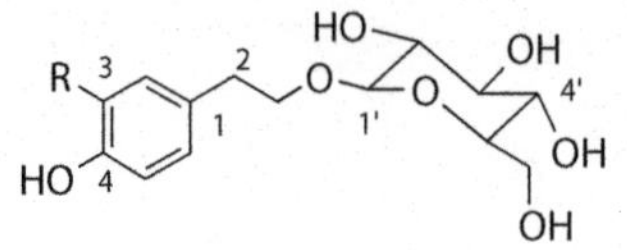

R = H: Tyrosol-1'-β-D-*O*-glucoside (**1**) (syn: salidroside)

R = OH: 3-Hydroxy-tyrosol-1'-β-D-*O*-glucoside (**2**)

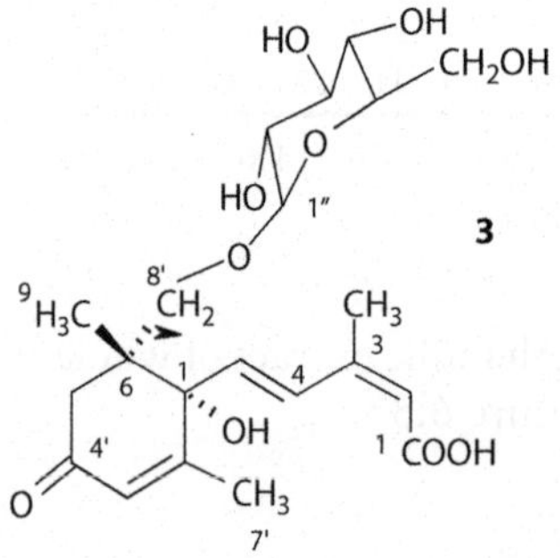

3

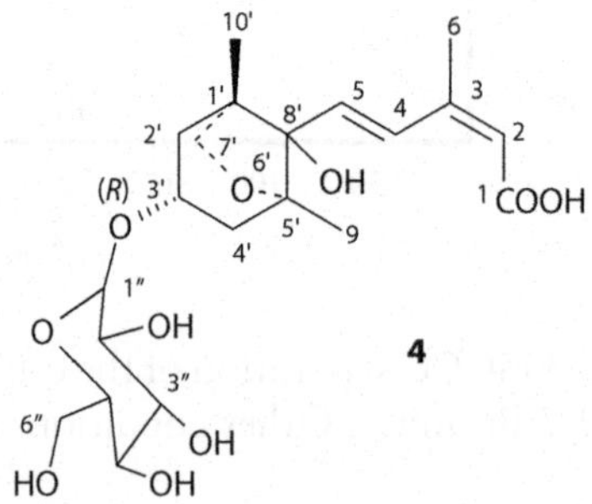

4

(1'*S*,6'*R*)-8'-Hydroxyabscisic acid β-D-glucoside (**3**)

(1'*R*,3'*R*,5'*R*,8'*S*)-*epi*-Dihydrophaseic acid β-D-glucoside (**4**)

FIGURE 6.7 Chemical structures of identified compounds from avocado seeds.

6.4 CONCLUSIONS

HSCCC is a versatile separation method for polar and nonpolar natural products. For polyphenols it has been successfully applied not only in the case of avocado seeds extracts but also, for example, in the case of anthocyanins from wine (Winterhalter 2007), stilbenes from *Vitis chunganeniss* (He et al. 2009), betalains from Opuntia fruits (Jerz et al. 2013), antioxidants from vegetables as from fruits *Luffa cylindrica* (Du et al. 2006).

Acknowledgments go to C. Balcke for the gentle corrections of this chapter. Many thanks are due to the Deutscher Akademischer Austauschdienst (DAAD) and Consejo Nacional de Ciencia y Tecnología (CONACYT) for financial support during M. Ramos-Jerz's PhD work.

REFERENCES

Adikaram, N. K. B., Ewing, D. F., Karunaratne, A. M., Wijeratne, E. M. K. 1992. Antifungal compounds from immature avocado fruit peel. *Phytochemistry* 31: 93–96.

Amiot, M. J., Fleuriet, A., Macheix, J. J. 1989. Accumulation of oleuropein derivatives during olive maturation. *Phytochemistry* 28: 67–69.

Berthod, A. (ed.) 2002. *Countercurrent Chromatography: The Support-Free Liquid Stationary Phase*. Comprehensive Analytical Chemistry, vol. 38. Amsterdam: Elsevier.

Conway, W. D. 1990. *Countercurrent Chromatography: Apparatus, Theory and Applications*. New York: VCH.

Conway, W. D. 1995. Overview on countercurrent chromatography. In *Modern Countercurrent Chromatography, ACS Symposium Series Vol. 593*, ed. W. D. Conway, R. J. Petroski, pp. 1–14. Washington, DC: American Chemical Society.

Di Tommaso, D., Calabrese, R., Rotilio, D. 1998. Identification and quantification of hydroxytyrosol in Italian wines. *J. High Resolut. Chromatogr.* 21: 549–553.

Du, Q., Xu, Y., Li, L., Zhao, Y., Jerz, G., Winterhalter, P. 2006. Antioxidant constituents in the fruits of *Luffa cylindrica* (L.) *Roem. J. Agric. Food Chem.* 54: 4186–4190.

Han, X., Zhang, T., Wie, Y., Cao, X., Ito, Y. 2002. Separation of salidroside from *Rhodiola crenulata* by high-speed counter-current chromatography. *J. Chromatogr. A* 971: 237–241.

He, S., Lu, Y., Jiang, L., Wu, B., Zhang, F., Pan, Y. 2009. Preparative isolation and purification of antioxidative stilbene oligomers from Vitis chunganeniss using high-speed counter-current chromatography in stepwise elution mode. *J. Sep. Sci.* 32: 2339–2345.

Ito, Y. 2005. Golden rules and pitfalls in selecting optimum conditions for high-speed counter-current chromatography. *J. Chromatogr.* 1065: 145–168.

Ito, Y. and Conway, W. D. 1996. *High-Speed Countercurrent Chromatography*. New York: Wiley & Sons, Inc.

Jerz, G., Elnakady, Y. A., Braun, A., Jäckel, K., Sasse, F., Al Ghamdi, A. A., Omar, M. O. M., Winterhalter, P. 2014. Preparative mass-spectrometry profiling of bioactive metabolites in Saudi-Arabian propolis fractionated by high-speed counter-current chromatography and off-line atmospheric pressure chemical ionization mass-spectrometry injection. *J. Chromatogr. A* 1347: 17–29.

Jerz, G., Klose, B., Wybraniec, S., Villanueva Rodriguez, S., Winterhalter, P. 2013. Preparative separation and pigment profiling of betalains from fruits of *Opuntia ficus* by ion-pair high-speed countercurrent chromatography (IP-HSCCC) and off-line LC-ESI-MS/MS. In *Tropical and Subtropical Fruits: Flavors, Color, and Health Benefits*, ed. B. S. Patil, G. K. Jayaprakasha, C. Osorio Roa, K. Mahattanatawee, pp. 3–27. ACS Symp. Series 1129. Washington, DC: American Chemical Society.

Kashman, Y., Neeman, I., Lifshitz, A. 1969. New compounds from avocado pear. *Tetrahedron* 25: 4617–4631.

Li, H.-B., Chen, F. 2001. Preparative isolation and purification of salidroside from the Chinese medicinal plant *Rhodiola sachalinensis* by high-speed counter-current chromatography. *J. Chromatogr. A* 932: 91–95.

Li, H.-B., Ge, Y.-K., Zheng, X.-X., Zhang, L. 2008. Salidroside stimulated glucose uptake in skeletal muscle cells by activating AMP-activated protein kinase. *Eur. J. Pharmacol.* 588: 165–169.

Mao, G.-X., Deng, H.-B., Yuan, L.-G., Li, D.-D., Yvonne Li, Y.-Y., Wang, Z. 2010. Protective role of salidroside against aging in a mouse model induced by D-galactose. *Biomed. Environ. Sci.* 23:161–166.

Qian, E. W., Ge, D. T., Kong, S. 2012. Salidroside protects human erythrocytes against hydrogen peroxide-induced apoptosis. *J. Nat. Prod.* 75: 531–537.

Ramos, M. del R., Jerz, G., Villanueva, S., López-Dellamary, F., Waibel, R., Winterhalter, P. 2004. Two glucosylated abscisic acid derivates from avocado seeds (Persea americana Mill. Lauraceae cv. Hass). *Phytochemistry* 65: 955–962.

Ramos-Jerz, M. del R. 2007. *Phytochemical analysis of avocado seeds (Persea americana Mill., c.v. Hass)*. Göttingen: Cuvillier Verlag.

Ramos-Jerz, M. del R., Villanueva, S., Jerz, G., Winterhalter, P., Deters, A. M. 2013. *Persea americana* mill. seed: Fractionation, characterization and effects on human keratinocytes and fibroblasts. *Evid. Based Complement. Alternat. Med.* Article ID 391247. http://www.ncbi.nlm.nih.gov/pmc/articles/PMC3863524/

Rosenblat, G., Kagan, H. M., Shah, M. A., Spiteller, G., Neeman, I. 1995. Chemical characterization of lysyl oxidase inhibitor from avocado seed oil. *J. Am. Oil Chem. Soc.* 72: 225–229.

Sutherland, I. A. 2002. Liquid stationary phase retention and resolution in hydrodynamic CCC. In *Countercurrent Chromatography—The Support-Free Liquid Stationary Phase*, Comprehensive Analytical Chemistry vol. 38, ed. A. Berthod, pp. 159–176. Amsterdam: Elsevier.

Trujillo, M., Mateos, R., Collantes de Teran, L., Espartero, J. L., Cert, R., Jover, M., Alcudia, F., Bautista, J., Cert, A., Parrado, J. 2006. Lipophilic hydroxytyrosol esters. Antioxidant activity in lipid matrices and biological systems. *J. Agric. Food Chem.* 54: 3779—3785.

Uccella, N. 2003. Olive biophenols in the MAC-Mediterranean aliment culture. In *Polyphenols 2002, Recent Advances in Polyphenols Research* (XXI International Conference on Polyphenols, Marrakesh, Morocco, September 9–12, 2002), ed. I. El Hadrami, and F. Daayf, pp. 98–162. Bordeaux: Secrétariat du Groupe Polyphenols.

Weisz, A., and Ito, Y. 2000. High-speed countercurrent chromatography. In *Encyclopedia of Separation Science*, vol. 6 (III), ed. I. D. Wilson, E. R. Adlard, M. Cooke, C. F. Poole, pp. 2588–2602. London: Academic Press.

Werman, M. J., Mokady, S., Neeman, I. 1990. Partial isolation and characterization of a new natural inhibitor of lysyl oxidase from avocado seed oil. *J. Agric. Food Chem.* 38: 2164–2168.

Winterhalter, P. 2007. Application of countercurrent chromatography (CCC) to the analysis of natural pigments. *Trends Food Sci. Technol.* 18: 507–513.

Winterhalter, P. 2009. Application of countercurrent chromatography for wine research and wine analysis. *Am. J. Enol. Vitic.* 60 (2): 123-129.

Zhang, J., Liu, A., Hou, R., Zhang, J., Jia, X., Jiang, W., Chen, J. 2009. Salidroside protects cardiomyocyte against hypoxia-induced death: A HIF-1α-activated and VEGF-mediated pathway. *Eur. J. Pharmacol.* 607: 6–14.

SECTION III

Different Groups of Phenolic Compounds Related to Foods

CHAPTER 7

Xanthones

Begoña de Ancos and Concepción Sánchez-Moreno

CONTENTS

7.1 INTRODUCTION

Xanthones are polyphenol compounds found in some higher plant families, fungi, and lichens (Negi et al. 2013). All xanthones possess the same oxygen–heterocycle backbone formed by a γ-pyron moiety condensed with two benzene rings (C_6–C_1–C_6) known as xanthone, xanthene-9-one, or dibenzo-γ-pyrone. This symmetric chemical structure makes the xanthone a very stable molecule (Figure 7.1; Pedraza-Chavarri et al. 2008).

Xanthones were first discovered and isolated in 1855 from pericarp of mangosteen fruit (*Garcinia mangostana* L.) by a German scientist performing research on dysentery. α-Mangostin was found among the major xanthones taken from the pericarps of mangosteen fruit (Figure 7.2) (Schmid 1855). It is a yellow-colored matter, and for that reason, these compounds were named *xanthones* from the Greek word for yellow, *xanthos*. α-Mangostin can be obtained from other parts of the plant, such as the dried sap, leaves, bark, seeds, and whole fruit. Later, the α-mangostin structure was elucidated (Dragendorff 1930; Murakami 1932) and the molecular formula and type and position of substituents of α-mangostin was established (Yates and Stout 1958; Figure 7.2). It has been discovered that the compound possesses a wide range of biological activities, such as anti-inflammatory, antitumor, antioxidant, cardioprotective, antidiabetic, antiobesity, antifungal, antiparasitic, antiviral, antiallergy, and antibacterial (Gutierrez-Orozco and Failla 2013; Ibrahim et al. 2016).

The biological activities of xanthones are associated with their tricycle structure but vary depending on the nature or position of the different substituents, such as isoprene,

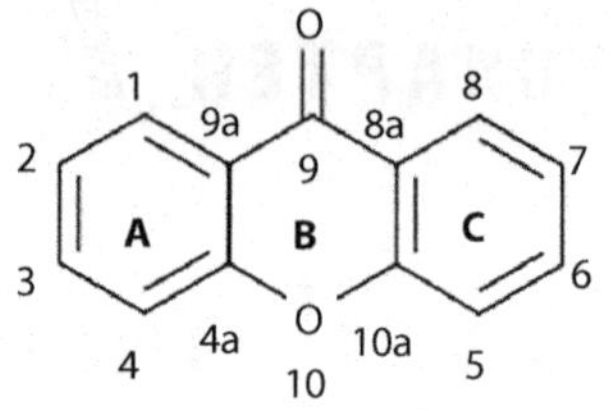

FIGURE 7.1 Xanthone backbone chemical structure.

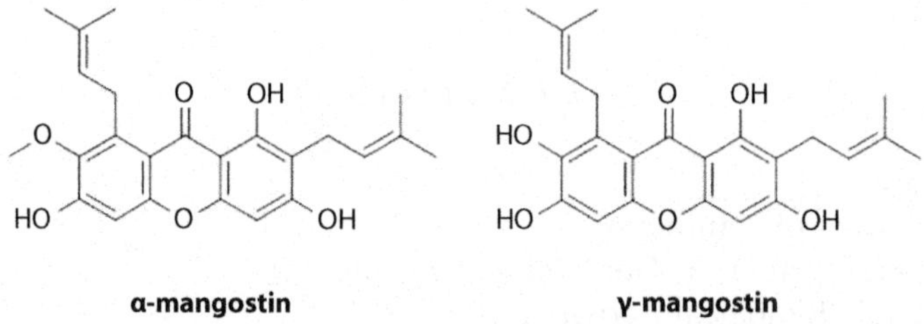

FIGURE 7.2 Chemical structures of the most abundant xanthones in mangosteen fruit (*Garcinia mangostana* L.).

methoxyl, and hydroxyl groups located at various positions of the A and B rings (Pedraza-Chavarri et al. 2008). Xanthones have been classified into five groups: (1) simple oxygenated xanthones, (2) xanthone glycosides, (3) prenylated xanthones, (4) xanthololignoids, and (5) miscellaneous xanthones (Negi et al. 2013).

In general, xanthones are biologically active polyphenolic compounds structurally very similar to flavonoids; their chromatographic behavior is also similar. Although flavonoids are frequently encountered in plant nature, xanthones are found in a limited number of families, such as *Gentianaceae*, *Guttiferae*, *Moraceae*, *Clusiaceae*, and *Polygalaceae*. The types, isolation, characterization, and biological applications of naturally occurring xanthones have been widely studied and reviewed (Obolsky et al. 2009; Negi et al. 2013; Gutierrez-Orozco and Failla 2013; Ibrahim et al. 2016). Naturally occurring xanthones have emerged as an important phytochemical in view of their remarkable pharmacological and other biological activities. It has been found that many plant products regularly used in traditional medicine as therapeutic agents contain xanthones as active constituents (Pinto et al. 2005; Na 2009; El-Seedi et al. 2010; Panda et al. 2013). In recent years, many chemical and pharmacological studies have been carried out to reveal a range of bioactivities of plants of the genus *Garcinia* (*Gentianaceae* family), such as anti-inflamation, antioxidation, anti-adipogenesis, anticancer, antimicrobial, anti-HIV, anticonvulsant, and antimalarial agents (Gutierrez-Orozco and Failla 2013; Jindarat 2014; Ibrahim et al. 2016). The pharmacological importance of xanthones have encouraged scientists not only to isolate these compounds from natural products but also to synthesize them as novel drug candidates (Shagufta and Ahmad 2016).

Related studies of xanthones have been gradually growing in the last years due to the important medicinal properties attributed to them. A search of available literature using *xanthone* as a term in the Web of Science retrieved 839 reports in the period 2002–2006 and 1815 reports in the period 2012–2016 (Figure 7.3). When using the terms *mangostana* and *xanthone* there is also a significant increase in the number of reports found in the last years, from 40 reports in the period of 2002–2006 to 284 in the period of 2012–2016 (Figure 7.4).

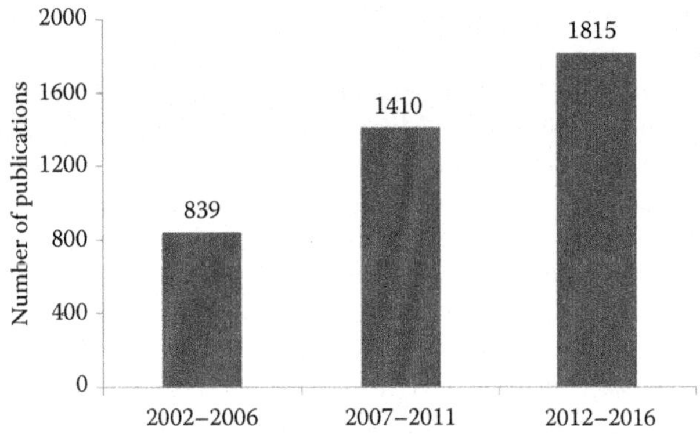

FIGURE 7.3 Number of publications related to xanthone from 2002 to 2016. (Search performed on September 14, 2016. Database: Web of Science.)

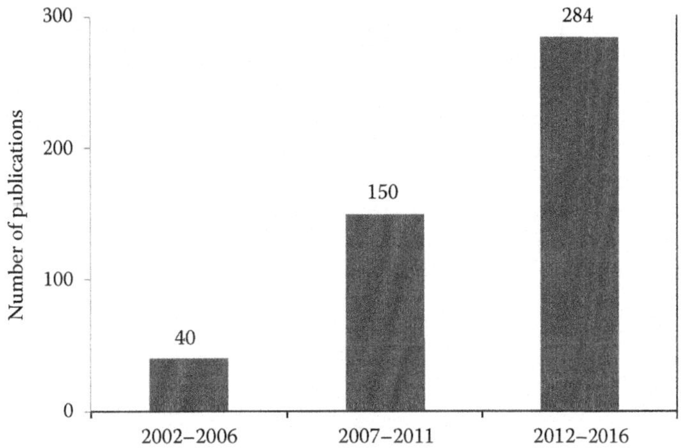

FIGURE 7.4 Number of publications related to mangosteen and xanthone from 2002 to 2016. (Search performed on September 14, 2016. Database: Web of Science.)

7.2 NATURAL SOURCES OF XANTHONES

The principal natural sources of xanthones are the mangosteen fruit (*Garcinia mangostana* L.) and the mango fruit (*Mangifera indica* L.). Mangosteen has a rich content of prenylated and oxygenated xanthones, the majority being α-mangostin. Mango fruit is characterized by the presence of C-glycosylated xanthones such as mangiferin (Berardini et al. 2005; Nack and Shahidi 2006; Jung et al. 2006; Walker 2007; Ji et al. 2007; Barreto et al. 2008; Zarena and Sankar 2009; Chaivisuthangkura et al. 2009). Tropical fruit are mostly consumed fresh but nowadays there is an increasing demand of fruit-derived products to be used in the beverage and dairy industry, among others (Ayala-Zavala et al. 2011). Processing of mangosteen and mango fruits generates a significant amount of by-products, such as the pericarp in mangosteen or the peel and seeds in mangos, which have shown to have very high levels of xanthones (Gutierrez-Orozco and Failla 2013; Juhurul et al. 2015).

7.2.1 Mangosteen Fruit (*Garcinia mangostana* L.)

Mangosteen (*Garcinia mangostana* L.,) is a tropical evergreen tree in the Southeast Asia region in places including Malaysia, Myanmar, Thailand, Philippines, Sri Lanka, and India. Mangosteen fruit is round and dark purple or reddish (Figure 7.5). The mangosteen fruit may be seedless or have one to five fully developed seeds with an obloid-oblong shape. The arils of the seeds are represented in four to eight triangular segments of white juice and soft flesh (pulp) that possesses a unique and delectable tropical flavor defined as a sweet, slightly acidic taste and a pleasant aroma for which it is known as the "queen of fruit" (Figure 7.5). The pericarp of mangosteen fruit has been used in traditional medicine in Southeast Asia for centuries to treat infections, dysentery, wounds, inflammation, fever, and so on (Pedraza-Chavarri et al. 2008; Gutierrez-Orozco and Failla 2013).

The phytochemical composition of *Garcinia mangostana* L. has been widely studied with the result of the identification of xanthones, benzophenones, flavonoids, phenolic acids, and triterpenoids (Obolskiy et al. 2009; Zadernowski et al. 2009). Among these compounds, prenylated and oxygenated xanthones are the major constituents. About 74 xanthones have been isolated and identified in the pericarp or hull, bark, roots, leaves, whole fruit, and other parts of *Garcinia mangostana* (Pedraza-Chavarri et al. 2008). The major constituents of the xanthone fraction of *G. mangostana* are α-mangostin and γ-mangostin (Figure 7.2) followed by β-mangostin and garcinone E (Figure 7.6).

Several reviews have summarized the chemical structures of xanthones found in *G. mangostana* fruit (Obolskiy et al. 2009), although continually new xanthones are isolated and characterized from different parts of this fruit, mainly pericarp (Mohamed et al. 2014; Zhou et al. 2015; Xu et al. 2016).

The majority of investigations are focused on the extraction and structure elucidation of xanthones from the pericarp of the mangosteen fruit that is mainly constituted

FIGURE 7.5 Mangosteen fruit (*Garcinia mangostana* L.).

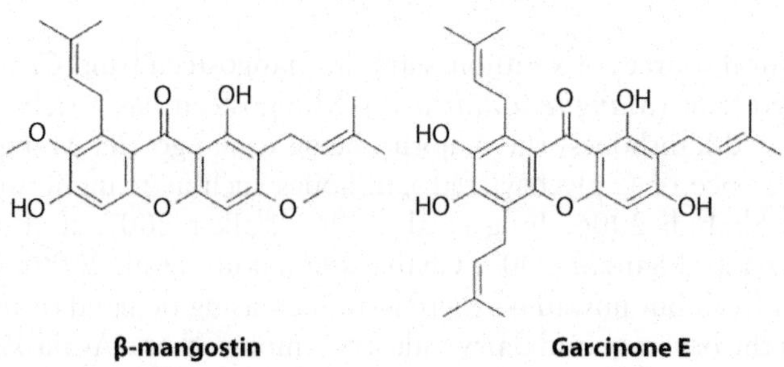

FIGURE 7.6 Chemical structure of the two second-most abundant xanthones in mangosteen fruit (*Garcinia mangostana* L.).

by α-mangostin and γ-mangostin (Figure 7.2). There are also studies dedicated to the extraction and structure characterization of xanthones from other parts of the mangosteen fruit such as arils, seeds, or whole fruit (with seeds), being also that α-mangostin and γ-mangostin are major xanthones (Obolskiy et al. 2009).

Consequently, Wittenauer et al. (2012) have described the isolation and characterization of major the xanthones of pericarp and aril segments of mangosteen fruit, besides a functional mangosteen beverage, by high-performance liquid chromatography coupled with mass spectrometry detection (HPLC–MS). HPLC are nowadays the best analytical approach to separate, identify, and quantify xanthones from mangosteen fruit and derived products (Jung et al. 2006; Ji et al. 2007; Walker 2007; Chaivisuthagkura et al. 2009; Zarena and Sankar 2011). Wittenauer et al. (2012) found a similar HPLC xanthone profile for pericarp and aril segments of mangosteen fruit, and also in the functional beverage made from whole mangosteen fruit (XanGo®) (Figure 7.7). Although in these three products the major compounds are α-mangostin and γ-mangostin, the pericarp showed the highest total xanthone content, followed by the aril segments and the functional beverage (Table 7.1). In fact, α-mangostin and total xanthone content were 29 and 16 times lower in the aril segments that in the pericarp of mangosteen fruit, respectively (Table 7.1).

Considering the consumption of the edible part of a mangosteen fruit (~30 grams fw of aril segments), it results in an intake of 12.2 and 2.5 milligrams of α-mangostin and γ-mangostin, respectively. Compared with the recommended daily dose of functional beverage (90 milliliters), the content of α-mangostin and γ-mangostin is similar to that of

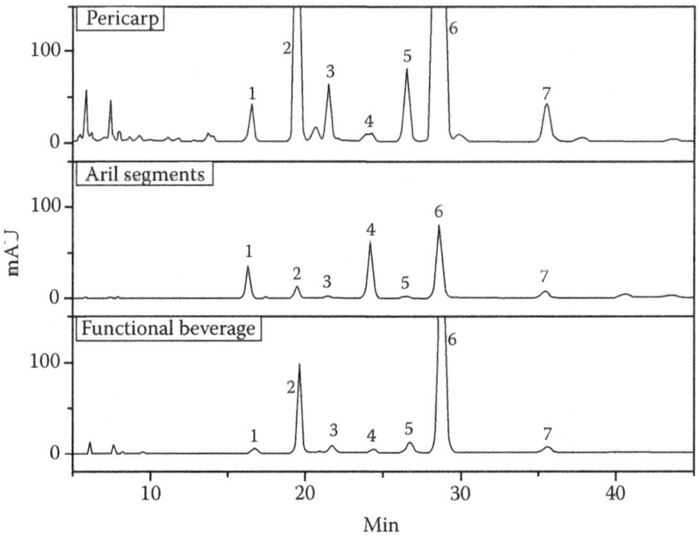

FIGURE 7.7 HPLC chromatogram (254 nm) of xanthones separated from pericarp and aril segments extracts (with methylene chloride) of mangosteen fruit and from a functional mangosteen beverage (XanGo®). Peaks: (1) 1,7-dihydroxy-3-methoxy-2-(3-methylbut-2-enyl)xanthone, (2) γ-mangostin, (3) 8-deoxygartanin, (4) 1,3,7-trihydroxy-2,8-di(3-methylbut-2-enyl)xanthone, (5) gartanin, (6) α-mangostin, (7) garcinone E. (Data from Wittenauer, J., Falk, S., Scheweiggert-Weisz, U., and Carle, R., *Food Chemistry* 134, 445–452, 2012.)

TABLE 7.1 Xanthone Content in Mangosteen Pericarp, Aril Segments, and Functional Beverage

Xanthone	Pericarp (mg/100 g fw)	Pericarp (mg/100 g dw)	Aril Segments (mg/100 g fw)	Aril Segments (mg/100 g dw)	Beverage (mg/100 mL)
(1) 1,7-Dihydroxy-3-methoxy-2-(3-methylbut-2-enyl)xanthone	35.27 ± 0.85	99.92 ± 2.41	12.52 ± 0.24	65.88 ± 1.28	0.48 ± 0.01
(2) γ-Mangostin	303.64 ± 6.16	860.17 ± 17.5	8.25 ± 0.24	43.39 ± 01.27	3.01 ± 0.06
(3) 8-Deoxygartanin	50.09 ± 0.80	141.90 ± 2.26	5.01 ± 0.09	26.39 ± 0.45	0.67 ± 0.01
(4) 1,3,7-Trihydroxy-2,8-di (3-methylbut-2-enyl)xanthone	19.12 ± 0.6	54.17 ± 1.58	25.76 ± 0.68	135.57 ± 3.56	0.53 ± 0.01
(5) Gartanin	70.41 ± 1.2	199.46 ± 3.50	5.49 ± 0.11	28.92 ± 0.58	0.85 ± 0.01
(6) α-Mangostin	1173.33 ± 29	3323.88 ± 82.2	40.59 ± 1.62	213.63 ± 8.53	12.53 ± 0.4
(7) Garcinone E	48.4 ± 1.9	137.09 ± 5.27	9.25 ± 0.17	48.70 ± 0.88	0.83 ± 0.0
Total Amount	1700.26 ± 40.5	*4816.59* ± 114.6	106.87 ± 3.15	*562.48* ± 16.56	18.79 ± 0.49

Source: Wittenauer, J., Falk, S., Scheweiggert-Weisz, U., and Carle, R., *Food Chemistry*, 134, 445–452, 2012.

aril segments of mangosteen fruit. Moreover, 90 milliliters of functional beverage have the same amount of major xanthones α-mangostin and γ-mangostin as 0.9 grams of pericarp, according to data showed in Table 7.1 (Wittenauer et al. 2012).

Traditional uses of mangosteen fruit are multifold. They are mainly eaten fresh as dessert. In this case, only the aril segments are eaten and the pericarp is discarded. However, the production of wine, preserves, jam, and purée from the arils and the whole fruit are further traditional food applications. A new potential market for puree obtained from the whole mangosteen fruit is the production of functional beverage (Wittenauer et al. 2012).

7.2.1.1 Biological Activity

Extensive literature has been reviewed concerning the biological activities of xanthone extracts of *G. mangostana* L. (Pedraza-Chaverri et al. 2008; Obolsky et al. 2009; Gutierrez-Orozco and Failla 2013) and, in particular, studies of the health-related properties of α-mangostin (Ibrahim et al. 2016), which is the main phytochemical of those extracts. A great number of studies indicate a high antioxidant capacity of the xanthone extracts of *G. mangostana* L. (Jung et al. 2006; Yu et al. 2007). Along with anti-inflammatory, antibacterial, antimalarial, antifungal, antiviral, anti-HIV, antiallergic, and antitumoral properties (Gutierrez-Orozco and Failla 2013; Shagufta and Ahmad 2016), extracts and xanthones from mangosteen fruit could be a great potential candidate for Alzheimer's disease (Wang et al. 2012) or for preventing and treating obesity (Liu et al. 2015).

Due to its potential role in promoting health, even more products derived from mangosteen fruit have been developed, such as Xango (from the United States), Verve (from the United States), or TriaXan (from Italy), which are commercialized as botanical healthy dietary supplements or functional beverages. Xango juice is a blend of mangosteen aril and pericarp purée with juice concentrates of eight other fruits: Apple, pear (juice and purée), grape, blueberry, raspberry, strawberry, cranberry, and cherry. Verve is an energy drink mainly formed with caffeine, vitamins, and a "superjuice" rich in phytochemicals of mangosteen and aloe. TriaXan is a liquid dietary supplement obtained from the whole mangosteen fruit (pericarp, pulp, and seeds). These beverages are included within the term *liquid botanical supplements*, which was coined for convenience products, mostly juice combinations of exotic fruits, linked to supposed health benefits from their traditional use of the original plant (Obolskiy et al. 2009). The shells of products derived from mangosteen fruits ranked 22nd among the top-selling dietary supplements in the United States, seven places ahead of green tea, and sales amounted to $176 million in the United States in 2014 (NBJ's Supplement Business Report, 2015).

Therefore, *G. mangostana* L. is an example of a medicinal and food plant that has now become widely available to obtain healthy dietary supplements (or nutraceuticals) and functional foods. Despite the success of these functional foods in the United States and Europe, the majority of the studies related to health-promoting activities of mangosteen fruit extracts and α-mangostin have been carried out using *in vitro* cellular models. Also, the antitumorigenic and anti-inflammatory activities of xanthones have been demonstrated with numerous studies using *in vivo* models with laboratory rodents (Gutierrez-Orozco and Failla 2013; Ibrahim et al. 2016). Controlled intervention trials to check the efficacy of xanthones in human volunteers, as well as investigations dedicated to study the absorption, metabolism, and elimination of xanthones or their metabolites are quite limited (Kondo et al. 2009; Udani et al. 2009; Tang et al. 2009; Ibrahim et al. 2016).

Only a few studies of dietary mangosteen xanthone bioavailability using *in vitro* assays or human studies have been found in the literature (Kondo et al. 2009;

Bumrungpert et al. 2009; Udani et al. 2009; Tang et al. 2009; Chitchumroonchokchai et al. 2012). Xanthone bioavailability studies carried out by Bumrungpert et al. (2009) were done with the pericarp and pulp of mangosteen fruit subjected to a simulated *in vitro* digestion combined with Caco-2 human intestinal cells. They found that the recovery of α-mangostin and γ-mangostin after a simulated digestion of pericarp and pulp exceeded 90 percent; meanwhile, bile salts were present but decreased to values <10 percent when bile salts were change by water. Therefore, the studies concluded that the absorption of α- and γ-mangostin is enhanced by dietary fat.

There are also several reports addressing the bioavailability of dietary xanthones in human subjects (Kondo et al. 2009; Chitchumroonchokchai et al. 2012). Kondo et al. (2009) designed a study with healthy human adults that consumed 59 milliliters of a xanthone-rich mangosteen juice containing 94.2 milligrams of xanthones. The maximum plasma concentration of α-mangostin (3.12 ± 1.47 ng/mL) was reached within one hour. Also, the plasma antioxidant capacity measured by the oxygen radical absorbance capacity (ORAC) increased by 18 percent two hours after ingestion of the mangosteen product compared with the subjects that ingested the placebo product. However, the contribution of xanthones to this increase of ORAC value is unknown due to the fact that his beverage also contained green tea; aloe vera; minerals; and vitamins A, B, C, D, and E (Kondo et al. 2009).

In a study carried out by Chitchumroonchokchai et al. (2012), healthy adults ingested 60 milliliters of 100 percent mangosteen juice (with pericarp particles) containing 130 milligrams of xanthones with a high-fat breakfast. Both free and glucuronidated/sulfated conjugates of the most abundant xanthones in mangosteen juice (α-mangostein, γ-mangostin, garcinones D and E, 8-deoxygatanin, and gartanin) were detected in serum and urine. The results of this study showed a great variability in maximum concentration of α-mangostin in serum (113 ± 107 mol/liter), as well as in the time to reach this maximum concentration (3.7 ± 2.4 hours). The absorption of xanthones from the juice was estimated to be 2 percent of the ingested dose as assessed by the quantity of total xanthones in urine collected from 24 hours. Results from simulated gastrointestinal digestion of mangosteen juice suggested that the apparently low bioaccessibility was due to inefficient release of xanthones from pericarp particles present in the juice. The transformation of ingested xanthones in bioactive metabolites by the gut bacteria, secretion of xanthones into bile, and retention in tissues beyond the 24-hour collection may underestimate the extent of absorption based on 24-hour urinary content. The results of this study clearly show that xanthones are absorbed and can be transformed into phase 2 metabolites. They also conclude that xanthones in mangosteen juice are better absorbed when ingested along with a high-fat meal (Chitchumroonchokchai et al. 2012).

Studies on the influence on health of consuming dietary xanthones are very scarce. There is some information about the anti-inflammatory activity of mangosteen xanthones in humans (Tang et al. 2009; Udani et al. 2009). The results were obtained in a randomized, double-blind, placebo-controlled trial with healthy adults who consumed 59 milliliters/day for 30 days of blended mangosteen juice (mangosteen juice, green tea, aloe vera, minerals, and multivitamins). At the end of the intervention study, a decrease of C-reactive protein (CRP) levels in serum was detected. However, other markers of inflammation were increased in subjects consuming the mangosteen product compared with a placebo such as cytokines IL-1α and IL-1β (Tang et al. 2009). Similar results were observed in other human studies with obese subjects. It was also reported that CRP levels in obese subjects consuming 18 ounces of a mangosteen juice blend per day during eight weeks were lower than those in the placebo group, but levels of the pro-inflammatory interferon-inducible protein 10 (IP-10) and macrophage inflammatory protein-1 β (MIP-1 β) increased (Udani et al. 2009). The results of these two studies (Tang et al. 2009; Udani et al. 2009) should be considered with caution because there is no way to discriminate the effects caused by other components of the mangosteen fruit juice blend.

Although no adverse effects were reported in the scarce studies with humans, the potential long-term toxicity of products containing mangosteen xanthones requires assessment. An important issue to consider is the potential toxicity of the mangosteen pericarp extract, even if the whole fruit, seeds, and seed oils did not provide evidence of toxic risks (Oblsikiy et al. 2009; Wittenauer et al. 2012).

The results obtained from human intervention studies and bioavailability in vitro studies of dietary xanthones of mangosteen products provide evidence that it is necessary to design future investigations to further elucidate the absorption, metabolism, and potential efficiency of dietary xanthones in human health without forgetting the safety studies of such products. Thus, it is necessary that more basic research studies on in vitro activity and, subsequently, human clinical trials are conducted to provide enough scientific evidence to support the use of mangosteen containing supplements or beverages as a health-promoting product.

There is relatively detailed phytochemical and pharmacological information on mangosteen fruit as a food supplement, but there are no clinical data available that would provide evidence of the efficacy of dietary mangosteen fruit xanthones in human health. Manufacturers of mangosteen products need clinical studies to prove their efficiency and in order to include nutrition and health claims on the labels or in the promotion of their products. Nutrition and health claim notifications are under the Regulation (EC) 1924/2006 in Europe (EC 2006). The Food and Drug Administration (FDA) in the United States requires that the manufacturer, packer, or distributors who wish to market a dietary supplement notify the FDA of certain information related to nutritional and health claims on the label of their products. These are summarized in the document titled "Guidance for Industry: A Food Labeling Guide" (FDA 2013).

7.2.2 Mango Fruit (*Mangifera indica* L.)

Mango (*Mangifera indica* L.) is one of the most popular tropical fruits. Production, trade, and consumption have increased significantly due to its attractive sensory properties and a growing recognition of its high nutritional and health-promoting properties (Figure 7.8; Masibo and He 2009; Rymbai et al. 2013).

Based on these properties, mango is the second-most traded tropical fruit in the world, after pineapple. The world production of mango (mangosteen and guavas included) is 43.93 million tons (MT) and is growing commercially in 111 countries, with India ranked first (18.00 MT) followed by China (4.62 MT), Thailand (3.14 MT), and

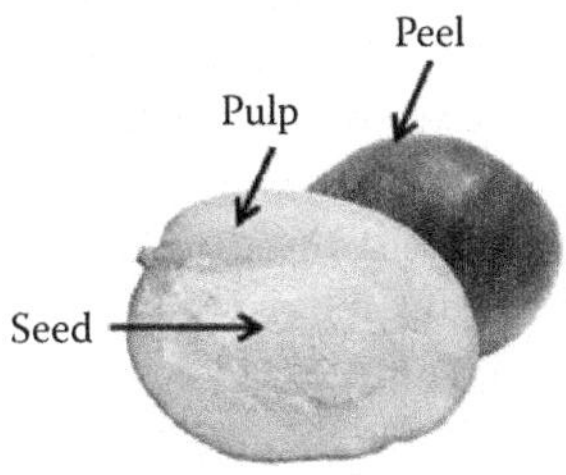

FIGURE 7.8 Mango fruit (*Mangifera indica* L.).

Indonesia (2.05 MT) (FAOSTAT 2016). There are three parts of interest in the mango fruit: the pulp, peel, and a central stone. The mango stone is composed of a woody outer shell encasing a kernel (seed proper). Depending on variety, the kernel or seed represents 45–85 percent of the stone and 20–60 percent of the mango fruit (Masibo and He 2009). Mango is generally consumed as a fresh fruit, and a wide range of foods can be prepared with the pulp. Demand for the concentrate of mango pulp is increasing due to its use as a base material in the beverage and dairy industry. The mango pulp, seed, and peel composition varies, depending on factors such as variety, locality, climate, and stage of maturity (Schieber et al. 2000; Berardine et al. 2005). There is an increased demand for mango products such as juice, nectar, canned mango slices in syrup, pickles, curries, dehydrated products, mango powder, chutneys, jams, and so on. Processing of mango fruits generates a significant amount of by-products, such as the peels and seeds that represent approximately 15–20 percent and 45–60 percent of the fresh fruit weight, respectively (Masibo and He 2009).

Extensive research on mango by-products has been performed in the last few years. These studies have revealed that mango by-products such as peels and seeds contain high concentrations of various health-promoting bioactive compounds, such as dietary fiber, vitamin C, carotenoids, and polyphenols (Ajila et al. 2007; Berardini et al. 2005; Sogi et al. 2013; Jahurul et al. 2015; Torres-León et al. 2016; Serna-Cock et al. 2016).

Within the group of polyphenols, the predominant compounds are mangiferin and quercetin-3-O-galactoside (Table 7.2). Mangiferin is a xanthone C-glycoside mainly found in the peel and seeds of mango fruit (Figure 7.9).

The mangiferin content of mango peel and seeds mainly depends of the cultivar and ripeness stage of the fruit. Thus, mangiferin is the major polyphenol compound in some cultivars (between 1263.2 mg/kg dw in Tommy Atkins and 13.9 mg/dw in Kent); meanwhile, in others is the flavonol quercetin 3-O-galactoside (Table 7.2). Compared with the peel, only small amounts of mangiferin were present in the pulp of few mango cultivars, going undetected in the majority of them (see Table 7.2; Berardini et al. 2005). Also, considerable variations in the amount of mangiferin in the peel and seeds of a mango fruit were found between cultivars, according to the results obtained by Barreto et al. (2008; see Table 7.3).

7.2.2.1 Biological Activity

There is a growing interest in the exploitation of the by-products of fruit processing that is rich in phytochemicals and that can be used as natural additives with different activities (antioxidant, anti-browning, antimicrobials, colorants, texturizers, etc.) but also as a source of bioactive compounds to obtain functional ingredients or functional foods (Masibo and He 2009: Ayala-Zavala et al. 2011; Jahurul et al. 2015; Torres-León et al. 2016; Serna-Cock et al. 2016; Rymbai et al. 2016). It is well documented that mango by-products, such as peels and seeds, present high antioxidant capacity and high levels of health-enhancing substances (dietary fiber, vitamin C, carotenoids, and polyphenols). Mango peels of different cultivars have shown a very high antioxidant capacity between 63–95 percent of DPPH inhibition (Serna-Cock et al. 2016). Mango peel and seeds can be dried and converted into powders and the powder can be used as a food supplement. Mango peel and seed powders can be used in bakery products (bread, cookies), ice cream, breakfast cereals, pasta products, beverages, extruded foods (snacks), meat, and so on (Juhurul et al. 2015; Serna-Cock et al. 2016).

Mango by-products are considered an important source of mangiferin. Mangiferin is a glucosylxanthone with a strong antioxidant capacity. Mangiferin isolated from

TABLE 7.2 Xanthones C-Glycosides and Flavonol Content in Peel, Pulp, and Kernel of Different Mango Cultivars (mg/kg dw)

	Tommy Atkins	**José**	**Kent**	**Haden**	***Ubá****
Peel					
Mangiferin	1263.2 ± 197.2	983.6 ± 50.1	13.9 ± 1.5	11.2 ± 0.1	199 ± 5.3
Isomangiferin	40.3 ± 0.8	45.5 ± 1.9	4.0 ± 0.3	21.0 ± 0.8	16.4 ± 2.9
Mangiferin gallate	87.3 ± 1.5	25.2 ± 2.0	ND	ND	28.0 ± 1.0
Isomangiferin gallate	12.3 ± 0.6	ND	ND	ND	26.9 ± 0.7
Quercetin-diglycoside	55.1 ± 0.7	40.3 ± 2.8	ND	ND	ND
Quercetin 3-O-galactoside	1217.3 ± 18.0	1467.3 ± 42.3	944.5 ± 38.3	1309.1 ± 26.0	151 ± 12.3
Quercetin 3-O-glucoside	882.0 ± 4.2	1045.3 ± 41.6	890.0 ± 39.8	912.7 ± 20.5	370 ± 25.6
Quercetin 3-O-xyloxide	239.5 ± 3.8	278.6 ± 8.4	150.7 ± 8.2	179.1 ± 4.5	84.4 ± 6.2
Quercetin 3-O-arabinofuranoside	163.5 ± 2.8	191.8 ± 7.5	91.6 ± 3.4	104.9 ± 5.1	64.8 ± 5.3
Quercetin 3-O-arabinopyranoside	152.4 ± 2.7	119.6 ± 3.5	84.8 ± 3.8	70.5 ± 0.8	35.0 ± 2.5
Quercetin 3-O-rhamnoside	38.2 ± 1.7	116.4 ± 4.3	58.1 ± 3.5	52.7 ± 0.6	15.8 ± 1.2
Kaempferol 3-O-glucoside	77.3 ± 5.3	171.7 ± 8.8	30.6 ± 1.8	43.7 ± 1.1	ND
Rhamnetin 3-O-galactoside	215.6 ± 4.9	374.4 ± 11.1	70.6 ± 3.4	228.6 ± 2.7	35.3 ± 2.7
Quercetin	ND	ND	3.3 ± 0.1	2.8 ± 0.0	64.1 ± 1.6
Total phenolics	4444.0 ± 198.3	4860.2 ± 80.0	2342.0 ± 56.4	2936.4 ± 33.9	1091 ± 67.3
Pulp					
Mangiferin	4.6 ± 0 1	19.4 ± 0.2	ND	16.2 ± 2.7	12.4 ± 0.3
Kernel					
Mangiferin					46.5 ± 4.7

Source: Berardini, N., Fezer, R., Conrad, J., Beifuss, U., Carle, R., and Schieber, A., *Journal of Agricultural and Food Chemistry*, 53, 1563–1570, 2005; Ribeiro, S. M. R., Barbosa, L. C. A., Queiroz, J. H., Knödler, M., and Schieber, A., *Food Chemistry*, 110, 620–626, 2008.

Abbreviations: ND, no detected; dw, dry weight.

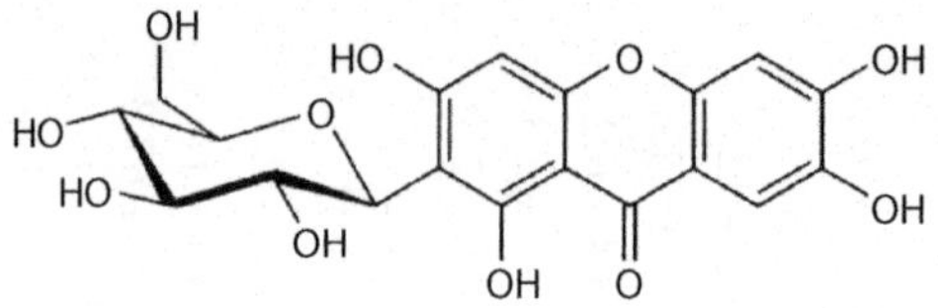

FIGURE 7.9 Mangiferin chemical structure.

TABLE 7.3 Mangiferin and Total Polyphenolic Compounds Content in Peels and Kernels of Different Mango Cultivars (g/kg dw)

	Van Dyke		Embrapa-141-Roxa	
	Peel	**Kernel**	**Peel**	**Kernel**
Mangiferin	4.94	6.40	15.23	8.98
Total polyphenolic	59.07	70.10	24.24	89.49

Source: Barreto, J. C., Trevisan, M. T. S., Hull, W. E. et al., *Journal of Agricultural and Food Chemistry*, 56, 5599–5610, 2008.

TABLE 7.4 Antioxidant Capacity of Mangiferin Purified from Peel and Seeds of Mango in the HX/XO, DPPH, FRAP, and ORAC Assays Systems

Substance	**HX/XO (mM) (IC_{50})**	**DPPH (μM) (IC_{50})**	**FRAP (μM) (EC_1)**	**ORAC (units)**
Mangiferin	0.47	0.59	1.03	12.85
Ascorbic acid	Pro-oxidant	50.74	81.82	1.07
Trolox	2.30	65.96	62.00	1.0

Source: Barreto, J. C., Trevisan, M. T. S., Hull, W. E. et al., *Journal of Agricultural and Food Chemistry*, 56, 5599–5610, 2008.

mango by-products has higher antioxidant capacity than well-known antioxidants such as ascorbic acid and Trolox (synthetic form of vitamin E) assayed by different test systems (hypoxanthine/xanthine oxidase, DPPH, FRAP, and ORAC assays; see Table 7.4; Barreto et al. 2008).

Many authors have studied mangiferin (Figure 7.9) as the possible active principle of mango peel and seeds and most of the biological activities of the mango by-products extracts have been attributed, in part, to this polyphenolic compound. Mangiferin was found to have the following activities: to protect mitochondrial membrane against lipid peroxidation, to participate in biological process related to apoptotic/necrotic cell death, to inhibit colon tumorigenesis in rats, and to inhibit immunoglobulin's G and various allergenic pathways. Mangiferin also has a very high antihyperlipidemic and anti-atherogenic activity, besides its antioxidant and antiviral properties. Mangiferin has also demonstrated efficacy as an anti-inflammatory, analgesic, immunomodulatory, and antioxidant, and also has high chemopreventive potential (Masibo and He 2009; Torres-León et al. 2016). It has also demonstrated effects against Alzheimer's disease (Sethiya and Mishra 2014). Vimang® is a formulation manufactured in Cuba that contains mangiferin as the main active ingredient (64 percent of total bioactive compounds isolated) and has presented protective effects against oxidative stress, has high chemopreventive potential, and has found to be effective in the management of patients with diabetes mellitus, benign prostate hyperplasia, dermatitis, and Lupus erythematosus (Masibo and He 2009; Sanchez et al. 2000).

7.3 EXTRACTION, ISOLATION, AND IDENTIFICATION OF XANTHONES

Xanthones are a class of polyphenolic compounds with a skeleton of xanthene-9-one that is substituted with a variety of isoprene, phenolic, and methoxyl groups, which give a large variety of possible structures. Natural xanthones are structurally similar to flavonoids and their chromatographic behavior is also similar (Naczk and Shahidi 2006). Procedures for the isolation and characterization of plant xanthones have been extensively reviewed (Obolskiy et al. 2009; Negi et al. 2013).

7.3.1 Extraction and Isolation

Extraction is a very important step in the isolation and characterization of polyphenolic compounds and there is no single or standard extraction method. Solvent extraction and extraction with supercritical fluids are the most common procedures used for the isolation of polyphenolic compounds. There are a large number of articles in the literature that focus on the extraction and analysis of polyphenols from plant material and derived products (Ignat et al. 2011).

The first step to obtain the xanthones' rich fraction is the grinding of the dried or lyophilized plant material with subsequent solvent extraction. Dorta et al. (2012) studied the effects of different drying methods on the polyphenol content and antioxidant capacity of mango peel and seed. They found that freeze-drying allowed the peel and seed to be stabilized without diminishing their polyphenol content and antioxidant capacity; meanwhile, the oven-drying treatment had the most negative effect on these characteristics.

Solvent extraction as a function of biomass status may be liquid–liquid extraction or solid–liquid extraction. Liquid–liquid extraction could be used for the isolation of xanthones from mangosteen fruit beverages commercialized as food supplements. Solid–liquid extraction has been widely employed for the xanthone extraction from different parts of mangosteen fruit and mango fruit (pericarp, peel, seeds, pulp, whole fruit, etc.; Ignat et al. 2011; Dorta et al. 2012, 2013a, 2013b, 2014; Zhou et al. 2015; Ibrahim et al. 2016).

Polyphenol extraction efficiency depends on the type of process employed and process conditions. Several factors affect the concentration of the determined compound in the extract, including particle size, solvent characteristics, temperature, solid–liquid ratio, extraction time, and number of steps of the extraction process (Ignat et al. 2011; Dorta et al. 2013a, 2013b, 2014).

The first step of the procedure typically involves the use of an organic or aqueous organic solvent. Methanol, ethanol, and acetone have been widely used as extracting solvents. For example, Yu et al. (2007) extracted the air-dried pericarp of mangosteen fruit with 70 percent methanol in the first extraction step. Chin et al. (2008) isolated and identified several xanthones from mangosteen powder fruit using methanol. Also, Mohamed et al. (2014) employed methanol for the extraction of xanthones from mangosteen fruit pericarp. Methanol also was used by Barreto et al. (2008) in the second extraction step of freeze-dried kernels and seeds of mango fruit after a first extraction with hexane in a soxhlet to remove lipids. Acetone has been employed for the extraction of xanthones from mangosteen pericarp (Ji et al. 2007). Also, mixtures of methanol/water (60:40 v/v) have been employed for the extraction of xanthones from mango peel and seed powder (Ribero et al. 2008) and 70 percent methanol has been used with air-dried mangosteen pericarp (Yun et al. 2007). In general, mixtures of ethanol/water or acetone/water (between 50 percent and 99.5 percent) are the most widely used extraction solvents because the food industry avoids the use of methanol due to its toxicity. For example, the extraction of xanthones from mangosteen pericarp was carried out with 95 percent ethanol by different authors (Zhan et al. 2010; Zhou et al. 2015) and mixtures of acetone/water (50:50 v/v and 80:20 v/v) was used to extract polyphenols, including xanthones from lyophilized mango seeds (Schieber et al. 2003; Chin et al. 2008; Ribero et al. 2008; Dorta et al. 2013b, 2014). Other solvents such as methylene chloride have been assayed for the extraction of xanthones from the arils and pericarp of mangosteen fruit (Wittenauer et al. 2012).

The isolation of xanthones from plant material is also influenced by the weight of plant material to solvent ratio. The most commonly reported ratios range between 1:1 and 1:10 (Yu et al. 2007; Chin et al. 2008; Dorta et al. 2014; Zhou et al. 2015). However, higher ratios between 1:20 and 1:100 have been also reported (Schieber et al. 2003; Berardini et al. 2005; Ji et al. 2007; Ribero et al. 2007; Dorta et al. 2014).

Temperature is an important factor in the extraction of xanthones from plant material. The conditions found in the literature ranged between room temperature (between three hours and three days; Berardini et al. 2005; Chin et al. 2008; Mohamed et al. 2014; Zhou et al. 2015) to 50–75°C during several hours (two to three hours; Yu et al. 2007; Wittenauer et al. 2012; Dorta et al. 2014). The extractions at room temperature could be combined with sonication for a short time (approximately 20 minutes; Ji et al. 2007), mechanical stirring for several hours (approximately three hours; Berardini et al. 2005), or maceration for one day (Chin et al. 2008). Microwave-assisted extraction of mango peels and seeds (one hour at 75°C) has been assayed with exit to obtain xanthone-rich extracts (Dorta et al. 2014). Temperature could be a critical factor for the extraction of determined xanthones. When mango peel is subjected to high temperatures, a slight decrease in mangiferin concentration has been detected, while the concentration of other xanthone derivatives increases significantly. These changes may be attributed to the formation of xanthones from bezophenone derivatives in mango peel, which are considered to be the precursors of xanthone *C*-glycosides (Berardini et al. 2005).

Today there is an increasing interest in the use of supercritical fluid extraction (SFE) because it is an environmentally beneficial alternative to the conventional organic solvent extraction. SFE is rapid, automatable, selective, and avoids the use of large amounts of toxic solvents. SFE utilizes the ability of certain chemicals to become excellent solvents for certain solutes under a combination of temperature and pressure. The most utilized fluid has been supercritical carbon dioxide ($SC–CO_2$). This solvent becomes supercritical when it is raised above its critical point of both pressure and temperature. In the case of carbon dioxide, the critical point is at 31.2°C and 7.38 MPa. The SFE of xanthones from mangosteen pericarp using $SC–CO_2$ (60°C/30 MPa) showed a much higher yield (7.56 percent) and xanthone content (65.93 percent) with an enrichment of xanthones 1.4–3.2 folds than the traditional ethanol extraction using a Soxtec™ (Zarena and Sankar 2011). Also, the xanthones *C*-glycosides from mango peel by-products have been successfully extracted with $SC–CO_2$ (García-Mendoza et al. 2015).

The separation of xanthones from the extract was commonly carried out by column chromatography on silica gel using different solvent mixtures with increasing polarity (for example, in order of addition into the column: n-hexane, chloroform, ethyl acetate, and methanol; Schieber et al. 2000; Yu et al. 2007; Mohamed et al. 2014). Also, the extract could be partitioned sequentially and treated with solvents of different polarity (hexane, dichloromethane, ethyl acetate, etc.; Chin et al. 2008). The majority of studies related to xanthone extraction procedures have been included in several reviews (Obolsky et al. 2009; Negi et al. 2013).

However, these procedures proved to be time-consuming and hardly applicable to routine analysis of a great number of samples. Therefore, other procedures have been assayed, such as the purification of polyphenol extract using solid-phase extraction with polyamide or Shephadex LH20 as solid phase previous to HPLC/DAD and HPLC/MS analysis (Schieber et al. 2003; Wittenauer et al. 2012; Dorta et al. 2014). Xanthones may be also separated and identified using thin layer chromatography (TLC) and HPLC by comparison with authentic standards (Negi et al. 2013).

7.3.2 Identification

Structural identification of all known xanthones has been established on the basis of ultraviolet visible spectroscopy (UV), infrared spectroscopy (IR), ^{1}H and/or ^{13}C nuclear magnetic resonance (NMR), x-ray crystallographic, and mass spectroscopy (MS) data (Chin et al. 2008; Obolskiy et al. 2009; Negi et al. 2013; Mohamed et al. 2014).

Ultraviolet visible spectroscopy (UV) technique is useful for locating free hydroxyl groups in xanthones, especially hydroxyl groups in position 3 (see Figure 7.1), which is easily detected by an addition of sodium acetate that results a bathochromic shift of 300–330 nm bands with increased intensity (Negui et al. 2013).

^{1}H and/or ^{13}C NMR spectroscopy is probably the most useful method in the structure elucidation of naturally occurring xanthones, and the NMR data of a great number of xanthones from *Garcinia mangostana* has been reported (Chin et al. 2008; Fu et al. 2013; Mohamed et al. 2014; Zhou et al. 2015; Xu et al. 2016). Different specific NMR techniques have been employed for the structural elucidation of more complex phenolic structures isolated from plant food–derived products, including naturally occurring xanthones. This includes ^{1}H and ^{13}C NMR, two dimensional homonuclear (^{2}D ^{1}H-^{1}H) correlated NMR spectroscopy (COSY), heteronuclear chemical shift correlation NMR (C-H HECTOR), totally correlated NMR spectroscopy (TOCSY), nuclear Overhauser effect in the laboratory frame (NOESY), and rotating frame of reference (ROESY; Obolskiy et al. 2009; Ignat et al. 2011; Mohamed et al. 2014; Zhou et al. 2015).

Proton nuclear magnetic resonance spectroscopy (^{1}H NMR) has been widely used to determine the chemical structure of xanthones. The integral signal of the ^{1}H NMR spectrum is proportional to the number of protons present in the xanthone structure. ^{1}H NMR gives information about the substitution pattern on each ring (Chin et al. 2008; Negi et al. 2013; Mohamed et al. 2014; Zhou et al. 2015).

Carbon nuclear magnetic resonance spectroscopy (^{13}C NMR) gives information about the total number of C atoms present in the molecule due to the number of signals of the ^{13}C. The NMR spectrum indicates the number of different types of C atoms (Chin et al. 2008; Negi et al. 2013; Mohamed et al. 2014; Zhou et al. 2015).

X-ray crystallography has been proven as a useful technique for determining the three-dimensional structures, including absolute configuration of xanthones. The main condition to use this technique is to obtain diffraction-quality crystals. The xanthones' crystal structures have revealed that the three-ring system of the xanthones core is mainly planar (Obolskiy et al. 2009).

Mass spectrometry (MS) is also a useful tool for elucidating the xanthones' chemical structure. The MS principle consists of ionizing chemical compounds to generate charged molecules or molecule fragments and measuring their mass-to-charge ratios. The main ionization systems used to analyze xanthones are fast atom bombardment (FAB); electrospray ionization (ESI); atmospheric pressure ionization (API), including atmospheric pressure chemical ionization (APPI); and, in parallel to electrospray advent, matrix-assisted laser desorption ionization (MALDI).

The analytical technique mainly used to separate, identify, and quantify polyphenol compounds of fruit and derived products is high-performance liquid chromatography (HPLC). The chromatographic conditions of the HPLC method include the use of reverse-phase C18 column and a binary solvent system containing acidified water solvent (solvent A) and polar organic solvent (solvent B). HPLC are combined with different detectors, such as UV-Vis diode array detector (HPLC–DAD), mass, or tandem mass spectrometry (HPLC–MS). Different authors have described rapid and efficient HPLC–DAD chromatographic systems to separate and identify xanthones from mangosteen fruit (Walker 2007; Chaivisuthangkura et al. 2009). One example of this efficient chromatographic approach is the use of HPLC–DAD for fingerprinting the main xanthones of mangosteen fruit using fourteen xanthone standards previously isolated from mangosteen pericarp. Figure 7.10 shows the HPLC–DAD (254 nm) chromatogram of the xanthones standards perfectly separated in a run of 60 minutes using a C18 column and mobile phase in a gradient preparing from acetonitrile (solvent A), 2 percent (v/v), acetic acid in water (solvent B), and n-butanol (solvent C). This HPLC–DAD fingerprinting method can be readily used as a suitable method for evaluation of the quality of mangosteen fruit and its derived products (Figure 7.11; Chaivisuthangkura et al. 2009).

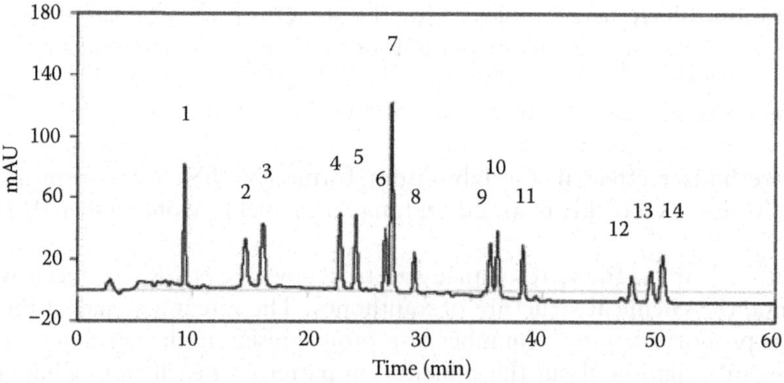

FIGURE 7.10 HPLC-DAD chromatogram detected at 254 nm of a mixture of standards of xanthones isolated from mangosteen fruit. Peak identification: (1) 11-hydroxy-1-isomangostin; (2) garcinone C; (3) garcinone D; (4) γ-mangostin; (5) 8-deoxygartanin; (6) gartanin; (7) α-mangostin; (8) garcinone E; (9) demethylcalabaxanthone; (10) 1,6-dihydroxy-7-methoxy-8-(3-methylbut-2-enyl)-6′,6′-dimethylpyrano(2′,3′:3.2)xanthone; (11) β-mangostin; (12) mangostenona; (13) calabaxanthone; (14) tovophyllin B. (Data from Chaivisuthangkura, A., Malaikaew, Y., Chaovanalikit, A. et al., *Chromatographia*, 69: 315–318, 2009.)

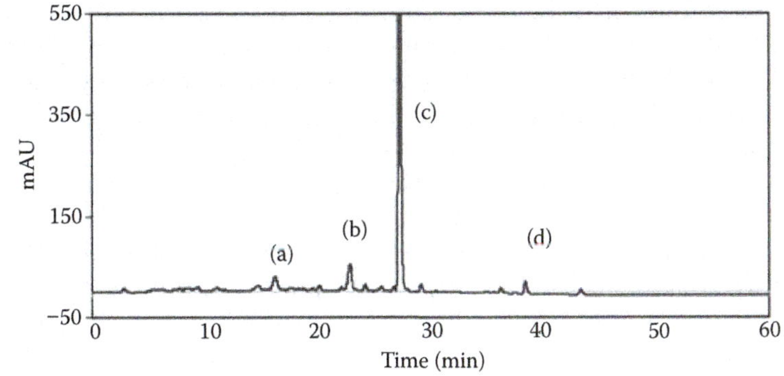

FIGURE 7.11 HPLC-DAD chromatogram detected at 254 nm of xanthones found in a crude extracts of mangosteen fruit hull. Peak identification: (a) garcinone D; (b) γ-mangostin; (c) α-mangostin; (d) β-mangostin. (Data from Chaivisuthangkura, A., Malaikaew, Y., Chaovanalikit, A. et al., *Chromatographia*, 69: 315–318, 2009.)

HPLC–MS techniques are nowadays the best analytical approach to study polyphenols from different biological sources, such as xanthones in mangosteen fruit (Jung et al. 2006; Ji et al. 2007; Zarena and Sankar 2011; Wittenauer et al. 2012). Moreover, for a more accurate characterization of phenolic compounds present in mangosteen fruit, mango fruits, and derived products, it is essential to use advanced and powerful techniques like high performance liquid chromatography coupled to electrospray ionization and quadrupole-time of flight–mass spectrometry (HPLC-ESI-QTOF-MS; Dorta et al. 2014). These techniques facilitate the exact mass measurement of both MS and MS/MS ions that enable the detection of hundreds of compounds within a single extract and provide essential information to characterize the structures of the phenolic compounds present in different plant derived products. A specific fragmentation pattern in the MS/MS mode has been observed to determine the main xanthone structures in mangosteen fruit and mango fruit. Thus, the elimination of prenyl moieties (–56 Da) in the allylic position may be an important reaction for the identification of xanthones. Also, the formation of significant fragment ions from the loss of OH, H_2O, and CHO are typical for xanthones and related compounds with a methoxy substituent near a carbonyl group. Also, xanthone C-glycoside, such as mangiferin, could be well identified by the elimination of the C-glucosidic moiety (–120 Da and 90 Da; Schieber et al. 2003; Berardini et al. 2005; Ji et al. 2007; Barreto et al. 2008; Ribeiro et al. 2008; Wittenauer et al. 2012; Negui et al. 2013; Dorta et al. 2014).

7.4 FINAL REMARKS

The principal natural sources of xanthones are the mangosteen fruit (*Garcinia mangostana* L.) and the mango fruit (*Mangifera indica* L.). Mangosteen has a rich content of prenylated and oxygenated xanthones, the majority of which are α-mangostin and γ-mangostin. Mango fruit is characterized by the presence of C-glycosylated xanthones such as mangiferin. The improvement in the extraction techniques, such as the use of supercritical fluid extraction (SFE) using supercritical fluids such as supercritical carbon dioxide (SC–CO_2) and the development of more accurate, advanced, and powerful identification techniques such as high performance liquid chromatography coupled to electrospray ionisation and quadrupole–time of flight–mass spectrometry (HPLC–ESI–QTOF–MS), allows the more accurate separation and identification of xanthones from different foods and biological matrices and the discovery of new ones.

Tropical fruit are mostly consumed fresh but nowadays there is an increasing demand for fruit-derived products to be used in the beverage and dairy industries, among others. Processing of mangosteen and mango fruits generates a significant amount of valuable by-products, such as the pericarp in mangosteen and the peel and seed in mangos, which have been shown to have very high levels of xanthones. Numerous studies have discovered that xanthones, mainly α-mangostin and mangiferin, have a wide range of biological activities such as anti-inflammatory, antitumor, antioxidant, cardioprotective, antidiabetic, antiobesity, antifungal, antiparasitic, antiviral, antiallergy, and antibacterial. Due to its potential role in promoting health, even more products derived from mangosteen and mango fruits have been developed and commercialized as botanical healthy dietary supplement or functional beverages. Detailed phytochemical and pharmacological information exists to support the biological activities of xanthones obtained mainly from in vitro or in vivo models, but there is scarce clinical data available that would provide evidence of efficacy of dietary mangosteen fruit xanthones in human health. Thus, there is the necessity for much more basic research studies to further elucidate the absorption, metabolism, and potential efficiency of dietary xanthones and, subsequently, human clinical trials to provide enough scientific evidence to support the use of mangosteen- and mango-containing supplements or beverages as healthy promoting products without forgetting safety studies of such products.

ACKNOWLEDGMENTS

The authors would like to thank the Spanish Ministry of Science and Innovation [project AGL2010-15910 (subprogram ALI)] and the Spanish Ministry of Economy and Competitiveness [(project AGL2013- 46326-R)].

REFERENCES

Ajila, C. M., Naidu, K. A., Bhat, S., and Prasada Rao, U. J. S. 2007. Bioactive compounds and antioxidant potential of mango peel extract. *Food Chemistry* 105: 982–988.

Ayala-Zavala, J. F., Vega-Vega, V., Rosas-Dominguez, C., Palafox-Carlos, H., Villa-Rodriguez, J. A., Wasim Siddiqui, M. W., Dávila-Aviña, F. E. and González-Aguilar, G. A. 2011. Agro-industrial potential of exotic fruit byproducts as a source of food additives. *Food Research International* 44: 1866–1874.

Barreto, J. C., Trevisan, M. T. S., Hull, W. E. et al. 2008. Characterization and quantitation of polyphenolic compounds in bark, kernel, leaves, and peel of mango (*Mangifera indica* L.). *Journal of Agricultural and Food Chemistry* 56: 5599–5610.

Berardini, N., Fezer, R., Conrad, J., Beifuss, U., Carle, R., and Schieber, A. 2005. Screening of mango (*Mangifera indica* L.) cultivars for their contents of flavonol O- and xanthone C-glycosides, anthocyanins, and pectin. *Journal of Agricultural and Food Chemistry* 53: 1563–1570.

Bumrungpert, A., Kalpravidh, R. W., Suksamrarn, S., Chaivisuthangkura, A., Chitchumroonchokchai, C., and Failla, M. L. 2009. Bioaccessibility, biotransformation, and transport of α-mangostin from *Garcinia mangostana* (Mangosteen) using simulated digestion and Caco-2 human intestinal cells. *Molecular Nutrition & Food Research* 53: S54–S61.

Chaivisuthangkura, A., Malaikaew, Y., Chaovanalikit, A. et al. 2009. Prenylated xanthone composition of *Garcinia mangostana* (mangosteen) fruit hull. *Chromatographia* 69: 315–318.

Chin, Y. W., Jung, H. A., Chai, H., Keller, W. J., and Kighorn, A. D. 2008. Xanthones with quinone reductase-inducing activity form the fruits of *Garcinia mangostana* (Mangostee). *Phytochemistry* 69: 754–759.

Chitchumroonchokchai, C., Riedl, K. M., Suksumrarn, S., Clinton, S. K., Kinghorn, A. D., and Failla, M. L. 2012. Xanthones in mangosteen juice are absorbed and partially conjugated by healthy adults. *The Journal of Nutrition* 142: 675–680.

Dorta, E., González, M., Globo, M. G., Sánchez-Moreno, C., and de Ancos, B. 2014. Screening of phenolic compounds in by-product extracts from mangoes (*Mangifera indica* L.) by HPLC-ESI-QTOF-MS and multivariate analysis for use as a food ingredient. *Food Research International* 57: 51–60.

Dorta, E., Lobo, M. G., and González, M. 2012. Using drying treatments to stabilize mango peel and seed: Effect on antioxidant activity. *LWT- Food Science and Technology* 45: 261–268.

Dorta, E., Lobo, M. G., and González, M. 2013a. Optimization of factors affecting extraction of antioxidants from mango seed. *Food and Bioprocess Technology* 6: 1067–1081.

Dorta, E., Lobo, M. G., and González, M. 2013b. Improving the efficiency of antioxidant extraction from mango peel by using microwave-assisted extraction. *Plant Foods for Human Nutrition* 68: 190–199.

Dragendorff, O. 1930. Über das Harz von *Garcinia Mangostana* L. *Liebigs Annalen* 482: 280–301.

EC. 2006. Regulation (EC) No 1924/2006 of the European Parliament and the Council of 20 of December of 2006 on Nutrition and Health Claims Made in Foods.

El Seedi, H. R., El-Ghorab, D. M. H., Bohlin, L., Borg-Karlson, A. K., Goransoon, U., and Verpoorte, R. 2010. Recent insights into the biosynthesis and biological activities of natural xanthones. *Current Medical Chemistry* 17: 854–901.

FAOSTAT. 2016. Mango, Mangosteen, and Guava 2013 World Production. http://faostat3.fao.org/download/Q/QC/E.

FDA. 2013. Guidance for Industry: A Food Labeling Guide (8. Claims). http://www.fda.gov/Food/GuidanceRegulation/GuidanceDocumentsRegulatoryInformation/LabelingNutrition/ucm064908.htm (last updated 8/20/2015).

Fu, M., Qiu, S. X., Xu, Y., Wu, J., Chen, Y., Yu, Y., and Xiao, G. 2013. A new xanthone from the pericarp of *Garcinia mangostana*. *Natural Products Communication* 8: 1733–1734.

García-Mendoza, M. P., Paula, J. T., Paviani, L. C., Cabral, F., and Martínez-Correa, H. A. 2015. Extracts from mango peel by-product obtained by supercritical CO2 and pressurized solvent processes. *LWT Food Science and Technology* 62: 131–137.

Gutierrez-Orozco, F., and Failla, M. L. 2013. Biological activities and bioavailability of mangosteen xanthones: A critical review of the current evidence. *Nutrients* 5: 3168–3183.

Ibrahim, N. Y., Hashim, N. M., Mariod, A. A. et al. 2016. α-Mangostin from *Garcinia mangostana* Linn: An updated review of its pharmacological properties. *Arabian Journal of Chemistry* 9: 317–329.

Ignat, I., Volf, I., and Popa, V. I. 2011. A critical review of methods for characterization of polphenolic compounds in fruits and vegetables. *Food Chemistry* 126: 1821–1835.

Ji, X., Avula, B., and Ikhlas, A. K. 2007. Quantitative and qualitative determination of six xanthones in *Garcinia mangostana* L. by LC-PDA and LC-ESI-MS. *Journal of Pharmaceutical and Biomedical Analysis* 43: 1270–1276.

Jindarat, S. 2014. Xanthones from mangosteen (*Garcinia mangostana*): Multitargeting pharmacological properties. *Journal of the Medical Association of Thailand* 97: S196–S201.

Juhurul, M. H. A., Zaidul, I. S. M., Ghafoor, K., Al-Juhaimi, F., Nyam, K. L., Norulaini, N. A. N., Sahena, F., and Omar, A. K. M. 2015. Mango (*Mangiferin indica* L.) by-products and their valuable components: A review. *Food Chemistry* 183: 173–180.

Jung, H. A., Su, N., Keller, W. J., Mehta, R. G., and Kinghorn, A. D. 2006. Antioxidant xanthones from the pericarp of *Garcinia mangostana* (mangosteen). *Journal of Agricultural and Food Chemistry* 54: 2077–2085.

Kondo, M., Zhang, L., Ji, H., Kou, Y., and Ou, B. 2009. Bioavailability and antioxidant activity of a xanthone-rich mangosteen (*Garcinia mangostana*) product in humans. *Journal of Agricultural and Food Chemistry* 57: 8788–8792.

Masibo, M., and He Q. 2009. Mango bioactive compounds and related nutraceutical properties-A review. *Food Research International* 25: 346–370.

Mohamed, G. A., Ibrahim, S. R. M., Shaaban, M. I. A., and Ross, S. A. 2014. Mangostanaxanthones I and II, new xanthones from the pericarp of *Garcinia mangostana*. *Fitoterapia* 98: 215–221.

Murakami, M. 1932. Über die constitution des mangostins. *Liebigs Annalen* 496: 122–151.

Na, Y. 2009. Recent cancer drug development with xanthone structures. *Journal of Pharmacy and Pharmacology* 61: 707–712.

Nack, M., and Shahidi, F. 2006. Phenolics in cereals, fruits and vegetables: Occurrence, extraction and analysis. *Journal of Pharmaceutical and Biomedical Analysis* 41: 1523–1542.

NBJ's Supplement Business Report, U.S. 2015. *Nutrition Business Journal.* http://www.transchem.com.au/wp-ontent/uploads/2015/09/NBJ_2015_Supplement_Report.pdf

Negi, J. S., Bisht, V. K., Singh, P., Rawat, S. M., and Joshi, G. P. 2013. Naturally occurring xanthones: Chemistry and biology. *Journal of Applied Chemistry* (2013): 1–9. http://dx.doi.org710.1155/2013/621459.pdf

Obolskiy, D., Pischel, I., Siriwatanametanon, N., and Heinrich, M. 2009. *Garcinia mangostana* L.: A phytochemical and pharmacological review. *Phytotherapy Research* 23:1047–1065.

Panda, S. S., Chand, M., Sakuja, R., and Jain, S. C. 2013. Xanthones as potential antioxidants. *Current Medical Chemistry* 20: 4481–4507.

Pedraza-Chaverri, J., Cárdenas-Rodríguez, N., Orozco-Ibarra, M., and Pérez-Rojas, J. M. 2008. Medicinal properties of mangosteen (*Garcinia mangostana*). *Food and Chemical Toxicology* 46: 3227–3239.

Pinto, M. M. M., Sousa, M. E., and Nascimiento, M. S. J. 2005. Xanthone derivatives: New insights in biological activities. *Current Medical Chemistry* 12: 2517–2538.

Ribeiro, S. M. R., Barbosa, L. C. A., Queiroz, J. H., Knödler, M., and Schieber, A. 2008. Phenolic compounds and antioxidant capacity of Brazilian mango (*Mangifera indica* L.) varieties. *Food Chemistry* 110: 620–626.

Rymbai, H., Roy, A. R., Deshmukh, N. A., Jha, A. K., Shimray, W., War, G. F. 2016. Analysis study on potential under utilized edible fruit genetic resources of the foothills track of Eastern Himalayas, India. *Genetic Sources and Crop Evolution* 63: 125–139.

Rymbai, H., Srivastav, M., Sharma, R. R., Patel, C. R., and Singh, A. K. 2013. Bioactive compounds in mango (*Mangifera indica* L.) and their roles in human health and plant defense: A review. *Journal of Horticultural Science & Biotechnology* 88: 369–379.

Sanchez, G. M., Re, L., Giuliani, A., Núnzez-Sellés, A. J., Davison, G. Pl, and León-Fernández, O. S. 2000. Protective effects of *Mangiferin indica* L. extract, mangiferin, and selected antioxidants against TPA-induced biolemolecules oxidation and peritoneal macrophage activation in mice. *Pharmacological Research: The Official Journal of Italian Pharmacological Society* 42: 565–573.

Schieber, A., Berardini, N., and Carle, R. 2003. Identification of flavonol and xanthone glycosides from mango (*Mangifera indica* L. cv. "Tommy Atkins") peels by high-performance liquid chromatography-electrospray ionization mass spectrometry. *Journal of Agricultural and Food Chemistry* 51: 5006–5011.

Schieber, A., Ullrich, W., and Carle, R. 2000. Characterization of polyphenols in mango puree concentrate by HPLC with diode array and mass spectrometric detection. *Innovative Food Science and Emerging Technology* 1: 161–166.

Schmid, W. 1855. Ueber das mangostin. *Liebigs Annalen der Chemie* 93: 83–89.

Serna-Cock, L., García-Gonzales, E., and Torres-León, C. 2016. Agro-industrial potential of the mango peel based on its nutritional and functional properties. *Food Reviews International* 32: 364–376.

Sethiya, N., and Mishra, S. 2014. Investigation of mangiferin, as a promising natural polyphenol xanthone on multiple targets of Alzheimer's disease. *Journal of Biologically Active Products from Nature* 4: 111–119.

Shagufta, and Ahmad, I. 2016. Recent insight into the biological activities of synthetic xanthone derivatives. *European Journal of Medicinal Chemistry* 116. 267–280.

Sogi, D. S., Siddip, M., Greiby, I, and Dolan, K. D. 2013. Total phenolics, antioxidant activity, and functional properties of "Tommy Atkins" mango peel and kernel as affected by drying methods. *Food Chemistry* 141: 2649–2655.

Tang, Y. P., Li, P. G., Kondo, M., Ji, H. P., Kou Y., and Ou, B. 2009. Effect of a mangosteen dietary supplement on human immune function: A randomized, double-blind, placebo-controlled trial. *Journal of Medicinal Food* 12: 755–763.

Torres-León, C., Rojas, R., Contreras-Esquivel, J. C., Serna-Cock, L., Belmares-Cerda, R. E., and Aguilar, C. N. 2016. Mango seed: Functional and nutritional properties. *Trends in Food Science and Technology* 55: 109–117.

Udani, J. K., Singh, B. B., Barret, M. L., and Singh, V. J. 2009. Evaluation of mangostee juice blend on biomarkers of inflammation in obese subjects: A pilot, dose finding study. *Nutrition Journal* 8: 48–54.

Walker, E. B. 2007. HPLC analysis of selected xanthones in mangosteen fruit. *Journal of Separation Science* 30: 1229–1234.

Wang, Y., Xia, Z., Xu, J. R., Wang, Y. X., Hou, L. N., Qiu, Y., and Chen, H. Z. 2012. α-Mangostin, a poyphenolic xanthone derivative from mangosteen, attenuates β-amyloid oligomers-induced neurotoxicity by inhibiting amyloid aggregation. *Neuropharmacology* 62: 871–881.

Wittenauer, J., Falk, S., Scheweiggert-Weisz, U., and Carle, R. 2012. Characterization and quantification of xanthones from the aril and pericarp of mangosteens (*Garcinia mangostana* L.) and mangosteen containing functional beverage by HPLC-DAD-MS^n. *Food Chemistry* 134: 445–452.

Xu, T., Deng, Y. X., Zhao, S., and Shao, Z. Y. 2016. A new xanthone from the pericarp of *Garcinia mangostana*. *Journal of Chemical Research* 40: 10–11.

Yates, P., and Stout, G. H. 1958. The structure of mangostin. *Journal of the American Chemical Society* 80: 1691–1700.

Yu, L., Zhao, M., Yang, B., Zhao, Q., and Jiang, Y. 2007. Phenolics from the hull of *Garcinia mangostana* fruit and their antioxidant activities. *Food Chemistry* 104: 176–181.

Zadernowski, R., Czplicki, S., and Naczk, M. 2009. Phenolic acid profiles of mangosteen fruits (Garcinia mangostana). *Food Chemistry* 112: 685–689.

Zarena, A. S. and Sankara, U. 2009. Supercritical carbon dioxide extraction of xanthones with antioxidant activity from Garcinia mangostana: Characterization by HPLC/LC-ESI-MS. *Journal of Supercritical Fluids* 49: 330–337.

Zarena, A. S., and Sankar, U. 2011. Xanthones enriched extracts from mangosteen pericarp obtained by supercritical carbon dioxide process. *Separation and Purification Technology* 80: 172–178.

Zhou, X., He, L., Wu, X., Zhong, Y. et al. 2015. Two new xanthones from the pericarp of *Garcinia mangostana*. *Natural Products Research* 29: 19–23.

CHAPTER 8

Stilbenes in Foods

Rubén Francisco González-Laredo, Nuria Elizabeth Rocha-Guzmán, José Alberto Gallegos-Infante, Martha Rocío Moreno-Jiménez, and Claudia Ivette Gamboa-Gómez

CONTENTS

8.1 INTRODUCTION

Among secondary metabolites commonly found in fruits, vegetables, grains, spices, herbs, and their derived foods and beverages, there are polyphenols, which are significant phytochemicals in human diet. These compounds at minor concentrations exert important antioxidant activity as radical scavengers in vitro against radical oxygen species (ROS), which is a distinctive property; thus, foods containing these metabolites are beneficial to sustaining good health. The consumption of foods with high contents of phenolic compounds at a ratio of one gram per day is highlighted as safe and helpful in the reduction of risks of many chronic diseases (Scalbert and Williamson 2000).

Additionally, it has been known for decades that polyphenols may protect plants against different types of biotic or abiotic stresses (drought, heat, UV radiation, insects and microorganisms, nutrients deficiency, etc.). In the plant kingdom there are reports of more than 8,000 phenolic compounds, which are produced from phenylalanine, a common intermediate, or from shikimic acid, a close precursor (Pandey and Rizvi 2009). Major groups of polyphenol compounds comprise flavonoids and non-flavonoid compounds (Figure 8.1). Flavonoids are mostly present in human diet foodstuffs, such as fruits and vegetables, and may be classified into subgroups, including flavanols, flavonols, flavones, isoflavones, flavanones, anthocyanidins, chalcones, and aurones (Manach et al. 2004). The non-flavonoid compounds may include phenolic acids, lignans, and stilbenes.

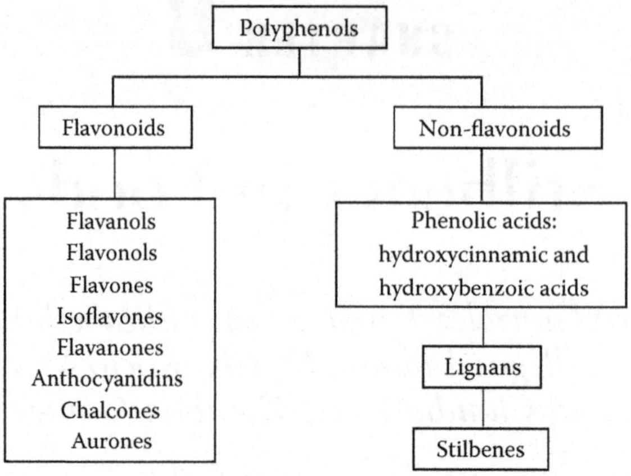

FIGURE 8.1 Stilbenes as a class of non-flavonoid polyphenols.

8.2 STILBENES

Stilbenes comprise a relatively small group among non-flavonoids phenolic compounds. Structurally, stilbenes are 1,2-diarylethenes: they are phenolic compounds holding two benzene rings separated by an ethane bridge, that is, a C_6–C_2–C_6 skeleton (Figure 8.2). Stilbenes exist as two possible isomers: The trans-1,2-diphenylethylene, called (E)-stilbene or trans-stilbene, and the cis-1,2-diphenylethylene, called (Z)-stilbene or cis-stilbene. The latter is sterically hindered and less stable due to steric interactions from the out-of-plane aromatic rings, which prevent them from chemical reaction and conjugation. Commonly, ring A is substituted by two hydroxyl groups in meta-position, while ring B holds one or several hydroxyls or a methoxy groups in ortho-, meta-, or para-positions.

Stilbene	**Precursor**	**R1**	**R2**	**R3**
Trans-resveratrol	*p*-coumaric acid	H	H	OH
Pinosylvin	Cinnamic acid	H	H	H
Piceatannol	Caffeic acid	H	OH	OH
Hydroxyresveratrol	2´,4´-dihydroxycinnamic acid	OH	H	OH
Rhapontigenin	Isoferulic acid	H	OH	OCH_3
Piceid	Trans-resveratrol	H	H	Oglu

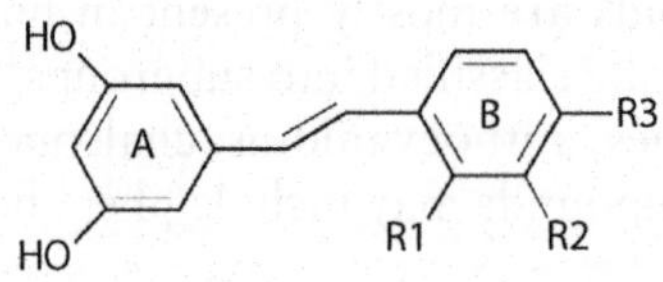

FIGURE 8.2 Resveratrol and related stilbene structures.

Usually stilbenes are present as monomers, oligomers, and polymers such as viniferines (Figure 8.3) in a diversity of herbal and vegetable sources (Ozcan et al. 2014).

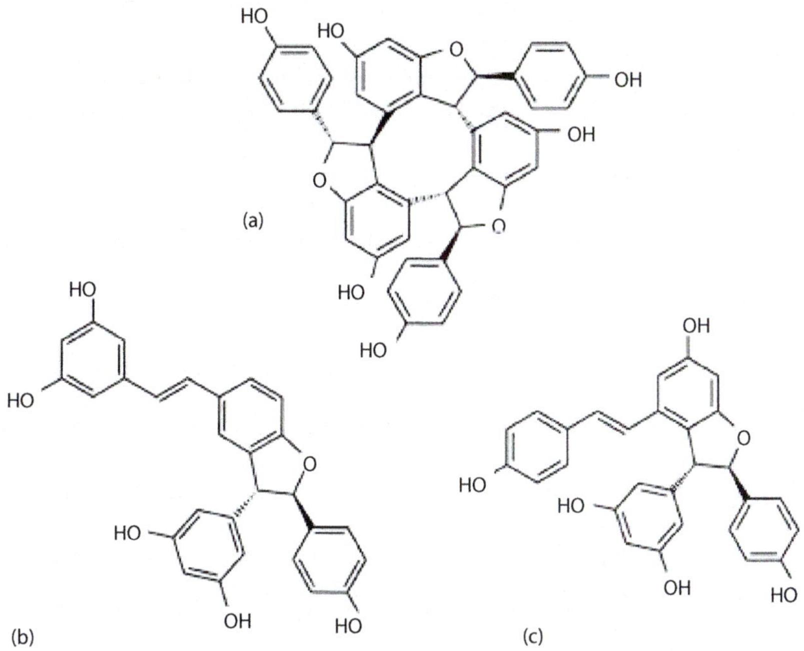

FIGURE 8.3 Structures of viniferins (resveratrol oligomers): (a) Alpha-viniferin, (b) delta-viniferin, and (c) epsilon-viniferin.

They may be present also in free phenolic form or conjugated, creating glycosylated molecules usually with glucose or xylose moieties (González-Laredo et al. 1997). Several basic sources of stilbenoids are reported in human diets and in more than 70 plant species and plant-derived foods (Cassidy et al. 2000), such as grapes (including peel and seeds), wines, some berries, soy, and peanuts—although it is a fact that the occurrence of stilbenes in the human diet is quite low. Due to stilbenoid bioactivity, their potential application either as phytochemical supplements or as active ingredients in nutraceutical and cosmetic preparations have been considered. Furthermore, stilbenes in plants are considered as antifungal phytoalexins. These are inducible metabolites excreted by plants as a chemical defense and as response to a variety of stress conditions, injuries, or infections by pathogens such as *Botrytis cinerea*, a necrotic fungus that uses wine grapes as hosts and degrades resveratrol owing to its laccase activity (Faravon et al. 2009). Stilbenes are also produced by plants as constitutive woody metabolites, thus the natural antimicrobial activity of these compounds suggests that their concentration in situ may be a good indication of stress levels and disease resistance.

Stilbenes are synthesized by a wide range of plant species, including Cyperaceae, Dipterocarpaceae, Fagaceae, Gnetaceae, Leguminosae, Liliaceae, Moraceae, Myrtaceae, Pinaceae, and Vitaceae, and are normally detected in barks, leaves, rhizomes, and roots. Nevertheless, stilbenoids are commonly found in nonedible plants. Therefore, current data available highlights grapes, grape juices and wine, peanuts, peanut oil, and peanut butter as the major dietary sources of stilbenes. In traditional medicine, we have used medicinal plants, such as the rhizomes and roots of *Veratrun formosanum* and *Polygonum cuspidatum*, which contain resveratrol and are used in traditional Chinese medicine (Chung et al. 1992).

8.2.1 Resveratrol in Grapes and Wines

The most studied naturally occurring stilbene is resveratrol (3,4',5-trihydroxystilbene), which is found mainly in grapes. Wine, as a product of grapes, and particularly red wine, also contains a significant amount of resveratrol. It was first isolated from the roots of white hellebore (*Veratrum grandiflorum* O. Loes) by Takaoka in 1939 and can be found in at least 72 plant species, including 12 families and 31 genera (Bhat and Pezzuto 2002).

In grapevines, the synthesis of trans-resveratrol takes place in leaves, roots, and particularly in grape skins; it may be absent or low in the fruit flesh. It may be a good marker for gray mold (*B. cinerea*) resistance and positively correlated to defense of the grapevine to cryptogamic diseases. Resveratrol is produced in response to UV radiation or microbial infection; when infestation in grapes is about 10 percent, the resulting wines have higher resveratrol content (Jeandet et al. 1991). Contrarily, its production also decreases naturally during the fruit-ripening process. Resveratrol production may be also provoked by chemical applications, which act as abiotic elicitors in fungicide and herbicide treatments.

Additionally to trans-resveratrol, its 3-O-glucoside (i.e., piceid) and its *cis* counterpart (i.e., cis-piceid) can similarly be detected in grape products. However, the cis-resveratrol isomer is released only as a by-product of fermentation from fresh grapes either by hydrolysis of their glucoside or by isomerization of the trans freeform (Mattivi et al. 1995). Also, grape genetics play an important role in stilbenes synthesis, as well as agronomical factors and climate conditions. The variety and the elevation (i.e., up to 320 m) influence the piceid production. Particularly, the year (i.e., temperature and moisture in the last phase of ripening) determines the cis-piceid content as a phytoalexin response to higher fungal pressure (Bavaresco et al. 2007).

As normal in natural products, the resveratrol content in wines depends on the agronomical, climatic, and ecological conditions of the harvesting. The metabolite levels change significantly from one variety to another, although similar trans-resveratrol contents were published for skins in red and white varieties of grapes (Okuda and Yokotsuka 1996). The resveratrol and piceid levels may be used as chemotaxonomic wine markers, allowing wines from different wineries to be grouped according to their common chemical content varieties (Romero-Pérez et al. 1996). The degree of soaking of grapes with seeds and skins during fermentation is the key factor that conditions the stilbenoid concentration in wines. Although considering other factors, such as aging time in oak barrels, yeast strains, fining agents, and process conditions, the longer the maceration of skins, the higher the extraction yield of resveratrol (Pezet and Cuenat 1996).

Other treatments during the wine-making process can modify stilbene extraction as the use of commercial pectolytic enzymes to break skin cells and release phenolic and flavor components. Also, filtering treatments and fining agents can reduce the resveratrol content in wines. In the first case, some membranes can adsorb more than half the resveratrol present. Meanwhile, in the second case, polyvinylpolypyrrolidone (PVPP) and gelatin may affect the levels of resveratrol and piceid in the wine (Cassidy et al. 2000).

The highest trans-resveratrol level is shown by red wines at about 8 milligrams/liter, depending on the grape variety. In rosé wines, the content may vary from 1.3 to 3.0 milligrams/liter. Meanwhile, white and sparkling wines have lower resveratrol concentrations of around 0.1 to 1.2 milligrams/liter due to the minimal skin contact related to white wine production. In respect to grape juices, the total resveratrol level in red juices is about 0.7 to 15 milligrams/liter; in white juices, it may be up to 1.5 milligrams/liter. The stilbene contents in red wines/juices seem on average to be tenfold higher than in their white counterparts. However, in the end, it is clear that differences in stilbenoid concentrations are mainly due to changes in processing technology (Okuda and Yokotsuka 1996).

8.2.2 Food Preservative Properties of Stilbenes

Besides the known and well-documented bioactive properties of resveratrol from grapes and grape seeds, there are products like Vineatrol® (www.vineatrol.com), which endorses the benefits from resveratrol and resveratrol oligomers. It has been labeled as a natural source of active stilbenes and as oligomers with strong antioxidant activities (Müller et al. 2009). This product is an extract from grapevine shoots and woody aerial parts of the grapevine, containing about 29 percent stilbenoids. It has demonstrated Botox- and retinoic-like activities, triggering SIRT1 expression, which protects genome integrity and cell lifespan. This extract is suggested as an active ingredient in formulations for anti-aging cosmetics as an option to retinoic acid in facial creams, for relaxation preparations of facial muscle contractions, and for wrinkle cream prescriptions.

Additionally, stilbenoids have shown promising properties as a red wine preservative. Traditionally, sulfur dioxide (SO_2) has been used as a preservative in the food industry, particularly for wines. SO_2 is maybe the most efficient and multipurpose additive ever used in winemaking due to its antioxidant and antiseptic properties, which inhibit the effects of oxidase enzymes and dissolved oxygen. Due to its accumulation in the organism, the reduction of SO_2 is a demand from consumers in favor of more natural and organic substitutes; thus, this may be the case of some polyphenols such as stilbenes.

The efficacy of Vineatrol® to preserve the quality of red wines made under traditional and Ganimede® winemaking systems was reported lately, at bottling and after one year of storage (Raposo et al. 2016). Sensory and olfactometric analysis, enological and color parameters, and volatile compounds composition were assessed. At bottling, treated wines with stilbenoids showed higher scores in sensory analysis and better color-related parameters than those samples treated with SO_2. Using SO_2 has increased the production of alcohol volatile and ester compounds compared with wines treated with the phenolic additive Vineatrol®. The olfactometric profile of wines was modified by the stilbene extract addition, showing two new odorant zones with high modified frequencies in wines treated with the stilbene preservative. After one year of storage, wines in bottles containing Vineatrol® showed oxidation characters in color parameters and sensory analysis, exposing the evolution during the storage in bottles as the weakest point of the process.

In other potential applications, resveratrol and related stilbenoids may be incorporated into biofilms for food packaging. As an example, different matrices have been experimented with, with relative success. Such is the case of polylactic acid (PLA) based films, obtained by a blow-extrusion process. Soto-Valdez et al. (2011) suggested that these films for diverse food, medical, and pharmaceutical applications act as antioxidant release membranes. Similarly, resveratrol was efficiently incorporated into chitosan and methylcellulose films. Pastor et al. (2013) reported some changes in the physicochemical and microstructural properties of the films, particularly when the highest concentration ratio of polymer:resveratrol was used (1:0.1). Even though the films resulted as opaque and less resistant and stretchable, the changes did not fundamentally affect the appearance and manipulation of films.

These composite films did not show antimicrobial activity (e.g., against *Botrytis cinerea* and *Penicillium italicum*), but exhibited constant antioxidant activity in close relation to the content of resveratrol as no evident losses of the activity were observed during film formation. Therefore, resveratrol-based films may be appropriate for coating food products, retarding oxidation processes, preserving nutritional value, and extending product shelf life. More studies should explore applications of these films on delicate food products.

8.2.3 Biological Activities of Resveratrol and Stilbenes

Resveratrol oligomers have been documented as fungal detoxification products of resveratrol metabolism. These oligomers have exhibited widely biological activities, such as antibacterial, antifungal, anticancer, anti-HIV, and antioxidant activities (Figure 8.4). Besides, during the last two decades, resveratrol has incited interest due to its preventive possibilities against aging, cancer, cardiovascular diseases, diabetes, and neurodegenerative diseases, the most prevalent human diseases of our time (Xue et al. 2014).

Resveratrol, as present in grapes and wine, shows scavenging activity in vitro for oxygen reactive species, OH radicals, and for lipid hydroperoxyl free radicals. This significant antioxidant activity may be correlated with the prophylactic effect of the moderated drinking of red wine against mental deterioration and senile dementia in mature people. It is accepted now that the consumption of fruit and vegetable juices rich in polyphenols three or more times per week may play an important role in delaying the beginning of conditions like Alzheimer's disease (Dai et al. 2006). Polyphenols from fruits and vegetables look like unique potential ingredients for neuroprotection against neurodegenerative diseases due to their capacity to influence and modulate several cellular processes such as signaling, proliferation, apoptosis, differentiation, and redox balance (Singh et al. 2008).

Resveratrol has a great influence on the carcinogenesis process, affecting cancer initiation and progression. Resveratrol has shown the capacity to inhibit through various mechanisms several human tumors cells by *in vitro* experiments and in different *in vivo* animal models, presenting no significant toxicity at relatively high doses of the stilbene.

When diet resveratrols are absorbed, they become metabolized quickly to resveratrol-3-O-glucuronides, to resveratrol-4-O-sulfates, and to other related metabolites. They may accumulate in relatively high levels in human tissues and plasma, but little is known about their consequent biological activity. However, resveratrol as a phytoestrogen has been reported to increase protein levels of Mn superoxide dismutase (MnSOD) and enhance its activity both in vitro and in vivo assays. This mitochondrial antioxidant enzyme confers cytoprotection and regulates cell cycle progression. Likewise, it was shown that natural resveratrol structural analogues such as piceid and pterostilbene (i.e., 3,5-dimethoxy-resveratrol) are able of eliciting similar effects and to slow cell growth

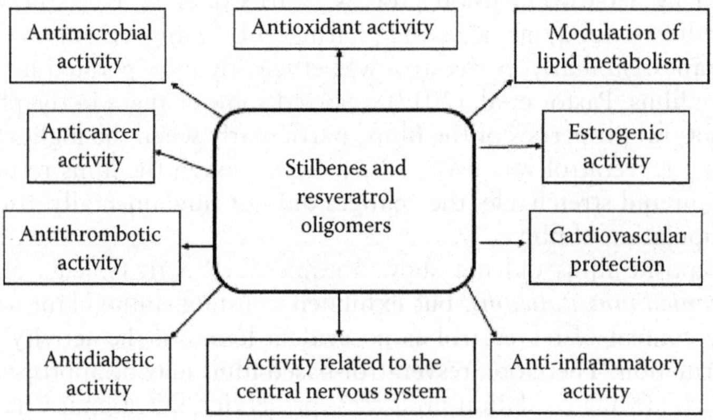

FIGURE 8.4 Bioactivities of stilbenes and resveratrol oligomers.

(Robb et al. 2014). In the end, it seems that these stilbenes activate the same mitochondrial response in mammalian cells, and therefore they might be as effective as resveratrol in modulating MnSOD expression and producing positive health effects in vivo.

8.2.4 Antidiabetic Activity of Stilbenes

Stilbenes and mainly resveratrol have been reported as antidiabetic phenolics, among many other biological activities (Figure 8.4). Although several mechanisms have been suggested to describe the antidiabetic action of resveratrol, the modulation of SIRT1, the enzyme that deacetylates proteins to contribute to cellular regulation and longevity, is one of the more accepted, improving the insulin sensitivity and the whole-body glucose homeostasis of diabetic rats (Milne et al. 2007). Also extracts from *Vitis vinifera* leaves have exhibited important antihyperglycemic effects as well as antioxidant activity equivalent to tolbutamide, a known hypoglycemic agent (Orhan et al. 2006).

Resveratrol treatment in diabetic rats stimulated their physiological processes such as glucose uptake by hepatocytes, adipocytes, and skeletal muscle and hepatic glycogen synthesis (Su et al. 2006). Another potential use of resveratrol is for the treatment of diabetic neuropathies. The administration of resveratrol to diabetic rats significantly tempered extreme pain conditions like cold allodynia and hyperalgesia, which are frequent medical findings from subjects with neuropathic pain (Sharma et al. 2006). More favorable effects have been observed when resveratrol was given in combination with insulin (Harikumar and Aggarwal 2008).

It has been claimed that oxidative stress and high glucose induced cytotoxicity were inhibited by grape seed phenolic extracts in cultured porcine kidney cells (LLC-PK1). Particularly, resveratrol prevents diabetes-induced changes in the kidney (i.e., diabetic nephropathy) and improves meaningfully oxidative stress and renal dysfunction in diabetic rats. Treatment with resveratrol also delayed the onset of insulin resistance and the decreasing of insulin secretion. A proposed mechanism was relating the inhibition of ATP-sensitive potassium channels in beta cells (Chen et al. 2007).

Siemann and Creasy (1992) suggested that red wine was rich in resveratrol and that long-term moderated red wine drinking might be associated to the lower prevalence of cardiovascular diseases observed in France and some Mediterranean countries. This would be confirmed later by many epidemiological studies, referring to it as the French paradox (Renaud and de Lorgeril 1992).

So far, numerous studies have proven the antiglycation capacity of resveratrol and related stilbenes (Yeh et al. 2017). That is, the capacity to inhibit the nonenzymatic glycation or Maillard reaction, which conducts a spontaneous post-translational modification of proteins and amino acids through reducing sugars. The products resulting from these reducing sugars are known as advanced glycation end products (AGE). In female gametocyte or germ cells involved in reproduction (oocytes), resveratrol inhibits the oxidative damage triggered by highly reactive carbonyl species (RCS) in the human body such as methylglyoxal (MG), which are dicarbonyl compounds produced by Maillard reactions in a key step for the generation of AGE. Likewise, resveratrol can protect cells from MG-induced mitochondrial dysfunction and oxidative stress, showing a significant reduction effect on diabetic complications. Resveratrol also protects against MG-induced injuries and may ease liver damage caused by type 2 diabetes mellitus, diminishing the expression of receptors of AGEs in the organ. Thus, resveratrol might moderate long-term diabetic complications, which might be associated with the inhibition of glycation factors, the accumulation of MG and AGE, and the expression of AGE receptors.

8.2.5 Anticarcinogenic Properties of Stilbenes

Actual epidemiological data suggest that there is a significant correlation between dietary intake and prevalence of many kinds of chronic and degenerative diseases such as cancers. On one hand, the incidence of cancer in people tends to rise every year in the current century due to changes in diet and modern lifestyles. On the other hand, present cancer chemotherapies are not as efficient as desired, demanding new anticancer drugs with less toxic and more effective outputs. Therefore, preventing or treating cancers with phytochemicals, particularly from diet, looks like a reasonable strategy. Additionally, in vitro and in vivo studies have shown that many dietary substances have anticancer properties such as some polyphenols, including stilbenes, resveratrol, and its oligomers (Fresco et al. 2006).

Polyphenols prompt the lessening of tumors, causing a protective effect on diverse human organs. Some stilbenes are among the more studied plant-derived therapeutics for chemotherapy and cancer prevention. Resveratrol, in particular, is known for having potent antioxidant and anti-inflammatory effects and for inhibiting platelet aggregation and the growth of a variety of cancer cells. Its chemotherapeutic and chemopreventive activities have been observed in all stages of cancer progress (initiation, promotion, and progression) in mice with chemical and UVB-induced skin carcinogenesis, and some murine models of human cancers. Resveratrol is reported as effective in most types of cancers, including breast, gastric, colorectal, lung, prostate, and skin, suppressing metastasis and angiogenesis. Extensive data in different human cell cultures indicate that this stilbene can modulate various targets and multiple pathways involved in cell growth, inflammation, and apoptosis (Athar et al. 2007).

However, resveratrol may act also as a pro-oxidant and its activity may look contradictory, depending on the cell line tested and its concentration. Particularly, Rossi et al. (2013) obtained unexpected results when they compared antioxidant protection from three resveratrol derivatives. Pterostilbene and 3,5,4′-trimethoxystilbene resulted in being more effective as antioxidants than resveratrol, at the same concentration. This is regardless of resveratrol having more hydroxyl groups in its structure. Consequently, the methoxylated stilbenes provide more protection than resveratrol against hydrogen peroxide mediated oxidative damage of DNA in Chinese hamster ovary (CHO) cells treated in vitro. As the mechanism of action involved is the scavenging of OH• radicals by the stilbenes, the trimethoxy derivative can act by an electron transfer process rather than a hydrogen transfer because there are not OH groups available in the molecule.

It has been compared to the antiproliferative and survival effects of some grape stilbenes extracts on leukemic and normal human lymphocytes (Billard et al. 2002). The trans-resveratrol epsilon-viniferin (Figure 8.3), a resveratrol dimer, and two different samples of Vineatrol® were tested. The two Vineatrol preparations were purified by preparative chromatrography, containing 10 and 25 percent of trans-resveratrol. The rich stilbenes samples were added to leukemic cells cultures from chronic B cell malignancies and to normal peripheral blood-derived mononuclear cells (PBMC) as controls. From the detected inhibition and the decline of cell recovery, the four different stilbenes samples showed anti-proliferative effects on the tested leukemic cells. Vineatrol at 10 percent resulted as the stronger treatment, while resveratrol and Vineatrol at 25 percent showed similar medium activities, and finally, the epsilon-viniferin sample exhibited just minor effects. The same relative order of effectiveness was perceived from their capacity to induce apoptosis in the leukemic B cells, while the same phenolic compounds showed minimum effects on the survival of PBMC, requiring higher concentrations to elicit cell death. The apoptotic effect of stilbenes in chronic leukemic B cells has been well-correlated with the activation of caspase 3, with the reductions on the inducible nitric oxide synthase (iNOS) and the anti-apoptotic protein bcl-2 expressions, and also with a decrease in the mitochondrial transmembrane potential. Extracts from vine shoots may be a convenient and natural source of crude stilbenes for the purification of resveratrol and other polyphenols of nutraceutical interest.

Resveratrol may interact with multiple molecular targets at relatively nontoxic dose models. However, the rapid metabolism of resveratrol to its glucuronate and sulfate conjugates has prompted a debate on the mechanisms underlying its bioactivity. Therefore, its bioavailability and pharmacological properties at different target tissues need to be improved, and the exact mechanisms of action that regulate its effectiveness and use to be understood. An example is the synthesis of resveratrol prodrugs and sulfated metabolites as promising candidates for anticancer drugs to overcome these problems (Falomir et al. 2016). Resveratrol glucosylated (RG) prodrugs have shown more cytotoxicity in HT-29 and MCF-7 cells than in the stilbene alone. Likewise, RG, and particularly, the resveratrol disulfate (RDS), reduced production of the vascular endothelial growth factor (VEGF) more significantly than resveratrol. Also, RG and RDS inhibited the expression of the human telomerase reverse transcriptase (hTERT) to a higher degree than resveratrol. Remarkably, resveratrol sulfated metabolites have displayed different biological activity in vivo, showing that they are not just bioactive resveratrol holders.

Finally, given the numerous effects of resveratrol even at its relatively low content in wine or other food sources, its nutraceutical benefits may depend on the synergistic mixtures that could be tested with other phytochemicals. It has been shown that resveratrol presents synergistic apoptotic effects in vitro with other polyphenols such as both ellagic acid and quercetin (Baur and Sinclair 2006). The preclinical studies of stilbenes and other phytochemicals in combined treatments are in progress. Thus, it is expected that these studies will collect data to sustain action mechanisms of resveratrol and provide basis for forthcoming preventive therapeutic trials and clinical research using this natural stilbene.

8.3 CONCLUSIONS

The diverse bioactivities shown by resveratrol and diet stilbenes promote their consumption as potent phytochemicals and endorse continuing research on their preventive possibilities against chronic diseases such as aging, cancer, cardiovascular diseases, diabetes, and neurodegenerative diseases. Also, the antiglycation and antiproliferative effects of resveratrol and related stilbene compounds drive scientific attention as pertinent anti-diabetic and anticarcinogenic nutraceutical agents. The elucidation of mechanisms of action, exploring the modulation of many different pathways, understanding physiological processes and the biochemistry behind the activity of stilbenes-rich plant extracts is in course. In addition, the study of synergistic effects of resveratrol and other bioactive polyphenols such as quercetin and gallic acid to enhance functional activities is an attractive and a prospective investigation topic, already at preclinical stages. Stilbenes as induced and constitutive secondary metabolites may be synthesized selectively in plant sources by abiotic elicitation for the enrichment in specific active phytochemicals.

In food technology, the utilization of stilbenes from grape extracts as antioxidant ingredients has promising applications; substituting SO_2 in wine bottling and adding it as an active ingredient in biofilms for food packaging are encouraging opportunities.

REFERENCES

Athar, M., J.H. Back, X. Tang, K.H. Kim, L. Kopelovich, D.R. Bickers, and A.L. Kim. 2007. Resveratrol: A review of preclinical studies for human cancer prevention. *Toxicol. Appl. Pharmacol.* 224:274–283.

Baur, J.A. and D.A. Sinclair. 2006. Therapeutic potential of resveratrol: The in vivo evidence. *Nat. Rev. Drug Discov.* 5:493–506.

Bavaresco, L., S. Pezzutto, M. Gatti, and F. Mattivi. 2007. Role of the variety and some environmental factors on grape stilbenes. *VITIS-Journal of Grapevine Research* 46:57–61.

Bhat, K.P.L. and J.M. Pezzuto. 2002. Cancer chemopreventive activity of resveratrol. *Annals of the New York Acad. Sci.* 957:210–229.

Billard, C., J.C. Izard, V. Roman, C. Kern, C. Mathiot, F. Mentz, and J.P. Kolb. 2002. Comparative antiproliferative and apoptotic effects of resveratrol, epsilon-viniferin and vine-shots derived polyphenols (vineatrols) on chronic B lymphocytic leukemia cells and normal human lymphocytes. *Leuk. Lymphoma* 43(10):1991–2002.

Cassidy, A., B. Hanley, and R.M. Lamuela-Raventos. 2000. Isoflavones, lignans and stilbenes–origins, metabolism and potential importance to human health. *J. Sci. Food & Agric.* 80(7):1044–1062.

Chen, W.P., T.C. Chi, L.M. Chuang, and M.J. Su. 2007. Resveratrol enhances insulin secretion by blocking K(ATP) and K(V) channels of beta cells. *Eur. J. Pharmacol.* 568:269–277.

Chung, M.I., C.M. Teng, K.L. Cheng, F.N. Ko, and C.N. Lin. 1992. An antiplatelet principle of Veratrum formosanum. *Planta Med.* 58:274–275.

Dai, Q., A.R. Borenstein, Y. Wu, J.C. Jackson, and E.B. Larson. 2006. Fruit and vegetable juices and Alzheimer's disease: The Kame Project. *Am. J. Medicine* 119:751–759.

Falomir, E., R. Lucas, P. Peñalver, R. Martí-Centelles, A. Dupont, A. Zafra-Gómez, and J.C. Morales. 2016. Cytotoxic, antiangiogenic and antitelomerase activity of glucosyl-and acyl-resveratrol prodrugs and resveratrol sulfate metabolites. *ChemBioChem.* 17:1343–1348.

Faravon, F., M. Luccetta, S. Odorizzi, A.T. Pais da Cunha, and L. Sella. 2009. The role of grape polyphenols on trans-resveratrol activity against *Botrytis cinerea* and of fungal laccase on the solubility of putative grape proteins. *J. Plant Pathol.* 91:579–588.

Fresco, P., F. Borges, C. Diniz, and M.P.M. Marques. 2006. New insights on the anticancer properties of dietary polyphenols. *Medicinal Res. Rev.* 26(6):747–766.

González-Laredo, R.F., J. Cháidez-González, A. Ahmed, and J.J. Karchesy. 1997. A stilbene xyloside from *Holodiscus discolor. Phytochem.* 46(1):175–176.

Harikumar, K.B. and B.B. Aggarwal. 2008. Resveratrol: A multitargeted agent for age-associated chronic diseases. *Cell Cycle* 7:1020–1035.

Jeandet, P., R. Bessis, and B. Gautheron. 1991. The production of resveratrol (3,5,4′-trihydroxystilbene) by grape berries in different development stages. *Am. J. Enol. Vitic.* 42:41–46.

Manach, C., A. Scalbert, C. Morand, C. Rémésy, and L. Jiménez. 2004. Polyphenols: Food sources and bioavailability. *Am. J. Clin. Nutr.* 79(5):727–747.

Mattivi, F., F. Reniero, and S. Korhammer. 1995. Isolation, characterization, and evolution in red wine vinification of resveratrol monomers. *J. Agric. Food Chem.* 43:1820–1823.

Milne, J.C., P.D. Lambert, S. Schenk, D.P. Carney, J.J. Smith, D.J. Gagne et al. 2007. Small molecule activators of SIRT1 as therapeutics for the treatment of type 2 diabetes. *Nature* 450:712–716.

Müller, C., K. Ullmann, A. Wilkens, P. Winterhalter, S. Toyokuni, and P. Steinberg. 2009. Potent antioxidative activity of Vineatrol_30 grapevine-shoot extract. *Biosci. Biotechnol. & Biochem.* 73(8):1831–1836.

Okuda, T. and K. Yokotsuka. 1996. Trans-Resveratrol concentrations in berry skins and wines from grapes grown in Japan. *Am. J. Enol. Vitic.* 47:93–99.

Orhan, N., M. Aslan, D.D. Orhan, F. Ergun, and E. Yesilada. 2006. In-vivo assessment of antidiabetic and antioxidant activities of grapevine leaves (*Vitis vinifera*) in diabetic rats. *J. Ethnopharmacol.* 108:280–286.

Ozcan, T., A. Akpinar-Bayizit, L. Yilmaz-Ersan, and B. Delikanli. 2014. Phenolics in human health. *Int. J. Chem. Eng. & Appl.* 5(5):393.

Pandey, K.B. and S.I. Rizvi. 2009. Plant polyphenols as dietary antioxidants in human health and disease. *Oxid. Med. & Cell. Long.* 2(5):270–278.

Pastor, C., L. Sánchez-González, A. Chiralt, M. Cháfer, and C. González-Martínez. 2013. Physical and antioxidant properties of chitosan and methylcellulose based films containing resveratrol. *Food Hydrocoll.* 30:272–280.

Pezet, R. and P. Cuenat. 1996. Resveratrol in wine: Extraction from skins during fermentation and post-fermentation standing of must from Gamay grapes. *Am. J. Enol. Vitic.* 47:287–290.

Raposo R., M.J. Ruiz-Moreno, T. Garde-Cerdán, B. Puertas, J.M. Moreno-Rojas, A. Gonzalo-Diago, R. Guerrero, V. Ortíz, and E. Cantos-Villara. 2016. Grapevine-shoot stilbene extract as a preservative in red wine. *Food Chem.* 197(Part B):1102–1111.

Renaud, S. and M. de Lorgeril. 1992. Wine, alcohol, platelets, and the French paradox for coronary heart disease. *Lancet* 339:1523–1526.

Robb, E.L., and J.A. Stuart. 2014. The stilbenes resveratrol, pterostilbene and piceid affect growth and stress resistance in mammalian cells via a mechanism requiring estrogen receptor beta and the induction of Mn-superoxide dismutase. *Phytochemistry* 98:164–173.

Romero-Pérez, A.I., R.M. Lamuela-Raventós, A.L. Waterhouse, and M.C. de la Torre-Boronat. 1996. Levels of cis- and transresveratrol and piceid isomers in rose and white *Vitis vinifera* wines. *J. Agric. Food Chem.* 44:2124–2128.

Rossi, M., F. Caruso, R. Antonioletti, A. Viglianti, G. Traversi, S. Leone, and R. Cozzi. 2013. Scavenging of hydroxyl radical by resveratrol and related natural stilbenes after hydrogen peroxide attack on DNA. *Chemico-Biological Interactions*. 206:175–185.

Scalbert, A. and G. Williamson. 2000. Dietary intake and bioavailability of polyphenols. *J. Nutr.* 130:2073S–2085S.

Sharma, S., M. Anjaneyulu, S.K. Kulkarni, and K. Chopra. 2006. Resveratrol, a polyphenolic phytoalexin, attenuates diabetic nephropathy in rats. *Pharmacol.* 76:69–75.

Siemann, E.H. and L.L. Creasy. 1992. Concentration of the phytoalexin resveratrol in wine. *Am. J. Enol. Vitic.* 43(1):49–52.

Singh, M., M. Arseneault, T. Sanderson, V. Murthy, and C. Ramassamy. 2008. Challenges for research on polyphenols from foods in Alzheimer's disease: Bioavailability, metabolism, and cellular and molecular mechanisms. *J. Agric. Food Chem.* 56:4855–4873.

Soto-Valdez, H., R. Auras, and E. Peralta. 2011. Fabrication of poly(lactic acid) films with resveratrol and the diffusion of resveratrol into ethanol. *J. Applied Polym. Sci.* 121:970–978.

Su, H.C., L.M. Hung, and J.K. Chen. 2006. Resveratrol, a red wine antioxidant, possesses an insulinlike effect in streptozotocin-induced diabetic rats. *Am. J. Physiol. Endocrinol. Metab.* 290:1339–1346.

Takaoka, M. 1939. Resveratrol, a new phenolic compound from *Veratrum grandiflorum*. *J. Chem. Soc. Japan* 60:1090–1100.

Xue, Y.-Q., J.-M. Di, Y. Luo, K.-J. Cheng, X. Wei, and Z. Shi. 2014. Resveratrol oligomers for the prevention and treatment of cancers. *Oxid. Med. & Cell. Long.* Vol 2014, Article ID 765832, 9pp.

Yeh, W.-J., S.-M. Hsia, W.-H. Lee, and C.-H. Wu. 2017. Polyphenols with antiglycation activity and mechanisms of action: A review of recent findings. *J. Food Drug Anal.* 25(1):84–92.

CHAPTER 9

Anthraquinones

Mireille Fouillaud, Yanis Caro, Mekala Venkatachalam, Isabelle Grondin, and Laurent Dufossé

CONTENTS

9.1 INTRODUCTION

Phenolics are of outstanding importance in many foods consumed by humans. Among the most popular foods containing phenolics are wine, olive oil, coffee, and tea. Among the phenolics, anthraquinones are less known, as well as the foods containing this class of molecules, and their properties.

This chapter describes anthraquinones, their structure, and their occurrence in foods such as plants, fermented products, insects, and so on. Uses of anthraquinones are closely linked to their properties such as colorant, antioxidant, antimicrobial, and so on, sometimes only relying on an empirical knowledge.

As anthraquinone structure is based on an anthracenedione chemical backbone, they have often been associated with undesirable properties. This Dr. Jekyll and Mr. Hyde aspect is therefore presented in this chapter.

9.2 ANTHRAQUINONES' MAIN STRUCTURES

Anthraquinones, also called anthracenediones or dioxoanthracenes, are important members of the quinone family, and constitute a large structural variety of compounds among the polyketide group. Anthraquinones are structurally built from an anthracene ring with a keto group on position 9, 10 as basic core and different functional groups such as -OH, $-CH_3$, $-OCH_3$, $-CH_2OH$, -CHO, -COOH, and so on may substitute at various positions (Figure 9.1). Anthraquinones and their derivatives, produced as secondary metabolites in plants, lichens, insects, and higher filamentous fungi, occur either in a free form or as glycosides. These glycosides are formed when one or more sugar molecules, mostly glucose or rhamnose, are bound to the aglycone by an *O*-glycoside linkage to a hydroxyl group. At times, other complexes linked by C- or O- in the side chain can also be synthesized (Gessler, Egorova, and Belozerskaya 2013, Caro et al. 2012). The electronic absorption spectra is a characteristic feature of the parent compound 9,10-anthraquinone and its dihydroxy- and diamino-derivatives, which permit understanding of the effects of the hydrogen bond, solvent polarity, and nature of substituents on the spectral shift (Khan 2012). These detailed studies are of great importance owing to the wide-ranging

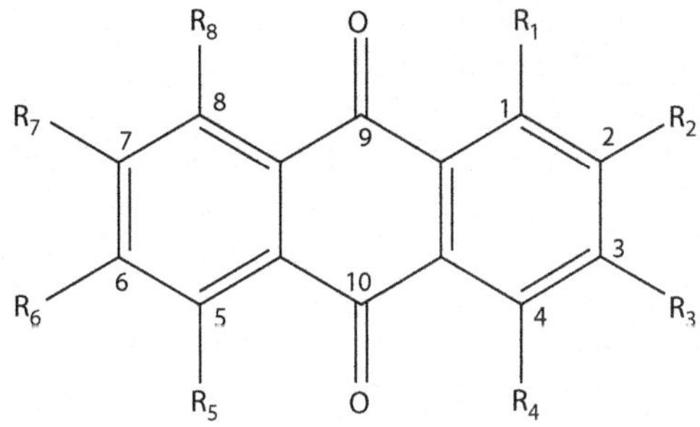

FIGURE 9.1 General structure of anthraquinone.

applications of anthraquinones in many fields, especially in the fields of dyeing. About 700 natural anthraquinoid pigments have already been identified from insects, plants, or microbes (Caro et al. 2012, Gessler, Egorova, and Belozerskaya 2013, Fouillaud et al. 2016). They are moreover considered "reactive dyes" as they can form covalent bonds, that is, with food components. Hydroxyanthraquinone derivatives can also easily form coordination complexes with several cations as metals. Opposite to direct dyes, their chemical structures are often much simpler, their absorption spectra show narrower absorption bands, and the dyeings are brighter (Hunger 2003).

Emodin- and Alizarin-Type Pigments. Based on the biosynthetic pathways, the pigment type is classified as an emodin or alizarin type. The emodin (6-methyl-1,3,8-trihydroxyanthraquinone)-type of anthraquinones shows a substitution on both aromatic rings and has a structure of 1,8-dihydroxyanthraquinone. Acetate-malonate pathway leads to the production of this type of anthraquinone pigments and contains basic aglycones such as emodin, aloe-emodin, physcion, and chrysophanol. Whereas the alizarin-type anthraquinones have one benzene ring unsubstituted with at least one hydroxyl group in position R1, and are typically synthesized by the shikimate-o-succinylbenzoate pathway (Sajc et al. 1999, Caro et al. 2012).

9.3 ANTHRAQUINONES NATURALLY OCCURRING IN FOODS

9.3.1 Anthraquinones in Edible Plants

In higher plants, anthraquinone derivatives are found in a wide range of species (Caro et al. 2012). Even if the distribution of these compounds in edible plants and vegetables is less extensive than that of other phenolic molecules (e.g., flavonoids), it cannot be discounted.

A number of anthraquinones derivatives found in higher plants, especially in *Rheum*, *Rumex*, *Rhamnus*, *Aloe*, and *Cassia* species, are excellent examples of acetate-derived structures formed through the (acetate-malonate)-polyketide pathway. These plants' anthraquinones show substitutions on both aromatic rings and have at least two hydroxyl groups in both the R1 and R8 positions. In contrast, in plants from the Rubiaceae family (e.g., *Morinda*, *Rubia*, and *Galium* species), the most common naturally occurring anthraquinones, such as alizarin, are synthesized via the chorismate/O-succinylbenzoic acid pathway (Caro et al. 2012). In plants, anthraquinones are not only present under their free form as aglyca, but often bound to sugars, forming water-soluble glycosides (Teuscher and Lindequist 1994, Thomson 1997, Lu et al. 1998, Derksen et al. 2003).

For ages, the plants containing anthraquinones have been mainly exploited like purgative drugs and consumed as such. Emodin, physcion, chrysophanol, aloe-emodin, or rhein form the basis of a range of natural anthraquinones' derivatives molecules found in purgative drugs of plant origin. The plant purgative drugs as senna (obtained from leaves and fruits of *Cassia angustifolia* and *C. senna*), cascara (obtained from bark extracts of *Rhamnus purshianus*), and frangula (obtained from bark extracts of *Rhamnus frangula*) are thus only suitable for medicinal and pharmaceutical uses. The free forms have little therapeutic activity in purgative drugs. They need to be under the form of water-soluble glycosides such as anthraquinone O- and C-glycosides, or dianthrone O-glycosides, in order to exert their action.

9.3.1.1 Rheum spp. *(Polygonaceae)*

The edible stem of the common rhubarb, that is, *Rheum rhaponticum* (also known as garden [English] rhubarb, is cultivated in various regions of Europe for culinary purposes (Figure 9.2).

In contrast, anthraquinone glycosides such as emodin-1-O-glycoside, chrysophanol-1-O-glycoside, emodin-8-O-glycoside, aloe-emodin-8-O-glycoside, rhein-8-O-glycoside, and chrysophanol-8-O-glycoside are a series of major active anthraquinoid compounds found in dried rhizome and root of some rhubarb species (*Rheum palmatum*, *R. officinale*, *R. tanguticum*, and *R. australe* or *R. emodi* wall [e.g., Meissn. also known as Himalayan rhubarb]). The aglycone structures, especially rhein, emodin, and chrysophanol, are also present in these "medicinal" rhubarb species.

The dianthrone derivatives, such as emodin dianthrone, physcion dianthrone, and sennosides A and B (i.e., dimers formed by oxidative coupling of two single anthraquinones), have also been characterized in these plants, especially in the roots (Teuscher and Lindequist 1994, Qhotsokoane-Lusunzi and Karuso 2001, Nunez Montoya, Agnese, and Cabrera 2006, Huang et al. 2007, Xiong et al. 2011).

Among these rhubarb species, *R. palmatum* extracts contained the highest amount of anthraquinones, for example, 34.0 milligrams/gram of dry material (Kosikowska Smolarz, and Malm 2010).

FIGURE 9.2 (a) *Rheum rhaponticum* leaves; (b) *Rheum rhaponticum* stems cooked in sugar syrup for pastry; (c) *Rheum tanguticum* rhizoma, common name Rhei Rhizoma in traditional Chinese medicine; and (d) *Rheum palmatum*.

Anyway, the variations in the anthraquinones' glycoside content are also due to the different altitudes where the plants are grown (Li, Sun, and Feng 2010, Wang et al. 2013).

9.3.1.2 Aloe spp. *(Liliaceae or Xanthorrhoeaceae)*

Aloe, which consists of dried juice from the leaves of various *Aloe* species from the Liliaceae family (i.e., *A. ferox* [Cape aloes], *A. barbadensis* [Curacao aloes], and *A. perryi* [Socotrine aloes]) contains from 10 to 30 percent anthracene derivatives. *Aloe* dried juice obtained from the leaves of *Aloe vera* (= *Aloe barbadensis* Miller), is largely used in food and beverages for the aromatic and bitter taste. The leaf of the *Aloe vera* plant consists of two main parts: An inner central leaf pulp that produces and stores *Aloe vera* gel, and an outer leaf pulp that produces and transports *Aloe vera* latex. According to the International *Aloe* Scientific Council, the *Aloe* leaf can be processed into two types of juices for commercial use: Inner leaf gel juice and decolorized whole leaf juice. Inner leaf gel juice is only produced from the gelatinous fillet of the leaf. Decolorized whole-leaf juice is produced by grinding the leaves. The grinding is followed by the treatment of the extracted juice with activated charcoal to remove aloe "latex." Approximately 80 phenolic anthraquinone's derivatives are produced by the pericyclic cells, located just below the epidermis. They can be found in some *Aloe vera* preparations. The main anthraquinone derivative of *Aloe*'s latex is aloin. This is a mixture of two diastereomers, termed aloin A and aloin B (Sehgal, Winters, Scott, David, et al. 2013; see Figure 9.3).

Aloin A, also called barbaloin, is the major C-glycoside anthraquinone in *Aloe*'s latex. When oxidized, it yields the free aglycone aloe-emodin. Aloinosides A and B, and the aglycone anthraquinone chrysophanol are also present in some *Aloe* varieties (Caro et al. 2012). Because of some adverse pharmacological effects of aloe constituents on consumers, the European Economic Community (EEC) listed aloin as a marker of aloe occurrence in food, and limited the amount of aloin to levels of 0.1 ppm in foods and beverages, and 50 ppm in alcoholic beverages (European Community Directive 88/388; EEC 1988). In contrast, the fresh mucilaginous gel in the leaves obtained from *Aloe vera* does not contain high level of anthraquinone derivatives. The fresh mucilaginous gel from *A. barbadensis* has been widely used for ages as a raw material or additive for health drinks, health foods, and health supplements. Recent studies indicated that commercial stabilized *Aloe* gel consumed as a beverage was not genotoxic or toxic in vivo (Sehgal, Winters, Scott

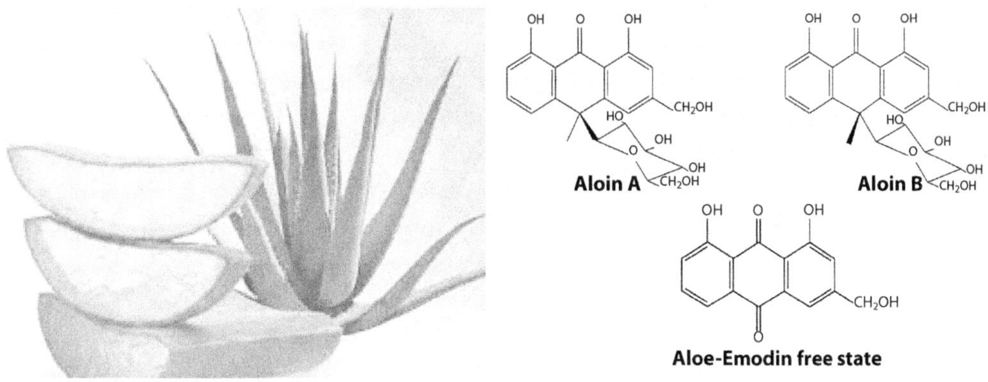

(a) (b)

FIGURE 9.3 (a) *Aloe vera* leaves; (b) main anthraquinones found in *Aloe vera*: Aloin A, aloin B, and the free aglycone form aloe-emodin.

et al. 2013, Sehgal, Winters, Scott, and Kousoulas 2013). Concerning *Aloe vera* preparations, the United States Food and Drug Administration (USFDA) then specified that anthraquinones' levels should be kept below 50 ppm, and Cosmetic Ingredient Review (CIR) concluded that a concentration of anthraquinones below 50 ppm in a product is adequately safe (CIR 2007). In 0.5% *Aloe vera* solution for oral consumption, aloin should not exceed 10 ppm. EEC foodstuff regulation allows a maximum of 0.1 ppm aloin to be used for flavoring purposes in food and drinks (Müller et al. 1996).

9.3.1.3 Morinda sp. (Rubiaceae)

Root of Indian mulberry (*Morinda citrifolia L.*, Rubiaceae) commonly known as noni, also potentially contains natural anthraquinone derivatives such as damnacanthal, morindone, rubiadin, and rubiadin-1-methyl ether (Deng et al. 2009, Bussmann et al. 2013) (Figure 9.4).

Noni products (fermented or unfermented juices or powders) have been used in traditional medicine and also as a nutritional supplement in foodstuffs. Noni's fruit puree (from which seeds had been removed), as well as food products derived from the puree, did not contain any detectable amount of anthraquinone derivatives (Bussmann et al. 2013). However, noni products that did contain seeds or leaf material did contain significant amounts of anthraquinone derivatives. To alleviate safety concerns for food uses, noni products should be derived only from fully ripe noni fruits. Therefore, any seed material needs to be removed during the production process (Bussmann et al. 2013).

9.3.1.4 Cassia spp. (Fabaceae)

Cassia gum, which comes from the purified flour from the endosperm of the seeds of *Cassia tora* and *Cassia obtusifolia*, is an authorized food additive (CAS Registry Number 11078-30-1) (Figure 9.5). It is mainly used as a thickener, emulsifier, foam stabilizer, moisture retention agent, or texturizing agent in diverse processed foods (cheese, frozen dairy desserts and mixes, meat products, and poultry products).

It has recently been evaluated according to its anthraquinone content: Rhein, emodin, aloe-emodin, and physcion where identified (Kim et al. 2004). Maximum use levels

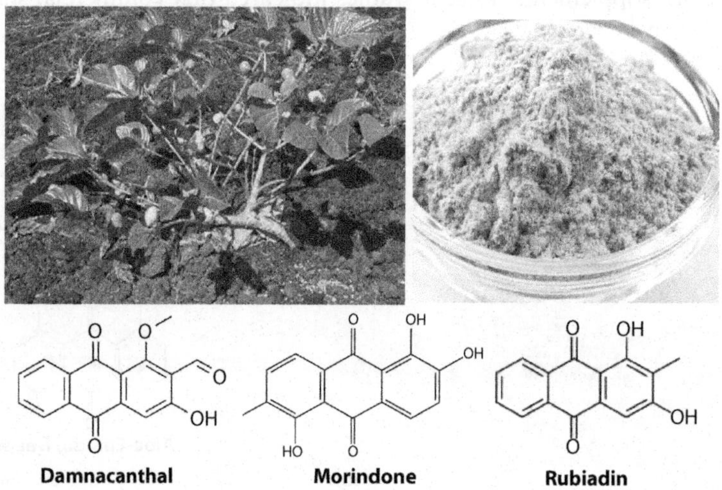

FIGURE 9.4 Indian mulberry (*Morinda citrifolia L.* plant and root's powder) and the main anthraquinones present in its roots.

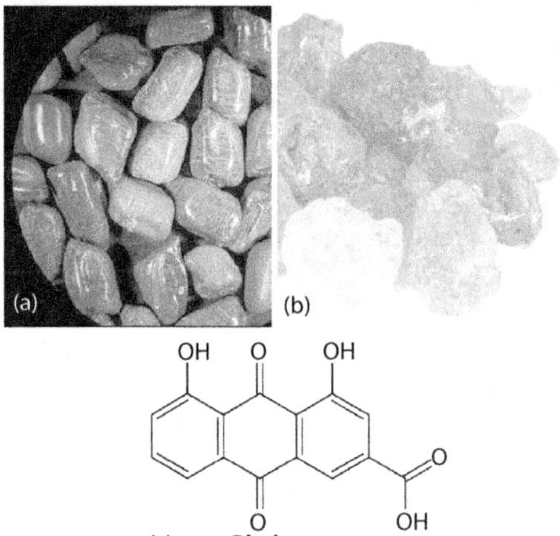

(c) **Rhein**

FIGURE 9.5 (a) *Cassia obtusifolia* seeds, (b) *Cassia* Gum (Gum Arabic), (c) structure of rhein, one major anthraquinone in *Cassia* species.

for cassia gum ranged from 2.5, grams/kilogram food in frozen desserts and 3 grams/kilogram food in cheeses to 3.5, grams/kilogram food in meat and poultry products. The concentration of anthraquinones in *Cassia* gum obtained from an isopropanol extraction purification step was below the 0.5 milligrams/kilogram detection limit. Moreover, traditional *Cassia* gum, containing approximately 70 milligrams/kilogram of total anthraquinones, was not mutagenic or clastogenic in mammalian cells.

Nevertheless, in a submission to the European Commission, use levels for *Cassia* gum only up to 2.5, grams/kilogram food were considered, with a maximum of 1.5, grams/kilogram food for processed meat and poultry products (EFSA 2006).

Cassia fistula Linn. ripe pods and leaves have been used for a long time in Thai traditional medicine (locally called "Khun"). They have been used as a laxative drug by boiling with water. Ripe pods and leaves of *Cassia* species contain several anthraquinones both in aglycone and glycoside forms such as rhein, aloe-emodin, chrysophanic acid, and sennosides (Dave and Ledwani 2012). The content of total anthraquinone glycosides in the dried leaves of *C. fistula* was evaluated at 0.36 percent w/w (average value). The laxative activity depends on the amount of total anthraquinone glycosides for which the Standard of ASEAN Herbal Medicine recommended should not be less than 0.5 percent of dried leaf raw materials. In the European Pharmacopoeia, the percentage of hydroxyanthracene glycosides in *C. angustifolia* dried leaves recommended should not be less than 2.5 percent. The recommended dose of hydroxyanthracene glycosides in the Senna leaf extract is 15–30 milligrams (Heilpflanzen-Welt 1993a,b, EMEA 2006). Thus, the dose of *C. fistula* decoction leaf extract equivalent to the dose of senna leaf extract should be 1–2 grams while the dose of the dried leaves should be 4–9 grams (Sakulpanich and Gritsanapan 2009).

9.3.1.5 Other Edible Vegetables

Anthraquinone derivatives can also be found at lower amounts in other types of vegetables and herbs. For example, a study has screened a variety of vegetables (cabbage lettuce, beans, and peas), herbs and herbal-flavored liquors for their content in the "free" (aglycone) anthraquinones emodin, chrysophanol, and physcion. The vegetables showed a large batch-to-batch variability, from 0.04 to 3.6, 5.9 and 36 milligrams total anthraquinones per kilogram fresh weight in peas, cabbage lettuce, and beans, respectively (Mueller et al. 1999). Physcion predominated in all vegetables. In herbs, grape vine leaves, couch grass root, and plantain herb, anthraquinones' contents ranged below 1 milligrams/kilogram (dry weight).

9.3.2 Microbial Consortia Producing Anthraquinones, Empirically Used in Asian Productions

Asia is often a precursor in using natural substances from microbes, based on empirical evidences. Well-known examples are alcoholic beverages made from rice (Japanese sake or Chinese huangjiu), red soya bean cheese, anka (red rice), processed meat (sausage, ham), fish paste, and so on (Dufossé 2006). Because of the renewed interest in natural food components in relation with health, anthraquinones have recently been carefully studied in some ancient processed foods.

9.3.2.1 Fuzhuan Brick-Tea

Fuzhuan brick-tea (*Camelia sinensis* var. *sinensis*), a traditional fermented Chinese drink, has been demonstrated to contain a mixed microscopic fungal population producing anthraquinones (mainly emodin and physcion; Figure 9.6). The strains are involved in the red color, the flavor, and certainly also in the health benefits, showing antidysenteric effects, anti-food born spoilage, and anti-pathogenic microorganisms (Anke et al. 1980, Anke, Kolthoum, and Laatsch 1980, Mo, Zhu, and Chen 2008, Mo et al. 2008, Ling et al. 2010, Singh et al. 2005). The manufacturing process implies several steps of tea leaves' treatments (mixing, grinding, steaming, cooling, etc.) followed by a solid-state fermentation step (15–17 days) (Mo, Zhu, and Chen 2008, Xu et al. 2011). The main

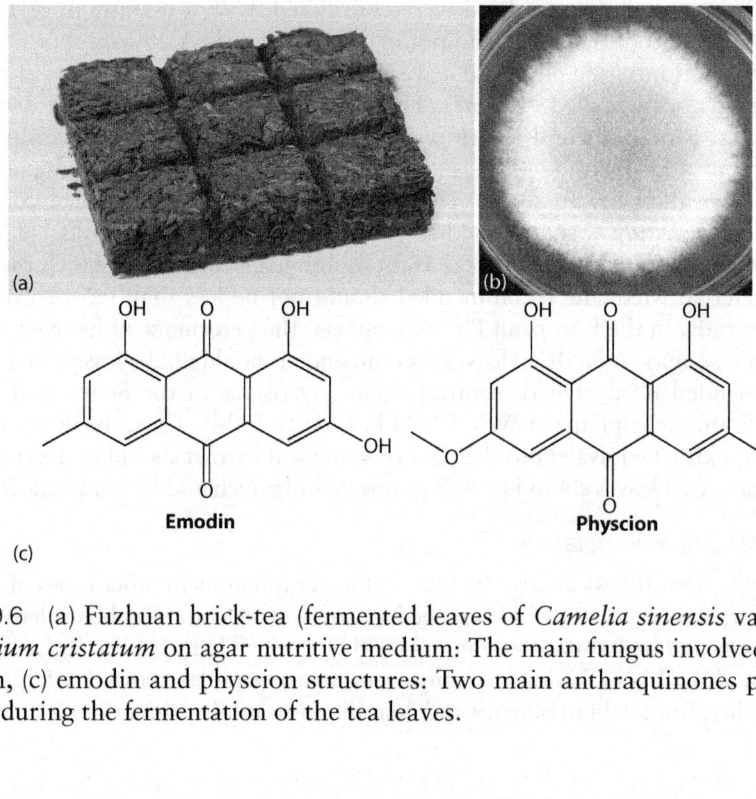

FIGURE 9.6 (a) Fuzhuan brick-tea (fermented leaves of *Camelia sinensis* var. *sinensis*), (b) *Eurotium cristatum* on agar nutritive medium: The main fungus involved in the fermentation, (c) emodin and physcion structures: Two main anthraquinones produced by the fungi during the fermentation of the tea leaves.

microorganisms involved are *Debaromyces*, *Aspergillus*, *Verticillium* and *Eurotium* spp. Several species of *Eurotium* were identified from the fermented material (10^5 CFU/g dry weight of readymade Fuzhuan brick-tea), but *Eurotium* sp. FZ (*Eurotium cristatum*) was the predominant strain characterized during the fermentation (Ge et al. 2016, Mo, Zhu, and Chen 2008, Xu et al. 2011, Qi and Sun 1990).

Further studies were conducted, aiming at standardizing Fuzhuan tea industrial production, through identifying and optimizing the synthesis of antioxidants and antimicrobial substances during the fermentation (Xu et al. 2011, Huang et al. 2010, Liu et al. 2010, Abe et al. 2008, Cao, Zhao, and Liu 1998). These research works stated that emodin was present in all dark tea samples but that physcion was only detectable in the teas fermented by *E. cristatum*. As they found that *Beauveria* sp. (an entomogenous fungus occuring in the stored brick-tea) was a part of the fungal community, probably acting as a protectant against insects, they assessed that a microbial consortia should be used as starter cultures to improve the quality of Fuzhuan tea fermentations.

9.3.2.2 Katsuobushi (Karebushi)

The Asian empirical knowledge about the involvement of non-mycotoxigenic strains of *Eurotium* (*E. rubrum*, *E. repens*, and *E. herbariorum*) in the production of natural anthraquinoid compounds in foods is illustrated through the fermentation of fishes. These fungi are already extensively used as starter cultures in Japanese manufactures of katsuobushi (or karebushi), that is, fermented slices from bonito (*Katsuwonus pelamis*) (Dimici and Wada 1994; Figure 9.7). These fungi were demonstrated to produce

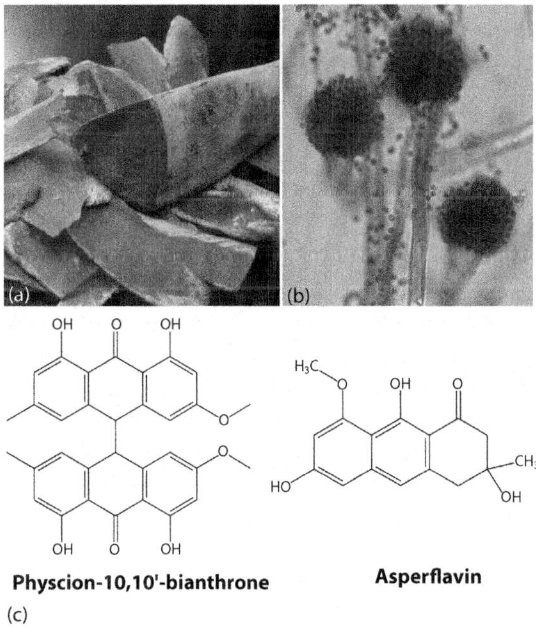

FIGURE 9.7 Anthraquinones in fermented *Katsuwonus pelamis*: (a) slices of katsuobushi (karebushi), (b) microscopic structures of *Eurotium* sp. (conidiophores colored with lactophenol blue) involved in anthraquinones production during the fermentation, (c) two of the main anthraquinones produced by *Eurotium herbariorum* in katsuobushi: Asperflavin and physcion-10,10'-bianthrone.

effective antioxidant extrolites, which participate in the suppression of lipid oxidation in fermented fish and therefore in the long shelf life. Their extrolites also participate in giving a deep red color to the final product (Manabe 2001, Pitt and Hocking 2009). Among them physcion, physcion-10,10'-bianthrone, catenarin, questin, asperflavin, and questinol were clearly identified (Miyake et al. 2014).

Another fungal species *A. glaucus*, already known to produce various anthraquinones such as 10,10'-dimer of emodin and physcion, catenarin, cynodontin, emodin, erythroglaucin, helminthosporin, physcion, questin, rubrocristin, tritisporin, variecolorquinone A (Fouillaud et al. 2016), is nominatively involved in katsuobushi fermentation (Nout and Aidoo 2010).

Obviously, a panel of secondary metabolites may occur in an uncontrolled manner, in a large majority of traditional indigenous fermented foods, as they naturally contain anthraquinone-producing microbial strains. Therefore, this is still a challenge to provide a warranty for the stability of the composition, the quality, and food safety.

9.4 ANTHRAQUINONES USED AS COLORANTS

9.4.1 Colors in Foods

We have known for a long time that colors are key factors for human and animal behaviors. Their attractive or repulsive effects are carefully studied as they highly influence the consumer's attitude. The "buying act," especially in foods, is undoubtedly based on the perception the consumer obtains from the color, as color and flavor are closely connected (Garber, Hyatt, and Starr 2000). In industrial fields, colorants are commonly used to enhance the product's natural color when its components are unable to provide a sufficient or attractive hue; they can also standardize the color and the appearance of products. Sometimes they are applied to restore the color that has been lost during processing. Until the middle of the nineteenth century, dyeing molecules used in cosmetics, drugs, and foods were of natural origin (mainly plants, animals, minerals). The situation rapidly changed when the first synthetic dyestuff was discovered (1856). The easier mass production conditions and the cheaper costs of chemical processes have made them the main sources of dyes in most industries (Hunger 2003). Nowadays about 7000 synthetic dyestuffs are known but consequently due to health risks, especially towards children (allergies, hyperactivity, risks on sexual development; Blendford 1995, Hunger and Sewekow 2003, Kobylewski 2010, FSA 2011), many regulations appeared in industrialized countries to regulate and control the introduction and the use (dose) of dyes in daily-use products. A quite recent awareness therefore led consumers to prefer foodstuffs, or more generally products, containing natural colorants and additives. Even if "natural" does not mean harmless, this undoubtedly applies a strong pressure on the market.

9.4.2 Anthraquinones and Derivatives

Anthraquinoid molecules show a wide range of nuances in shades of brown, purple, red, orange, and yellow. Their structures are relatively stable and they demonstrate a superior brightness compared to azopigments (Caro et al. 2012). In food industries, only few anthraquinones are already manufactured and marketed, either coming from insects (carminic acid), plants (alizarin from European madder roots), or microbes (Arpink Red® or Natural red™). These molecules present a great interest in the field of dyeing, even if they are sometimes described as "Dr. Jekyll or Mr. Hyde" (Dufossé 2014b).

9.4.3 Legislation and Use

Today, each country has its own regulations and European, American, and Asian standards highly diverge about authorizations and conditions of use for dyes.

9.4.3.1 European Standards

In the European Union, the directive referred to as the EU Directive 94/36/EC (Color Directive), then 89/107/EEC, has been implemented throughout national legislations that locally rule the specific uses of dyes (Mapari et al. 2009). About 100 authorized colorants used in foodstuffs are given E-numbers and tested for biosafety before commercialization. About 40 percent of them are from natural origins (Caro et al. 2012). A quite recent CEE regulation (CE n°1223/2009) (EEC 2009) intends to unify the rules concerning colorants used in cosmetics, but the work is presently in progress.

The European Food Safety Authority (EFSA) first fixed in the UE a detection limit of anthraquinones below 0.5 milligrams/kilogram (EFSA 2006). Specific maximum residue limits for food were then defined with a new regulation (EC), No. 1146/2014, effective from May 18, 2015 (EEC 2014). A European maximum residue limit of 0.01 milligrams/kilogram was applicable for the presence of anthraquinones in food, irrespective of the origin of the foodstuff. Taking into account state-of-the-art research, the specific standard values were adapted to the limit of quantification. According to matrix, these maximum residue limits may deviate from the common 0.01 milligrams/kilogram and are listed for anthraquinones in annex V of regulation (EC) 396/2005 (EEC 2005).

9.4.3.2 American Rules

In the United States, the specific use of colors is outlined by the United States Food and Drug Administration (USFDA) in the Code of Federal Regulation, Title 21 (21 CFR, titles 73-82) (USFDA 2001). Twenty-nine "exempt colorants" (synthetic or natural), are available in the United States for foods without being submitted to the rigorous requirements applied to all the certified ones. Indeed, a list of nine molecules was termed "the permitted list" including dyestuffs, which achieved certification procedures and were proved to be more or less harmless. However, very few of the dyeing molecules had therefore been extensively tested for safety and the harmlessness of synthetic colorants has been replete with controversies and contradictions (Francis 2002). This is particularly true for azoïc dyes in red hues, presently authorized over the world, but exhibiting the above-mentioned negative effects on human health (Greenhawt and Baldwin 2009, USFDA 2011, Weiss 2012, Yilmaz, Ergun, and Yilmaz 2014). The conclusions of the Southampton study (McCann et al. 2007) thus led to the obligation for food companies to apply a label mentioning that some "azo-dyes (i.e., synthetic dyes) may have an adverse effect on activity and attention in children" (EFSA 2008).

Today it is noteworthy that, under the pressure of regulatory agencies, now more than ever, extensive, lengthy and costly toxicity studies are required for the commercialization of new dyes.

9.4.3.3 In Asia

The concern in Asian countries about the impact of food additives on health is relatively recent. As an example, European madder root extract (i.e., madder color) has been long accepted for use as a food additive in Japan and South Korea. It was present among food additives that were already marketed or used on the date of the amendment of the Japanese Food Sanitation Law in 1995. Thus, it appeared in the List of Existing Food Additives. As a coloring agent, its food use was limited to wakame, kelp, meat, fresh fishes, shellfish, whale meat, tea, beans, and vegetables (Dufossé 2014b). Nevertheless, due to its extensive use as a food colorant, its safety has been recently studied in Japan. The analysis of the biological effects of the coloring compounds extracted from madder roots clearly indicated that this dyestuff exerts a carcinogenic potential in animals organs and cells (Inoue,Yoshida, Takahashi, Fujimoto et al. 2009, Inoue, Yoshida, Takahashi, Shibutani, et al. 2009, Ishii et al. 2014). The Japanese Ministry of Health, Labor, and Welfare of Japan then concluded that no acceptable daily intake (ADI) could be established for this substance. Madder color was then delisted and prohibited for use in foods (JFAEC 2004).

9.4.4 Colored Anthraquinones from Plants

Historically, plants have been used for the extraction of a majority of natural colorants (pepper, red beet, grapes, and saffron) before being replaced by synthetic dyes. A renewed interest in natural colorants has increased their commercial availability. The most common plant pigments from edible plants, fruit, and vegetables are carotenoids, chlorophylls, anthocyanins, and betalains. Nevertheless, anthraquinones' derivatives are common aromatic compounds in plant pigments. They are the largest group of plant quinones, better known for their use as mordant dyes as well as bird repellants. The anthraquinone derivatives occur in many different higher plants and are generally present as anthraquinones glycosides in young plants. The anthraquinone-based pigmented compounds present in plants are often under 5 percent (dry weight) (Caro et al. 2012).

Various plant parts, including roots, leaves, twigs, stems, heartwood, bark, wood shavings, flowers, fruits, rinds, hulls, and husks, can serve as natural colorant sources. However, the dried roots of higher plants are usually used to extract the dyestuff containing anthraquinone derivatives.

9.4.4.1 Madder Root (Rubia tinctorum Linn., Rubiaceae)

The mixture of color compounds extracted from dried roots of European madder (*Rubia tinctorum* Linn., Rubiaceae) is one of the oldest red dyes used throughout the history in Europe, Asia, and Northern and Southern America. European madder roots contain from 2 to 3.5 percent (dry weight) of anthraquinones glycosides (Caro et al. 2012, Dufossé et al. 2014, Dufossé 2014b). The roots' bark contains a higher amount of dyestuff than the wooden parts. During storage, hydrolysis of some glycosides occurs, which is completed under acidic conditions. The color shades of madder vary from scarlet, carmine red, pink (high content of pseudopurpurin or purpurin, called pink madder or rose madder), to red with a bluish tint (alizarin lakes) (Figure 9.8). European madder roots contain an impressive number of anthraquinone derivatives; a total of more than 36 have been identified in *Rubia tinctorum* roots, even if a part of these compounds is believed to be artifacts formed during extraction or drying of the dyestuff. Fifteen anthraquinones' derivatives from *Rubia tinctorum* roots play an important role in dyeing and are grouped together

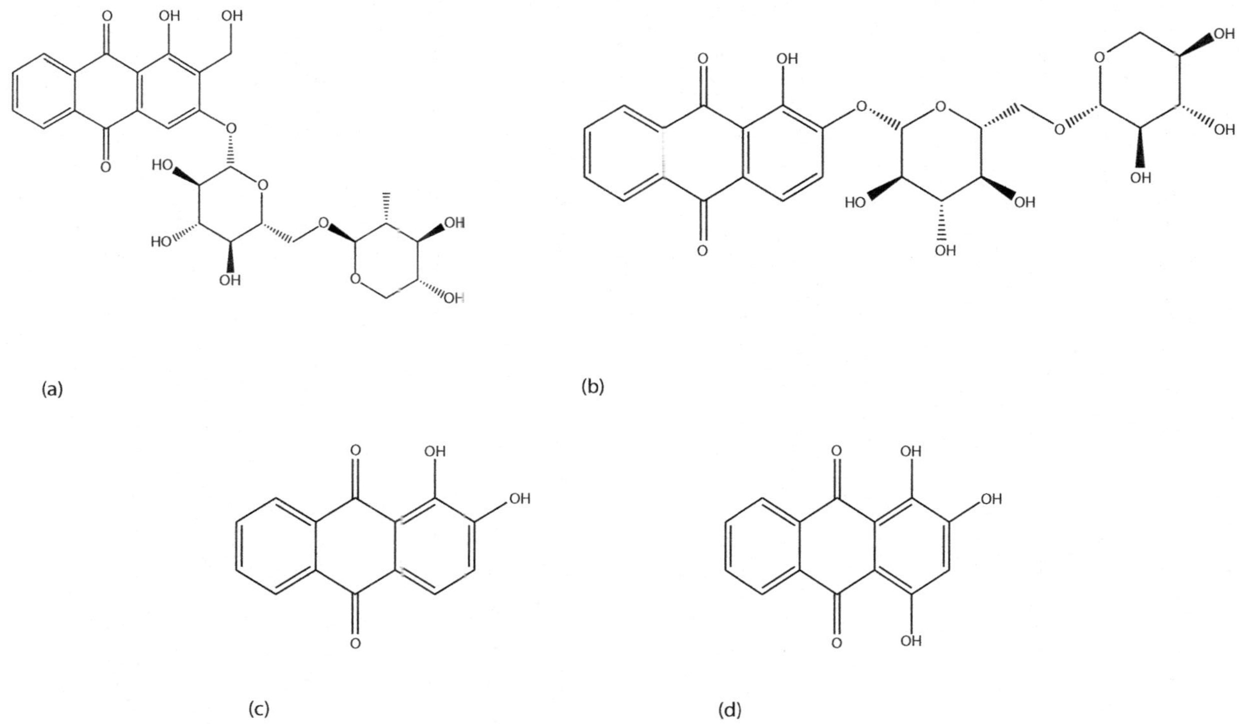

FIGURE 9.8 The main structures of glycones (a, b) and aglycones (c, d) present in madder root: (a) Lucidin primeveroside, (b) ruberythric acid, (c) alizarin, and (d) purpurin.

in the Color Index as C. I. Natural Red 8. The yield of *Rubia tinctorum* roots from the three-year-old plant is between 3–5 tons per hectare producing about 150–200 kilograms of dye (Saxena and Raja 2014).

The composition of the extracted anthraquinone derivatives differs between the varieties of *Rubia*. Other madder plants yielding anthraquinone red dyes include Indian madder (*Rubia cordifolia Linn*) and Naga madder (*R. sikkimensis*).

The main anthraquinone derivatives isolated from plants in *Rubia* spp., and also in *Galium* spp. (another Rubiaceae) are usually alizarin (orange-red), purpurin (dark red), pseudopurpurin (orange), lucidin-primeveroside (red), nordamnacanthal (orange), rubiadin (yellow), and munjistin (orange-red) (Kawasaki, Goda, and Yoshihira 1992, Westendorf, Pfau, and Schulte 1998, Caro et al. 2012). In some *Rubia* species, the anthraquinone alizarin is bound to the disaccharide primeverose (6-O-β-D-xylopyranosyl-β-D-glucose) to build up the anthraquinone glycoside ruberythric acid (golden-yellow) (Derksen et al. 2003). Surprisingly, these phenol anthraquinone derivatives are not formed via the polyketide pathway, but through a more elaborate sequence, involving shikimate and an isoprene unit through the chorismate/O-succinylbenzoic acid pathway. Such structures, like the potent mutagenic-alizarin, rubiadin, and lucidin pigments, do not contain the characteristic meta-oxygenation pattern of phenol anthraquinone, and often have oxygenation in only one aromatic ring (Caro et al. 2012).

The roots of the plant are rich in the highly colored, naturally occurring glycosidic anthraquinoid compounds ruberythric acid and lucidin-primeveroside. Alongside, the corresponding aglycones can be readily formed by deglycosylation, particularly during extraction of the pigments, and free aglycones or glycosides can also be extracted, depending on the polarity of the solvent used for the process. For the production of a commercially useful dye extracted from European madder, the glycoside ruberythric acid has to be hydrolyzed to the water-soluble aglycone alizarin. Alizarin (1,2-dihydroxyanthraquinone) is the main red dye found in commercial madder color. It is also known as Pigment Red 83 or C. I. Mordant Red 11. An intrinsic problem is the simultaneous hydrolysis of the glycoside lucidin-primeveroside to the unwanted lucidin and rubiadin aglycones proved to be mutagenic (Kawasaki, Goda, and Yoshihira 1992, Westendorf, Pfau, and Schulte 1998). Purpurin (1,2,4-trihydroxyanthraquinone; C. I. Natural Red 16) is a minor component in the European madder roots, but is the main dye (bright red crystals) in addition with munjistin (orange-red crystals) extracted from Indian madder (*Rubia cordifolia*). Both alizarin and purpurin contained in European madder and Indian madder, respectively, are only sparingly soluble in water, but are freely soluble in alcohol, ether, acetone, and alkaline solutions. It has been demonstrated that alizarin can be extracted from the roots of *R. tinctorum* with methanol at 25°C with an extraction yield of 2.9 grams/kilogram of dried material (De Santis and Moresi 2007). This yield can be increased to 4.0 grams/kilogram by means of microwave assisted extraction, with purpurin being extracted from *R. tinctorum* at a yield of 2.1 grams/kilogram by microwave-assisted extraction (Dabiri et al. 2005). It has also been demonstrated that extraction of *R. tinctorum* in methanol/water mixtures can be conducted at lower temperatures and in shorter times to obtain similar yields by application of ultrasound assisted extraction (Cuoco et al. 2009).

9.4.4.2 Other Plants

Several other species, although producing colored anthraquinones, are not considered viable contributors to the natural dye market. This includes *Anchusa tinctoria*, *Lithospermum* spp. (Boraginaceae), *Carthamus tinctoria* (Asteraceae), and *Galium* species (Rubiaceae). *Galium tinctorium*, *Galium mullugo* (great lady's bedstraw or wild madder), *Galium verum* (yellow lady's bedstraw), and *Galium aperine* (goosegrass or cleavers) are, however, considered to produce inferior dyes compared with the red pigments obtained from European madder (Rymbai, Sharma, and Srivastav 2011).

9.4.5 Colored Anthraquinone from Insects: Carminic Acid

Carminic acid, carmine, cochineal extract are produced in Peru, Bolivia, Mexico, Chile, and Spain (Canary Islands), from the dried bodies of female cochineal insects (*Dactylopius coccus*), primarily grown on *Opuntia cacti*. The pigments can create red, orange, purple, and pink shades, depending on formulation.

These dyes are allowed by most of the food laws in different countries, such as the Food and Drug Administration (FDA) of the United States, and the European Union, where food additive identification code is E120 (Müller-Maatsch and Gras 2016).

The chemical structure of carminic acid, the main pigment of cochineal, consists of a glucose unit, which is attached to an anthraquinone (Figure 9.9).

Carminic acid is soluble in water, alcohol, acid, and alkaline solutions. It presents good light stability and its color varies depending on pH. Because of its carbonyl and hydroxyl groups, carminic acid is ideally suited to coordination bonding with metals, creating carmine. Some cationic metal complexes can form lakes, giving precipitates of different colors (Borges et al. 2012).

The coloring is currently used in a variety of products such as ice creams, yogurts, fruit drinks, candies, alcoholic drinks, and meat products.

While carmine is considered as a safe and effective natural alternative to synthetic red color FD&C Red #40, manufacturers have faced pressure to replace it for vegans, vegetarians, shoppers seeking kosher and halal products, plus those suffering from the "ick" factor (Watson 2013). This "ick" factor is the main consumer issue for carminic acid, carmine, and cochineal extract and it started when these colorings were implicated in adverse reactions, that is, anaphylactic shock reaction in a small number of people due to impurities in the preparation, not due to the pigment itself. In 1998, it was reported that IgE-mediated allergy might be caused by the consumption of carmine due to the presence of protein or protein-derived residues. In another case an anaphylactic reaction

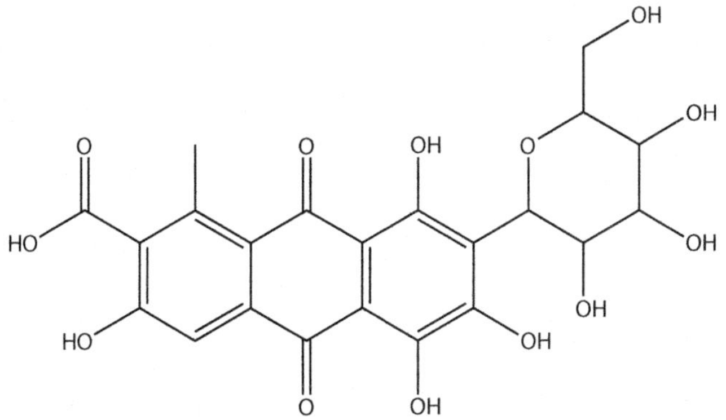

FIGURE 9.9 Carminic acid.

occurred in a 35-year-old woman after she ingested mixed-fruit yogurt colored with carmine. Acute allergic reactions after the ingestion of orange beverages, strawberry milk, and red-colored cocktail, all containing carmine, were also mentioned (Beaudouin et al. 1995, Baldwin, Chou, and Solomon 1997, Acero et al. 1998). A major 38-kd cochineal allergen was cloned, expressed, and characterized by Ohgiya et al. in cochineal extract, protein allergen that could be a phospholipase or related enzyme, which are both known to be allergens in other insects (Ohgiya et al. 2009).

Companies producing carminic acid, carmine, and cochineal extract should better communicate to researchers and consumers these observations (e.g., improvements made in extraction processes, in order to minimize or to suppress the "protein" or "peptidic" content). Consumer lobbying groups are now requesting/looking for replacement products and companies producing food colorants are selling other pigments as alternatives to what they both now call "crushed bug juice."

Another drawback of carmine products is not scientific nor technical but linked to the market as prices are highly volatiles (from a stable level of 15 USD per kg it surged in 2010–2011 up to 120 USD per kilogram and moderated down again to 15 USD per kilgogram—a previous price peak occurred during the 1995–1996–1997 years). This fact also prompted manufacturers to seek alternatives, such as natural red color from tomatoes, beetroots, grape skins, and purple carrots. The world's largest food color company, Chr. Hansen, which sources one-third of global carmine production, decided in 2011 to explore whether it would be commercially viable to produce carmine with a controlled fermentation process (proof of concept test). The genome of the cochineal insect has been sequenced and candidate genes identified (Watson 2013). The host organism for industrial application is unknown up to now (genetically modified organism [GMO] could be a yeast or a filamentous fungi), as are the future regulatory status of that carmine and the consumer perception of such a colorant (GMO-derived; insect genes inserted in an eukaryotic microorganism). Another issue not addressed up till now by scientists from universities, research centers, and private companies is also important: Who is truly producing the anthraquinones present in *Dactylopius coccus*: The insect itself, as hypothesized above, or the symbiotic microflora living in the insect? (Ramirez-Puebla et al. 2010). Isolation, screening, and cultivation of these microorganisms in conditions that mimic symbiosis should be investigated.

As a conclusion, the Dr. Jekyll aspect of carminic acid and derivatives is the excellent stability in food formulations (between 2004 and 2009 a 76 percent increase in new European food product launches listing carmine as an ingredient was observed; the increase was also linked to the consequences of the "Southampton six" study, warning for hyperactivity in children related to the occurrence in food of six artificial colorants, including three sulfonated mono-azo red dyes—E122 carmoisine/azorubine, E124 Ponceau 4R, E129 Allura Red AC), whereas the Mr. Hyde aspects include (1) allergenicity in some cases, (2) not vegan-vegetarian-kosher-halal, (3) price versatility.

9.4.6 Colored Anthraquinones from Microbes: Arpink Red®, Natural Red™

Anthraquinone compounds have been considered among the most abundant microbial natural pigments. If few bacteria species have been proven to produce anthraquinones (*Streptomyces* spp. [Balachandran et al. 2016, Duraipandiyan, Al-Dhabi, and Ignacimuthu 2016]; *Photorhabdus temperata* [Ahn et al. 2013]), anthraquinoid molecules are widespread among fungi and lichens, giving color to spores, sclerotia, sexual bodies, and other developmental structures (Yu and Keller 2005, Fouillaud et al. 2016). However, only rare experiences have been up to date, successful in industrial production of microbial anthraquinones.

9.4.6.1 *Arpink Red®*

Arpink Red® is the first anthraquinoid-type pigment that has been produced from microbes (fungus *Penicillium oxalicum* var *armeniaca*) and used in an industrial setting. This natural colorant, initially manufactured and commercialized by the Czech company Ascolor Biotech s.r.o., received in 2004 a two-year temporary approval by the EU for distribution as a food additive, exclusively in the Czech Republic (Dufossé 2006) (Figure 9.10).

In 2006 the file for homologation was still under progress at the European Food Safety Authority (EFSA). The situation is still not clear as Ascolor Biotech s.r.o. or the new Biomedical s.r.o. did not send the required data to authorities till later on (WHO 2006).

9.4.6.1.1 Arpink Red® This colorant was the first one produced by fermentation and bioprocess engineering using the strain *Penicillium oxalicum* var *armeniaca* CCM 8242, obtained from soil (the variety was morphologically described). The fungus produces a pigment (C25H26014, MW= 550 Da) up to 2 grams/liter of culture medium, providing a raspberry-red color in an aqueous solution. It is stable at pH over 3,5. Neutral solutions are even stable after 30 minutes of boiling and the color shade does not change in relation with pH (WO 9950434; CZ 285721; EP 1070136; US 6340586, cited in Sardaryan [2004]). The liquid culture conditions allow the crystallization of a red powder including carminic acid (Sardaryan 1999). Spectral analysis of the red-colored mixture obtained through the extraction process showed to contain no more than 52 percent (dry weight) of colored substances. Toxicological data about the pigment, produced by Ascolor Biotech s.r.o, examined oral toxicity in mice, 90-day subchronical toxicology, dermal irritation/corrosion, eye irritation/corrosion, anti-tumor effectiveness, micronucleus tests in mice, AMES tests, antibiotic activity, and presence of mycotoxins. The results allowed its acceptance by the Codex Alimentarius Commission (Rotterdam meeting, March 11–15, 2002) and its safety assessment during the Joint FAO/WHO Expert Committee on Food Additives (JECFA, 63rd meeting in Geneva, June 8–17, 2004). A specific formulation of the colorant was moreover patented as a food supplement, supposed to develop prophylactic and therapeutic anticancer activities (patents n° WO 2002011563 A1 [Sardaryan 2002, 2006]).

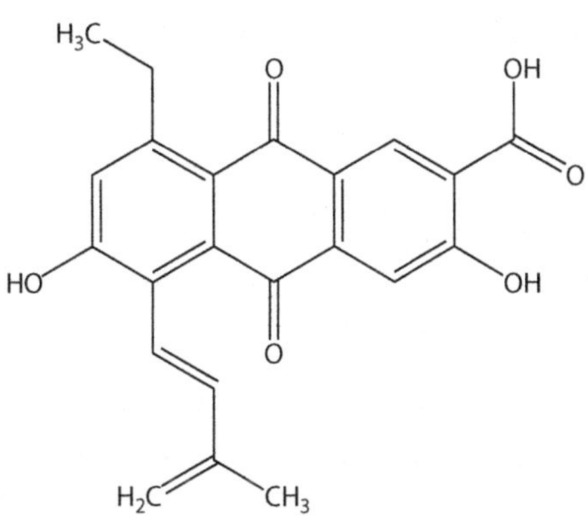

FIGURE 9.10 General structure of Arpink Red®.

9.4.6.2 Natural Red™

In 2009, a new patent (CZ patent n° 302696) was filled in by the Czech company Biomedical s.r.o., dealing with the fungus *Penicillium oxalicum* var *armeniaca* CCM 8374. The strain produced an exogenous red pigment of anthraquinone type (Sardaryan 2009). The organism was obtained from genetic modifications (site-directed mutagenesis) applied to the former strain *P. oxalicum* var *armeniaca* CCM8242, coupled with culture medium optimization and guidance of the biosynthetic pathways. A molecular weight of 378.32 Da was determined for the major chromophore and the production of the anthraquinone type pigment was improved up to 5–10 grams/liter.

The recommended use of Arpink Red® was 100 milligrams/kilogram in meat products and in non-alcoholic drinks, 200 milligrams/kilogram in alcoholic drinks, 150 milligrams/kilogram in milk products including ice creams, and 300 milligrams/kilogram in confectionery products (Codex Alimentarius 2002).

According to the patent filled in 2009 (Natural Red™), ethanol–water solutions of the dye (2 to 50 percent) can be prepared from aqueous solutions. The product can be used at concentrations between 50 and 400 milligrams/kilogram. It seems that antibiotics are also produced along with the pigment. However, no significant effect was found when testing bacteria (G+ or G-). All toxins detected in the mix were in low concentrations. Therefore, in the range of recommended quantities, it was asserted that the use of this product did not present any inconvenience, according to the presented tests.

The formulation of the final mixture commercialized for dyeing foodstuffs may include maltodextrine (around 65 percent) along with the chromophore (around 35 percent) (Dufossé 2014a).

9.5 ANTHRAQUINONES AND FOOD PROCESSING

Investigation of the stability of anthraquinones is important for the evaluation of health foods, cosmetics, and pharmaceuticals containing these secondary phenolic metabolites. Several factors such as pH, temperature, light, water, or oxidants have been recognized to affect their stability. However, the available information concerning the effect of food processing on the degradation of anthraquinones is more limited. On the other hand, many studies have been carried out to evaluate their stability in model systems.

As little is known about the decrease in the content of anthraquinones during food processing, forced degradation studies provide an insight and information about the storage and intrinsic stability conditions of food products containing anthraquinones.

9.5.1 Stability in Model Systems

Temperature shows a clear effect on anthraquinone stability. It has been established that molecular structure of quinones is a determining factor for their thermal stability and strongly influences their thermal decomposition (Sousa et al. 2012). These authors found that 9,10-anthraquinone starts to decompose above 215°C. A clear effect of the different substituted groups attached to the skeleton of anthraquinones has been mentioned. Zhang et al. investigated the thermal behavior of five free anthraquinones having the same skeleton (1,8-dihydroxyanthraquinone), obtained from rhubarb, using thermogravimetry (TG), differential thermogravimetry (DTG), and different thermal analysis technique (Zhang et al. 2009). Similar TG and DTG curve shapes were observed for chrysophanol, emodin, physcion, aloe-emodin, and rhein. But due to the different substituted groups, they exhibit different mass loss features. For example, chrysophanol, emodin, and physcion displaying different groups R3 showed different thermal behavior, showed different thermal behavior. It was found that the decomposition point of chrysophanol is near to physcion and their mass loss occur between 189–293°C and 200–317°C, but the emodin is stable before 281°C and its temperature range is 245–352°C due to the substituted hydroxyl group.

Concerning the stability of aloin A, Ding et al. evaluated the effects of varying pH, temperature, and light conditions, usually encountered in processing, and characterized its major degradation products (Ding et al. 2014). Aloin (also named barbaloin) is an anthraquinone-C glycoside which occurs naturally as a mixture of two diastereoisomers: Aloin A (configuration at C10, C1:S, S) and aloin B (C10, C1:R, S). The effect of pH on aloin A stability was evaluated at pH 2.0, 3.0, 5.0, 7.0, and 8.0. The thermal stability of aloin A was studied at four different temperatures (4, 30, 50, and 70°C) at the same pH values (pH 7.0), all involved in industrial treatments. The photostability test occurred in protection from light, under natural light, and strong light (4000 lx) conditions (at pH 7.0 and room temperature). It was reported that aloin A decomposed quickly at high temperature or under neutral-basic conditions, but light did not promote the degradation of aloin A. The main degradation products of aloin A were identified as aloe-emodin, elgonica-dimers A and B at pH 5.0 or below, and elgonica-dimers were mainly formed at 4°C as well. In addition, it was shown that 10-hydroxyaloins A and B were found under any condition except at pH 2.0 and 3.0, and they were mainly formed under high temperature, neutral-basic and any light conditions. Several studies have been previously carried out with aloin. For example, Zonta et al. (1995) have mentioned a remarkable decrease of aloin A content with the increase in pH value and temperature. Compositional variations of aloin have been studied by Chang et al. (2006) with an instability when dissolved in methanol. Establishment of preservation conditions and manufacturing process of aloe-based products must consider this degradation of aloin (Ramachandra and Rao 2006). Pellizzoni et al. (2011) have evaluated the stability of the main *Aloe* fractions and aloe-based commercial products under different storage conditions. It was mentioned that aloin stability was not increased by ascorbate nor by the antimicrobial agents used.

Narayanan et al. (2015) carried out forced degradation studies on aloe-emodin and emodin by HPTLC. Various degradation parameters were evaluated such as oxidation (6 percent v/v hydrogen peroxide for three hours), acid (0.1 N HCl for two hours) and alkaline (0.1 N NaOH for two hours) hydrolysis, photolysis (sunlight or UV-254 nm for eight hours), hydrolytic and thermal degradation (dry heat at 105°C for eight hours). A significant degradation of aloe-emodin and emodin under acid hydrolysis was reported, whereas these anthraquinones were found less susceptible to base degradation. A moderate thermal degradation was also mentioned. Both anthraquinones underwent moderate oxidative and photolytic degradation. In constrast, aloe-emodin was found more susceptible to hydrolytic degradation than emodin.

Ali et al. (2014) have evaluated a simultaneous determination of diacerein, rhein, and emodin using an accurate, sensitive and selective thin-layer chromatography–densitometry method. Different stress conditions, including hydrolysis, oxidation, and photolytic degradation of diacerein, were analyzed in bulk powder and different pharmaceutical formulations. These authors observed that diacerein was degraded under hydrolytic and oxidative degradation conditions to give one degradation product, rhein, whereas the drug was stable upon exposure to photolytic degradation conditions. The thermal stability of anthraquinones has also been studied in course of extraction processing. Several studies have been carried out with pressurized hot water extraction (PHWE). Barrera Vasquez et al. (2015) evaluated the effect of temperature, pressure, and water flow rate on the extraction yield of four anthraquinones (soranjidiol, rubiadin, rubiadin 1-methyl ether and 2-hydroxy-3-methyl anthraquinone) from aerial parts of *Heterophyllaea pustulata* Hook f. It was mentioned that extractions at higher temperature (220°C) gave lower yields of anthraquinones, apparently due to the thermal decomposition of these compounds. Investigating PHWE water extraction of anthraquinones of *Morinda citrifolia* roots, using alizarin or 1,2-dihydroxy anthraquinone as a standard compound, Shotipruk et al. (2004) have determined the effects of extraction temperature (110, 170, and 220°C) and water flow rate on extraction yield and rate of extraction, and reported that alizarin was stable up to 220°C. Studying PHWE water extraction of the anti-cancer damnacanthal (3-hydroxy-1-methoxyanthraquinone-2-aldehyde) from roots of *Morinda citrifolia*, Anekpankul et al. (2007) mentioned its decomposition at higher temperature than 170°C. It has therefore been established that anthraquinones usually decompose at high temperatures. It is also appropriate to pay particular attention to matters of hydrolytical degradation of glycoside anthraquinones.

Wianowska demonstrated hydrolytical unstability of glycoside forms of hydroxyanthraquinones during their extraction from *Rumex crispus* roots in different pressurized liquid extraction conditions using a methanol/water mixture as an extractant (Wianowska 2014). Different solvent compositions, extraction temperatures, pressures, and static extraction times were investigated, examining concentration changes of some monoglycosides (emodin-8-O-β-D-glucopyranoside, chrysophanol-8-O-β-D-glucopyranoside, and physcion-8-O-β-D-glucopyranoside) and their aglycones (emodin, chrysophanol, and physcion). A gradual concentration increase of all the examined aglycones was observed with temperature increase (50–150°C), followed by a decrease of aglycones concentration above 150°C, attributed to their thermal decomposition.

Several studies were previously dedicated to the barbaloin stability. Yasuda et al. studied its concentration in 15 products containing aloe. A decrease was observed in liquefied products up to 50 percent after storage for one month in a cold, dark place. These authors also explored the stability and degradation pathways of barbaloin suspension in an aqueous solution over the pH range of 1.1–8.4 (Yasuda et al. 1997, 2000). It was found a 1.15 days half-life of barbaloin at pH 7.2 (20°C), and five days at pH 3.4 (20°C). It was mentioned that barbaloin is converted to dimers and then to trimers during storage.

Gutterman and Chauser-Volfson (2006) compared the decrease in the content of three secondary phenolic metabolites (two C-glycosides: barbaloin, aloeresin, and one O-glycoside: aloenin). They used (1) a suspension of *Aloe arborescens* powder in water after storage for up to 45 days, and (2) harvested leaves stored for up to 3.5 months at 4°C in darkness. A rapid degradation of aloenin was observed during storage in water, while a gradual and slow degradation was measured for the two C-glycosides, compounds known for their resistance towards hydrolysis. On the other hand, it was shown that the variation in the relative amounts of these three glycosides was quite similar in stored leaves. Several years' stability was mentioned for these three secondary phenolic metabolites when stored as dry powder.

9.5.2 Stability in Food Processing

Concerning the anthraquinones stability in food processing, McDougall et al. (2010) have evaluated the effects of several cooking methods on the polyphenolic composition of garden rhubarb *Rheum rhapontigen*. Total polyphenolic content, anthocyanin content, and total antioxidant capacity were studied. Products were analyzed by liquid chromatography–mass spectrometry, leading to a putative identification of 40 polyphenol compounds, including anthraquinone, stilbene, anthocyanin, and flavonol derivatives. Four cooking regimes were developed: Blanching (boiling water), slow cooking (70–80°C), fast cooking (100°C), and baking (180°C), from 2 to 30 minutes. Most cooking regimes, except blanching, increased total polyphenol content and overall antioxidant capacity, compared to the raw material. The authors found a yield increase of all components but the initial increase in the content of some anthraquinone aglycones was followed by their destruction with increasing cooking time. For example, a dramatic reduction in the relative amounts of the anthraquinone aglycones between 5 and 10 minutes of baking, accompanied by a contents decrease of the anthraquinone glycoside derivatives were observed. It was suggested a breakdown of the anthraquinone dimer derivatives forms anthraquinone monomer glycosides.

Yen and Chung (1999) investigated the effects of heating on water extracts from *Cassia tora* L. seeds, prepared under different degrees of roasting. It was found that the total content of anthraquinones in water extracts was in the order of unroasted > 150°C roasted > 200°C roasted, indicating that anthraquinones were degraded by thermal treatments. Wu and Yen (2004) have analyzed the contents of chrysophanol, emodin, and rhein in *C. tora* seeds, showing also that the unroasted samples contained the highest anthraquinones content. The authors have mentioned that anthraquinoness were degraded to a free form (aglycon) by roasting treatment.

9.6 DR. JEKYLL AND MR. HYDE? BIOLOGICAL EFFECTS OF ANTHRAQUINONES

As anthraquinones are not yet widely applied as dietary supplements or food colorants, research work needs to extend to knowledge concerning their potential roles in human and animal health. In recent years, anthraquinones are increasingly attracting the attention of the pharmaceutical community as they include a wide diversity of pharmacologically active compounds (Xie et al. 2010, Zhang et al. 2010, Matsuda et al. 2001, Riecken et al. 1990, Firuzi et al. 2011, Zhou and Chen 1988, Zhou et al. 2006, Izhaki 2002). The review of Fouillaud et al. (2016) highlights some selected bioactive effects of a large panel of anthranoid molecules (Fouillaud et al. 2016). Their positive or negative effect(s) due to the 9,10-anthracenedione structure and its substituents are still not clearly understood and their potential roles or effects on human health are today strongly discussed among scientists. Extending the knowledge about these widespread molecules may help to open doors toward innovative and useful natural substances, potentially usable in daily diets.

9.6.1 Benefits

9.6.1.1 Anti-Tumor

Cancer development largely results from an uncontrolled growth of malignant cells in which cell proliferation surpasses cell death. Deregulation of apoptosis, occurring in a majority of cancer types, has since become a non-negligible target for anticancer strategies and pro-apoptotic compounds are thus under active investigations (Xie et al. 2010).

Emodin. Huang et al. (2007) and other teams clearly demonstrated that anthraquinones, such as emodin, aloe-emodin, and rhein, inhibit the growth and proliferation of various cancer cells, such as lung adenocarcinoma, myelogenous leukemia, neuroblastoma, hepatocellular carcinoma, bladder cancer, and others through cell death and survival's modulation (Olsen, Bjorling-Poulsen, and Guerra 2007, Chen et al. 2014, Meggio et al. 2004). Emodin also demonstrated its capacity to reduce toxicity and to enhance efficacy in combination chemotherapy with standard drugs (arsenic trioxide and docosahexaenoic acid, or gemcitabine) against tumor cells (Srinivas et al. 2007, Brown, Bellon, and Nicot 2007, Guo et al. 2012). Emodin might also suppress the growth of cancers by reducing tumor neovascularization and decreasing macrophages' migration inhibitory factor expression. It also attenuates tumor cell-induced metastasis (Zhang, Hu, and Chen 2015, Ma et al. 2015). Following the pharmaceutical hits with emodin, studies about anthraquinones were expanded with the addition of bromo, nitro, amino or bromoacetamido groups, and other compounds such as citreorosein (ω-hydroxyemodin) were also proved to be active (Lu et al. 2012, Lim et al. 2014).

Chrysophanol, found in rhubarb, is chemically closely related to emodin. It stimulates reactive oxygen species (ROS) production, mitochondrial dysfunction, loss of ATP, and DNA damage in J5 human liver cancer cells, which leads to necrotic cell death (Lu et al. 2010, Pandith et al. 2014).

Physcion, the orange pigment found in the roots of curled dock (*Rumex crispus*; yellow dock in the United States) and also in rhubarb, has antitumor and antifungal properties. Recent research suggests that physcion effectively inhibited a part of the pentose–phoshate pathway responsible for constructing the cellular building blocks necessary for rapid growth of cancer cells. The inhibition of cancer cell proliferation and tumor growth in nude mice xenografts takes place without obvious toxicity (Lin et al. 2015).

Rhein (4,5-dihydroxyanthraquinone-2-carboxylic acid), used in clinical studies on animal disease models or functional cells, exerted multiple functions including anticarcinogenesis, antioxidant, anti-inflammation, and immunosuppression (Hu et al. 2015, Tsang and Bian 2015, Huang, Chu, and Chao 1991, Chang et al. 1996, Zhang et al. 2005).

Damnacanthal, present in noni plants, targets several tyrosine kinases and also proved its antitumor effects (Garcia-Vilas, Quesada, and Medina 2015).

Purpurin. The antigenotoxic effect of purpurin against a range of environmental carcinogens has previously been observed in *Drosophila melanogaster.* The compound also clearly inhibited the formation of hepatic DNA adducts in mice exposed to carcinogens (Marczylo et al. 1999, Marczylo, Sugiyama, and Hayatsu 2003, Takahashi et al. 2007).

More recently, the bioactivity of the major constituents of *R. cordifolia* roots has been explored by Biswas et al. (2015). The study indicated that the most bioactive fraction of the plant extract and purpurin showed primarily monophenolase inhibition and to a lesser extent diphenolase inhibitory activity. In addition, results of enzyme kinetic analysis showed they reversibly inhibited tyrosinase in a competitive manner. Molecular docking results implied that the possible inhibitory mechanisms might be attributed to purpurin interaction with copper ion, coordinating three histidine residues of tyrosinase. Authors concluded that this finding could be of importance in prevention of the undesirable enzymatic browning reaction of food products, as well as hyper-pigmentation of human skin (Biswas et al. 2015).

9.6.1.2 Antimicrobial, Antiviral, Antiparasitic

To date, most of the anthraquinones studied, isolated from various sources (plants, microbes), exhibited more antibacterial than antifungal activities. One aspect of the mechanism was elucidated by Daly et al. (2015). The polyhydroxyanthraquinone ω-hydroxyemodin (OHM) was identified as a suppressor of quorum sensing (QS) which is controlling the production of a virulence factor, essential for causing tissue infections by *Staphylococcus aureus* (through agr inhibition f. i.). Decreased dermonecrosis with OHM treatment was associated with enhanced bacterial clearance and reductions in inflammatory cytokine transcription and expression at the site of infection. Furthermore, OHM treatment enhanced the immune cell killing of *S. aureus in vitro* in an agr-dependent manner. These data suggest that bacterial disarmament through the suppression of *S. aureus* QS may bolster the host innate immune response and limit inflammation.

Several anthraquinones are able to inhibit the replication of viruses, or even directly kill enveloped or un-enveloped strains. Alizarin, quinalizarin, rhein, hypericin, and protohypericin, but also other anthraquinones derivatives as emodin, aloe-emodin, emodin anthrone, emodin bianthrone chrysophanic acid, and hypericin showed activity against several strains of human or animal viruses, clearly distinguishable from cytotoxic effects on cells (Barnard et al. 1992, Li et al. 2014, Sydiskis et al. 1991, Lin et al. 2008, Semple et al. 2001, Kubin et al. 2005, Shuangsuo et al. 2006). Aloe-emodin, moreover, showed dose-dependent inhibition of virus-induced cytopathic effects. A significant anti-leishmanial activity has also been demonstrated for 1,8-dihydroxy-3-methoxy-6-methyl-anthraquinone against *Leishmania major* (Awaad et al. 2014).

9.6.1.3 Antioxidant and Chelation Properties

Oxidative stress contributes to free radical–mediated diseases such as aging, atherosclerosis, cancer, ischemic heart disease, diabetes, hyperlipidaemia, hepatotoxicity, and neurodegenerative diseases. Natural and synthetic anthraquinones and their derivatives (Emodin, aloe-emodin, alizarin, physcion, etc.) clearly demonstrated their antioxidant potential (Li, Li, and Wang 2009, Kosalec et al. 2013, Firuzi et al. 2011, Brash and Havre 2002, Fiorentino et al. 2007, Heo et al. 2008, Zargar et al. 2011). From their quinonoid structures, they are bound to participate in redox reactions, exhibiting antioxidant or pro-oxidant properties. According to Yen et al. (2000), the basic anthrone chemical structure exhibited the role of electron acceptor, and the hydroxy substituent accompanied with methylations are multifunctional antioxidants, combining both chain-breaking and metal-chelating properties.

9.6.1.4 Excretion Functions: Laxative, Diuretic Activities

Anthranoid laxatives of natural origin, mainly extracted from plants (emodin, aloe-emodin, and chrysophanol) are widely used, even since ancient times (Evans and Evans 2002, Bruneton 2009, Van Gorkom et al. 1999). Senna, cascara, frangula, rhubarb, and aloe are commonly used for their laxative effects (IARC 2002). It is believed that the presence of hydroxyl groups, in position 1 and 8 or 9 of the aromatic ring system, are essential for the purgative action (Paneitz and Westendorf 1999). Because of their chemical structure, emodin glycosides (and other anthraquinones) are carried unabsorbed to the large intestine in mammals, where metabolism to the active aglycones takes place by intestinal bacterial flora. The aglycone exerts its laxative effect by damaging epithelial cells, which leads directly and indirectly to changes in absorption, secretion and motility (Van Gorkom et al. 1999, Mueller et al. 1999). One main target is the inhibition of the ion transport (Cl^--channels) across colon cells, contributing to the laxative effect (Izhaki 2002, Rauwald 1998). Moreover Na^+/K^+-ATPase (pump) was inhibited by those 1,8-dihydroxyanthrones/anthraquinones that bear an additional phenolic hydroxyl group. Interference with oxidative ATP production, as an additional effect, may explain the known synergistic action described for the combination of different anthrones/anthraquinones or anthranoid drugs, respectively (Rauwald 1998).

The diuretic action of emodin and aloe-emodin is probably due to this Na^+-K^+-ATPase inhibition (Zhou and Chen 1988). 1,3,6,8-trihydroxymethylanthraquinone has been used in a patented laxative preparation for intravascular injection, active by stimulating the neuromuscular junction of the bowel wall (Mobley 1991).

Nevertheless, studies in humans have also suggested tumor-promoting activities for these laxatives. Although the short-term use of these substances is generally safe, long-term utilization cannot be recommended.

9.6.1.5 Other Identified Biological Activities

Effects on lipid and glucose metabolism. Recent findings about the therapeutic potential against diabetes mellitus of several naturally occurring anthraquinones and their derivatives (including emodin, physcion, cascarin, catenarin, chrysophanol, and rhein) were highlighted in Chien et al. (2015) and Mishra et al. (2014). Emodin, for example, demonstrated a dose-dependent antidiabetic effect (reductions in blood glucose) and lipid-modulating effects (serum total cholesterol, triglycerides, free fatty acids, and malonaldehyde) that involve, in part, upregulation of L-type calcium channel expression in the pancreas and heart of dyslipidaemic-diabetic rats (Zhao et al. 2009). Emodin also caused dose-dependent increases in the plasma superoxide dismutase activity.

Estrogenic activity. Insufficiency of endogenous estrogen secretion is known to cause several physical disorders in postmenopausal women, such as osteoporosis, hypercholesteremia, and symptoms of menopause. Synthetic estrogen-replacement therapy has been reported to be effective for these diseases. Emodin was mentioned for its high estrogenic activity. Conversely to aloe-emodin and chrysophanol, the compound is able to bond with human ERα and Erβ, competing with 17β-estradiol. Concerning the structure–activity relationships of anthraquinones regarding the estrogenic activity, it is quite clear that the unchelated hydroxyl group is essential for a strong competency (Matsuda et al. 2001, Fain, Zaitsev, and Ryabov 2004).

Vasorelaxant or contractile effects. Emodin dose-dependently relaxed isolated vascular rings of several vessels in animal and humans (Huang, Chu, and Chao 1991, Huang et al. 1991). Emodin can also induce muscle contracture, simultaneously depressing twitch amplitude. It seemed to be caused myogenically and it suggests that muscle contraction induced by emodin was dose-dependent (Cheng and Kang 1998).

9.6.2 Risks: Cytotoxicity, Carcinogenic Effects

9.6.2.1 Aloe Constituents

Several studies have attempted to determine whether or not *Aloe vera* is toxic to animals or humans. Many of the adverse effects of *Aloe vera* preparations should be related to the anthraquinones content and more particularly to the aloin level. Recently, the U.S. Department of Health and Human Services, in the National Toxicology Program (NTP/NCTR), has demonstrated a dose-dependent increase in large intestinal tumors in rats, chronically exposed to *Aloe vera* non-decolorized whole-leaf extract (in daily drinking water containing 60 ppm of aloin for nearly their entire lifetime; NTP Technical Report, 2013). In the study, the increased incidence of colon adenomas and carcinomas was related to intake of non-decolorized *Aloe vera* leaf extracts (unpurified, high anthraquinone level), supporting the notion that preparations containing aloe latex phenolic compounds, such as anthraquinones, are responsible mediators of the adverse effects on the colon. A more recent study has confirmed that *Aloe vera* whole-leaf extract is an intestinal irritant in rats and mice and a carcinogen of the large intestine in rats (Boudreau et al. 2013). Concerning the compound aloe-emodin in *Aloe vera*, it was reported that it induced micronucleus frequencies in *in vitro* micronucleus test in mouse lymphoma L5178Y cells (Müller et al. 1996). There are thus concerns for various adverse side effects usually related to oral intake of *Aloe* latex rather than *Aloe* gel (Dell'Agli et al. 2007).

9.6.2.2 Madder Root Compounds

A number of long-term genotoxicity studies on rats demonstrated positive results for madder color, suggesting that the carcinogenicity is based on genotoxicity. Madder root causes DNA adducts in the kidneys, livers, and colons of rats and provide clear evidence that madder color exerts unequivocal carcinogenicity (Westendorf, Pfau, and Schulte 1998, Inoue, Yoshida, Takahashi, Fujimoto et al. 2009, Yokohira et al. 2008, Inoue, Yoshida, Takahashi, Shibutani et al. 2009). Rubiadin, the major contributor to madder color, plays the role of an initiator as well as a promoter of carcinogenic effects.

Indeed, rubiadin aglycones and lucidin are found to be positive to bacterial mutagenicity tests, as well a number of other anthraquinone compounds like 1-hydroxy-2-methylanthraquinone, lucidin-ω-methylether, lucidin-ω-ethylether, xanthoprupurin, 7-hydroxy-2-methyl-anthraquinone, and lucidin-primeveroside (Kawasaki, Goda, and Yoshihira 1992, Westendorf, Pfau, and Schulte 1998, Yasui and Takeda 1983, Ishii et al. 2014). Alizarin from madder color also exerts promotor potential in the kidney, but the effects are much weaker than with rubiadin. From structure mutagenicity studies, it was concluded that 1,3-dihydroxyanthraquinones that bear a methyl (CH_3) or hydroxymethyl (CH_2OH) group in position R2, for example, rubiadin and lucidin aglycones from madder color, respectively, are mutagenic. For direct mutagenicity an oxygenated state of the benzylic carbon-2 is required. Mutagenic studies about lucidin, more particularly, showed that a reactive compound is formed from the metabolism of the pigment, which then reacts with DNA and possibly other macromolecules to form covalent adducts with adenine and guanidine under physiological conditions. Other 1,3-dihydroxyanthraquinones that do not possess a methyl or hydroxymethyl group in position R2, such as the orange pigment nordamnacanthal and the orange-red munjistin pigment, are not found to be mutagenic, since the dehydration to the exomethylenic compound is not possible under physiological conditions (Kawasaki, Goda, and Yoshihira 1992, Westendorf, Pfau, and Schulte 1998).

In conclusion, rubiadin, and more generally madder color, can induce carcinogenicity and should be dealt with carefully as a significant carcinogen against humans (Inoue, Yoshida, Takahashi, Fujimoto et al. 2009).

9.6.2.3 Common Vegetables

The genotoxicity of several anthraquinones found in a variety of vegetables (cabbage lettuce, beans, and peas) was investigated in the comet assay, the micronucleus test, and the mutation assay in mouse lymphoma cells (Mueller and Stopper 1999). Emodin was genotoxic, whereas chrysophanol and physcion showed no effects. Indeed, pure emodin has toxic and direct gene mutagenic properties to *Salmonella typhimurium* TA1537 (Fullbeck et al. 2005). Another study mentioned that emodin was clearly genotoxic in mouse lymphoma cells, but also inhibits cell invasiveness in human cancer cells (Huang, Shen, and Ong 2004). This is probably due to the dose-dependent action. However, complete vegetable extract on its own did not show any effect in the micronucleus test. Taking into consideration the measured concentrations of anthraquinones, estimated daily intakes, the genotoxic potency, as well as protective effects of the food matrix, authors concluded that the analyzed constituents do not represent a high priority genotoxic risk in a balanced human diet (Mueller et al. 1999).

9.6.2.4 Senna Ingredients

1,8-Dihydroxyanthraquinone, the aglycone moiety of the laxative ingredient of senna, was formerly marketed as a laxative under the trade name Dantron®, but human drug products containing Dantron (IARC, 1990) were withdrawn from commerce in the United States in 1987 after it was shown to cause intestinal tumors in experimental animals (IARC 1990, 2002).

9.7 IMPROVING INDUSTRIAL SCALE PRODUCTION OF ANTHRAQUINONES FOR FUTURE APPLICATIONS

Since the dawn human history, plants have been a well-known, renewable, and almost fully controllable source of food additives. However, to match the high increase of the world's demand and to compete with the efficacy of chemical synthesis, the raw material productions, thus the agricultural yields, should be incredibly enhanced and improved. Thus, the need for extended agricultural surfaces, huge water volumes used for the plants' growth, and today's financial strategies (short-term profits) cause plants to be an expensive way of producing useful chemicals for food or daily products. This high cost is bearable when the product has a high added-value (as pharmaceutical drugs or luxury cosmetics), but it cannot be easily supported if the final compound is of low price on the global market. In our societies, where we search for what's cheap right now, the plant-based productions are facing increasing difficulties.

9.7.1 Plants

New solutions appeared with the development of plant cells cultures, allowing the in vitro production of biometabolites and ensuring uniform quality and continuous delivery (Rymbai, Sharma, and Srivastav 2011). These processes have already been applied at a laboratory scale with the culture of callus tissues and cells of *Frangula alnus*, *Frangula rupestris*, or *Rhamnus purshiana* (Rhamnaceae) in order to produce anthraquinoid drugs (emodin, aloe-emodin, chrysophanol, and physcion) (Van den Berg and Labadie 1984, 1988, 1989, Van den Berg, Radema, and Labadie 1988a,b, Suzuki and Matsumoto 1988, Van den Berg 1991, Sajc et al. 1999). Unfortunately, the productivity was not high enough at the moment, whatever the improvement factors they tried (maximum 0.5 percent of anthraquinoid glycosides w/w dry weight). However, the cultured tissues accumulated higher amounts of free anthraquinone aglycones compared to corresponding plants. In vitro shoot multiplication of *Frangula alnus* was then obtained on woody plant medium with indole-3-acetic acid and 6-benzylaminapurine. The highest anthraquinone production was in the shoots grown on the Murashige and Skoog medium (MS medium) with addition of 1-naphthaleneacetic, thidiazuron and chitin (Dörnenburg and Knorr 1994, Namdeo 2007). Good results of elicitation came from a study from Komaraiah et al. (2005) on *Morinda citrifolia* cells suspension cultures (noni fruit). Enhancement of accumulation of anthraquinones in plant cell cultures was accomplished by treatment with elicitors such as polyunsaturated fatty acids, methyl jasmonate, salicylate, and nitric oxide, coupled with ultrasonication and a controlled feeding of the carbon source in the growth medium. The anthraquinone production was increased up to 16.74 milligrams/gram of dry weight, which was more than a four-fold increase above the control cultures (Komaraiah et al. 2005). However, in vitro culture of plants for industrial production is undoubtedly still in its infancy. Strategies need to develop information based on cellular and molecular levels. Cell cultures should then provide new continuous and reliable means for the commercial processing of even rare plants, and the chemicals they provide. This is the basis for the production of commercially acceptable levels of compounds for health benefits.

9.7.2 Microbes

Since the food company DSM has gained EU approval for food use of fungal originated β-carotene, produced from the fermentation of *Blakeslea trispora* in 2000 (EEC 2000), industrial interest in microbial metabolites has been revived, and new investigations have been ongoing to develop stable, continuous, and cost effective microbial products ever since. Indeed, the past decade was a period of great improvement for microbial metabolites synthesis and knowledge about different ways to increase yields have been greatly extended. Anthraquinones in microbes belong to secondary metabolites, the pathways of which generate a great diversity of compounds, arising from a few key intermediates. Therefore, the biotechnological approach is the royal road to improve anthraquionoid metabolites production.

Four major fronts are currently ongoing:

- Overall analysis of gene expression, that is, genomics, proteomics, metabolomics, fluxomics, and transcriptomics to better understand the production pathways and general metabolisms as well as the genes and the molecules involved.
- Molecular techniques to carry out metabolic engineering, to modify and improve particular biosynthetic pathways. Further metabolic engineering to optimize already existing or exogenous biosynthetic pathways coupled with the use of powerful computational algorithms, and databases based on the above mentioned « omic » sciences (genomics, proteomics, and metabolomics) (Chen and Nielsen 2013).

- Production of interesting new metabolites in alternative hosts that have already been given GRAS status by the U.S. Food and Drug Administration (Generally Recognized as Safe), to be used in the food industry (*Penicillium roquefortii, Aspergillus oryzae, A. sojae, A. japonicus, Mortierella vinaceae, M. alpina, Fusarium monoliforme, F. veneratum, Saccharomyces cerevisiae*, etc.; Duran, De Conti, and Teixeira 2009, USFDA 2015a,b). Some long-ago studied bacteria or fungi (*Escherichia coli, Corynebacterium glutamicum, Bacillus subtilis, Aspergillus niger, A. oryzae, Penicillium chrysogenum, Saccharomyces cerevisiae*) are already operated in industry for enzymes, nutraceuticals, or pharmaceuticals. Indeed, the numerous years of research done on the selected strains led to high robustness and remarkable tolerances against various stresses under industrial conditions. This is the guarantee of stable and efficient production levels.
- Extensive use of Design of Experiment (DOE) is also of great interest to improve the conditions of metabolites production, combining the main optimal physicochemical parameters: Temperature, oxygen, carbon, nitrogen and other nutrient sources, pH regulation, light exposure, and physiological stage of the fungi. A side goal is to decrease the total production costs using by-products of agroindustrial origin as low-cost alternative substrates for microbial metabolites production (Sánchez 2009).

9.8 CONCLUSION

As it was presented throughout the chapter, anthraquinones constitute a large group of natural compounds that occur in many foods, from plants such as *Aloe* to more elaborated products such as Asian fermented tea or tuna. They are able to bring color to food, and the biological properties described for some anthraquinones are broad, for example, antimicrobial, antiviral, antiparasitic, antitumor, antioxidant, chelatant, diuretic, laxative, and so on.

Legislation is also an important point to address when using anthraquinones for food use. Many countries already set values of maximum limits, some (Europe) being less permissive than others (Asia).

As a concluding remark, the global biological effect of anthraquinones and derivatives formed during food processing or in the human body still need to be studied in order to have a clear picture of these compounds.

REFERENCES

Abe, M., N. Takaoka, Y. Idemoto, C. Takagi, T. Imai, and K. Nakasaki. 2008. Characteristic fungi observed in the fermentation process for Puer tea. *Int. J. Food Microbiol.* 124 (2):199–203. doi: http://dx.doi.org/10.1016/j.ijfoodmicro.2008.03.008.

Acero, S., A. I. Tabar, M. J. Alvarez, B. E. Garcia, J. M. Olaguibel, and I. Moneo. 1998. Occupational asthma and food allergy due to carmine. *Allergy* 53 (9):897–901.

Ahn, J. Y., J. Y. Lee, E. J. Yang, Y. J. Lee, K. B. Koo, K. S. Song, and K. Y. Lee. 2013. Mosquitocidal activity of anthraquinones isolated from symbiotic bacteria *Photorhabdus* of entomopathogenic nematode. *J. Asia-Pac. Entomol.* 16 (3):317–320. doi: http://dx.doi.org/10.1016/j.aspen.2013.04.005.

Ali, N. W., N. S. Abdelwahab, M. Abdelkawy, and A. A. Emam. 2014. Validated stability indicating TLC-densitometric method for the determination of diacerein. *J. Chromatogr. Sci.* 52 (1):5–11. doi: 10.1093/chromsci/bms197.

Anekpankul, T., M. Goto, M. Sasaki, P. Pavasant, and A. Shotipruk. 2007. Extraction of anti-cancer damnacanthal from roots of *Morinda citrifolia* by subcritical water. *Sep. Purif. Technol.* 55 (3):343–349. doi: 10.1016/j.seppur.2007.01.004.

Anke, H., I. Kolthoum, and H. Laatsch. 1980. Metabolic products of microorganisms. 192. The anthraquinones of the *Aspergillus glaucus* group. II. Biological activity. *Arch. Microbiol.* 126 (3):231–236. doi: 10.1007/BF00409925.

Anke, H., I. Kolthoum, H. Zähner, and H. Laatsch. 1980. Metabolic products of microorganisms. 185. The anthraquinones of the *Aspergillus glaucus* group. I. Occurrence, isolation, identification and antimicrobial activity. *Arch. Microbiol.* 126 (3):223–230. doi: 10.1007/BF00409924.

Awaad, A. S., H. M. Al-Zaylaee, S. I. Alqasoumi, M. E. Zain, E. M. Aloyan, A. M. Alafeefy, E. S. Awad, and R. M. El-Meligy. 2014. Anti-leishmanial activities of extracts and isolated compounds from *Drechslera rostrata* and *Eurotium tonpholium*. *Phytother. Res.* 28 (5):774–780. doi: 10.1002/ptr.5096.

Balachandran, C., D. Veeramuthu, A. Yuvaraj, S. Balachandran, E. Nobuhiko, A. D. Naif Abdullah, I. Savarimuthu, I. Yoko, O. Akinao, and T. P. Paramasivan. 2016. Isolation and characterization of 2-hydroxy-9,10-anthraquinone from *Streptomyces olivochromogenes* (ERINLG-261) with antimicrobial and antiproliferative properties. *Rev. Bras. Farmacogn.* 26 (3):285–295. doi: http://dx.doi.org/10.1016/j.bjp.2015.12.003.

Baldwin, J. L., A. H. Chou, and W. R. Solomon. 1997. Popsicle-induced anaphylaxis due to carmine dye allergy. *Ann. Allergy Asthma Immunol.* 79 (5):415–9. doi: 10.1016/s1081-1206(10)63035-9.

Barnard, D. L., J. H. Huffman, J. L. Morris, S. G. Wood, B. G. Hughes, and R. W. Sidwell. 1992. Evaluation of the antiviral activity of anthraquinones, anthrones and anthraquinone derivatives against human cytomegalovirus. *Antiviral Res.* 17 (1):63–77.

Barrera Vázquez, M. F., L. R. Comini, J. M. Milanesio, S. C. Núñez Montoya, J. L. Cabrera, S. Bottini, and R. E. Martini. 2015. Pressurized hot water extraction of anthraquinones from *Heterophyllaea pustulata* Hook f. (Rubiaceae). *The Journal of Supercritical Fluids* 101:170–175. doi: http://dx.doi.org/10.1016/j.supflu.2015.02.029.

Beaudouin, E., G. Kanny, H. Lambert, S. Fremont, and D. A. Moneret-Vautrin. 1995. Food anaphylaxis following ingestion of carmine. *Ann. Allergy Asthma Immunol.* 74 (5):427–30.

Biswas, R., P. Mukherjee, M. Dalai, P. Mandal, and M. Nag. 2015. Tyrosinase inhibitory potential of purpurin in *Rubia cordifolia*—A bioactivity guided approach. *Industrial Crops and Products* 74 (15):319–326.

Blendford, D. 1995. Colouring consumers' perceptions. *Food Ingredients Analysis* 17 10–15.

Borges, M. E., R. L. Tejera, L. Díaz, P. Esparza, and E. Ibáñez. 2012. Natural dyes extraction from cochineal (*Dactylopius coccus*). New extraction methods. *Food Chem.* 132 (4):1855–1860. doi: http://dx.doi.org/10.1016/j.foodchem.2011.12.018.

Boudreau, M. D., F. A. Beland, J. A. Nichols, and M. Pogribna. 2013. Toxicology and carcinogenesis studies of a nondecolorized [corrected] whole leaf extract of *Aloe barbadensis* Miller (*Aloe vera*) in F344/N rats and B6C3F1 mice (drinking water study). *Natl. Toxicol. Program Tech. Rep. Ser.* 577 (0888-8051 (Print)):1–266.

Brash, D. E., and P. A. Havre. 2002. New careers for antioxidants. *Proc. Natl. Acad. Sci. USA* 99 (22):13969–71. doi: 10.1073/pnas.232574399.

Brown, M., M. Bellon, and C. Nicot. 2007. Emodin and DHA potently increase arsenic trioxide interferon-α–induced cell death of HTLV-I–transformed cells by generation of reactive oxygen species and inhibition of Akt and AP-1. *Blood* 109 (4):1653–1659. doi: 10.1182/blood-2006-04-015537.

Bruneton, J. 2009. *Pharmacognosie, phytochimie, plantes médicinales*. Edited by Tec & Doc., 4 ed. Paris: Lavoisier.

Bussmann, R. W., L. Hennig, A. Giannis, J. Ortwein, T. M. Kutchan, and X. Feng. 2013. Anthraquinone Content in Noni (*Morinda citrifolia* L.). *J. Evidence-Based Complementary Altern. Med.* 2013:208378. doi: 10.1155/2013/208378.

Cao, J., Y. Zhao, and J. W. Liu. 1998. Safety evaluation and fluorine concentration of Pu'er brick-tea and Bianxiao brick-tea. *Food Chem. Toxicol.* 36 (12):1061–3.

Caro, Y., L. Anamale, M. Fouillaud, P. Laurent, T. Petit, and L. Dufosse. 2012. Natural hydroxyanthraquinoid pigments as potent food grade colorants: An overview. *Nat. Prod. Bioprospect.* 2 (5):174–193. doi: 10.1007/s13659-012-0086-0.

Chang, C. H., C. C. Lin, J. J. Yang, T. Namba, and M. Hattori. 1996. Anti-inflammatory effects of emodin from *Ventilago leiocarpa*. *Am. J. Chin. Med.* 24 (2):139–42. doi:10.1142/s0192415x96000189.

Chang, X. L., C. Wang, Y. Feng, and Z. Liu. 2006. Effects of heat treatments on the stabilities of polysaccharides substances and barbaloin in gel juice from *Aloe vera* Miller. *J. Food Eng.* 75 (2):245–251. doi: http://dx.doi.org/10.1016/j.jfoodeng.2005.04.026.

Chen, Y., and J. Nielsen. 2013. Advances in metabolic pathway and strain engineering paving the way for sustainable production of chemical building blocks. *Curr. Opin. Biotechnol.* 24 (6):965–972. doi: http://dx.doi.org/10.1016/j.copbio.2013.03.008.

Chen, Y., J. Li, J. Hu, J. Zheng, Z. Zheng, T. Liu, Z. Lin, and M. Lin. 2014. Emodin enhances ATRA-induced differentiation and induces apoptosis in acute myeloid leukemia cells. *Int. J. Oncol.* 45 (5):2076–84. doi: 10.3892/ijo.2014.2610.

Cheng, Y. W., and J. J. Kang. 1998. Emodin-induced muscle contraction of mouse diaphragm and the involvement of Ca2+ influx and Ca2+ release from sarcoplasmic reticulum. *Br. J. Pharmacol.* 123 (5):815–20. doi: 10.1038/sj.bjp.0701677.

Chien, S. C., Y. C. Wu, Z. W. Chen, and W. C. Yang. 2015. Naturally occurring anthraquinones: Chemistry and therapeutic potential in autoimmune diabetes. *Evid. Based Complement. Alternat. Med.* 2015:13. doi: 10.1155/2015/357357.

CIR. 2007. Final report on the safety assessment of *Aloe andongensis* Extract, *Aloe andongensis* leaf juice, *Aloe arborescens* leaf extract, *Aloe arborescens* leaf juice, *Aloe arborescens* leaf protoplasts, *Aloe barbadensis* flower extract, *Aloebarbadensis* leaf, *Aloe barbadensis* leaf extract, *Aloe barbadensis* leaf juice, *Aloe barbadensis* leaf polysaccharides, *Aloe barbadensis* leaf water, *Aloe ferox* leaf extract, *Aloe ferox* leaf juice, and *Aloe ferox* leaf juice extract. *Int. J. Toxicol.* 26 (2):1–50. doi: 10.1080/10915810701351186.

Codex Alimentarius. 2002. Report of the 34th Session of the Codex Committee of food additives and contaminants, Rotterdam, March 2002. In *67 FR 4233* edited by Codex Alimentarius Commission: Office of the Federal Register, National Archives and Records Administration.

Cuoco, G., P. M. C. Fau-Archier, F. A. P. Fau-Chemat, C. C. F. Fau-Vieillescazes, and C. Vieillescazes. 2009. A multivariate study of the performance of an ultrasound-assisted madder dyes extraction and characterization by liquid chromatography-photodiode array detection. *Ultrason Sonochem.* 16 (1350-4177 (Print)):75–82.

Dabiri, M., S. Salimi, A. Ghassempour, A. Rassouli, and M. Talebi. 2005. Optimization of microwave-assisted extraction for alizarin and purpurin in Rubiaceae plants and its comparison with conventional extraction methods. *J. Sep. Sci.* 28 (4):387–96. doi:10.1002/jssc.200400041.

Daly, S. M., B. O. Elmore, J. S. Kavanaugh, K. D. Triplett, M. Figueroa, H. A. Raja, T. El-Elimat, H. A. Crosby, J. K. Femling, N. B. Cech, A. R. Horswill, N. H. Oberlies, and P. R. Hall. 2015. Omega-Hydroxyemodin limits *Staphylococcus aureus* quorum sensing-mediated pathogenesis and inflammation. *Antimicrob. Agents Chemother.* 59 (4):2223–35. doi: 10.1128/aac.04564-14.

Dave, H., and L. Ledwani. 2012. A review on anthraquinones isolated from *Cassia* species and their applications. *Indian J. Nat. Prod. Resour.* 3 (3):291–319.

De Santis, D., and M. Moresi. 2007. Production of alizarin extracts from *Rubia tinctorum* and assessment of their dyeing properties. *Ind. Crops Prod.* 26 (2):151–162. doi: http://dx.doi.org/10.1016/j.indcrop.2007.02.002.

Dell'Agli, M., F. Giavarini, P. Ferraboschi, G. Galli, and E. Bosisio. 2007. Determination of aloesin and aloeresin A for the detection of aloe in beverages. *J Agric. Food. Chem.* 55 (9):3363–7. doi: 10.1021/jf070182h.

Deng, S., B. J. West, C. J. Jensen, S. Basar, and J. Westendorf. 2009. Development and validation of an RP-HPLC method for the analysis of anthraquinones in noni fruits and leaves. *Food Chem.* 116 (2):505–508. doi: http://dx.doi.org/10.1016/j.foodchem.2009.02.070.

Derksen, G. C., M. Naayer, T. A. van Beek, A. Capelle, I. K. Haaksman, H. A. van Doren, and A. de Groot. 2003. Chemical and enzymatic hydrolysis of anthraquinone glycosides from madder roots. *Phytochem. Anal.* 14 (3):137–44. doi:10.1002/pca.694.

Dimici, L., and S. Wada. 1994. Lipid changes in bonito meat in the katsuobushi processing and quality assessment of the commercial product based on lipid composition. *J. Jpn. Oil Chem. Soc.* 43 (6):470–477.

Ding, W. J., X. F. Wu, J. S. Zhong, and J. Z. Wan. 2014. Effects of temperature, pH and light on the stability of aloin A and characterisation of its major degradation products. *Int. J. Food Sci. Technol.* 49 (7):1773–1779. doi: 10.1111/ijfs.12500.

Dörnenburg, H., and D. Knorr. 1994. Research report: Elicitation of chitinases and anthraquinones in *Morinda citrifolia* cell cultures. *Food Biotech.* 8 (1):57–65. doi: 10.1080/08905439409549868.

Dufossé, L. 2006. Microbial production of food grade pigments. *Food Technol. Biotechnol.* 44 (3):313–321.

Dufossé, L. 2014a. Personal communication about Natural Red™.

Dufossé, Laurent. 2014b. Anthraquinones, the Dr. Jekyll and Mr. Hyde of the food pigment family. *Food Res. Int.* 65, Part B:132–136. doi: http://dx.doi.org/10.1016/j.foodres.2014.09.012.

Dufossé, L., M. Fouillaud, Y. Caro, S. A. S. Mapari, and N. Sutthiwong. 2014. Filamentous fungi are large-scale producers of pigments and colorants for the food industry. *Curr. Opin. Biotechnol.* 26:56–61. doi: http://dx.doi.org/10.1016/j.copbio.2013.09.007.

Duraipandiyan, V., N. A. Al-Dhabi, and S. Ignacimuthu. 2016. New antimicrobial anthraquinone 6,61-bis (1,5,7-trihydroxy-3-hydroxymethylanthraquinone) isolated from *Streptomyces* sp. isolate ERI-26. *Saudi J. Biol. Sci.* In Press. doi: http://dx.doi.org/10.1016/j.sjbs.2016.02.008.

Duran, N., R. De Conti, and M. F. S. Teixeira. 2009. Pigments from fungi: Industrial perspective. In *Advances in Fungal Biotechnology*, edited by Mahendra Rai, 185–225. New Delhi, India: I.K. International Publishing House Pvt. Ltd.

EEC. 1988. Directive 88/388/CEE du Conseil du 22 juin 1988 relative au rapprochement des législations des États membres dans le domaine des arômes destinés à être employés dans les denrées alimentaires et des matériaux de base pour leur production edited by European Economic Community. Brussels, Belgium: Journal officiel n° L 184 du 15/07/1988.

EEC. 2000. Opinion of the scientific committee on food on ß-carotene from *Blakeslea trispora*–correction. In *SCF/CS/ADD/COL 158 Final –correction*, edited by the European Commission—Health & Consumer Protection. Brussels, Belgium: Directorate C–Scientific Opinions.

EEC. 2005. Regulation (EC) no 396/2005 of the european parliament and of the council of 23 February 2005 on maximum residue levels of pesticides in or on food and feed of plant and animal origin and amending Council Directive 91/414/EEC. In *L70*, edited by The European Parliament and the Council of the European Union. Brussels, Belgium: Official Journal of the European Union.

EEC. 2009. Règlement (CE) no 1223/2009 du parlement européen et du conseil du 30 novembre 2009 relatif aux produits cosmétiques (refonte). In *Journal officiel de l'Union européenne*, edited by Parlement Européen et Conseil de l'Union Européenne. Brussels, Belgium.

EEC. 2014. Amending Annexes II, III, IV and V to Regulation (EC) No 396/2005 of the European Parliament and of the Council as regards maximum residue levels for anthraquinone, benfluralin, bentazone, bromoxynil, chlorothalonil, famoxadone, imazamox, methyl bromide, propanil and sulphuric acid in or on certain products. In *Commission regulation (EU) no 1146/2014 of* 23 October 2014, edited by European Parliament and Council. Brussels, Belgium: Official Journal of the European Union.

EFSA. 2006. Opinion of the scientific panel on food additives, flavourings, processing aids and materials in contact with food on a request from the commission related to an application on the use of *Cassia* gum as a food additive (Question N° EFSA-Q-2003-134). *The EFSA Journal* 389:1–16.

EFSA. 2008. Scientific opinion of the panel on food additives, flavourings, processing aids and food contact materials (AFC) on a request from the commission on the results of the study by McCann et al. on the effect of some colours and sodium benzoate on children's behaviour. *The EFSA Journal* 660:1–54.

EMEA. 2006. Community herbal monograph on Senna leaf (*Sennae folium*). Edited by European Medicines Agency. London: Committee on Herbal Medicinal Products.

Evans, W. C., and D. Evans. 2002. *Trease and Evans Pharmacognosy*, 15th ed. London, UK: Elsevier.

Fain, V. Ya, B. E. Zaitsev, and M. A. Ryabov. 2004. Metal complexes with 1,5- and 1,8-Dihydroxy-9,10-anthraquinones: Electronic absorption spectra and structure of ligands. *Russ. J. Coord. Chem.* 30 (5):360–364. doi: 10.1023/B:RUCO.0000026007.04814.ad.

Fiorentino, A., B. D'Abrosca, S. Pacifico, G. Cefarelli, P. Uzzo, and P. Monaco. 2007. Natural dibenzoxazepinones from leaves of *Carex distachya*: Structural elucidation and radical scavenging activity. *Bioorg. Med. Chem. Lett.* 17 (3):636–9. doi:10.1016/j.bmcl.2006.11.002.

Firuzi, O., R. Miri, M. Tavakkoli, and L. Saso. 2011. Antioxidant therapy: Current status and future prospects. *Curr. Med. Chem.* 18 (25):3871–3888.

Fouillaud, M., M. Venkatachalam, E. Girard-Valenciennes, Y. Caro, and L. Dufossé. 2016. Anthraquinones and derivatives from marine-derived fungi: Structural diversity and selected biological activities. *Mar. Drugs* 14 (4):64.

Francis, F. J. 2002. Food colorings. In *Colour in Food, Improving Quality*, edited by D. B. MacDougall, pp.297–330. Boca Raton, Florida: Woodhead Publishing Limited & CRC Press.

FSA. 2011. Food Standard Agency Advice to Parents on Food Colours and Hyperactivity. http://www.food.gov.uk/safereating/chemsafe/additivesbranch/colours/hyper.

Fullbeck, M., X. Huang, R. Dumdey, C. Frommel, W. Dubiel, and R. Preissner. 2005. Novel curcumin- and emodin-related compounds identified by in silico 2D/3D conformer screening induce apoptosis in tumor cells. *BMC Cancer* 5:97. doi: 10.1186/1471-2407-5-97.

Garber, L. L., E. M. Hyatt, and R. G. Starr. 2000. The effects of food color on perceived flavor. *Journal of Marketing Theory and Practice* 8 (4):59–72. doi:10.1080/10696679.2000.11501880.

Garcia-Vilas, J. A., A. R. Quesada, and M. A. Medina. 2015. Damnacanthal, a noni anthraquinone, inhibits c-Met and is a potent antitumor compound against Hep G2 human hepatocellular carcinoma cells. *Sci Rep* 5:8021. doi: 10.1038/srep08021.

Ge, Y., Y. Wang, Y. X. Liu, Y. Tan, X. Ren, X. P. Zhang, D. K. Hyde, Y. X. Liu, and Z. Liu. 2016. Comparative genomic and transcriptomic analyses of the Fuzhuan brick-tea-fermentation fungus *Aspergillus cristatus*. *BMC Genomics* 17 (1):1–13. doi:10.1186/s12864-016-2637-y.

Gessler, N. N., A. S. Egorova, and T. A. Belozerskaya. 2013. Fungal anthraquinones. *Appl. Biochem. Microbiol.* 49 (2):85–99. doi: 10.1134/s000368381302004x.

Greenhawt, M. J., and J. L. Baldwin. 2009. Carmine dye and cochineal extract: Hidden allergens no more. *Ann. Allergy, Asthma, Immunol.* 103 (1):73–75. doi:10.1016/s1081-1206(10)60146-9.

Guo, H. C., H. Q. Bu, J. Luo, W. T. Wei, D. L. Liu, H. Chen, H. F. Tong, Z. H. Wang, H. Y. Wu, H. H. Li, M. M. Zuo, W. Li, and S. Z. Lin. 2012. Emodin potentiates the antitumor effects of gemcitabine in PANC-1 pancreatic cancer xenograft model in vivo via inhibition of inhibitors of apoptosis. *Int. J. Oncol.* 40 (6):1849–57. doi:10.3892/ijo.2012.1389.

Gutterman, Y., and E. Chauser-Volfson. 2006. Changes in secondary phenolic metabolites during storage as an aqueous suspension in comparison with the content in harvested *Aloe arborescens* leaves. *Int. J. Food Sci. Technol.* 41 (6):662–666. doi:10.1111/j.1365-2621.2005.01131.x.

Heilpflanzen-Welt. 1993a. List of German commission E monographs (Phytotherapy): Senna leaf. Special Expert Committee of the German Federal Institute for Drugs and Medical Devices. Accessed September 2016 from http://buecher.heilpflanzen-welt.de/BGA-Commission-E-Monographs/0336.htm.

Heilpflanzen-Welt. 1993b. List of German Commission E Monographs (Phytotherapy): Senna pod. Special Expert Committee of the German Federal Institute for Drugs and Medical Devices. Accessed September 2016 from http://buecher.heilpflanzen-welt.de/BGA-Commission-E-Monographs/0337.htm.

Heo, S. J., J. P. Kim, W. K. Jung, N. H. Lee, H. S. Kang, E. M. Jun, S. H. Park, S. M. Kang, Y. J. Lee, P. J. Park, and Y. J. Jeon. 2008. Identification of chemical structure and free radical scavenging activity of diphlorethohydroxycarmalol isolated from a brown alga, *Ishige okamurae*. *J. Microbiol. Biotechnol.* 18 (4):676–81.

Hu, F., F. Xing, G. Zhu, G. Xu, C. Li, J. Qu, I. Lee, and L. Pan. 2015. Rhein antagonizes P2X7 receptor in rat peritoneal macrophages. *Sci. Rep.* 5:14012. doi:10.1038/srep14012.

Huang, H. C., S. H. Chu, and P. D. Chao. 1991. Vasorelaxants from Chinese herbs, emodin and scoparone, possess immunosuppressive properties. *Eur. J. Pharmacol.* 198 (2–3):211–3.

Huang, H. C., C. R. Lee, P. D. Chao, C. C. Chen, and S. H. Chu. 1991. Vasorelaxant effect of emodin, an anthraquinone from a Chinese herb. *Eur. J. Pharmacol.* 205 (3):289–94.

Huang, H., H. Zheng, B. Su, and Y. Huang. 2010. Research on the Metabolites of Eurotium in Fuzhuan Brick-Tea I. Liquid Culture of Eurotium. *Tea Communication* 21 (2):15–17, 21.

Huang, Q., G. Lu, H. M. Shen, M. C. M. Chung, and C. N. Ong. 2007. Anti-cancer properties of anthraquinones from rhubarb. *Med. Res. Rev.* 27 (5):609–630. doi:10.1002/med.20094.

Huang, Q., H.M. Shen, and C. N. Ong. 2004. Inhibitory effect of emodin on tumor invasion through suppression of activator protein-1 and nuclear factor-κB. *Biochem. Pharmacol.* 68 (2):361–371. doi: http://dx.doi.org/10.1016/j.bcp.2004.03.032.

Hunger, K. 2003. *Industrial Dyes: Chemistry, Properties, Applications.* Kelkheim, Germany: Wiley-VCH Verlag GmbH & Co. KGaA.

Hunger, K., and Sewekow, U. 2003. Health and safety aspects. In *Industrial Dyes*, edited by K. Hunger, pp. 625–641. Darmstadt, Federal Republic of Germany: Wiley-VCH Verlag GmbH & Co. KGaA, Weinheim.

IARC. 1990. *Dantron (Chrysazin; 1,8-Dihydroxyanthraquinone).* Edited by International Agency for Research on Cancer. Vol. 50, *IARC Monographs on the Evaluation of Carcinogenic Risks to Humans.* Lyon, France: IARC Press.

IARC. 2002. *Some Traditional Herbal Medicines, Some Mycotoxins, Naphthalene and Styrene.* Edited by International Agency for Research on Cancer. Vol. 82, *IARC Monographs on the Evaluation of Carcinogenic Risks to Humans.* Lyon, France: IARC Press.

Inoue, K., M. Yoshida, M. Takahashi, H. Fujimoto, M. Shibutani, M. Hirose, and A. Nishikawa. 2009. Carcinogenic potential of alizarin and rubiadin, components of madder color, in a rat medium-term multi-organ bioassay. *Cancer Sci.* 100 (12):2261–2267. doi: 10.1111/j.1349-7006.2009.01342.x.

Inoue, K., M. Yoshida, M. Takahashi, M. Shibutani, H. Takagi, M. Hirose, and A. Nishikawa. 2009. Induction of kidney and liver cancers by the natural food additive madder color in a two-year rat carcinogenicity study. *Food Chem. Toxicol.* 47 (1):184–91. doi:10.1016/j.fct.2008.10.031.

Ishii, Y., S. Takasu, K. Kuroda, K. Matsushita, A. Kijima, T. Nohmi, K. Ogawa, and T. Umemura. 2014. Combined application of comprehensive analysis for DNA modification and reporter gene mutation assay to evaluate kidneys of gpt delta rats given madder color or its constituents. *Anal. Bioanal. Chem.* 406 (9):2467–2475. doi:10.1007/s00216-014-7621-2.

Izhaki, I. 2002. Emodin—A secondary metabolite with multiple ecological functions in higher plants. *New Phytol.* 155 (2):205–217. doi: 10.1046/j.1469-8137.2002.00459.x.

JFAEC, Japanese Food Additives Expert Committee. 2004. Madder color. Evaluation Report, 8, edited by Food Safety Commission. Japan.

Kawasaki, Y., Y. Goda, and K. Yoshihira. 1992. The mutagenic constituents of *Rubia tinctorum. Chem. Pharm. Bull. (Tokyo)* 40 (6):1504–9.

Khan, M. S. 2012. *Electronic Spectroscopy of Amino and HydroxyAnthraquinones: Absorption Spectra,* Theoretical Interpretation and Solvent Effect on Absorption Spectra, 192. Germany: Lap Lambert Academic Publishing GmbH and Co, Saarbrucken.

Kim, Y. M., C. H. Lee, H. G. Kim, and H. S. Lee. 2004. Anthraquinones isolated from *Cassia tora* (Leguminosae) seed show an antifungal property against phytopathogenic fungi. *J. Agric. Food Chem.* 52 (20):6096–6100. doi: 10.1021/jf049379p.

Kobylewski, S. 2010. Food dyes: A rainbow of risks. Ph.D, Center for Science in the Public Interest, University of California, Los Angeles.

Komaraiah, P., P. B. Kavi Kishor, M. Carlsson, K. E. Magnusson, and C. F. Mandenius. 2005. Enhancement of anthraquinone accumulation in *Morinda citrifolia* suspension cultures. *Plant Sci.* 168 (5):1337–1344.

Kosalec, I., D. Kremer, M. Locatelli, F. Epifano, S. Genovese, G. Carlucci, M. Randic, and M. Zovko Koncic. 2013. Anthraquinone profile, antioxidant and antimicrobial activity of bark extracts of *Rhamnus alaternus, R. fallax, R. intermedia* and *R. pumila*. *Food Chem.* 136 (2):335–41. doi: 10.1016/j.foodchem.2012.08.026.

Kosikowska, U., H. Smolarz, and A. Malm. 2010. Antimicrobial activity and total content of polyphenols of *Rheum* L. species growing in Poland. *Open Life Sci.* 5 (6). doi: 10.2478/s11535-010-0067-4.

Kubin, A., F. Wierrani, U. Burner, G. Alth, and W. Grunberger. 2005. Hypericin—The facts about a controversial agent. *Curr. Pharm. Des.* 11 (2):233–53.

Li, D. L., X. M. Li, and B. G. Wang. 2009. Natural anthraquinone derivatives from a marine mangrove plant-derived endophytic fungus *Eurotium rubrum:* Structural elucidation and DPPH radical scavenging activity. *J. Microbiol. Biotechnol.* 19 (7):675–680. doi:10.4014/jmb.0805.342.

Li, L., P. Sun, and C. Q. Feng. 2010. Comparison of the contents of anthraquinones between cultivated and wild *Rheum palmatum. Lishizhen Medicine and Materia Medica Research (Shizhen Guoyi Guoyao)* 21:2251–2253.

Li, S. W., T. C. Yang, C. C. Lai, S. H. Huang, J. M. Liao, L. Wan, Y. J. Lin, and C. W. Lin. 2014. Antiviral activity of aloe-emodin against influenza A virus via galectin-3 up-regulation. *Eur. J. Pharmacol.* 738:125–32. doi: 10.1016/j.ejphar.2014.05.028.

Lim, K. H., J. H. Kwon, K. Kim, J. Y. Noh, S. Kang, J. M. Park, M. Y. Lee, O. N. Bae, and J. H. Chung. 2014. Emodin inhibits tonic tension through suppressing PKCδ-mediated inhibition of myosin phosphatase in rat isolated thoracic aorta. *Br. J. Pharmacol.* 171 (18):4300–10. doi: 10.1111/bph.12804.

Lin, C. W., C. F. Wu, N. W. Hsiao, C. Y. Chang, S. W. Li, L. Wan, Y. J. Lin, and W. Y. Lin. 2008. Aloe-emodin is an interferon-inducing agent with antiviral activity against Japanese encephalitis virus and enterovirus 71. *Int. J. Antimicrob. Agents* 32 (4):355–9. doi: 10.1016/j.ijantimicag.2008.04.018.

Lin, R., S. Elf, C. Shan, H. B. Kang, Q. Ji, L. Zhou, T. Hitosugi, L. Zhang, S. Zhang, J. H. Seo, J. Xie, M. Tucker, T. L. Gu, J. Sudderth, L. Jiang, M. Mitsche, R. J. DeBerardinis, S. Wu, Y. Li, H. Mao, P. R. Chen, D. Wang, G. Z. Chen, S. J. Hurwitz, S. Lonial, M. L. Arellano, H. J. Khoury, F. R. Khuri, B. H. Lee, Q. Lei, D. J. Brat, K. Ye, T. J. Boggon, C. He, S. Kang, J. Fan, and J. Chen. 2015. 6-Phosphogluconate dehydrogenase links oxidative PPP, lipogenesis and tumour growth by inhibiting LKB1-AMPK signaling. *Nat. Cell Biol.* 17 (11):1484–1496. doi: 10.1038/ncb3255 http://www.nature.com/ncb/journal/v17/n11/abs/ncb3255.html-supplementary-information.

Ling, T. J., X. C. Wan, W. W. Ling, Z. Z. Zhang, T. Xia, D. X. Li, and R. Y. Hou. 2010. New triterpenoids and other constituents from a special microbial-fermented tea—Fuzhuan brick-tea. *J. Agric. Food Chem.* 58 (8):4945–4950. doi:10.1021/jf9043524.

Liu, Z. Y., A. Q. Xu, Z. J. Li, and Y. L. Wang. 2010. Research progress on *Eurotium cristataum* and its metabolites in Fuzhuan tea. *J. Tea Sci., Tea Comm.* 37 (1):23–26.

Lu, C. C., J. S. Yang, A. C. Huang, T. C. Hsia, S. T. Chou, C. L. Kuo, H. F. Lu, T. H. Lee, W. G. Wood, and J. G. Chung. 2010. Chrysophanol induces necrosis through the production of ROS and alteration of ATP levels in J5 human liver cancer cells. *Mol. Nutr. Food Res.* 54 (7):967–76. doi: 10.1002/mnfr.200900265.

Lu, Y., S. J. Suh, X. Li, J. L. Liang, M. Chi, K. Hwangbo, O. Kwon, T. W. Chung, C. H. Kwak, K. M. Kwon, M. Murakami, Y. Jahng, C. H. Kim, J. K. Son, and H. W. Chang. 2012. Citreorosein inhibits production of proinflammatory cytokines by blocking mitogen activated protein kinases, nuclear factor-kappaB and activator protein-1 activation in mouse bone marrow-derived mast cells. *Biol. Pharm. Bull.* 35 (6):938–45.

Lu, Y., P. J. Xu, Z. N. Chen, and G. M. Liu. 1998. Anthraquinone glycosides from *Rhynchotechum vestitu* Mfn2. *Phytochem.* 49 (4):1135–1137. doi: http://dx.doi.org/10.1016/S0031-9422(98)00079-X.

Ma, J., H. Lu, S. Wang, B. Chen, Z. Liu, X. Ke, T. Liu, and J. Fu. 2015. The anthraquinone derivative Emodin inhibits angiogenesis and metastasis through downregulating Runx2 activity in breast cancer. *Int. J. Oncol.* 46 (4):1619–28. doi:10.3892/ijo.2015.2888.

Manabe, M. 2001. Fermented foods and mycotoxins. *J. Jpn Assoc. Mycotoxicol.* 51:25–29.

Mapari, S. A. S., A. S. Meyer, U. Thrane, and J. C. Frisvad. 2009. Identification of potentially safe promising fungal cell factories for the production of polyketide natural food colorants using chemotaxonomic rationale. *Microb. Cell Fact.* 8. doi:10.1186/1475-2859-8-24.

Marczylo, T. H., T. Hayatsu, S. Arimoto-Kobayashi, M. Tada, K. Fujita, T. Kamataki, K. Nakayama, and H. Hayatsu. 1999. Protection against the bacterial mutagenicity of heterocyclic amines by purpurin, a natural anthraquinone pigment. *Mutat. Res.* 444 (2):451–61.

Marczylo, T., C. Sugiyama, and H. Hayatsu. 2003. Protection against Trp-P-2 DNA adduct formation in C57bl6 mice by purpurin is accompanied by induction of cytochrome P450. *Journal of Agricultural and Food Chemistry* 51 (11):3334–3337.

Matsuda, H., H. Shimoda, T. Morikawa, and M. Yoshikawa. 2001. Phytoestrogens from the roots of *Polygonum cuspidatum* (Polygonaceae): Structure-requirement of hydroxyanthraquinones for estrogenic activity. *Bioorg. Med. Chem. Lett.* 11 (14):1839–1842. doi: http://dx.doi.org/10.1016/S0960-894X(01)00318-3.

McCann, D., A. Barrett, A. Cooper, D. Crumpler, L. Dalen, K. Grimshaw, E. Kitchin, K. Lok, L. Porteous, E. Prince, E. Sonuga-Barke, J. O. Warner, and J. P. Stevenson. 2007. Food additives and hyperactive behaviour in 3-year-old and 8/9-year-old children in the community: A randomised, double-blinded, placebo-controlled trial. *The Lancet* 370 (9598):1560–1567. doi: 10.1016/S0140-6736(07)61306-3.

McDougall, G. J. P. Dobson, and N. Jordan-Mahy. 2010. Effect of different cooking regimes on rhubarb polyphenols. *Food Chem.* 119 (2):758–764. doi:10.1016/j.foodchem.2009.07.030.

Meggio, F., M. A. Pagano, S. Moro, G. Zagotto, M. Ruzzene, S. Sarno, G. Cozza, J. Bain, M. Elliott, A. D. Deana, A. M. Brunati, and L. A. Pinna. 2004. Inhibition of protein kinase CK2 by condensed polyphenolic derivatives. An in vitro and in vivo study. *Biochem.* 43 (40):12931–6. doi: 10.1021/bi048999g.

Mishra, S. K., S. Tiwari, A. Shrivastava, S. Srivastava, G. K. Boudh, S. K. Chourasia, U. Chaturvedi, S. S. Mir, A. K. Saxena, G. Bhatia, and V. Lakshmi. 2014. Antidyslipidemic effect and antioxidant activity of anthraquinone derivatives from *Rheum emodi* rhizomes in dyslipidemic rats. *J. Nat. Med.* 68 (2):363–71. doi:10.1007/s11418-013-0810-z.

Miyake, Y., C. Ito, T. Kimura, A. Suzuki, Y. Nishida, and M. Itoigawa. 2014. Isolation of aromatic compounds produced by *Eurotium herbariorum* NU-2 from Karebushi, a Katsuobushi, and their dpph-radical scavenging activities. *Food Sci. Technol. Int., Tokyo* 20 (1):139–146. doi: 10.3136/fstr.20.139.

Mo, H., H. Zhang, Y. Li, and Y. H. Zhu. 2008. Antimicrobial activity of the indigenously microbial fermented Fuzhuan brick-tea. *J. Biotechnol.* 136, Supplement:S722. doi: http://dx.doi.org/10.1016/j.jbiotec.2008.07.1719.

Mo, H., Y. H. Zhu, and Z. Chen. 2008. Microbial fermented tea—A potential source of natural food preservatives. *Trends Food Sci. Technol.* 19 (3):124–130. doi: http://dx.doi.org/10.1016/j.tifs.2007.10.001.

Mobley, J. R. 1991. 1,3,6,8-trihydroxymethylanthraquinone: Patent US 5039707 A.

Mueller, S. O., M. Schmitt, W. Dekant, H. Stopper, J. Schlatter, P. Schreier, and W. K. Lutz. 1999. Occurrence of emodin, chrysophanol and physcion in vegetables, herbs and liquors. Genotoxicity and anti-genotoxicity of the anthraquinones and of the whole plants. *Food Chem. Toxicol.* 37 (5):481–491. doi: 10.1016/s0278-6915(99)00027-7.

Mueller, S. O., and H. Stopper. 1999. Characterization of the genotoxicity of anthraquinones in mammalian cells. *Biochimica et Biophysica Acta (BBA) – General Subjects* 1428 (2–3):406–414. doi: http://dx.doi.org/10.1016/S0304-4165(99)00064-1.

Müller, S. O., I. Eckert, W. K. Lutz, and H. Stopper. 1996. Genotoxicity of the laxative drug components emodin, aloe-emodin and danthron in mammalian cells: Topoisomerase II mediated? *Mutat. Res., Genet. Toxicol.* 371 (3):165–173. doi: http://dx.doi.org/10.1016/S0165-1218(96)90105-6.

Müller-Maatsch, J., and C. Gras. 2016. 18—The "carmine problem" and potential alternatives." In *Handbook on Natural Pigments in Food and Beverages*, 385–428. Woodhead Publishing.

Namdeo, A. 2007. Plant cell elicitation for production of secondary metabolites: A review. *Pharmacognosy Reviews* 1 (1):69–79.

Narayanan, S., A. P. Jadhav, and V. J. Kadam. 2015. Forced degradation studies of Aloe emodin and emodin by HPTLC. *Indian J. Pharm. Sci.* 77 (6):795–8.

Nout, M. J. R., and K. E. Aidoo. 2010. Asian fungal fermented food. In *Industrial Applications*, edited by Martin Hofrichter, 29–58. Berlin, Heidelberg: Springer Berlin Heidelberg.

Nunez Montoya, S. C., A. M. Agnese, and J. L. Cabrera. 2006. Anthraquinone derivatives from *Heterophyllaea pustulata*. *J. Nat. Prod.* 69 (5):801–3. doi: 10.1021/np050181o.

Ohgiya, Y., F. Arakawa, H. Akiyama, Y. Yoshioka, Y. Hayashi, S. Sakai, S. Ito, Y. Yamakawa, S. Ohgiya, Y., Z. Ikezawa, and R. Teshima. 2009. Molecular cloning, expression, and characterization of a major 38-kd cochineal allergen. *J. Allergy Clin. Immunol.* 123 (5):1157–1162. e4. doi: http://dx.doi.org/10.1016/j.jaci.2008.12.1111.

Olsen, B. B., M. Bjorling-Poulsen, and B. Guerra. 2007. Emodin negatively affects the phosphoinositide 3-kinase/AKT signalling pathway: A study on its mechanism of action. *Int. J. Biochem. Cell Biol.* 39 (1):227–37. doi: 10.1016/j.biocel.2006.08.006.

Pandith, S. A., A. Hussain, W. W. Bhat, N. Dhar, A. K. Qazi, S. Rana, S. Razdan, T. A. Wani, M. A. Shah, Y. S. Bedi, A. Hamid, and S.K. Lattoo. 2014. Evaluation of anthraquinones from Himalayan rhubarb (*Rheum emodi* Wall. ex Meissn.) as antiproliferative agents. *S. Afr. J. Bot.* 95:1–8. doi: 10.1016/j.sajb.2014.07.012.

Paneitz, A., and J. Westendorf. 1999. Anthranoid contents of rhubarb (*Rheum undulatum* L.) and other *Rheum* species and their toxicological relevance. *Eur. Food Res. Technol.* 210 (2):97–101. doi: 10.1007/s002170050542.

Pellizzoni, M., G. P. Molinari, and L. Lucini. 2011. Stability of the main Aloe fractions and Aloe-based commercial products under different conditions. *Agrochimica* LV (5):288–296.

Pitt, J. I., and A. D. Hocking. 2009. *Fungi and Food Spoilage*. United States: Springer.

Qhotsokoane-Lusunzi, M. A., and P. Karuso. 2001. Secondary metabolites from Basotho medicinal plants. I. *Bulbine narcissifolia*. *J. Nat. Prod.* 64 (0163-3864 (Print)):1368–1372.

Qi, Z. T., and Z. M. Sun. 1990. Identification of predominant species in brick-tea. *Acta Mycol. Sin. (Chen Chun Hsueh Pao)* 9:176–179.

Ramachandra, C. T., and P. S. Rao. 2006. Processing of *Aloe vera* leaf gel: A review. *Am. J. Agric. Biol. Sci.* 3 (2):502–510. doi: 10.3844/ajabssp.2008.502.510.

Ramirez-Puebla, S. T., M. Rosenblueth, C. K. Chavez-Moreno, M. C. de Lyra, A. Tecante, and E. Martinez-Romero. 2010. Molecular phylogeny of the genus *Dactylopius* (Hemiptera: Dactylopiidae) and identification of the symbiotic bacteria. *Environ. Entomol.* 39 (4):1178–83. doi: 10.1603/en10037.

Rauwald, H. W. 1998. Herbal laxatives: Influence of anthrones, anthraquinones on energy metabolism and ion transport in a model system. In *Phytomedicines of Europe*, 97–116. Washington, DC, USA: American Chemical Society.

Riecken, E. O., M. Zeitz, C. Emde, R. Hopert, L. Witzel, R. Hintze, U. Marsch-Ziegler, and J. C. Vester. 1990. The effect of an anthraquinone laxative on colonic nerve tissue: A controlled trial in constipated women. *Zeitschrift fur Gastroenterologie* 28 (12):660–664.

Rymbai, H., R. R. Sharma, and Manish Srivastav. 2011. Biocolorants and its implications in health and food industry—A review. *Int. J. PharmTech Res.* 3 (4):2228–2244.

Sajc, L., N. Kovačević, D. Grubišić, and G. Vunjak-Novaković. 1999. Frangula species: In vitro culture and the production of anthraquinones. In *Medicinal and Aromatic Plants XI*, edited by Y. P. S. Bajaj, 157–176. Berlin, Heidelberg: Springer-Verlag.

Sakulpanich, A., and W. Gritsanapan. 2009. Determination of anthraquinone contents in *Cassia fistula* leaves for alternative source of laxative drugs. *Planta Med.* 75 (09):PG11. doi: 10.1055/s-0029-1234665.

Sánchez, C. 2009. Lignocellulosic residues: Biodegradation and bioconversion by fungi. *Biotechnol. Adv.* 27 (2):185–194. doi: http://dx.doi.org/10.1016/j.biotechadv.2008.11.001.

Sardaryan, E. 1999. Strain of the microorganism *Penicillium oxalicum* var. armeniaca and its application. Patent WO1999050434 A1.

Sardaryan, E. 2002. Food supplement. Patent WO 2002011563 A1.

Sardayan, E. 2004. Food Supplement. Patent US 20040105864 A1.

Sardaryan, E. 2006. Food supplement. Patent US 20060247316 A1.

Sardaryan, E. 2009. Strain of *Penicillium oxalicum var. Armeniaca* microscopic fungus and process for preparing magenta. In *Industrial Property Office: Patents and Utility Models*, edited by G.E.S. Biomedical s.r.o. Ceska Republika.

Saxena, S., and A. S. M. Raja. 2014. Natural dyes: Sources, chemistry, application and sustainability issues. 37–80. doi: 10.1007/978-981-287-065-0_2.

Sehgal, I., W. D. Winters, M. Scott, A. David, G. Gillis, T. Stoufflet, A. Nair, and K. Kousoulas. 2013. Toxicologic assessment of a commercial decolorized whole leaf *Aloe vera* juice, lily of the desert filtered whole leaf juice with aloesorb. *J. Toxicol.* 2013:802453. doi: 10.1155/2013/802453.

Sehgal, I., W. D. Winters, M. Scott, and K. Kousoulas. 2013. An in vitro and in vivo toxicologic evaluation of a stabilized *Aloe vera* gel supplement drink in mice. *Food Chem. Toxicol.* 55:363–370. doi: http://dx.doi.org/10.1016/j.fct.2013.01.012.

Semple, S. J., S. M. Pyke, G. D. Reynolds, and R. L. Flower. 2001. In vitro antiviral activity of the anthraquinone chrysophanic acid against poliovirus. *Antiviral Res.* 49 (3):169–78.

Shotipruk, A., J. Kiatsongserm, P. Pavasant, M. Goto, and M. Sasaki. 2004. Pressurized hot water extraction of anthraquinones from the roots of Morinda citrifolia. *Biotechnol. Prog.* 20 (6):1872–5. doi: 10.1021/bp049779x.

Shuangsuo, D., Z. Zhengguo, C. Yunru, Z. Xin, W. Baofeng, Y. Lichao, and C. Yan'an. 2006. Inhibition of the replication of hepatitis B virus in vitro by emodin. *Med. Sci. Monit.* 12 (9):Br302-6.

Singh, N. P., A. P. Gupta, A. K. Sinha, and P. S. Ahuja. 2005. High-performance thin layer chromatography method for quantitative determination of four major anthraquinone derivatives in *Rheum emodi. J. Chromatogr. A.* 1077 (2):202–206. doi: http://dx.doi.org/10.1016/j.chroma.2005.03.130.

Sousa, E. T., M. M. Da Silva, S. J. De Andrade, M. P. Cardoso, L. A. Silva, and J. B. De Andrade. 2012. Evaluation of thermal stability of quinones by thermal analysis techniques. *Thermochim. Acta* 529:1–5. doi: 10.1016/j.tca.2011.11.012.

Srinivas, G., S. Babykutty, P. P. Sathiadevan, and P. Srinivas. 2007. Molecular mechanism of emodin action: Transition from laxative ingredient to an antitumor agent. *Med. Res. Rev.* 27 (5):591–608. doi: 10.1002/med.20095.

Suzuki, H., and T. Matsumoto. 1988. Anthraquinone: Production by plant cell culture. In *Medicinal and Aromatic Plants I*, edited by Y. P. S. Bajaj, 237–250. Berlin, Heidelberg: Springer Berlin Heidelberg.

Sydiskis, R. J., D. G. Owen, J. L. Lohr, K. H. Rosler, and R. N. Blomster. 1991. Inactivation of enveloped viruses by anthraquinones extracted from plants. *Antimicrob. Agents Chemother.* 35 (12):2463–6.

Takahashi, E., S. Arimoto, K. Okamoto, and T. Negishi. 2007. Enhancement of phase II enzyme activity by purpurin resulting in the suppression of MeIQx-DNA-adduct formation in mice. *Mutat. Res., Genet. Toxicol.* 626 (1-2):128–34. doi:10.1016/j.mrgentox.2006.09.011.

Teuscher, E., and U. Lindequist. 1994. *Biogene gifte*. 2nd ed. Stuttgart: Wiss. Verl.-Ges.

Thomson, R. H. 1997. Anthraquinones. In *Naturally Occurring Quinones IV: Recent Advances*, 309–483. Dordrecht: Springer Netherlands.

Tsang, S. W., and Z. X. Bian. 2015. Anti-fibrotic and anti-tumorigenic effects of rhein, a natural anthraquinone derivative, in mammalian stellate and carcinoma cells. *Phytotherapy Res.* 29 (3):407–414. doi: 10.1002/ptr.5266.

USFDA. 2001. Code of Federal Regulation. Title 21, Volume I, Part 1 to 99, edited by US Food and Drugs Administration. Washington DC, USA: US Government Publishing Office.

USFDA. 2011. Certified Color Additives in Food and Possible Association with Attention Deficit Hyperactivity Disorder in Children. USFDA Center for Food Safety. http://www.fda.gov/downloads/AdvisoryCommittees/CommitteesMeetingMaterials/FoodAdvisoryCommittee/UCM248549.pdf.

USFDA. 2015a. Microorganisms and microbial-derived ingredients used in food (partial list). U.S. Government Publishing Office. Accessed September 2015 from http://www.fda.gov/Food/IngredientsPackagingLabeling/GRAS/MicroorganismsMicrobialDerivedIngredients/default.htm.

USFDA. 2015b. Title 21—Food and Drugs, Chapter I-Food and Drug Administration Department of Health and Human Services, Subchapter B-food for human consumption (continued) Part 184—Direct food substances affirmed as generally recognized as safe. U.S. Government Publishing Office. Accessed September 2015 from http://www.ecfr.gov/cgi-bin/text- idx?SID=e956d645a8b4e6b3e34e4e5d1b690209&mc=true&node=pt21.3.184&rgn=di v5.

Van den Berg, A. J. J., and R. P. Labadie. 1988. Rhamnus spp.: In vitro production of anthraquinones, anthrones, and dianthrones. In *Medicinal and Aromatic Plants I*, edited by Y. P. S. Bajaj, 513–528. Berlin, Heidelberg: Springer.

Van den Berg, A. J. J., and R. P. Labadie. 1989. Quinones. In *Methods in Plant Biochemistry*, edited by P. M. Dey and J. B. Harborne, 413–491. London: Academic Press.

Van den Berg, A. J. J., M. H. Radema, and R. P. Labadie. 1988a. Anthra-derivatives in suspension cell cultures of *Rhamnus frangula. Plant Sci.* 56 (2):123–127. doi: http://dx.doi.org/10.1016/0168-9452(88)90025-8.

Van den Berg, A. J. J., M. H. Radema, and R. P. Labadie. 1988b. Effects of light on anthraquinone production in *Rhamnus purshiana* suspension cultures. *Phytochem.* 27 (2):415–417. doi: http://dx.doi.org/10.1016/0031-9422(88)83110-8.

Van den Berg, A. J., and R. P. Labadie. 1984. Anthraquinones, anthrones and dianthrones in callus cultures of *Rhamnus frangula* and *Rhamnus purshiana. Planta Med.* 50 (5):449–51. doi: 10.1055/s-2007-969765.

Van den Berg, A. J. J. 1991. Biotechnology and biosynthesis of quinones. *Pharmaceutisch Weekblad* 13 (2):74–77. doi: 10.1007/bf01974984.

Van Gorkom, B. A., E. G. De Vries, A. Karrenbeld, and J. H. Kleibeuker. 1999. Review article: Anthranoid laxatives and their potential carcinogenic effects. *Aliment. Pharmacol. Ther.* 13 (4):443–52.

Wang, Z., P. Ma, L. Xu, C. He, Y. Peng, and P. Xiao. 2013. Evaluation of the content variation of anthraquinone glycosides in rhubarb by UPLC-PDA. *Chem. Cent. J.* 7 (1):1–11. doi: 10.1186/1752-153x-7-170.

Watson, E. 2013. An alternative to crushed bugs? Chr. Hansen explores producing carmine via controlled fermentation process. FOODnavigator-usa.com. Accessed September from http://www.foodnavigator-usa.com/Suppliers2/An-alternative-to-crushed-bugs-Chr.-Hansen-explores-producing-carmine-via-controlled-fermentation-process.

Weiss, B. 2012. Synthetic food colors and neurobehavioral hazards: The view from environmental health research. *Environ. Health Perspect.* 120 (1):1–5. doi:10.1289/ehp.1103827.

Westendorf, J., W. Pfau, and A. Schulte. 1998. Carcinogenicity and DNA adduct formation observed in ACI rats after long-term treatment with madder root, *Rubia tinctorum* L. *Carcinogenesis* 19 (0143-3334 (Print)):2163–2168.

WHO. 2006. FAO/WHO Expert Committee on food additives: Evaluation of certain food additives. In *WHO Technical Report Series n° 934*, edited by WHO. Geneva, Switzerland.

Wianowska, D. 2014. Hydrolytical instability of hydroxyanthraquinone glycosides in pressurized liquid extraction. *Anal. Bioanal. Chem.* 406 (13):3219–27. doi:10.1007/s00216-014-7744-5.

Wu, C. H., and G. C. Yen. 2004. Antigenotoxic properties of *Cassia tea* (*Cassia tora* L.): Mechanism of action and the influence of roasting process. *Life Sci.* 76:85–101. doi:10.1016/j.lfs.2004.07.011

Xie, G., X. Zhu, Q. Li, M. Gu, Z. He, J. Wu, J. Li, Y. Lin, M. Li, Z. She, and J. Yuan. 2010. SZ-685C, a marine anthraquinone, is a potent inducer of apoptosis with anticancer activity by suppression of the Akt/FOXO pathway. *Br. J. Pharmacol.* 159 (3):689–697. doi: 10.1111/j.1476-5381.2009.00577.x.

Xiong, H. R., J. Luo, W. Hou, H. Xiao, and Z. Q. Yang. 2011. The effect of emodin, an anthraquinone derivative extracted from the roots of *Rheum tanguticum*, against herpes simplex virus in vitro and in vivo. *J. Ethnopharmacol.* 133 (2):718–723. doi: http://dx.doi.org/10.1016/j.jep.2010.10.059.

Xu, A., Y. Wang, J. Wen, P. Liu, Z. Liu, and Z. Li. 2011. Fungal community associated with fermentation and storage of Fuzhuan brick-tea. *Int. J. Food Microbiol.* 146 (1):14–22. doi: http://dx.doi.org/10.1016/j.ijfoodmicro.2011.01.024.

Yasuda, K., S. Uehara, T. Shindo, I. Takano, and M. Nishijima. 2000. Stability of barbaloin in aqueous solution. *J. Japan Association Food Preservation Scientists (Nihon Shokuhin Hozou Kagaku Kaishi)* 26:85–90.

Yasuda, K., K. Yokoyama, H. Ushiyama, H. Ogawa, and Y. Kawai. 1997. Hygienic studies on health food IV. Barbaloin content and stability in foods containing *Aloe*. *Food Hygiene and Safety Science (Shokuhin Eiseigaku Zasshi)* 38 (5):335–340.

Yasui, Y., and N. Takeda. 1983. Identification of a mutagenic substance, in *Rubia tinctorum* L. (madder) root, as lucidin. *Mutation Res. Lett.* 121 (3):185–190. doi: http://dx.doi.org/10.1016/0165-7992(83)90201-4.

Yen, G. C., and D. Y. Chung. 1999. Antioxidant effects of extracts from *Cassia tora* L. prepared under different degrees of roasting on the oxidative damage to biomolecules. *J. Agric. Food Chem.* 47 (4):1326–32.

Yen, G. C., P. Duh, and D. Chuang. 2000. Antioxidant activity of anthraquinones and anthrone. *Food Chem.* 70 (4):437–441. doi: 10.1016/S0308-8146(00)00108-4.

Yilmaz, U. T., F. Ergun, and H. Yilmaz. 2014. Determination of the food dye carmine in milk and candy products by differential pulse polarography. *J. Food Drug Anal.* 22 (3):329–335. doi: http://dx.doi.org/10.1016/j.jfda.2013.12.002.

Yokohira, M., Yamakawa K., K. Hosokawa, Y. Matsuda, T. Kuno, K. Saoo, and K. Imaida. 2008. Promotion potential of madder color in a medium-term multi-organ carcinogenesis bioassay model in F344 rats. *J. Food Sci.* 73 (1750–3841 (Electronic)):T26-32.

Yu, J. H., and N. Keller. 2005. Regulation of secondary metabolism in filamentous fungi. *Annu. Rev. Phytopathol.* 43:437–58. doi: 10.1146/annurev.phyto.43.040204.140214.

Zargar, B. A., M. H. Masoodi, B. Ahmed, and S. A. Ganie. 2011. Phytoconstituents and therapeutic uses of *Rheum emodi* wall. ex Meissn. *Food Chem.* 128 (3):585–589. doi: http://dx.doi.org/10.1016/j.foodchem.2011.03.083.

Zhang, J., Z. Hu, and Z. Chen. 2015. Inhibitory effect of emodin on the growth of cervical cancer in tumor-transplanted mice and underlying mechanism. *Xi Bao Yu Fen Zi Mian Yi Xue Za Zhi* 31 (3):350–4.

Zhang, J. Y., L. Y. Tao, Y. J. Liang, L. M. Chen, Y. J. Mi, L. S. Zheng, F. Wang, Z. G. She, Y. C. Lin, K. K. To, and L. W. Fu. 2010. Anthracenedione derivatives as anticancer agents isolated from secondary metabolites of the mangrove nndophytic fungi. *Mar. Drugs* 8 (4):1469–1481. doi: 10.3390/md8041469.

Zhang, L. M., W. G. Xie, T. T. Wen, and X. Zhao. 2009. Thermal behavior of five free anthraquinones from rhubarb. *J. Therm. Anal. Calorim.* 100 (1):215–218. doi: 10.1007/s10973-009-0013-8.

Zhang, X. P., Z. F. Li, X. G. Liu, Y. T. Wu, J. X. Wang, K. M. Wang, and Y. F. Zhou. 2005. Effects of emodin and baicalein on rats with severe acute pancreatitis. *World J. Gastroenterol.* 11 (14):2095–100.

Zhao, X. Y., G. F. Qiao, B. X. Li, L. M. Chai, Z. Li, Y. J. Lu, and B. F. Yang. 2009. Hypoglycaemic and hypolipidaemic effects of emodin and its effect on L-type calcium channels in dyslipidaemic-diabetic rats. *Clin. Exp. Pharmacol. Physiol.* 36 (1):29–34. doi: 10.1111/j.1440-1681.2008.05051.x.

Zhou, X. M., and Q. H. Chen. 1988. Biochemical study of chinese rhubarb. XXII. Inhibitory effect of anthraquinone derivatives on Na^+-K^+-ATPase of the rabbit renal medulla and their diuretic action. *Yaoxue Xuebao* 23 (1):17–20.

Zhou, X., B. Song, L. Jin, D. Hu, C. Diao, G. Xu, Z. Zou, and S. Yang. 2006. Isolation and inhibitory activity against ERK phosphorylation of hydroxyanthraquinones from rhubarb. *Bioorg. Med. Chem. Lett.* 16 (3):563–568.

Zonta, F., P. Bogoni, P. Masotti, and G. Micali. 1995. High-performance liquid chromatographic profiles of aloe constituents and determination of aloin in beverages, with reference to the EEC regulation for flavouring substances. *J. Chromatogr. A.* 718 (1):99–106.

CHAPTER 10

Flavonoids

Marilena Antunes-Ricardo and Janet Alejandra Gutiérrez-Uribe

CONTENTS

10.1 INTRODUCTION

Flavonoids are a group of secondary metabolites that can be found in fruits, vegetables, grains, roots, stems, flowers, tea, and wine. These compounds are synthesized in the cytoplasm and are transported to the vacuoles (Pinelo et al. 2006; Markham et al. 2001). Besides their function as the pigments of flowers and fruits, flavonoids are used by plants to attract pollinators and seed dispersers. They are also involved in plants' resistance to diseases and have been associated with pollen germination and fertility (Schijlen et al. 2004, Treutter 2005). On the other hand, flavonoids, including anthocyanins, are involved in plant resistance to UV radiation. Plants subjected to UV radiation artificially respond by producing changes in the flavonoid biosynthetic pathways such as by increasing the activities of phenylalanine ammonia-lyase (PAL), which plays a pivotal role in the biosynthesis of the phenylpropanoid skeleton and, therefore, in the production of phenolic compounds. Also, PAL is responsible for plant defense strategies in difficult environments (Nazi et al 2012, Makoi et al. 2010, Roze et al. 2011). Recently, flavonoids have attracted special attention due to their multiple biological activities, such as antioxidant, anti-inflammatory, anticancer, antiviral, antibacterial, and hepatoprotective activity, among others (Kumar and Pandey 2013).

10.2 CHEMISTRY OF FLAVONOIDS

Despite the wide variety of flavonoids, a limited number of them are abundant in foods commonly consumed by humans. Table 10.1 summarizes the classification of flavonoids according to their chemical structure, examples of most important flavonoids in each class, and some food sources of them.

Many flavonoid derivatives can be found in plants due to hydrogenation reactions, hydroxylation, malonylation, glycosylation, acylation and methylation (Tapas et al. 2008). Glycosylation, methylation, and acylation have been the most-studied flavonoid

TABLE 10.1 Chemical Skeletons of the Main Flavonoid Classes

Flavonoid Class	Structure Backbone	Examples	Food Sources	Reference
Flavan-3-ol	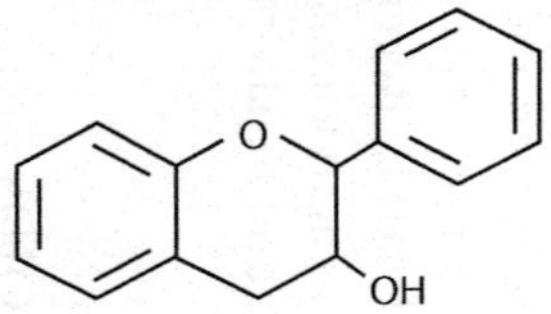	Catechin, epicatechin, epigallocatechin	Grapes, apple, pear, cherries, apricots, strawberries, green tea, red wine, blackberries, broad beans, peaches, raspberries, chocolate.	(Kumar and Pandey 2013; Gadkari and Balaraman 2015; López-Miranda et al. 2016; Zhang et al. 2014; Santos-Buelga et al. 2016)
Flavanones	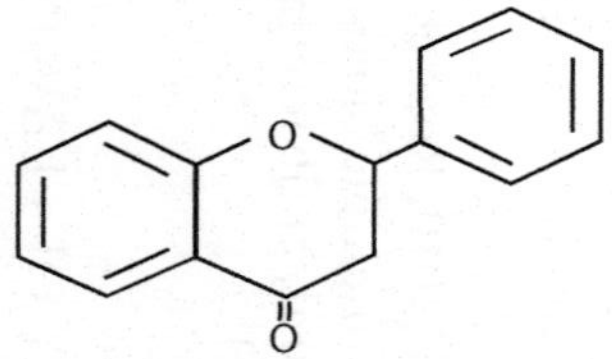	Hesperidin, taxifolin, naringenin, naringin, eriodictyol.	Citrus fruits, grapes, lemon, mandarins, oranges, bergamot, potato.	(Kumar and Pandey 2013; Santos-Buelga et al. 2016)
Isoflavones	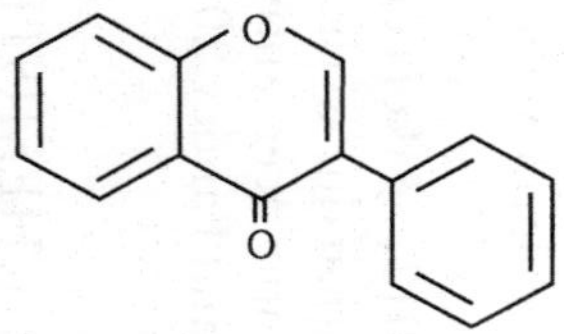	Genistein, daidzein, biochaninA, puerarin	Soybeans, chickpeas.	(Kumar and Pandey 2013; Santos-Buelga et al. 2016; Guardado-Félix et al. 2017)
Flavones	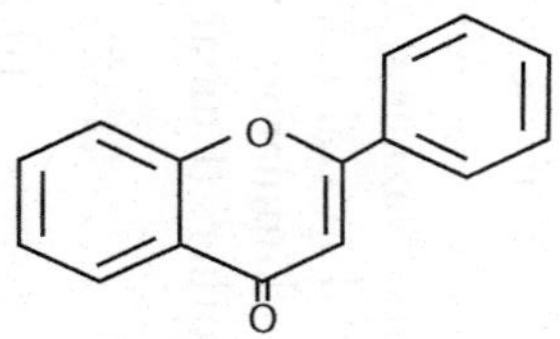	Apigenin, luteolin,chrysin, chrysoeriol	Fruit skins, red wine, buckwheat, red pepper, tomato, honey, sage, wava, garlic, broccoli, celery.	(Kumar and Pandey 2013; Santos-Buelga et al. 2016; Miean and Mohamed 2001)

(Continued)

TABLE 10.1 (CONTINUED) Chemical Skeletons of the Main Flavonoid Classes

Flavonoid Class	Structure Backbone	Examples	Food Sources	Reference
Flavonols	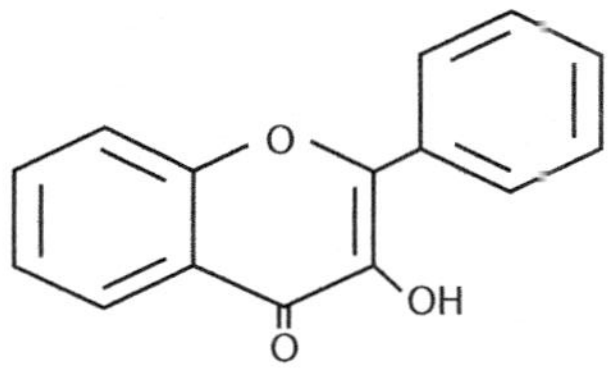	Quercetin, kaempferol, myricetin, isorhamnetin	Apple, onion, papaya, black beans, cactus.	(Kumar and Pandey 2013; Santos-Buelga et al. 2016; Miean and Mohamed 2001; Guajardo-Flores et al. 2014; Santos-Zea et al. 2011)
Anthocyanins	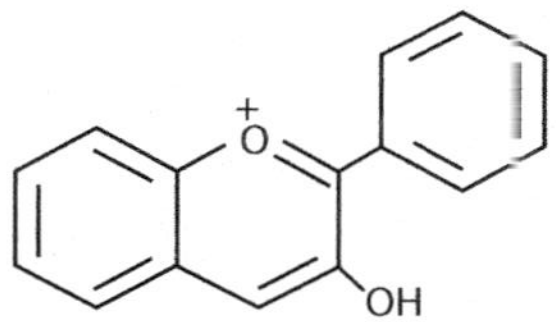	Cyanidin, delphinidin, malvidin, pelargonidin, petunidin, peonidin	Blue maize, berries, black bean, cabbage, red onions.	(Kumar and Pandey 2013; Santos-Buelga et al. 2016)
Chalcones	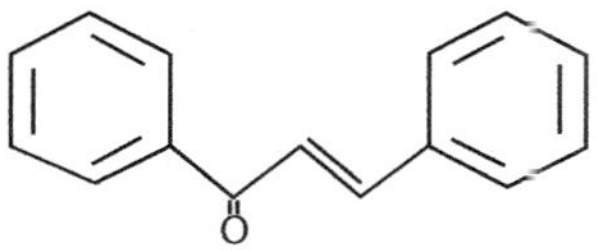	Naringenin chalcone, phloretin, arbutin	Apple, tomato skin, oriental pear.	(Kumar and Pandey 2013; Shao et al. 2008; Tomás-Barberán and Clifford 2000)

modifications since these reactions dramatically affect the polarity of the molecule, their biological properties, and increase the molecular weight of the flavonoid (Aherne and O'Brien 2002). Figure 10.1 shows the most common substituents of flavonoids found in nature, including methyl groups, sugar moieties or glycoside groups, and acyl groups.

Glycosylated flavonoids possess sugar residues attached to their structure while flavonoids without sugar residues are known as *aglycones*. Glucose is the most commonly encountered sugar, followed by galactose, rhamnose, xylose, and arabinose, whereas glucuronic and galacturonic acids are rare.

With the exception of the catechins, flavonoids are not found in nature in the form of aglycone; they are generally found in plants as glycosylated derivatives (Survay et al. 2011). Glycosylation may influence both the stability and solubility of molecules. Quercetin-3-O-glucosyl-rhamnoside had higher stability in aqueous solution at 100°C under air perfusion conditions compared to quercetin (Buchner et al. 2006; Plaza et al. 2014). Mäkilä et al. (2016) evaluated the stability of different hydroxycinnamic acid derivatives, flavonol glycosides, and anthocyanins contained in black currant juice at ambient temperature (in light and in dark conditions) and at 4°C for a year. O-glucosides of hydroxycinnamic acid compounds stored at low temperatures and excluding light were the most stable. Likewise, it has been demonstrated that the stability of flavonoids increases with the number of glycosylated hydroxyl groups (Plaza et al. 2014).

On the other hand, a sugar moiety increases the solubility of the flavonoids in water compared to the corresponding aglycones, or well, decreases their lipophilicity (Plaza et al. 2014). The position and the number of sugar moieties play an important role in flavonoid solubility as shown by Rothwell et al. (2005), who reported that the water solubility of quercetin glycosides decreased in the following order: Quercetin-3,4′-diglucoside > quercetin-3-glucoside > quercetin-4′-glucoside.

Also, flavonoid glycosides have shown to be more resistant to thermal processing than their corresponding aglycones, as observed with apigenin-7-O-apiosyl-glucoside, apigenin 7-O-glucoside. Apigenin-7-O-apiosyl-glucoside resists β-glucosidase activity, but was converted to apigenin 7-O-glucoside at pH 2.7 when processed at 100°C for 90 minutes; it could be then further deglycosylated to apigenin. Apigenin aglycone was most stable at pH 3 but progressively degraded at pH 5 or 7 at 100°C

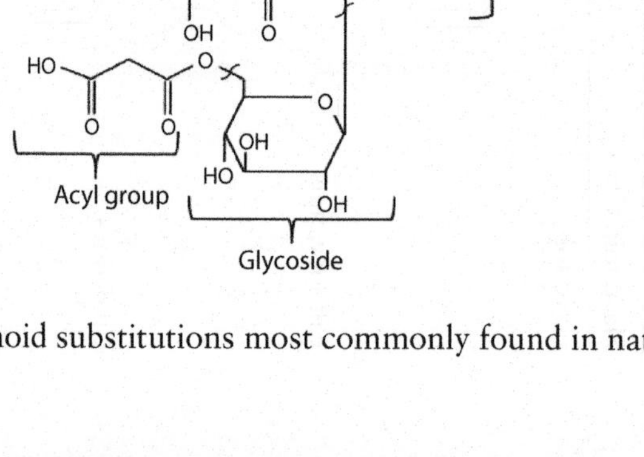

FIGURE 10.1 Flavonoid substitutions most commonly found in nature.

(Hostetler et al. 2013). Flavonol glycosides were the more stable compounds after baking muffins enriched with strawberry, blackcurrant, raspberry, and sour cherry pomace at three different temperatures (140, 180, and 220°C). These compounds showed losses from 0 percent to 21 percent; meanwhile, anthocyanins exhibited losses of 36 percent to 97 percent after the baking process (Górnaś et al. 2016).

Glycosides of pelargonidin and cyanidin showed high stability in strawberry juices stored at 4°C with a degradation degree about 69 percent. However, differences in degradation degrees of anthocyanins glycosides were observed. Degradation degrees of pelargonidin-3-glucoside and pelargonidin-3-malonyl-glucoside (69 percent and 73 percent, respectively) were higher than cyanidin-3-glucoside and cyanidin-3-malonylglucoside (68 percent and 56 percent, respectively) (Teleszko et al. 2016). Likewise, quercetin monoglycosides showed differential stability after drying using microwave vacuum drying, freeze drying, or air drying. Results showed that both microwave vacuum drying and freeze drying maintained high quercetin contents while air drying induced a significant loss of quercetin glycosides and browning reactions. The stability of quercetin glycosides, compared to their initial concentration in apple slices, was in descending order: quercetin-3-O-galactoside > quercetin-3-O-glucoside > quercetin-3-O-rhamnoside > quercetin-3-O-xyloside > quercetin-3-O-arabinoside (Schulze et al. 2014).

Flavonoids, aglycones, and glycosides have shown different biological effects observing that bioactivity of the last ones depend on the type and position of sugar residue linked to the aglycone (Antunes-Ricardo et al. 2015; Kathirvel and Richards 2009). For example, quercetin-3-O-rhamnoside exhibited greater *in vitro* antioxidant effect than quercetin-3-O-glucoside (Hopia and Heinonen 1999). Likewise, the number of sugar residues affects the flavonol antioxidant effect. Quercetin monosaccharides showed to be more active than quercetin disaccharides and this could be partly explained by different solubility (Hopia and Heinonen 1999).

The O-methylation of flavonoids is a common xenobiotic transformation occurring in plants, microbes, and mammals from high selective enzymatic systems, the O-methyl transferases (Bernini et al. 2011; Wen and Walle 2006). A large number of methyl groups increase the aqueous solubility of flavonoids, improving their intestinal permeability and therefore intestinal absorption (Wen and Walle 2006). An example of this was showed by Walle (2009), who observed that 5,7-dimethoxyflavone (5,7-DMF) exhibited solubility in water or pH 7.4 Hank's buffered salt solution up to 100 µM; meanwhile, chrysin reached maximum solubility at around 20 µM despite its two free hydroxyl groups.

In addition, methylation of free hydroxyl groups in flavonoids also increases their metabolic stability and enhances biological effects on health. Koirala et al. (2016) reported that 7-hydroxyflavone, 7,4′-dihydroxyflavone, and 5,7-dihydroxyflavone (chrysin) were undetectable in tissue levels after administration to rats, whereas the corresponding methylated derivatives reached high tissue levels. These data suggest that methylated flavonoids could be bioactive for a longer period of time than their non-methylated analogs.

Acylation also increases the stability of flavonoids by the intra- and intermolecular interactions, between the flavonoid skeleton and the aromatic ring in the acyl moiety, reducing the degradation index of the molecule and, in some cases, improving its bioactivity (Plaza et al. 2014). Acylation occurs mainly at the C-6 position of the glucose molecule by esterification with acetic, p-coumaric, or caffeic acids, although they have also been reported with lactic acid in grape anthocyanins (Monagas and Bartolomé 2009). The majority of the known anthocyanins are acylated by organic acids via ester bonds, which is referred to as *anthocyanin glycosyl acylation* (Zhao et al. 2017). Anthocyanins acylated with p-coumaric, sinapic, caffeic, ferulic or sinapic acids, and cyanidin were identified in broccoli sprouts (Moreno et al. 2010). More than 70 types of anthocyanins have been reported in *Raphanus sativus* (L.), of which about 80 percent to 90 percent were acylated (Matera et al. 2012). Likewise, acylation of flavonoids enhanced the capacity to inhibit xanthine oxidase and the antiproliferative activity of isorhamnetin-3-O-glucoside (Salem et al. 2011).

10.3 FLAVONOID BIOAVAILABILITY

A challenge with flavonoids is their low bioavailability, a key step to ensure their bioefficacy. Bioavailability is a complex process that involves different stages: liberation, absorption, distribution, metabolism, and elimination (LADME). Numerous factors affect the bioavailability of dietary flavonoids, such as chemical structure and molecular weight, glycosylation, esterification, and natural source (Thilakarathna and Rupasinghe 2013).

The absorption of flavonoids mainly depends on their solubility and permeability. The flavonoid aglycones can easily permeate through the intestinal monolayer due to their high lipophilicity and low molecular weight. However, flavonoid glycosides showed limited permeability, possibly due to higher hydrophilicity and larger molecular weight. The aglycones and simple hydrophilic flavonoid glucoside are transported into epithelial cells across the small intestine by the intestinal Na^+-dependent glucose cotransporter (SGLT1) to be hydrolyzed by the cytosolic β-glucosidase (CBG) or can be hydrolyzed by lactase phloridzin hydrolase (LPH), a β-glucosidase on the outside of the brush border membrane of the small intestine. Then the liberated aglycone can be absorbed across the small intestine. Complex glycosides are transported toward the colon to be hydrolyzed by bacterial enzymes and subsequently liberate flavonoid aglycones (Kumar and Paney 2013). Felgines et al. (2000) evaluated the absorption kinetics of naringenin and its glycosides in rats. The absorption kinetics of naringenin and naringenin-7-glucoside were similar, whereas naringenin-7-rhamnoglucoside exhibited a delay in its intestinal absorption, resulting in decreased bioavailability. After naringenin-7-glucoside feeding, no glucoside was found in the cecum. Data suggested that these flavanones were efficiently absorbed after being fed to rats and that their bioavailability was related to their glycosidic moiety. Other reports about the bioavailability of quercetin and its glycosides after oral administration to rats found that the aglycone bioavailability was lower (2.0 percent) than the observed for quercetin-3-O-glucoside and quercetin-3-O-maltoside that showed 12 percent and 30 percent of bioavailability, respectively (Makino et al. 2009). Likewise, some flavonoid glycosides such as puerarin 7-O-glucoside and puerarin 7-O-isomaltoside showed higher concentration levels in plasma and have longer residence time in the blood than aglycone after intravenous administration (Xiao et al. 2014). A single-dose pharmacokinetic trial was conducted in 10 healthy adults younger than 50 years old to evaluate the acute (24-hour) absorption and excretion of flavonoids, phenolic acids, and proanthocyanidins from a low-calorie cranberry juice cocktail (54 percent juice). The main phenolics in the test beverage were the anthocyanins peonidin-3-galactoside and 3-arabinoside, followed by cyanidin-3-arabinoside and 3-galactoside, and the flavonols hyperoside and quercetin. The sum total of phenolics including phenolic acids, flavonols, and flavanols detected in plasma reached a peak concentration of 34.2 µg/mL at 8–10 hours, and the highest contributors to this were protocatechuic acid, quercetin, and vanillic acid (McKay et al. 2015).

Transportation and tissue distribution of dietary flavonoids depend on their affinities toward plasma proteins such as hemoglobin (Ding et al. 2012). Methylated flavonoids have shown 2–16 times more protein affinity than the non-methylated due to the increase in hydrophobic interactions with human serum albumin and ovoalbumin (Cao et al. 2013).

After absorption, flavonoids are treated by the body as xenobiotics as they are rapidly removed from the bloodstream through hepatic reactions of glucuronidation, sulfation, or methylation, or may be metabolized to smaller phenolic compounds (Del Rio et al. 2010). These reactions seem to be more efficient than P450-mediated oxidation and might significantly contribute to the effects of these dietary constituents (Xiao and Hogger 2013). Five principal quercetin metabolites, quercetin-3′-O-sulfate, quercetin-3-O-glucuronide, isorhamnetin-3-O-glucuronide, a quercetin-O-glucuronide-O-sulfate, and a quercetin-O-diglucuronide were detected in plasma 24 hours after consuming 270 grams of lightly fried onions with a total of 275 μmol of flavonol glucosides with 4′-O-glucoside and 3,4′-O-diglucoside as the main constituents (Crozier et al. 2010). Likewise, Hosoda et al. (2010) observed the presence of five isoflavone metabolites, being daidzein-4′-O-glucuronide, genistein-4′-O-glucuronide, daidzein-7-O-glucuronide, genistein-7-O-glucuronide, daidzein-4′-O-sulfate, genistein-4′-O-sulfate, daidzein-7-O-sulfate, and genistein-7-O-sulfate in plasma in the 1–7 hour period post-consumption of 50 grams of baked soya bean powder containing 66 μmol of daidzein, 106 μmol of genistein, 120 μmol of diadzein-7-glucoside, and 205 μmol of genistein-7-O-glucoside suspended in cow milk.

During absorption, flavonoid metabolites may suffer conjugation to generate more stable structures. The longer plasma $t_{1/2}$ value of isorhamnetin-3-O-glucuronide could reflect post-absorption 3′-O-methylation of quercetin-3-O-glucuronide in the liver. Meanwhile, the delayed $t_{1/2}$ of the quercetin-O-glucuronide-O-sulfate could be a consequence of post-absorption sulfation of quercetin-3-O-glucuronide or glucuronidation of quercetin-3′-O-sulfate. Likewise, it has been demonstrated that rutinose sugar moiety of quercetin-3-O-rutinoside is not cleaved by the action of either CBG or LPH during passage through the small intestine and, as a consequence, the release of quercetin aglycone occurs by the action of colonic bacterial enzymes. Therefore, a lower production and absorption of methylated and glucuronidated quercetin metabolite takes place in the large intestine (Crozier et al. 2010). Similarly, green tea flavan-3-ols and chlorogenic acids in coffee are subject to extensive metabolism prior to absorption, initially in the small intestine and in the large intestine where the colonic microbiota produces phenolic acids.

The colon plays an important role in the bioavailability and metabolism of dietary phenolic and polyphenolic compounds generating new catabolites that affect colonic microbiota (Bianchi et al. 2010). A high level of flavonoids metabolites in urine indicates that substantial quantities of these compounds were metabolized by colonic microbiota (Crozier et al. 2010). In this sense, Felgines et al. (2000) reported that recovery of glucoside or rhamnoglucoside forms of naringenin in urine was lower (31 percent and 14 percent) than aglycone (14 percent), indicating that a higher concentration (>70 percent) of the flavonoid could be available to exert a specific biological effect.

10.4 CONCLUSIONS

Flavonoids have been widely studied due to their biological activities and multiples applications in food industry. The chemical structure of flavonoids determines their phytochemical characteristics, and even more, their activity of functionality. Conjugations like glycosylation, methylation, and acylation confers to flavonoids more stability to processing conditions and also to intestinal metabolism. On the other hand, effective dose and active metabolite profiles depend on flavonoid absorption across biological membranes. More studies are needed to understand clearly the relationship between conjugation patterns of flavonoids and their bioavailability and biological activities.

REFERENCES

Aherne, S., and O'Brien, N. 2002. Dietary flavonols: Chemistry, food content, and metabolism. *Nutrition* 18:75–81.

Antunes-Ricardo, M., Gutiérrez-Uribe, J.A., López-Pacheco, F., Alvarez, M.M., and Serna-Saldivar, S.O. 2015. *In vivo* anti-inflammatory effects of isorhamnetin glycosides isolated from *Opuntia ficus-indica* (L.) Mill cladodes. *Industrial Crops and Products* 76:803–808.

Bernini, R., Crisante, F., and Ginnasi, M.C. 2011. A convenient and safe O-methylation of flavonoids with dimethyl carbonate (DMC). *Molecules* 16:1418–1425.

Bianchi, M.A., Scazzina, F., Del Rio, D. et al. 2010. Ability of a high-total antioxidant capacity diet to increase stool weight and bowel antioxidant status in human subjects. *British Journal of Nutrition* 104:1500–1507.

Buchner, N., Krumbein, A., Rohn, S., and Kroh, L.W. 2006. Effect of thermal processing on the flavonols rutin and quercetin. *Rapid Communications in Mass Spectrometry* 20:3229–3235.

Cao, H., Jing, X., Wu, D., and Shi, Y. 2013. Methylation of genistein and kaempferol improves their affinities for proteins. *International Journal of Food Sciences and Nutrition* 64:437–443.

Crozier, A., Del Rio, D., and Clifford, M.N. 2010. Bioavailability of dietary flavonoids and phenolic compounds. *Molecular Aspects of Medicine* 31:446–467.

Del Rio, D., Borges, G., and Crozier, A. 2010. Berry flavonoids and phenolics: Bioavailability and evidence of protective effects. *British Journal of Nutrition* 104:S67–S90.

Ding, F., Sun, Y., Diao, J.X. et al. 2012. Features of the complex of food additive hesperidin to hemoglobin. *Journal of Photochemistry and Photobiology B* 106:53–60.

Felgines, C., Texier, O., Morand, C. et al. 2000. Bioavailability of the flavanone naringenin and its glycosides in rats. *American Journal of Physiology-Gastrointestinal and Liver Physiology* 279:G1148–G1154.

Gadkari, P.V., and Balaraman, M. 2015. Catechins: Sources, extraction and encapsulation: A review. *Food and Bioproducts Processing* 93:122–138.

Górnaś, P., Juhņeviča-Radenkova, K., Radenkovs, V. et al. 2016. The impact of different baking conditions on the stability of the extractable polyphenols in muffins enriched by strawberry, sour cherry, raspberry or black currant pomace. *LWT-Food Science and Technology* 65:946–953.

Guajardo-Flores, D., Serna-Guerrero, D., Serna-Saldívar, S.O., and Jacobo-Velázquez, D.A. 2014. Effect of germination and UV-C radiation on the accumulation of flavonoids and saponins in black bean seed coats. *Cereal Chemistry* 91:276–279.

Guardado-Félix, D., Serna-Saldivar, S.O., Cuevas-Rodríguez, E.O., Jacobo-Velázquez, D.A., and Gutiérrez-Uribe, J.A. 2017. Effect of sodium selenite on isoflavonoid contents and antioxidant capacity of chickpea (*Cicer arietinum* L.) sprouts. *Food Chemistry* 226:69–74.

Hopia, A., and Heinonen, M. 1999. Antioxidant activity of flavonol aglycones and their glycosides in methyl linoleate. *Journal of the American Oil Chemists' Society*, 76:139–144.

Hosoda, K., Furuta, T., Yokokawa, A., and Ishii, K. 2010. Identification and quantification of daidzein-7-glucuronide-4′-sulfate, genistein-7-glucuronide-4′-sulfate and genistein-4′,7-diglucuronide as major metabolites in human plasma after administration of kinako. *Analytical and Bioanalytical Chemistry* 397:1563–1572.

Hostetler, G.L., Riedl, K.M., and Schwartz, S.J. 2013. Effects of food formulation and thermal processing on flavones in celery and chamomile. *Food Chemistry* 141:1406–1411.

Kathirvel, P., and Richards, M.P. 2009. Mechanisms by which flavonol aglycones inhibit lipid oxidation better than glycosylated flavonols in comminuted muscle tissue. *Food Chemistry* 117:75–82.

Koirala, N., Thuan, N.H., Ghimire, G.P., Van Thang, D., Sohng, J.K. 2016. Methylation of flavonoids: Chemical structures, bioactivities, progress and perspectives for biotechnological production. *Enzyme and Microbial Technology* 86:103–116.

Kumar, S., and Pandey, A.K., 2013. Chemistry and biological activities of flavonoids: An overview. *The Scientific World Journal* 2013:1–16.

López-Miranda, S., Serrano-Martínez, A., Hernández-Sánchez, P. et al. 2016. Use of cyclodextrins to recover catechin and epicatechin from red grape pomace. *Food Chemistry* 203:379–385.

Mäkilä, L., Laaksonen, O.A., Alanne, A.L.M., Kortesniemi, M.K., Kallio, H.P., and Yang, B. 2016. Stability of hydroxycinnamic acid derivatives, flavonol glycosides and anthocyanins in black currant juice. *Journal of Agricultural and Food Chemistry* 64:4584–4598.

Makino, T., Shimizu, R., Kanemaru, M., Suzuki, Y., Moriwaki, M., and Mizukami, H. 2009. Enzymatically modified isoquercitrin, alpha-oligoglucosyl quercetin 3-O-glucoside, is absorbed more easily than other quercetin glycosides or aglycone after oral administration in rats. *Biological and Pharmaceutical Bulletin* 32:2034–2040.

Makoi, J.H., Belane, A.K., Chimphango, S.B., and Dakora, F.D. 2010. Seed flavonoids and anthocyanins as markers of enhanced plant defence in nodulated cowpea (*Vigna unguiculata* L. Walp.). *Field Crops Research* 118:21–27.

Markham, K.R., Gould, K.S., and Ryan, K.G. 2001. Cytoplasmic accumulation of flavonoids in flower petals and its relevance to yellow flower colouration. *Phytochemistry* 58:403–413.

Matera, R., Gabbanini, S., De Nicola, G.R., Iori, R., Petrillo, G., and Valgimigli, L. 2012. Identification and analysis of isothiocyanates and new acylated anthocyanins in the juice of *Raphanus sativus* cv. Sango sprouts. *Food Chemistry* 133:563–572.

McKay, D.L., Chen, C.Y.O., Zampariello, C.A., and Blumberg, J.B. 2015. Flavonoids and phenolic acids from cranberry juice are bioavailable and bioactive in healthy older adults. *Food Chemistry* 168:233–240.

Miean, K.H., and Mohamed, S. 2001. Flavonoid (myricetin, quercetin, kaempferol, luteolin, and apigenin) content of edible tropical plants. *Journal of Agricultural and Food Chemistry* 49:3106–3112.

Monagas, M. and Bartolomé, B. 2009. Anthocyanins and anthocyanin-derived compounds. In *Wine Chemistry and Biochemistry*, ed. V.M. Moreno-Arribas and C.M. Polo, pp. 439–462. New York: Springer.

Moreno, D.A., Pérez-Balibrea, S., Ferreres, F., Gil-Izquierdo, Á., and García-Viguera, C. 2010. Acylated anthocyanins in broccoli sprouts. *Food Chemistry* 123:358–363.
Nazi, N., Ali, A., Javad, H., and Shahram, P. 2012. Phenylalanin ammonia-lyase activity, total phenolics and flavonoids contents in flowers, leaves, hulls and kernels of three pistachio (*Pistacia vera* L.) cultivars. *Bulletin of Botanical Research* 32:519–525.
Pinelo, M., Arnous, A., and Meyer, A.S. 2006. Upgrading of grape skins: Significance of plant cell-wall structural components and extraction techniques for phenol release. *Trends in Food Science & Technology* 17:579–590.
Plaza, M., Pozzo, T., Liu, J., Gulshan Ara, K.Z., Turner, C., and Nordberg Karlsson, E. 2014. Substituent effects on *in vitro* antioxidizing properties, stability, and solubility in flavonoids. *Journal of Agricultural and Food Chemistry* 62:3321–3333.
Rothwell, J.A., Day, A.J., and Morgan, M.R. 2005. Experimental determination of octanol-water partition coefficients of quercetin and related flavonoids. *Journal of Agricultural and Food Chemistry* 53:4355–4360.
Roze, L., Chanda, A., and Linz, J. 2011. Compartmentalization and molecular traffic in secondary metabolism: A new understanding of established cellular processes. *Fungal Genetics and Biology* 48:35–48.
Salem, J.H., Chevalot, I., Harscoat-Schiavo, C., Paris, C., Fick, M., and Humeau, C. 2011. Biological activities of flavonoids from *Nitraria retusa* (Forssk.) Asch. and their acylated derivatives. *Food Chemistry* 124:486–494.
Santos-Buelga, C., and González-Paramás, A.M. 2016 Flavonoids: Functions, metabolism and biotechnology. In *Industrial Biotechnology of Vitamins, Biopigments, and Antioxidants*, ed. E.J. Vandamme and J.L. Revuelta, pp. 469–496. Weinheim, Germany: Wiley-VCH Verlag GmbH & Co.
Santos-Zea, L., Gutierrez-Uribe, J., and Serna-Saldivar, S. 2011. Comparative analyses of total phenols, antioxidant activity, and flavonol glycoside profile of cladode flours from different varieties of Opuntia spp. *Journal of Agricultural and Food Chemistry* 59:7054–7061.
Schijlen, E.G., De Vos, C.R., van Tunen, A.J., and Bovy, A.G. 2004. Modification of flavonoid biosynthesis in crop plants. *Phytochemistry* 65:2631–2648.
Schulze, B., Hubbermann, E.M., and Schwarz, K. 2014. Stability of quercetin derivatives in vacuum impregnated apple slices after drying (microwave vacuum drying, air drying, freeze drying) and storage. *LWT-Food Science and Technology* 57:426–433.
Shao, X., Bai, N., He, K. et al. 2008. Apple polyphenols, phloretin and phloridzin: New trapping agents of reactive dicarbonyl species. *Chemical Research in Toxicology* 21:2042–2050.
Survay, N.S., Upadhyaya, C.P., Kumar, B. et al. 2011. New genera of flavonols and flavonol derivatives as therapeutic molecules. *Journal of the Korean Society for Applied Biological Chemistry* 54:1–18
Tapas, A.R., Sakarkar, D., and Kakde, R. 2008. Flavonoids as nutraceuticals. A review. *Tropical Journal of Pharmaceutical Research* 7:1089–1099.
Teleszko, M., Nowicka, P., and Wojdyło, A. 2016. Effect of cultivar and storage temperature on identification and stability of polyphenols in strawberry cloudy juices. *Journal of Food Composition and Analysis* 54:10–19.
Thilakarathna, S.H., and Rupasinghe, H.P. 2013. Flavonoid bioavailability and attempts for bioavailability enhancement. *Nutrients* 5:3367–3387.
Tomás-Barberán, F.A., and Clifford, M.N. 2000. Flavanones, chalcones and dihydrochalcones–nature, occurrence and dietary burden. *Journal of the Science of Food and Agriculture* 80:1073–1080.

Treutter, D. 2005. Significance of flavonoids in plant resistance and enhancement of their biosynthesis. *Plant Biology* 7:581–591.

Walle, T. 2009. Methylation of dietary flavones increases their metabolic stability and chemopreventive effects. *International Journal of Molecular Sciences* 10:5002–5019.

Wen, X., and Walle, T. 2006. Methylated flavonoids have greatly improved intestinal absorption and metabolic stability. *Drug Metabolism and Disposition* 34:1786–1792.

Xiao, J., and Hogger, P. 2013. Metabolism of dietary flavonoids in liver microsomes. *Current Drug Metabolism* 14:381–391.

Xiao, J., Chen, T., and Cao, H. 2014. Flavonoid glycosylation and biological benefits. *Biotechnology Advances* 32:1145–1156.

Zhang, H., Tang, B., and Row, K. 2014. Extraction of catechin compounds from green tea with a new green solvent. *Chemical Research in Chinese Universities* 30:37–41.

Zhao, C.L., Yu, Y.Q., Chen, Z.J. et al. 2017. Stability-increasing effects of anthocyanin glycosyl acylation. *Food Chemistry* 214:119–128.

CHAPTER 11

Lignans

Alessandra Durazzo

CONTENTS

11.1 INTRODUCTION

Within phytochemicals, phenolic compounds called lignans have attracted the interest of food chemists and nutrition researchers over the years. Lignans are vascular plant secondary metabolites, with widespread occurrence in the plant kingdom, and which are ascribed a wide range of physiological functions and beneficial properties (Adlercreutz 2007; Muir 2010; Peterson et al. 2010; Touré and Xueming 2010).

11.2 DIETARY LIGNANS AND CHEMICAL STRUCTURAL DIVERSITY

Lignans belong to the group of diphenolic compounds derived from the combination of two phenylpropanoid C6-C3 units at the β and β' carbon atoms and sometimes by additional ether, lactone, or carbone bonds, and have a chemical structure like the 1,4-diarylbutan (Lewis and Davin 1999). They are derived from the shikimic acid biosynthetic pathway (Ayres and Loike 1990; Mazur and Adlercreutz 1998; Imai et al. 2006). Lignans, optically active compounds, may exist as two enantiomers, that is, the right- and left-handed forms (Umezezawa 2003; Hemmati et al. 2010). The range relative to structurally different forms of lignans and biological activities is broad. In plant physiology, lignans are probably involved in defense and growth regulation mechanisms for the development of plants, besides their antifungal, antimicrobial, antiviral, insecticidal, and antifeeding properties (Harmatha and Dinan 2003; Arts and Hollman 2005; Saleem et al. 2005).

Plant lignans are widely distributed in many plant species where they occur as aglycones, glycosides, esterified glycosides, or bio-oligomers (Bambagiotti-Alberti et al. 1994; Kamal-Eldin et al. 2001). Numerous compounds have been discovered in several parts of various plants, that is, wooden parts, roots, leaves, flowers, fruits, and seeds (Pan et al. 2009). Lariciresinol, matairesinol, pinoresinol, and secoisolariciresinol (Figure 11.1) represent the most commonly detected plant lignans in foods (Milder et al. 2005a; Thompson et al. 2006). Other lignans such as medioresinol, syringaresinol (Figure 11.1), and sesamin were reported in various kinds of foods (Peñalvo et al. 2005a, 2008; Smeds et al. 2007, 2009). In the last year, investigations have focused their attention towards the isolation and structure elucidation of new lignan compounds (Woo et al. 2011; Huang et al. 2015).

11.3 LIGNANS IN FOODS

The sources of dietary lignans are represented by oilseeds (e.g., flax, soy, rapeseed, and sesame), whole-grain cereals (e.g., wheat, oats, rye, and barley), legumes, various vegetables and fruit (particularly berries), as well as beverages such as coffee, tea, and wine (Milder et al. 2005a; Thompson et al. 2006; Peñalvo et al. 2005a, 2008; Smeds et al. 2007, 2009).

Amongst edible products, flaxseed and sesame seeds represent rich sources of lignans (Thompson et al. 1997; Mazur 1998; Muir and Westcott 2003; Milder et al. 2005a; Peñalvo et al. 2005a), beside wood knots in coniferous trees, especially Norway spruce, are identified as the most concentrated lignan sources known so far (Holmbom et al. 2003).

However, the content of some lignans as well as their degree of esterification glycosides could vary in relation to different growing conditions, geographic location, climate, and genetic characteristics.

11.4 METABOLISM OF LIGNANS

Plant lignans are precursors of the compounds, generally called *enterolignans* due to their colonic origin (named also mammalian lignans), enterolactone, and enterodiol (Figure 11.2).

Enterolignans are biologically active metabolites due to their similar structure to the human hormone estrogen, so they may have estrogenic/anti-estrogenic effects (Axelson et al. 1982; Setchell et al. 1983); indeed, health-supporting effects of these metabolites have been investigated with a focus on phytoestrogen activity (Adlercreutz, 2002, 2007).

Lignans are stored in plants predominantly as glycosides. A major fraction of some dietary lignans are deglycosylated and partly converted to the mammalian lignans, enterodiol, and enterolactone by colonic bacteria; enterodiol is readily oxidized to enterolactone (Stitch et al. 1980; Borriello et al. 1985; Setchell and Adlercreutz 1998; Rowland et al. 2000; Wang et al. 2000). Moreover, it should be mentioned that, after ingestion, a small part of the dietary lignans could be absorbed in unchanged form in the small intestine (Glitso et al. 2000).

Presumably enterodiol and enterolactone are absorbed in the colon and reconjugated to glucuronides and sulfates in the intestinal wall and in the liver (Adlercreutz et al. 1995).

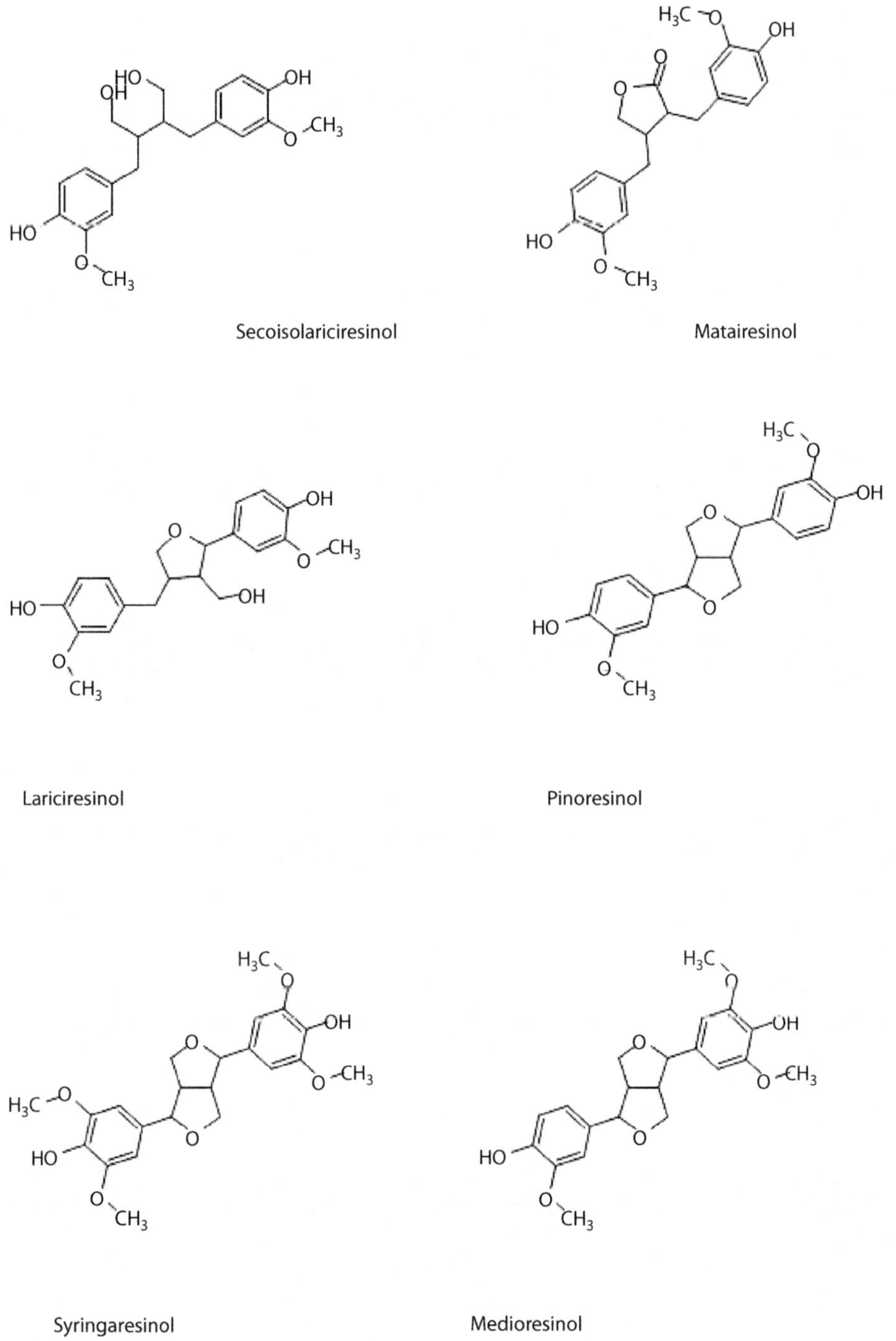

FIGURE 11.1 The chemical structures of some common dietary lignans.

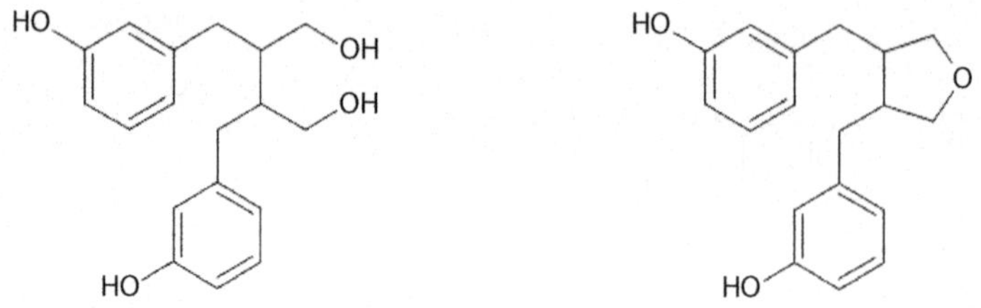

Enterodiol Enterolactone

FIGURE 11.2 The chemical structure of enterolignans.

Some of the metabolites may also undergo enterohepatic circulation. Lignans are excreted in bile and urine as conjugated glucuronides and in feces in the unconjugated form (Heinonen et al. 2001; Peñalvo et al. 2004).

The degree of conversion from common dietary lignans to enterolignans varies as described by Heinonen et al. (2001): secoisolariciresinol, matairesinol, pinoresinol, and lariciresinol have a high degree of conversion, from 55 percent to 100 percent. Several researchers have indicated as precursors of mammalian lignans, secoisolaresinol, matairesinol, pinoresinol, lariciresinol (Mazur et al. 2000; Heinonen et al. 2001; Xie et al. 2003a), as well as syringaresinol (Heinonen et al. 2001), 7-hydroxymatairesinol (Saarinen et al. 2000), arctiin (Xie et al. 2003b), and sesamin (Peñalvo et al. 2005b). Recently, several new precursors have been identified (Jin and Hattori 2011).

Interesting is the study of Bartkiene et al. (2011), which investigated the enterolactone and enterodiol formation from their precursors by the action of intestinal microbiota and their relationship with non-starch polysaccharides in various berries and vegetables: the enterolignan production from various berries ranged from 7.8 to 382.8 nmol/g as well as from vegetables from 10.5 till 91.2 nmol/g. Several studies have been focused on elucidation and description of enzymatic reactions and microorganisms implicated (Landete et al. 2015).

Kuijsten (2007) reported that the usual plasma concentration of enterolignans in Western populations is in the range of 10–30 nM. In addition, low concentrations of plant lignans respect to those observed for enterolignans was found in plasma (Peñalvo et al. 2004).

11.5 BIOLOGICAL ACTIVITIES

Enterolignans and some of their plant precursors have been indicated as having antitumorigenic (Kangas et al. 2002; Saarinen et al. 2010), anticarcinogenic (Bergman-Jungestrom et al. 2007; Velentzis et al. 2009), estrogenic or anti-estrogenic (Aehle et al. 2011), and antioxidant (Kitts et al. 1999; Prasad 2000; Durazzo et al. 2013a) properties.

For example, regarding antioxidant properties, studies of pure lignans have been carried out (Kitts et al. 1999; Eklund et al. 2005; Durazzo et al. 2013a). Niemeyer and Metzler (2001) have shown differences in the antioxidant activities between plant lignans (secoisolariciresinol, matairesinol) and enterolignans, presumably related to the methoxy group. Eklund et al. (2005) monitored the radical scavenging activity by DPPH (2,2-diphenyl-1-picrylhydrazyl): compounds with catechol (3,4-dihydroxyphenyl) moieties showed the highest radical scavenging activity.

Durazzo et al. (2013a) also evaluated the antioxidant activity of pure single lignans, their mixtures, and enterolignans by means of the FRAP assay: plant lignans have significantly higher FRAP values when compared with enterolignans. Matairesinol had the highest antioxidant properties and no synergistic effect was underlined in the mixture of plant lignans compared to single plant lignans.

Beside the antioxidant properties attributed to pure lignan compounds, data on antioxidant effects of lignan-rich food extracts are available (Willför et al. 2003; Hu et al. 2007; Durazzo et al. 2014a).

In the last years, biological activity of newly discovered compounds belonging to the lignan group have been tested and the spectrum of potential biological properties attributed to lignans is being enlarged (Cui et al. 2014; Kong et al. 2014; Wang et al. 2014; Yu et al. 2014; Zhang et al. 2014; Gnabre et al. 2015; Su and Wink 2015).

11.6 LIGNAN-RICH DIETS

Several investigations have demonstrated a potential protective effect of lignan-rich diets against the development of various diseases, particularly hormone-dependent cancer and cardiovascular diseases (Lemay et al. 2002; Webb and McCullough 2005; Lampe et al. 2006; Adlercreutz 2007; Bergman-Jungestrom et al. 2007; Saarinen et al. 2007; Bloedon et al. 2008; Dodin et al. 2008; Cornish et al. 2009; Prasad 2009; Velentzis et al. 2009; Adolphe et al. 2010; Buck et al. 2010; Peterson et al. 2010; Ward et al. 2010; Buck et al. 2011; Peñalvo and Lopez-Romero 2012; Guglielmini et al. 2012; Zamora-Ros et al. 2012; Durazzo et al. 2014b).

11.7 METHODS FOR IDENTIFICATION OF LIGNANS IN FOOD

Different methods have been reported for the extraction and identification of lignans (Obermeyer et al. 1995; Mazur et al. 1996; Nilsson et al. 1997; Meagher et al. 1999; Johnsson et al. 2000; Muir and Westcott 2000; Kraushofer and Sontag 2002). The extraction method from the food matrix represents a key issue and in particular the type of hydrolysis step. Lignans have been extracted using organic solvents, alkaline, acid hydrolysis, enzymatic hydrolysis, or a mixture of them.

Acid hydrolysis may also cause changes in the molecular structure, leading to interconversions between lignans and so forming artifacts, whereas alkaline hydrolysis may not be efficient enough in some cases, for example in strong matrices such as fiber-rich foods, causing an underestimation of matairesinol (Smeds et al. 2007; Peterson et al. 2010).

An interesting method is that developed by Penãlvo et al. (2005a): an alkaline hydrolysis was used as the step prior an enzymatic hydrolysis; under alkaline conditions, ester-linked oligomers of lignan were hydrolyzed to give the lignan monomer and then aglycone forms were obtained by enzymatic hydrolysis using *Helix pomatia* β-glucuronidase/sulfatase.

Lignans were detected using LC-MS/MS and quantified against individual stable $^{13}C_3$-labeled lignans. This method also incorporates isotope dilution to ensure accuracy and precision. Analytical values using HPLC as well as either gas or liquid chromatography–mass spectrometry were developed and carried out. In the last years, different authors have summarized different methodology approaches of lignan quantification (Schwartz and Sontag 2011).

11.8 LIGNAN DATABASES

A complete and comprehensive database on the content of lignans in foods is useful for assessing their dietary intake and diet formulations, and to be used in observational studies as key elements for healthy nutritional patterns (Blitz et al. 2007).

Knowledge of the dietary intake of lignans is needed for understanding the relationship between a lignan-rich diet and potential lower risk of development of various diseases, that is, hormone-related cancers, heart diseases, menopausal symptoms, and osteoporosis. Accurate information on dietary lignan in foods is crucial to determine exposure and to investigate health effects in vivo.

The structural diversity of the compounds, the large number of dietary sources, the great variability in content for a specified source as well as the use of different extraction techniques and analytical methods, and, in some instances, the lack of suitable analytical methods besides several studies focused only on a few compounds, represent a limitation for this purpose (Scalbert et al. 2011). More analytical data on lignans are now available and useful for updating and expanding food composition databases (Milder et al. 2005a; Peñalvo et al. 2005a, 2008; Thompson et al. 2006; Smeds et al. 2007, 2009, 2012; Kuhnle et al. 2008a, 2008b, 2012; Moreno-Franco et al. 2011; Durazzo et al. 2013a, 2013b, 2013c, 2014a).

Milder et al. (2005a) constructed a database with lignan contents (secoisolariciresinol, matairesinol, lariciresinol, pinoresinol) of 83 solid foods and 26 beverages commonly consumed in the Netherlands: the lignan contents of grain products, vegetables, fruits, and legumes varied mostly between 50 and 200 μg/100 grams of fresh edible weight and in beverages ranged from 0 for cola to 91 μg/100 milliliters for red wine. By using the previously mentioned database, Milder et al. (2005b) estimated lignan intake in a representative sample of 4660 Dutch adults (Dutch Food Consumption Survey, carried out in 1997–1998): lariciresinol and pinoresinol contributed 75 percent to lignan intake, whereas secoisolariciresinol and matairesinol contributed 25 percent; the major food sources of lignans were beverages (37 percent), followed by vegetables (24 percent), nuts and seeds (14 percent), bread (9 percent), and fruits (7 percent) (Milder et al. 2005b).

Thompson et al. (2006) developed a lignan database of foods consumed in Canada: decreasing amounts (on weight basis per 100 grams) of total lignans (secoisolariciresinol, matairesinol, lariciresinol, pinoresinol) are found in the following order: nuts and oilseeds, cereals and breads, legumes, fruits, vegetables, soy products, meat products, other processed foods, alcoholic, and nonalcoholic beverages. Matairesinol was the least-concentrated lignan in most studied foods whereas secoisolariciresinol reached the highest concentration in 63 foods, lariciresinol in 44 foods, and pinoresinol in 14 foods (Thompson et al. 2006).

Peñalvo et al. (2008) have described the content of six plant lignans (secoisolariciresinol, matereisinol, lariciresinol, pinoresinol, syringaresinol, medioresinol) in 86 food items commonly consumed in Japan: the amounts of plant lignans in the items selected for the study varied from 0 to 1724 μg/100 grams (wet basis).

Moreno-Franco et al. (2011) developed the Aligna databases, a comprehensive compilation of values for alkylresorcinols and lignans (secoisolariciresinol, matereisinol, lariciresinol, pinoresinol, syringaresinol, medioresinol) publically accessible that reported the lignan content of 593 commonly foods and beverages, with particular attention to foods particularly consumed in Spain and by estimating the intake of lignans in Spain, 0.76 milligrams/day; in addition, the major contributors to lignan intake in Spain were oils and fats (33 percent), fruits and vegetables (30 percent), bread (14 percent), and wine and beer (10 percent) (Moreno-Franco et al. 2011).

It is important to mention the development of Phenol-Explorer, the first comprehensive database on polyphenol content in foods, including lignans (Neveu et al. 2010): the data derived from the continuous systematic collection of original content values in scientific publications. The availability of new and appropriate lignan food composition data will also facilitate further nutrition-related studies and will likely encourage the consumption of foods rich in key nutrients.

11.9 EFFECT OF PROCESSING

Only small amounts of foods are consumed raw, while most of them are consumed after cooking or processing depending on the type of food matrices and the eating habits of the consumers.

Several researches have documented how variations in lignan content are correlated to way of cooking, as well as to phytochemical structure and concentration, food matrix, and so on (Brenes et al. 2002; Milder et al. 2005a; Peñalvo et al. 2008; Cerretani et al. 2009; Durazzo et al. 2013b; Gerstenmeyer et al. 2013). For example, in vegetables, Milder et al. (2005a) reported a 25 percent decrease of lignan content (on dry weight) after boiling and frying some vegetables. Peñalvo et al. (2008) studied the effect of boiling on some vegetables commonly consumed in Japan: in most cases, losses of lignans probably occur due to degradation of chemical structure or diffusion of lignans in water; in some cases, an increase of lignans occur attributable to release from food matrix (Peñalvo et al. 2008).

Considering an oil matrix, Brenes et al. (2002) described a high stability of pinoresinol in olive oil after thermal treatments below 180°C; Cerretani et al. (2009), by studying the effect of microwave heating of extra virgin olive oil, olive oil, and pomace olive oil, showed how pinoresinol and 1-acetoxy pinoresinol had a slight decrease after 15 minutes of microwave treatment.

By studying sesame oil, Wu (2007) found that the content of lignans didn't change after heating at 180°C for four minutes and the level of sesamol increased after heating at 180°C for 20 minutes; moreover, heating at 200°C for 20 minutes led to significant loss of sesamolin and sesamol, whereas sesamin was relatively heat-stable (Wu 2007).

A total lignan content of 16, 7, and 40 μg/100 gram wet basis, respectively, was found for cooked macaroni, cooked white rice, and cooked whole rice as shown in a study by Milder et al. (2005a).

Hirawan and Beta (2011) reported that whole grain wheat spaghetti exhibited a higher content of secoisolariciresinol digloside (SDG) than regular spaghetti due to the higher proportion of bran in whole grain wheat.

Durazzo et al. (2013b) studied the effect of cooking on lignans content in mixed-cereal pastas made by adding 60 percent of whole-grain flours of different cereals (wheat, oat, rye, barley, and rice) to durum wheat semolina: in raw mixed-cereal pastas, the total lignan content varied in a range between 94.91 and 485.62 μg/100 grams dry weight and after cooking total lignan losses of about 35.5 percent, 18.31 percent, and 5.46 percent were reported in oat-, rye-, and rice-added pastas, while increases of 5.74 percent and 13.62 percent were observed in barley-added and whole durum wheat pastas.

Several authors reported that lignan content was stable in breadmaking (Muir and Westcott 2000; Hyvärinen et al. 2006a; Simbalista et al. 2012; Turfani et al. 2017). Meija et al. (2013) reported a lignan content for selected Latvian breads in the range from 85.4 μg/100 grams wet weight for white bread to 10044 μg/100 grams wet weight for wheat bread with seeds as well as for Finnish breads in the range from 88.3 µg/100 grams wet weight for wheat breads and 3838.3 µg/100 grams wet weight for mixed flour bread with seeds.

Other processing techniques, that is, milling, heat treatment, and parboiling, have been used to make a desirable product for consumers and to improve product quality. For seeds, Murkovic et al. (2004) reported that secoisolariciresinol was destroyed after 20 minutes during roasting process. Gerstenmeyer et al. (2013), by studying the effect of thermal heating on some lignans in flax seeds, sesame seeds, and rye, concluded how the water content and the applied temperatures should be considered and optimized for preserving lignan content during processing: a moderate heating at 100°C did not degrade the lignan aglycones and glycosides in dry foods, whereas if samples with high moisture content were heated, the degradation of the lignans in sesame seeds and rye was observed already at 100°C.

Concerning cereal grains, Durazzo et al. (2009), by studying the effect of micronization and classification (turboseparation) on the lignan content of two Italian cultivars of barley, found different levels of total lignan enrichment. The coarse fractions, obtained both from the first and second air classification, were the richest ones. In addition, the major contribution to the enrichment was due to lariciresinol in both cultivars (Durazzo et al. 2009).

Durazzo et al. (2009), by studying wheat milling products, found a 4.87-, 3.88-, and 5.18-fold enrichment from the grains to the bran fractions calculated for the soft wheat cv. Bologna and for the durum wheat cv. Simeto and Creso; a selective enrichment of secoisolariciresinol, lariciresonol, and pinoresinol from grain to bran was also underlined (Durazzo et al. 2009). These results suggest how processing factors influence, in a different manner, structurally different compounds belonging to the same chemical group.

A distribution of lignans (secoisolariciresinol, matairesinol, hydroxymatairesinol, lariciresinol, pinoresinol, syringaresinol) in wholegrain rye and its fractions prepared by sieved and air classifications have been studied by Pihlava et al. (2015). As for many other phenolic compounds, lignans were mainly found in bran.

Interesting is the study of Hyvärinen et al. (2006b) on the effect of processing and storage on the stability of purified, flaxseed-derived secoisolariciresinol diglucoside (SDG) added to dairy products: for example, in edam cheese manufacture, most of the added SDG was retained in the whey fraction and 6 percent was found in the cheese curd. SDG was also relatively stable in edam cheese during ripening of six weeks at 9°C.

11.10 CONCLUSIONS

Further studies on the evaluation of the effects of food processing on lignan content in different food matrices might increase the reliability of lignan intake estimations. It is particularly important to use analytical techniques that facilitate the analysis in the food matrix to study their natural release during digestion in vivo and the possible interactions with intestinal microbiota.

REFERENCES

Adlercreutz, H. 2002. Phytoestrogen and cancer. *The Lancet Oncology* 3:364–373.
Adlercreutz, H. 2007. Lignans and human health. *Critical Reviews in Clinical Laboratory Sciences* 44:483–525.

Adlercreutz, H., Vanderwildt, J., Kinzel, J. et al. 1995. Lignan and isoflavonoid conjugates in human urine. *The Journal of Steroid Biochemistry and Molecular Biology* 52:97–103.

Adolphe, J.L., Whiting, S.J., Juurlink, B.H., Thorpe, L.U., and Alcorn, J. 2010. Health effects with consumption of the flax lignan secoisolariciresinol diglucoside. *British Journal Nutrition* 103:929–938.

Aehle, E., Müller, U., Eklund, P.C., Willför, S.M., Sippl, W., and Dräger, B. 2011. Lignans as food constituents with estrogen and antiestrogen activity. *Phytochemistry* 72:2396–2405.

Arts, I.C.W., and Hollman, P.C.H. 2005. Polyphenols and disease risk in epidemiological studies. *American Journal Clinical and Nutrition* 81:317s–325s.

Axelson, M., Sjoevall, J., Gustafsson, B., and Setchell, K. 1982. Origin of lignans in mammals and identification of a precursor from plants. *Nature* 298:659–660.

Ayres, D.C., and Loike, J.D. 1990 Lignans' chemical, biological and clinical properties. In *Chemistry & Pharmacology of Natural Products*, ed. J.D. Phillipson, D.C. Ayres, and H. Baxter, p. 402. Cambridge: Cambridge University Press.

Bambagiotti-Alberti, M., Coran, S.A., Ghiara, C., Giannellini, V., Raffaelli, A. 1994. Revealing the mammalian lignan precursor secoisolariciresinol di glucoside in flax seed by ionspray mass spectrometry. *Rapid Communication Mass Spectrometry* 8:595–598.

Bartkiene, E., Juodeikiene, G., Basinskiene, L., Liukkonen, K.H., Adlercreutz, H., and Kluge, H. 2011. Enterolignans enterolactone and enterodiol formation from their precursors by the action of intestinal microflora and their relationship with non-starch polysaccharides in various berries and vegetables. *LWT - Food Science and Technology* 44:48–53.

Bergman-Jungestrom, M., Thompson, L.U., and Dabrosin, C. 2007. Flaxseed and its lignans inhibit estradiol-induced growth, angiogenesis, and secretion of vascular endothelial growth factor in human breast cancer xenografts *in vivo*. *Clinical Cancer Research* 13:1061–1067.

Blitz, C.L., Murphy, S.P., and Au, D.L.M. 2007. Adding lignan values to a food composition database. *Journal of Food Composition and Analysis* 20:99–105.

Bloedon, L.T., Balikai, S., Chittams, J. et al. 2008. Flaxseed and cardiovascular risk factors: Results from a double blind, randomized, controlled clinical trial. *The Journal of the American College of Nutrition* 27:65–74.

Borriello, S.P., Setchell, K.D.R., Axelson, M., and Lawson, A.M.J. 1985. Production and metabolism of lignans by the human faecal flora. *The Journal of Applied Bacteriology* 58:37–43.

Brenes, M., Garcia, A., Dobarganes, M.C., Velasco, J., and Romero, C. 2002. Influence of thermal treatments simulating coking processes on the polyphenol content of virgin olive oil. *Journal of Agriculture and Food Chemistry* 50:5962–5967.

Buck, K., Vrieling, A., Zaineddin, A.K. et al. 2011. Serum enterolactone and prognosis of postmenopausal breast cancer. *Journal of Clinical Oncology* 29:3730–3738.

Buck, K., Zaineddin, A.K., Vrieling, A., Linseisen, J., and Chang-Claude, J. 2010. Meta-analyses of lignans and enterolignans in relation to breast cancer risk. *American Journal of Clinical and Nutrition* 92:141–153.

Cerretani, L., Bendini, A., Rodriguez-Estrada, M.T., Vittadini, E., and Chiavaro, E. 2009. Microwave heating of different commercial categories of olive oil: Part I. Effect on chemical oxidative stability indices and phenolic compounds. *Food Chemistry* 2:1381–1388.

Cornish, S.M., Chilibeck, P.D., and Paus-Jennsen, L. 2009. A randomized controlled trial of the effects of flaxseed lignan complex on metabolic syndrome composite score and bone mineral in older adults. *Applied Physiology, Nutrition and Metabolism* 34:89–98.

Cui, H., Xu, B., Wu, T., Xu, J., Yuan, Y., and Gu, Q. 2014 Potential antiviral lignans from the roots of *Saururus chinensis* with activity against Epstein–Barr virus lytic replication. *Journal of Natural Products* 77:100–110.

Dodin, S., Cunnane, S.C., Masse, B. et al. 2008. Flaxseed on cardiovascular disease markers in healthy menopausal women: A randomized, double-blind, placebo-controlled trial. *Nutrition* 24:23–30.

Durazzo, A., Azzini, E., Raguzzini, A. et al. 2009. Influence of processing on the lignans content of cereal based foods. *Tecnica Molitoria International* 60:163–173. ISSN: 0040-1862.

Durazzo, A., Turfani, V., Azzini, E., Maiani, G., and Carcea, M. 2013a. Phenols, lignans and antioxidant properties of legume and sweet chestnut flours. *Food Chemistry* 140:666–671.

Durazzo, A., Azzini, E., Turfani, V., Polito, A., Maiani, G., and Carcea, M. 2013b. Effect of cooking on lignans content in wholegrain pasta made with different cereals and other seeds. *Cereal Chemistry* 90:169–171.

Durazzo, A., Zaccaria, M., Polito, A., Maiani, G., and Carcea, M. 2013c. Lignan content in cereals, buckwheat and derived foods. *Foods* 2:53–63.

Durazzo, A., Turfani, V., Narducci, V., Azzini, E., Maiani, G., and Carcea, M. 2014a. Nutritional characterization and bioactive components of commercial carobs flours. *Food Chemistry* 153:109–113.

Durazzo, A., Carcea, M., Adlercreutz, H. et al. 2014b. Effects of consumption of whole grain foods rich in lignans in healthy postmenopausal women with moderate serum cholesterol: A pilot study. *International Journal of Food Science and Nutrition* 65:637–645.

Eklund, P.C., Långvik, O.K., Wärnå, J.P., Salmi, T.O., Willför, S.M., and Sjöholm, R.E. 2005. Chemical studies on antioxidant mechanisms and free radical scavenging properties of lignans. *Organic & Biomolecular Chemistry* 3:3336–3347.

Gerstenmeyer, E., Reimer, S., Berghofer, E., Schwartz, H., and Sontag, G. 2013. Effect of thermal heating on some lignans in flax seed, sesame seeds and rye. *Food Chemistry* 138:1847–1855.

Glitso, L.V., Mazur, W.M., Adlercreutz, H. et al. 2000. Intestinal metabolism of rye lignans in pigs. *British Journal of Nutrition* 84:429–437.

Gnabre, J., Bates, R., and Huang, R.C. 2015. Creosote bush lignans for human disease treatment and prevention: Perspectives on combination therapy. *Journal of Traditional and Complementary Medicine* 5:119–126.

Guglielmini, P., Rubagotti, A., and Boccardo, F. 2012. Serum enterolactone levels and mortality outcome in women with early breast cancer: A retrospective cohort study. *Breast Cancer Research and Treatment* 132:661–668.

Harmatha, J., and Dinan, L. 2003. Biological activities of lignans and stilbenoids associated with plant-insect chemical interactions. *Phytochem Reviews* 2:321–330.

Heinonen, S., Nurmi, T., Liukkonen, K. et al. 2001. *In vitro* metabolism of plant lignans: New precursors of mammalian lignans enterolactone and enterodiol. *Journal of Agricultural & Food Chemistry* 49:3178–3186.

Hemmati, S., Heimendahl, C.B., Klaes, M., Alfermann, A.W., Schmidt, T.J., and Fuss, E. 2010. Pinoresinol-lariciresinol reductases with opposite enantiospecificity determine the enantiomeric composition of lignans in the different organs of Linum usitatissimum L. *Planta Medica* 76:928–934.

Hirawan, R., and Beta, T., 2011. C-Glysocylflavone and lignan diglucoside contents of commercial, regular and whole-wheat spaghetti. *Cereal Chemistry* 88:338–343.

Holmbom, B., Eckerman, C., Eklund, P. et al. 2003. Knots in trees—A new rich source of lignans. *Phytochemistry Reviews* 2:331–340.

Hu, C., Yuan, Y.V., and Kitts, D.D. 2007. Antioxidant activities of the flaxseed lignan secoisolariciresinol diglucoside, its aglycone secoisolariciresinol and the mammalian lignans enterodiol and enterolactone *in vitro*. *Food and Chemical Toxicology* 45:2219–2227.

Huang, X.Y., Feng, Z.M., Yang, Y.N., Jiang, J.S., and Zhang, P.C. 2015. Four new neolignan glucosides from the fruits of Arctium lappa. *Journal of Asian Natural Products Research* 17:504–511.

Hyvärinen, H.K., Pihlava, J.M., Hiidenhovi, J.A., Hietaniemi, V., Korhonen, H.J.T., and Ryhänen, E.L. 2006a. Effect of processing and storage on the stability of flaxseed lignan added to bakery products. *Journal of Agricultural and Food Chemistry* 54:48–53.

Hyvärinen, H.K., Pihlava, J.M., Hiidenhovi, J.A., Hietaniemi, V., Korhonen, H.J.T., and Ryhänen, E.L. 2006b. Effect of processing and storage on the stability of flaxseed lignan added to dairy products. *Journal of Agricultural and Food Chemistry* 54:8788–8792.

Imai, T., Nomura, M., and Fukushima, K. 2006. Evidence for involvement of the phenylpropanoid pathway in the biosynthesis of the norlignan agatharesinol. *Journal of Plant Physiology* 163:483–487.

Jin, J.S., and Hattori, M. 2011. A new mammalian lignan precursor, asarinin. *Food Chemistry* 124:895–899.

Johnsson, P., Kamal Eldin, A., Lundgren, L.N., and Aman, P. 2000. HPLC method for analysis of secoisolariciresinol diglucoside in flaxseed. *Journal of Agricultural and Food Chemistry* 48:5216–5219.

Kamal-Eldin, A., Peerlkamp, N., Johnsson, P. et al. 2001. An oligomer from flaxseed composed of secoisolariciresinoldiglucoside and 3-hydroxy-3-methyl glutaric acid residues. *Phytochemistry* 58:587–90.

Kangas, L., Saarinen, N., Mutanen, M. et al. 2002. Antioxidant and antitumor effect of hydroxymatairesinol (HM-3000, HMR), a lignan isolated from the knots of spruce. *European Journal of Cancer Prevention* 11:S48–S57.

Kitts, D.D., Yuan, Y.V., Wijewickreme, A.N., and Thompson, L.U. 1999. Antioxidant activity of the flaxseed lignan secoisolariciresinol diglycoside and its mammalian lignan metabolites enterodiol and enterolactone. *Molecular and Cellular Biochemistry* 202:91–100.

Kong, P., Zhang, L., Guo, Y., Lu, Y., and Lin, D. 2014. Phillyrin, a natural lignan, attenuates tumor necrosis factor α-mediated insulin resistance and lipolytic acceleration in 3T3-L1 adipocytes. *Planta Medica* 80:880–886.

Kraushofer, T., and Sontag, G. 2002. Determination of matairesinol in flaxseed by HPLC with coulometric electrode array detection. *Journal of Chromatography B* 777:61–66.

Kuhnle, G.G.C., Dell'Aquila, C., Aspinall, S.M., Runswick, S.A., Mulligan, A.A., and Bingham, S.A. 2008a. Phytoestrogen content of beverages, nuts, seeds, and oils. *Journal of Agricultural and Food Chemistry* 56: 7311–7315.

Kuhnle, G.G.C., Dell'Aquila, C., Aspinall, S.M., Runswick, S.A., Mullingan, A.A., and Bingham, S.A. 2008b. Phytoestrogen content of foods of animal origin: Dairy products, eggs, meat, fish, and seafood. *Journal of Agricultural and Food Chemistry* 56:10099–10104.

Kuhnle, G.G.C., Dell'Aquila, C., Aspinall, S.M., Runswick, S.A., Mulligan, A.A., and Bingham, S.A. 2012. Phytoestrogen content of cereals and cereal based foods consumed in the UK. *Nutrition and Cancer* 61:302–309.

Kuijsten, A. 2007. Bioavailability of enterolignans and their relation with chronic diseases. PhD thesis. Wageningen: Wageningen University.

Lampe, J.W., Atkinson, C., and Hullar, M.A. 2006. Assessing exposure to lignans and their metabolites in humans. *Journal of AOAC International* 89:1174–1181.

Landete, J.M., J Arqués, J., Medina, M., Gaya, P., De La Rivas, B., and Muñoz, R. 2015. Bioactivation of phytoestrogens: Intestinal bacteria and health. *Critical Reviews in Food Science and Nutrition* 56:1826–1843.

Lemay, A., Dodin, S., Kadri, N., Jacques, H., and Forest, J.C. 2002. Flaxseed dietary supplement versus hormone replacement therapy in hypercholesterolemic menopausal women. *Obstetrics Gynecology* 100:495–504.

Lewis, N.G., and Davin, L.B. 1999. Lignans: Biosynthesis and function. In *Comprehensive natural products chemistry. Polyketides and other secondary metabolites including fatty acids and their derivatives*, ed. D. Barton, K. Nakanishi, O. Meth-Cohn, and U. Sankawa, pp. 639–712. Amsterdam: Elsevier.

Mazur, W. 1998. Phytoestrogen content in foods. *Baillière's Clinical Endocrinology Metabolism* 12:729–742.

Mazur, W.M., and Adlercreutz, H. 1998. Natural and anthropogenic environmental estrogens: The scientific basis for risk assessment; naturally occurring estrogens in food. *Pure and Applied Chemistry* 70:1759–1776.

Mazur, W., Fotsis, T., Wahala, K., Ojala, S., Salakka, A., and Adlercreutz, H. 1996. Isotope dilution gas chromatographic-mass spectrometric method for the determination of isoflavonoids, coumestrol, and lignans in food samples. *Analytical Biochemistry* 233:169–180.

Mazur, W., Uehara, M., Wahala, K., and Adlercreutz, H. 2000. Phyto-oestrogen content of berries, and plasma concentrations and urinary excretion of eterolactone after a single strawberry-meal in human subjects. *British Journal Nutrition* 83:381–387.

Meagher, L.P., Beecher, G.R., Flanagan, V.P., and Li, B.W. 1999. Isolation and characterization of the lignan, isolariciresinol and pinoresinol in flaxseed meal. *Journal of Agricultural and Food Chemistry* 47:3173–3180.

Meija, L., Söderholm, P.P., Krams, I., and Adlercreutz, H. 2013. Dietary intake and major sources of plant lignans in Latvian men and women. *International Journal of Food Science and Nutrition* 64:622–630.

Milder, I.E., Arts, I.C., van de Putte, B., Venema, D.P., and Hollman, P.C. 2005a. Lignan contents of Dutch plant foods: A database including lariciresinol, pinoresinol, secoisolariciresinol and matairesinol. *British Journal of Nutrition* 93:393–402.

Milder, I.E.J., Feskens, E.J.M., Arts, I.C.W., Bueno-de-Mesquita, H.B., Hollman, P.C.H., and Kromhout, D. 2005b. Intake of the plant lignans secoisolariciresinol, matairesinol, lariciresinol and pinoresinol in Dutch men and women. *Journal of Nutrition* 135:1202–1207.

Moreno-Franco, B., Garcia-Gonzalez, A., Montero-Bravo, A.M. et al. 2011. Dietary alkylresorcinols and lignans in the Spanish diet: Development of the Alignia database. *Journal of Agricultural and Food Chemistry* 59:9827–9834.

Muir, A.D. 2010. Flax lignans: New opportunities for functional foods. *Food Science and Technology Bulletin – Functional Foods* 6:61–79.

Muir, A.D., and Westcott, N.D. 2000. Quantitation of the lignan secoisolariciresinol diglucoside in baked goods containing flax seed or flax meal. *Journal of Agricultural and Food Agriculture* 48:4048–4052.

Muir, A.D., and Westcott, N.D. 2003. Flaxseed constituents and human health. In *Flax: The Genus Linum*, ed. A.D. Muir, and N.D. Westcott, pp. 243–251. London, UK: Taylor & Francis.

Murkovic, M.P., Piironen, V., Lampi, A.M., Kraushofer, T., and Sontag, G. 2004. Change in chemical composition of pumpkin seeds during the roasting process for production of pumpkin seed oil (Part 1: Nonvolatile compounds). *Food Chemistry* 84:359–365.

Neveu, V., Perez-Jiménez, J., Vos, F. et al. 2010. Phenol-Explorer: An online comprehensive database on polyphenol contents in foods. Accessed November 20, 2015 from http://www.phenol-explorer.eu.

Niemeyer, H.B., and Metzler, M. 2001. Differences in the antioxidant activity of plant and mammalian lignans. *Journal of Food Engineering* 56:255–256.

Nilsson, M., Åman, P., Härkönen, H. et al. 1997. Content of nutrients and lignans in roller milled fractions of rye. *Journal of the Science of Food and Agriculture* 73:143–148.

Obermeyer, W.R., Musser, S.M., Betz, J.M., Casey, R.E., Pohland, A.E., and Page, S.W. 1995. Chemical studies of phytoestrogens and related compounds in dietary supplements: Flax and chaparral. *Proceedings of the Society for Experimental Biology and Medicine* 208:6–12.

Pan, J.Y., Chen, S.L., Yang, M.H., Wu, J., Sinkkonen, J., and Zou, K. 2009. An update on lignans: Natural products and synthesis. *Natural Product Reports* 26:1251–1292.

Peñalvo, J.L., Nurmi, T., Haajanen, K., Al-Maharik, N., Botting, N., and Adlercreutz, H. 2004. Determination of lignans in human plasma by liquid chromatography with coulometric electrode array detection. *Analytical Biochemistry* 332:384–393.

Peñalvo, J.L., Haajanen, K.M., Botting, N., and Adlercreutz, H. 2005a. Quantification of lignans in food using isotope dilution gas chromatography/mass spectrometry. *Journal of Agricultural and Food Chemistry* 53:9342–9347.

Peñalvo, J.L., Heinonen, S.M., Aura, A.M., and Adlercreutz, H. 2005b. Dietary sesamin is converted to enterolactone in humans. *Journal Nutrition* 135:1056–1062.

Peñalvo, J.L., Adlercreutz, H., Uehara, M., Ristimaki, A., and Watanabe, S. 2008. Lignan content of selected foods from Japan. *Journal of Agricultural and Food Chemistry* 56:401–409.

Peñalvo, J.L., and Lopez-Romero, P. 2012. Urinary enterolignan concentrations are positively associated with serum HDL cholesterol and negatively associated with serum triglycerides in U.S. adults. *Journal of Nutrition* 142:751–756.

Peterson, J., Dwyer, J., Adlercreutz, H., Scalbert, A., Jacques, P., and McCullough, M.L. 2010. Dietary lignans: Physiology and potential for cardiovascular disease risk reduction. *Nutrition Reviews* 68:571–603.

Pihlava, J.M., Nordlund, E., Heinio, R.L., Hietaniemi, V., Lehtinen, P., and Poutanen, K. 2015. Phenolic compounds in wholegrain rye and its fractions. *Journal of Food Composition and Analysis* 38:89–97.

Prasad, K. 2000. Antioxidant activity of secoisolariciresinol diglucoside-derived metabolites, secoisolariciresinol, enterodiol, and enterolactone. *The International Journal of Angiology* 9:220–225.

Prasad, K. 2009. Flaxseed and cardiovascular health. *Journal of Cardiovascular Pharmacology* 54:369–377.

Rowland, I., Wiseman, H., Sanders, T., Adlercreutz, H., and Bowey, E. 2000. Interindividual variation in metabolism of isoflavonoids and lignans: The role of the gut microflora and habitual diet. *Nutrition and Cancer* 36:27–32.

Saarinen, N.M., Warri, A., Makela, S.I. et al. 2000. Hydroxymatairesinol, a novel enterolac-tone precursor with antitumor properties from coniferous tree (*Picea abies*). *Nutrition and Cancer* 36:207–216.

Saarinen, N.M., Warri, A., Airio, M., Smeds, A., and Makela, S. 2007. Role of dietary lignans in the reduction of breast cancer risk. *Molecular Nutrition Food Research* 51:857–866.

Saarinen, N.M., Tuominen, J., Pylkkänen, L., and Santti, R. 2010. Assessment of information to substantiate a health claim on the prevention of prostate cancer by lignans. *Nutrients* 2:99–115.

Saleem, M., Kim, H.J., Ali, M.S., and Lee, Y.S. 2005. An update on bioactive plant lignans. *Natural Product Reports* 22:696–716.

Scalbert, A., Andres-Lacueva, C., Arita, M. et al. 2011. Databases on food phytochemicals and their health-promoting effects. *Journal of Agricultural and Food Chemistry* 59:4331–4348.

Schwartz, H. and Sontag, G. 2011. Analysis of lignans in food samples-impact of sample preparation. *Current Bioactive Compounds* 7:156–171.

Setchell, K.D., Lawson, A.M., McLaughlin, L.M., Patel, S., Kirk, D.N., and Axelson, M. 1983. Measurement of enterolactone and enterodiol, the first mammalian lignans, using stable isotope di-lution and gas chromatography mass spectrometry. *Biomedical Mass Spectrometry* 10:227–235.

Setchell, K.D.R., and Adlercreutz, H. 1998. Mammalian lignans and phytoestrogens: Recent studies on their Formation, metabolism and biological role in health and disease. In *Role of Gut Flora in Toxicity and Cancer*, ed. I.R. Rowland, pp. 325–345. San Diego, CA: Academic Press.

Simbalista, R.L., Frota, K., Soares, R.A.M., and Arêas, J.A.G. 2012. Effect of storage and processing of Brazilian flaxseed on lipid and lignan contents. *Ciência e Tecnologia de Alimentos* 32:374–380.

Smeds, A.I., Eklund, P.C., Sjöholm, R.E. et al. 2007. Quantification of a broad spectrum of lignans in cereals, oilseeds, and nuts. *Journal of Agricultural and Food Chemistry* 55:1337–1346.

Smeds, A.I., Jauhiainen, L., Tuomola, E., and Peltonen-Sainio, P. 2009. Characterization of variation in the lignan content and composition of winter rye, spring wheat and spring oat. *Journal of Agricultural and Food Chemistry* 57:5837–5842.

Smeds, A.I., Eklund, P.C., and Willför, S.M. 2012. Content, composition, and stereochemical characterisation of lignans in berries and seeds. *Food Chemistry* 134:1991–1998.

Stitch, S.R., Toumba, J.K., Groen, M.B. et al. 1980. Excretion, isolation and structure of a new phenolic constituent of female urine. *Nature* 287:738–740.

Su, S., and Wink, M. 2015. Natural lignans from Arctium lappa as antiaging agents in *Caenorhabditis elegans*. *Phytochemistry* 117:340–350.

Thompson, L.U., Richard, S.E., Cheung, F., Kenaschuk, E.O., and Obermeyer W.R. 1997. Variability in anticancer lignan levels in flaxseed. *Nutrition and Cancer* 27:26–30.

Thompson, L.U., Boucher, B.A., Liu, Z., Cotterchio, M., and Kreiger, N. 2006. Phytoestrogen content of foods consumed in Canada, including isoflavones, lignans and coumestan. *Nutrition and Cancer* 54:184–201.

Touré, A., and Xueming, X. 2010. Flaxseed lignans: Source, biosynthesis, metabolism, antioxidant activity, bio-active components, and health benefits. *Comprehensive Reviews in Food Science and Food Safety* 9:261–269.

Turfani, V., Narducci, V., Durazzo, A., Galli, V., and Carcea, M. 2017. Technological, nutritional and functional properties of wheat bread enriched with lentil or carob flours. *LWT - Food Science and Technology* 78:361–366.

Umezezawa, T. 2003. Phylogenetic distribution of lignin producing plants. *Wood Reseach* 90:27–110.

Velentzis, L.S., Cantwell, M.M., Cardwell, C., Keshtgar, M.R., Leathem, A.J., and Woodside, J.V. 2009. Lignans and breast cancer risk in pre- and post-menopausal women: Meta-analyses of observational srudies. *British Journal of Cancer* 100:1492–1498.

Wang, L.Q., Meselhy, M.R., Li, Y., Qin, G.W., and Hattori, M. 2000. Human intestinal bacteria capable of transforming secoisolariciresinol digluco- side to mammalian lignans, enterodiol and enterolactone. *Chemical and Pharmaceutical Bulletin* 48:1606–1610.

Wang, O., Cheng, Q., Liu, J. et al. 2014. Hepatoprotective effect of *Schisandra chinensis* (Turcz.) Baill. lignans and its formula with *Rubus idaeus* on chronic alcohol-induced liver injury in mice. *Food & Function* 5:3018–3025.

Ward, H.A., Kuhnle, G.G., Mulligan, A.A. et al. 2010. Breast, colorectal, and prostate cancer risk in the European prospective investigation into cancer and nutrition-norfolk in relation to phytoestrogen intake derived from an improved database. *American Journal of Clinical Nutrition* 91:440–448.

Webb, A.L., and McCullough, M.L. 2005. Dietary lignans: Potential role in cancer prevention. *Nutrition and Cancer* 51:117–131.

Willför, S.M., Ahotupa, M.O., Hemming, J.E. et al. 2003. Antioxidant activity of knotwood extractives and phenolic compounds of selected tree species. *Journal of Agricultural and Food Chemistry* 51:7600–7606.

Woo, K.W., Choi, S.U., Park, J.C., and Lee, K.R. 2011. A new lignan glycoside from *Juniperus rigida*. *Archives of Pharmacal Research* 34:2043-2049.

Wu, W.H. 2007. The contents of lignans in commercial sesame oils of Taiwan and their changes during heating. *Food Chemistry* 104:341–344.

Xie, L.H., Ahn, E.M., Akao, T., Abdel-Hafez, A.A., Nakamura, N., and Hattori, M. 2003b. Transformation of arctiin to estrogenic and antiestrogenic substances by human intestinal bacteria. *Chemical and Pharmaceutical Bulletin* 51:378–384.

Xie, L.H., Akao, T., Hamasaki, K., Deyama, T., and Hattori, M. 2003a. Biotransformation of pinoresinol diglucoside to mammalian lignans by human intestinal microflora, and isolation of Enterococcus faecalis strain PDG-1 responsible for the transformation of (+)-pinoresinol to (+)-lariciresinol. *Chemical and Pharmaceutical Bulletin* 51:508–515.

Yu, H.Y., Chen, Z.Y., Sun, B. et al. 2014. Lignans from the fruit of *Schisandra glaucescens* with antioxidant and neuroprotective properties. *Journal of Natural Products* 77:1311–1320.

Zamora-Ros, R., Agudo, A., Lujan-Barroso, L. et al. 2012. Dietary flavonoid and lignan intake and gastric adenocarcinoma risk in the European Prospective Investigation into Cancer and Nutrition (EPIC) study. *American Journal of Clinical Nutrition* 96:1398–1408.

Zhang, C., Mao, X., Zhao, X. et al. 2014. Gomisin N isolated from *Schisandra chinensis* augments pentobarbital-induced sleep behaviors through the modification of the serotonergic and GABAergic system. *Fitoterapia* 96:123–130.

CHAPTER 12

Lignins in Food

Esther Pérez-Carrillo and Erick Heredia Olea

CONTENTS

12.1 INTRODUCTION

The chemical basis of hardening is related to modifications of cell wall composition and traditionally has been associated with lignification (Salunkhe and Desai 1984; Smith and Stanley 1987; Everson et al. 1992). The cell wall is a complex system composed of different polymers like cellulose, hemicellulose, and lignin. Cellulose is the most abundant polymer in cell wall (35–50 percent weight dry matter basis) and is composed of D-glucopyranose linked linearly by β-1-4 links (Agbogbo et al. 2006, Park et al. 2010). Hemicellulose is a heterogeneous polymer made of C5 and C6 sugars by xylose, arabinose, mannose, glucose, and glucuronic acid being the major components. Lignin is also a heterogeneous polymer as hemicellulose, but its fundamental blocks are phenolic alcohols. Lignin gives strength and structural support to cell wall, avoids the microbial degradation, and helps to the vascular system for water transport in plant cells (Ralph and Hatfield 1997, Mokochinski et al. 2015). The effect of lignin on cell wall properties is mediated by its hydrophobicity and its incrustation and attachment to structural polysaccharides and proteins (Grabber et al. 1996). Lignin and hemicellulose are covalently linked through acetyl groups between the hemicellulose arabinose branches and hydroxyl groups of the aromatic alcohols.

12.2 LIGNIN STRUCTURE

Lignin is a polymer with a three-dimensional network. A structural model was usually used to present the structure of lignin that only describes a hypothetical structure inferred from the average (Figure 12.1). It is an aromatic biopolymer that contains around 40 oxygenated phenylpropane units built up by oxidative coupling of three mayor C_6–C_3 (phenylpropanoid) units (hydroxyl-cinnamyl alcohols), namely sinapyl alcohol (M1S, syringyl alcohol), coniferyl alcohol (M1G, guayacyl alcohol), and *p*-coumaryl alcohol (M1H, *p*-hydroxyphenyl propanol), which forms a randomized structure in a tridimensional network inside cell walls. The major interunit linkage is an aryl–aryl ether type. Besides the 20 different types of bonds within the lignin, it seems to be particularly associated with the hemicellulosic polysaccharides (Xiao et al. 2001, Boerjan et al. 2003, Dhingra et al. 2011). They give rise to the *p*-hydroxyphenyl (H), guaiacyl (G), and syringyl (S) units of the lignin polymer, respectively (Mokochinski et al. 2015). The monolignols are C6–C3 phenylpropanoids and differ from each other only by their degree of methoxylation. In lignin, the C6–C3 units are interconnected by several types of ether and carbon–carbon linkages, but their relative number is not clearly established because different analytical methods give distinct results. The phenylpropane units are joined together both with C–O–C linkages (ether) approximately 70 percent or more and C–C linkages. The most common linkage is the β-O-4 (β-aryl ether) linkage (Boerjan et al. 2003).

12.2.1 Lignin Classification

Previously, lignin was divided into softwood, hardwood, and grass lignin. Although it only has three structures, the quantity proportions of these structures vary greatly in different families of plants, softwood, hardwood, and grass lignin. Lignin has no specific molecular weight, and due to the different monomers, it can be divided into three types, according to the structure elements (Chen 2014).

- Guaiacyl (G) lignin: It is formed through dehydrated oligomerization of coniferyl alcohol, and its structure is homogeneous. Usually, gymnosperm mainly contains G lignin, that is almost all softwoods.
- Guaiacyl–syringyl (GS) lignin: It is the result of the dehydrated oligomerization of coniferyl alcohol and sinapyl alcohol. The ratio varies from 4:1 to 1:2 for the two monomeric units. Hardwood and the dicotyledon mainly contain GS lignin.
- Guaiacyl–syringyl–hydroxy–phenyl (GSH) lignin: It is the *p*-hydroxyphenyl lignin, compression wood, that has a high proportion of phenylpropane units of the *p*-hydroxyphenyl type in addition to the normal guaiacyl units. Monocotyledons mainly contain this kind of lignin, for example, grass.

Hydroxyls and many polar groups exist in the lignin structure, resulting in strong intramolecular and intermolecular hydrogen bonds that make it insoluble in any solvent. Only a fraction of the lignin polymer is solubilized by most of the chemical degradation methods, and isolation lignin parts always cause chemical changes in the structure (Monties 1988). Condensation or degradation is divided into soluble and insoluble lignin (Chen 2014).

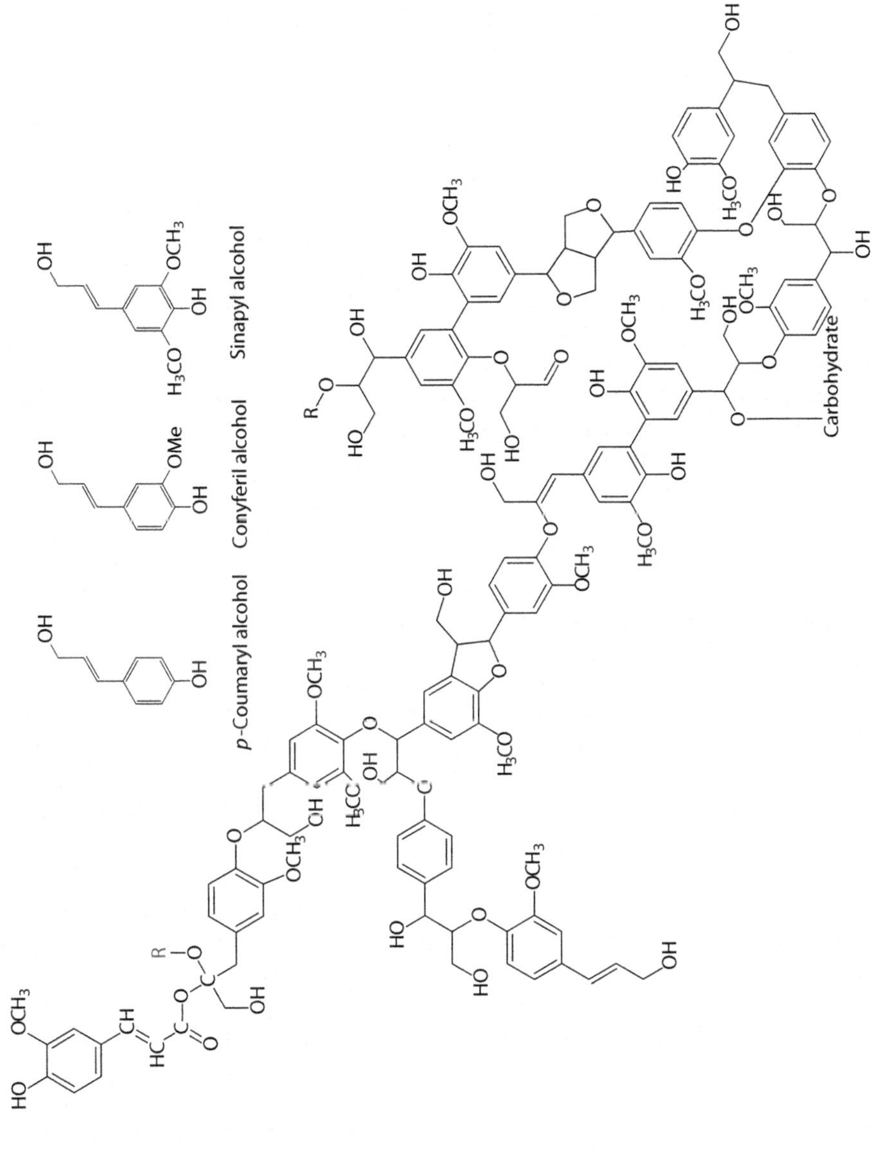

FIGURE 12.1 Lignin structure and its three fundamental monomers: *p*-coumaryl alcohol, coniferyl alcohol, and sinapyl alcohol.

The presence of phenolic hydroxyl and carboxyl make the lignin able to be dissolved in alkaline solution. Separated Brauns lignin (unhydrolyzable by acids, easily oxidizable, soluble in hot alkali and bisulfite, and condenses readily with phenols and thio compounds) and organosolv lignin can be dissolved in dioxane, dimethyl sulfoxide (DMSO), methanol, ethanol, acetone, methyl cellosolve, and pyridine. Alkali lignin and lignosulfonate usually can be dissolved in a diluted alkali, water, and salt solution. Brauns lignin, phenol lignin, and many organosolv lignins can be completely dissolved in dioxane. Acid lignin is not soluble in any solvents. The best solvents for lignin separation are acetyl bromide and hexafluoroisopropanol in acetic acid (Chen 2014).

12.3 LIGNIN SYNTHESIS

The main importance of the adaptation of plants from the aqueous media to the land was the ability to synthesize lignin (Boerjan et al. 2003). Many aspects of the chemistry of lignin still remain unclear, even though it has been studied for more than a century. The lignin biosynthesis starts with the deamination of the phenylalanine or tyrosine by phenylalanine ammonia-lyase or tyrosine ammonia-lyase and involves successive hydroxylation reactions of the aromatic ring, followed by phenolic *O*-methylation and conversion of the side-chain carboxyl to an alcohol, producing the first cinnamic acids (Sun 2010).

The content of lignin change with the developmental stage, variety, and external factors. Tricin was recently discovered in lignin preparations from wheat (*Triticum aestivum*) straw and subsequently studied in maize. To provide proof that tricin is involved in lignification and establish the mechanism by which it incorporates into the lignin polymer, the 4′-*O*-β-coupling products of tricin with the monolignols (*p*-coumaryl, coniferyl, and sinapyl alcohols) were synthesized along with the trimer that would result from its 4′-*O*-β-coupling with sinapyl alcohol and then coniferyl alcohol. Tricin was also found to cross couple with monolignols to form tricin-(4′-*O*-β)-linked dimers in biomimetic oxidations using peroxidase/hydrogen peroxide or silver (I) oxide. These findings suggest that tricin is fully compatible with lignification reactions, it is an authentic lignin monomer, and, because it can only start a lignin chain, works as a nucleation site for lignification in monocots. This initiation role helps resolve a long-standing dilemma that monocot lignin chains do not appear to be initiated by monolignol homodehydrodimerization as they are in dicots that have similar syringyl–guaiacyl compositions. The term *flavonolignin* is recommended for racemic oligomers and polymers of monolignols that start from tricin (or incorporate other flavonoids) in the cell wall, in analogy with the existing term *flavonolignan* that is used for the low-molecular mass compounds composed of flavonoid and lignan moieties (Lan et al. 2015; Lan et al. 2016 a,b). In case of the apple, ligin is gradually synthesized with the beginning of the cell enlargement period. Lignin content in the peel is positively correlated with the development time of the fruit and the activity of peroxidase. In contrast, the lignin content in the yellow apple peel is slightly more than that of the red apple. In addition, a low temperature in the field has a significant impact on the lignin content in the peel. The harvest time also affects the lignin content: the lignin content in the late-harvested Laiyang pears was the highest because many phenolics would be converted to lignin during the period of fruit development. Also, the conversion rate in the flesh is higher than that in the pit tissue. Bagging also affects the lignin content. Phenylalanine ammonia-lyase (PAL) is a photo-induced enzyme; its activity is positively correlated with light. However, after bagging the fruit, the activity of PAL decreases, and the lignin content in peel and pulp can both be reduced by 32 percent (Chen 2014).

12.4 LIGNIN CONTENT IN FOODS

Foods high in lignin include flaxseed, root vegetables (carrots, parsley, horseradish), wheat bran, edible seeds (in berries and tomatoes), vegetables with edible stems (cabbage, broccoli), green beans, peas, peaches, apples, and Brazil nuts (Table 12.1). The amount and type of lignin vary from plant to plant according to the genotype and the influence of biotic or abiotic stress factors. Lignification in fruit and vegetables occurs in response to different abiotic stresses, such as physical impact, wounding, and long-term low temperature storage.

Lignin is the cell wall component most frequently associated with hardening and therefore related to the changes in texture of vegetables such asparagus (*Asparagus officinalis*). Asparagus lignin content increased from the top to the bottom of the spear. The differences

TABLE 12.1 Typical Lignin Content of Whole Grain and Fresh Fruits and Vegetables

Food	Lignin (g/100 g raw sample)	Reference
Apple	1.70	Bunzel et al. 2005
Arum	0.43	Rahim et al. 2006
Asparagus	11.06–5.6	Bunzel et al. 2005; Jaramillo-Carmona et al. 2008
Bengal gram	ND	Rahim et al. 2006
Black gram	ND	Rahim et al. 2006
Carrot	3.2	Bunzel et al. 2005
Curly kale	2.5	Bunzel et al. 2005
Green gram	ND	Rahim et al. 2006
Jam Alu	0.42	Rahim et al. 2006
Kaon	ND	Rahim et al. 2006
Khesari	0.45	Rahim et al. 2006
Kiwi	8.30	Bunzel et al. 2005
Kohlrabi	1.30–3.80	Rahim et al. 2006; Bunzel et al. 2005
Lentil	ND	Rahim et al. 2006
Loquat	1.28–301	Cai et al. 2006
Maize	ND	Rahim et al. 2006
Mangosteen	0.10–0.60	Kamdee et al. 2014
Peach	0.50–4.00	Gabotti et al. 2015
Pear	23.5	Bunzel et al. 2005
Potato	0.32	Rahim et al. 2006
Radish	3.40	Bunzel et al. 2005
Rhubarb	9.00	Bunzel et al. 2005
Rice	ND	Rahim et al. 2006
Rye	0.70–1.12	Alijošius et al. 2016
Shak Alu	1.03	Rahim et al. 2006
Small radish	3.40	Bunzel et al. 2005
Spinach	3.3	Bunzel et al. 2005
Sweet Potato	2.87	Rahim et al. 2006
Triticale	0.73–1.07	Alijošius et al. 2016
Turnip	0.92	Rahim et al. 2006
Wheat	0.41–1.26	Alijošius et al. 2016; Rahim et al. 2006

in cell wall lignin compositions affect the mechanical properties of asparagus tissue (Jaramillo-Carmona et al. 2008). In case of peach (*Prunus persica*), lignin contributes to flesh texture; the concentration was different in the fruit of three cultivars and decreased with ripening (Gabotti et al. 2015). In loquat (*Eriobotrya japonica* L.), increased flesh firmness is related to incresed levels of lignin (Cai et al. 2006). Similarly, in mangosteen (*Garcinia mangostana* L.), increases in firmness and lignification occur in the pericarp of mechanically damaged fruit (Dangcham et al. 2008). Lignification was the main chilling injury symptom in kiwifruit (*Actinidia deliciosa*), and it occured not only in the outer pericarp but also in the inner pericarp. The occurrence of lignification in kiwifruit began from the stem end and gradually spread to the calyx end. The lignified cells mostly occurred around vascular bundles first and then the outer pericarp of kiwifruit. At the end of the cold storage, sclerenchyma under the outer pericarp became thicker than those observed on the harvest day. Epidermal cells appeared to be bronze-colored, which was darker than lignified cell groups in the core. Lignified cells emerged around vascular bundles and were smaller than surrounding parenchymal cells, occurring in clusters (Li et al. 2017).

In the case of minimally processed products, jicama (*Pachyrhizus erosus* L. Urban) cylinders had higher lignin content in the external tissue due to mechanical damage caused by cutting. During storage, it increased 3.1 times and was related to tissue browning (Aquino-Bolaños and Mercado-Silva 2004). Inversely, in carrots, lignin content increased during storage at 2°C with ethylene, but the product was discolored (Howard and Griffin 1993).

Lignin is an important component of dietary fiber. The National Academy of Science in 2002 included lignin among non-digestible carbohydrates (Dhingra et al. 2011). Lignin is generally considered an inert compound in the human gastrointestinal tract. However, it can regulate food intake and digestion, improve bowel function, and alleviate constipations in patients. However, recent studies by Niemi and other (2013) concluded that lignin did not suppress the conversion activity of gut microbiota and that several lower molecular weight metabolites were formed during the fermentation, which suggests partial degradation of lignin in the colon. Moreover, Zhang et al. (2014) observed pancreatic lipase inhibition by lignin reducing fat absorption.

12.5 LIGNIN CHARACTERIZATION

Lignin is the most abundant source of renewable aromatic units and therefore, detailed characterization to unveil its chemical properties is a critical step in its utilization.

12.5.1 Isolation and Purification

The first step needed for processing is an appropriate milling. A lot of woody material contains high lignin levels, and it is very difficult to cut and chop. Lignified materials are good abrasives, wearing the iron part of the mill. Only a fraction of the lignin polymer is solubilized by most of the chemical degradation methods, and isolation always causes chemical changes in the structure (Monties 1988). Enzymatic degradation is a common technique used to separate and solubilize lignin. Some enzymes such as peroxidases, laccases, polyphenol oxidase, and coniferyl alcohol oxidase have the potential to dehydrogenate the monolignol radicals. The use of hydrogen peroxide is needed to perform the degradation of monolignols used by these enzymes (Boerjan et al. 2003). A solution of 5 percent NaOH and temperatures at 130–170°C are usually used to dissolve the lignin in an aqueous liquor. With these treatments, a purity of 60–70 percent could be reached (Sun 2010). Besides enzymes, microbial secretomes have been successfully used to partially depolymerize solids with high contents of lignin (Salvachúa et al. 2016).

Another option of purification is the use of organosolvs with solutions of methanol, ethanol, 1-propanol, n-butanol, dioxane, or acetone in water. In the organosolv method, lignocelluloses are treated with organic solvents like ethanol or ethanol/water mixtures at mild conditions to obtain lignin with molecular weights between 1000 and 3000 Da (Singh and Dhepe 2016). The organosolv isolation is less difficult compared with the Kraft process in terms of hazardous residue, but the low recuperation yields of lignin have been a problem to find a market for purified lignin. Interestingly, organosolv fractionation is a good technique to produce lignin with different physicochemical characteristics (Tao et al. 2016).

12.5.2 Quantification Techniques

To elucidate the structural features and chemical components of lignin, two methods of chemical degradation and nondestructive techniques are used, each providing partial but complementary information. In the former method, the polymers are degraded to their constituent building blocks and the resulting products are analyzed by gas chromatography (GC) or high-performance liquid chromatography (HPLC). For example, acid hydrolysis has been used to estimate the sugar components and content of uronic acids. Meanwhile, alkaline nitrobenzene oxidation has been used to estimate the extent of uncondensed units in lignins based on the yield of vanillin, syringaldehyde, and *p*-hydroxybenzaldehyde, resulting from three constitutive monomeric lignin units guaiacyl, syringyl, and *p*-hydroxyphenyl, respectively. In the latter approaches, several physical nondestructive techniques used to analyze the polymers, such as Fourier transform infrared (FT-IR), carbon-13 nuclear magnetic resonance spectroscopies (^{13}C-NMR), and gel permeation chromatography (GPC), enable structural information to be derived from the intact macromolecule and avoid the possibility of degradation artifacts (Xiao et al. 2001).

12.5.2.1 Gravimetric Method

Goering and Van Soest (1970) proposed the first analytic method for quantification of cellulose, hemicellulose, and lignin. This method treats the sample with a neutral detergent solution leaving a fraction called neutral detergent fiber (NDF). The second step is to treat a new batch of the same sample with a solution composed of H_2SO_4 (1N) and detergent to obtain a new fraction called acid detergent fiber (ADF), followed by hydrolysis with concentrated H_2SO_4 (72% v/v), and finally the solid residue after the hydrolysis is burned at 525°C leaving only ashes. The amount of hemicellulose is calculated by subtracting the amount of ADF to NDF, the cellulose is obtained by subtracting the amount of acid hydrolyzed sample to the ADF, and the lignin, is calculated by subtracting the ashes to the acid hydrolyzed sample. Even if this technique is used for the quantification of the lignocellulose components, it underestimates the total lignin, and it is dependent on a huge amount of gravimetric measurements, especially in food rich in soluble fiber (Dhingra et al. 2011).

Nowadays, the Klason method is the most typical lignin determination procedure. The procedure separates lignin as an insoluble material by depolymerization of cellulose and hemicellulose in 72 percent sulfuric acid (SA) followed by hydrolysis of the dissolved polysaccharides in boiling 3 percent SA. However, part of the lignin is dissolved in the filtrate as so-called acid-soluble lignin (ASL). Because lignin possesses oxygen-containing functional groups at benzylic positions, it is sensitive to acidic media and therefore undergoes considerable changes during the lignin determination procedure. Chemical changes in 72 percent SA are due to reactions such as condensation between aromatic and benzylic carbons, cleavage of syringyl ether linkage, and rearrangement of diaryl ether units. The main reaction in boiling 3 percent SA is probably hydrolysis of the depolymerized polysaccharides to soluble monosaccharides. The higher ASL content of syringyl lignin-rich woods and the higher reactivity of the syringyl nucleus to SA in comparison with the guaiacyl nucleus suggest an important relation between ASL and syringyl lignin. Syringyl nuclei, furthermore, has higher reactivity during the condensation reaction with carbohydrates, in 72 percent SA, yielding glycosides with carbon-to-carbon linkages (C-glycosides) (Yasuda et al. 2001).

12.5.2.2 Spectrophotometric Method

Fourier transform–near infrared (FT-NIR) has been used to determine the structure and composition of lignin. This technique can identify the monolignols G and S and their interactions with other molecules. With this technique it is easy to compare lignin from wood samples against that found in soft grasses (Mokochinski el al. 2015).

12.5.2.3 Chromatographic Methods

High performance liquid chromatography has been used to determine the composition of lignin. The main problem with the technique is the type of detectors and the availability of quantification standards. The degradation products of lignin polymer can be identified with an LSD detector (Mokochinski el al. 2015). HPLC has a great advantage over GC since it does not need the sample derivatization. The limitation of HPLC quantification is the wide amount of lignans produced after lignin hydrolysis, limited by the low number of standards available for research. The best alternative to avoid this problem is the coupling of HPLC pumps to the mass detectors.

12.5.2.4 Mass Characterization

The proportion of S/G lignin subunits could be quantified using different analytical techniques. Nuclear magnetic resonance (NMR), pyrolysis gas chromatography-mass spectrometry, even GC–MS spectrometry has been used to quantify the hydrolyzed portions of lignin (Mokochinski et al. 2015). Naron et al. (2017) studied the differences between common methods like wet chemical methods, Fourier transformed infrared spectroscopy, and gel permeation chromatography, and concluded that Thermo Gravimetric analysis for lignin devolatilization, the capture of released volatile compounds in thermal desorption tubes and the quantification of the captured phenols by GC–MS, is a good alternative to estimate the monomers' content.

12.6 LIGNIN USES IN FOOD

Lignin is present in many different food by-products, such as sugarcane bagasse, bean dregs, sweet potato residues, sunflower seed shells, and aged bamboo shoots. Developing valuable products from lignin can improve the economics for the use of by-products (Azadfar et al. 2015). Lignin could become the main renewable aromatic source for the chemical industry in the future and substitute phenol for most of its industrial applications, such as phenolic resins, surfactants, epoxy resins, adhesives, or polyester (Agrawal et al. 2014).

Nonetheless, features in the lignin macromolecule and its characterization suggests the predominant presence of antioxidant monomers in the lignin. Lignin compounds that contain more phenolic hydroxyl groups, fewer aliphatic hydroxyl groups, low molecular weight, and narrow polydispersivity are reported to have higher antioxidant activity. Treatment with the ozone and soaking aqueous ammonia generated a lignin with antioxidant activity (Azadfar et al. 2015). It also has been used as a prebiotic after a multienzymatic treatment (Reddy and Krishnan, 2010). A lot of potentials to be used as an antioxidant material in a nanoscale has also been reported (Ge et al. 2014). Therefore, it could be used in foods, cosmetics, and other chemical products.

REFERENCES

Agbogbo, F.K., Coward-Kelly, G., Torry-Smith, M., and Wenger, K.S. 2006. Fermentation of glucose/xylose mixtures using *Pichia stipitis*. *Process Biochemistry* 41:2333–2336.

Agrawal, A., Kaushik, N., and Biswas, S. 2014. Derivatives and applications of lignin – An insight. *The Scitech J.* 1(7)30–36.

Alijošius, S., Švirmickas, G.J., Bliznikas, S. et al. 2016. Grain chemical composition of different varieties of winter cereals. *Zemdirbyste-Agriculture* 103(3):273–280.

Aquino-Bolaños, E.A., and Mercado-Silva, E. 2004. Effects of polyphenol oxidase and peroxidase activity, phenolics and liginin content on the browning of cut jicama. *Postharv. Biol. Techn.* 33:275–283.

Azadfar, M., Gao, A.H., Bule, M.V., and Chen, S. 2015. Structural characterization of lignin: A potential source of antioxidants guaiacol and 4-vinylguaiacol. *Int. J. Biol. Macromol.* 75:58–66.

Boerjan, W., Ralph, J., and Baucher, M. 2003. Lignin biosynthesis. *Annual Rev. Plant Biol.* 54:519–546.

Cai, C., Xu, C., Li, X., Ferguson, I., and Chen, K.S. 2006. Accumulation of lignin in relation to change in activities of lignification enzymes in loquat fruit flesh after harvest. *Postharv. Biol. Techn.* 40:163–169.

Chen, H. 2014. Chemical composition and structure of natural lignocellulose. In *Biotechnology of Lignocellulose: Theory and Practice*. doi: 10.1007/978-94-007-6898-7. Beijing: Chemical Industry Press, and Dordrecht: Springer Science Business Media.

Dhingra, D., Michael, M., Rajput, H., and Patil, R.T. 2011. Dietary fibre in foods, a review. *J. Foods Sci. Technol.* 49(3):255–266.

Everson, H.P., Waldron, K.W., Geeson, J.D., and Browne, K.M. 1992. Effects of modified atmospheres on textural and cell wall changes of asparagus during shelf-life. *International Journal of Food Science and Technology* 127:187–199.

Gabotti, D., Negrini N., Morgutti, S., Nocito, F.F., and Cocucci M. 2015. Cinnamyl alcohol dehydrogenases in the mesocarp of ripening fruit of *Prunus persica* genotypes with different flesh characteristics: Changes in activity and protein and transcript levels. *Physiologia Plantarum* 154:329–348.

Ge, Y., Wei, Q., and Li, Z. 2014. Preparation and evaluation of free radical scavenging activities of nanoscale lignin materials. *Bioresources* 9(4):6699–6706.

Goering, H.K., and Van Soest, P.J. 1970. Forage fibre analysis. U.S. Department of Agriculture, Washington.

Howard, L.R., and Griffin, L.E. 1993. Lignin formation and surface discoloration of minimally processed carrot sticks. *J. Food Sci.* 58(5):1065–1067.

Jaramillo-Carmona, S., Fuentes-Alventosa, J.M., Rodríguez-Gutiérrez, G. et al. 2008. Characterization of asparagus lignin by HPLC. *J. Food Sci.* 73(7):C526–532.

Lan, W., Lu, F., Regner, M. et al. 2015. Tricin, a flavonoid monomer in monocot lignification. *Plant physiology* 167(4):1284–1295.

Lan, W., Morreel, K., Lu, F. et al. 2016a. Maize tricin-oligolignol metabolites and their implications for monocot lignification. *Plant Physiology.* doi: http://dx.doi.org/10.1104/pp.16.02012

Lan, W., Rencoret, J., Lu, F. et al. 2016b. Tricin-lignins: Occurrence and quantitation of tricin in relation to phylogeny. *The Plant Journal* 88(6):1046–1057.

Li, J., Suo, J., Han, Y. et al. 2017. The effect of 1-methylcyclopropene, methyl jasmonate and methyl salicylate on lignin accumulation and gene expression in postharvest "Xuxiang" kiwifruit during cold storage. *Postharv. Biol. Techn.* 124:107–118.

Mokochinski, J.B., Bataglion, G.A., Kiyota, E. et al. 2015. A simple protocol to determine lignin S/G ratio in plants by UHPLC-MS. *Anal. Bioanal. Chem.* 407:7221–7227.

Monties, B. 1988. Preparation of dioxane lignin fractions by acidolysis. *Methods in Enzymology* 161:31–35.

Naron, D.R., Collard, F.X., Tyhoda, L., and Görgens, J.F. 2017. Characterization of lignins from different sources by appropriate analytical methods: Introducing thermogravimetric analysis-thermal desorption-gas chromatography–mass spectroscopy. *Industrial Crops and Products* 101:61–74.

Park, S., Baker, O.O., Himmel, M.E., Parilla, P.A., and Johnson, D.K. 2010. Cellulose crystallinity index: Measurement techniques and their impact on interpreting cellulase performance. *Biotechnology for Biofuels* 3(10):1–10.

Rahim, M.A., Akter, S., and Ahmed, N. 2006. Enzymatic-gravimetric estimation of the content and composition of insoluble dietary fiber in selected cereals, pulses, and starchy roots and tubers. *Bangladesh Journal of Nutrition* 18–19:69–75.

Ralph, J. and Hatfield, R.D. 1997. Abnormal Lignins in Antisense-CCR and Antisense-CAD Plants. US Dairy Forage Research Center, 1, 38–41.

Reddy, S.S. and Krishnan, C. 2010. Production of prebiotics and antioxidants as health food supplements from lignocellulosic materials using multienzymatic hydrolysis. *Int. J. Chem. Sci.* 8(3):S535–S549.

Salunkhe, D.K. and Desai, B.B. 1984. Asparagus and artichoke. In *Postharvest Biotechnology of Vegetables*, pp. 117–127. Boca Raton, FL: CRC Press.

Salvachúa, D., Katahira, R., Cleveland, N.S. et al. 2016. Lignin depolymerization by fungal secretomes and a microbial sink. *Green Chemistry* 18(22):6046–6062.

Singh, S.K., and Dhepe, P.L. 2016. Isolation of lignin by organosolv process from different varieties of rice husk: Understanding their physical and chemical properties. *Bioresource Technology* 221:310–317.

Smith, J.L., and Stanley, D.W. 1987. Nonenzymatic lignification of asparagus. *Texture Studies* 18:339–358.

Sun, R.-C. 2010. *Cereal Straw as a Resource for Sustainable Biomaterials and Biofuels: Chemistry, Extractives, Lignins, Hemicelluloses and Cellulose*, 1st ed. Oxford, UK: Elsevier.

Tao, J., Hosseinaei, O., Delbeck, L. et al. 2016. Effects of organosolv fractionation time on thermal and chemical properties of lignins. *RSC Advances* 6(82):79228–79235.

Xiao, B., Xun, X.F., and Sun, R.C. 2001. Chemical, structural, and thermal characterization of alkali-soluble lignins and hemicelluloses, and cellulose from maize stems, rye straw, and rice straw. *Polymer Degradation and Stability* 74:307–319.

Yasuda, S., Fukushima, K., and Kakehi, A. 2001. Formation and chemical structure of acid-soluble lignin I: Sulfuric acid treatment time and acid-soluble lignin content of hardwood. *Journal of Wood Science* 47:69–72.

CHAPTER 13

Tannins

Rocio Campos-Vega, B. Dave Oomah, Angélica María Hernández-Arriaga, Norma Julieta Salazar-López, and Kenia Vázquez-Sánchez

CONTENTS

13.1 INTRODUCTION

Tannins have played a pivotal role in human evolution. They enabled the transformation of animal hide into leather "tanning" during the Stone Age using tannin extracts from wild teak (*Pterocarpus angolensis*) bark or treatment with tannins and tannic acid derived from tannin-rich pods and barks (Rifkin 2011). The interaction of tannins with proteins and lipids strongly influence and assist cell–cell interaction and signal transduction responsible for anticarcinogenic, vascular, and cardioprotective activities associated with the evolution of the human brain and intelligence (Tarahovsky 2008). Tannin and sugar contents predict food choices of nonhuman primates in the wild, suggesting sensitivity and taste thresholds for these components (Wobber, Hare, and Wrangham 2008). However, these functionalities changed in human evolution after the discovery of fire cooking, thereby improving tannin or plant polyphenol consumption (Tarahovsky 2008).

Tannins are a critical component in animal husbandry by regulating feed intake, particularly of browsers and small ruminants, and at the same time controlling gastrointestinal parasites (Yoshihara, Minho, and Yamamura 2013). Tannins represent an effective alternative strategy against gastrointestinal nematode parasitism compared with drug resistant synthetic anthelmintics in the farming of small ruminants in tropical countries (Oliveira et al. 2011). In plants, tannins localized inside vacuoles and linked to cell walls (Dalzell and Kerven 2002) exert their defensive biochemical mechanisms from biological antagonists and natural pathogens. The strong binding ability of tannins defines their biological and physiological functions that until recently were considered adverse and tagged as antinutritional.

Structurally, tannins are oligomers or polymers of heterogeneous molecules containing aromatic rings varying in number and position of hydroxyl groups that can interact with proteins, carbohydrates, lipids, and metals (Jakobek 2015; Serrano et al. 2009). Several studies have highlighted the importance of tannins in nutraceutical foods and medicine with antioxidant (Pedan et al. 2017; Chen, Mcclung, and Bergman 2016), cytotoxic (Zarin et al. 2016), hypocholesterolemic (Tebib, Besancon, and Rouanet 1994), and antimicrobial activities (Santos et al. 2017) that are critical in modulating obesity, diabetes, and cardiovascular diseases. For example, a tannin extract with castalagin as the main constituent exerted antioxidant activity and inhibited α-glucosidase activity that contributed to the development of antidiabetic agents (Muccilli et al. 2017). Similarly, proanthocyanidins oligomers from pecan shell can limit carbohydrate and lipid hydrolysis in a simulated human digestive model by inhibiting α-amylase and pancreatic lipase activities (Vazquez et al. 2017).

Investigation of tannins has resulted in the development of emulsifying agents (Figueroa, Zafimahova, and Maldonado 2015), radiation protection agents (Zhou et al. 2016), adhesives (Ping et al. 2012), pigments in leathers (Carsote et al. 2016), and water treatment agents for metal absorption (Tondi et al. 2008; Anirudhan and Suchithra 2008). The potential use of tannins in these diverse areas depends on their structural characteristics and degree of polymerization; therefore, it is necessary to understand the specific methods of their analysis. This chapter updates the literature on the structural characteristics, analytical methods, and sources of tannins and their effects on major metabolic diseases.

13.2 TANNIN CLASSIFICATION

Tannins are ubiquitous with extremely high diversity, resulting in complex classification sometimes based on their biological effects differentiated by solubility/absorptivity or chemical structure. Complex unabsorbable tannin structures display binding properties that may produce local gastrointestinal effects (antioxidant, radical scavenging, antimicrobial, antiviral, antimutagenic, and antinutrient effects). Absorbable tannins, generally low molecular weight, exert systemic effects in various organs (Serrano et al. 2009). According to the Haslam–Bate–White definition, all "vegetable tannins" commonly referred as hydrolyzable tannins (HTs) are characterized as water soluble, have molecular weight between 500 and 5,000 Da or higher, react with phenol, and are able to precipitate alkaloids and protein (Buzzini et al. 2008). However, it is the structural characteristics of tannins that enable their binding or complexing with animal oils and iron-rich mineral ingredients in transforming hide into leather since the Middle Stone Age (Rifkin 2011).

1. Classification based on chemical structure.

   Tannins are commonly classified as proanthocyanidins, hydrolyzable tannins, phlorotannins, and complex tannins based on their chemical structure (Serrano et al. 2009; Khanbabaee and Ree 2001; Table 13.1). Proanthocyanidins (PAs), often referred to as condensed tannins, are oligomers and polymers with 2 to more than 200 monomers of flavan-3-ol units, with constitutive units usually catechins and epicatechins linked through the C4–C8 or C4–C6 (type B PAs) or linkages between C4–C8/C6 and C2–C7 (type A PAs) (Chen, Mcclung, and Bergman 2016; Serrano et al. 2009). Type B PAs are further segregated into procyanidins, prodelphidins, propelargonidins, proteracacinidins, promelacacinidins, procassinidins, probutinidins, and other non-PAs molecules with flavan-3-ol constituent units based on the hydroxylation patterns of the chain extender (Ferreira and Slade 2002).

   Hydrolyzable tannins (HTs) were the first group of plant polyphenols subjected to analytical research more than 200 years ago (Arapitsas 2012). Gallic acid linked to sugar molecules (mainly glucose, fructose, xylose, or saccharose) is the predominant structural characteristic of hydrolyzable tannins (Serrano et al. 2009). Hydrolyzable tannins are further categorized into gallotannins and ellagitannins; gallotannins or galloyl glucoses (GGs) are formed by gallic acid esters and glucose, whereas hexahydroxydiphenic acid esters and glucose or quinic acid are the linkages for ellagitannins (Jakobek 2015). Over 1000 HTs have been identified, from the simple glucogallin (332 Da) to the pentameric ellagitannins (> 5000 Da) well-known because of their astringency (Arapitsas 2012).

   Phlorotannins (PHT), oligomers, or polymers of phloroglucinol (1,3,5-trihydroxybenzene) are produced by brown algae (*Phaeophycea*) (Leyton et al. 2016). PHT are subclassified based on interlinkage types into four groups: fuhalos and phlorethols (ether linkage), fucols (phenyl linkage), fucophlorethols (ether and phenyl linkage), eckols and carmalols (dibenzodioxin linkage) (Corona et al. 2016; Kim et al. 2013).

   Complex tannins (CXT) reflect the structural complexity with different groups consisting of gallotannin or ellagitannin units bound to catechin or epicatechin (i.e., Acutissimin A found in red wine aged in oak barrels) (Serrano et al. 2009; Saucier et al. 2006; Khanbabaee and Ree 2001).

2. Classification based on extraction.

   The solubility of tannins reflects their extractability; soluble or free tannins that can be extracted with water or aqueous organic solvents such as ethanol, methanol, or acetone and non-extractable, insoluble, or bound tannins (Serrano et al. 2009). Soluble tannins are oligomeric or relatively low molecular weight hydrolyzable tannins (500–3000 Da), whereas insoluble tannins generally are of high molecular weight due to their complex formation with proteins or cell wall polysaccharides (Serrano et al. 2009; Haslam 2007; Dalzell and Kerven 2002). Insoluble tannins can be extracted after hydrolysis yielding two fractions: Hydrolyzable and non-hydrolyzable tannins (Pérez and Saura 2015).

TABLE 13.1 Tannins Classification Based on Their Chemical Structure

Basic Units		Source and Example		Reference
Proanthocyanidins or condensed tannins				
Flavan-3-ol 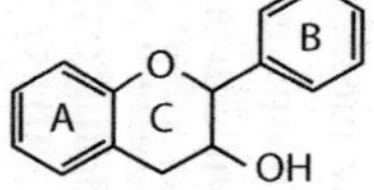	(+)-Catechin 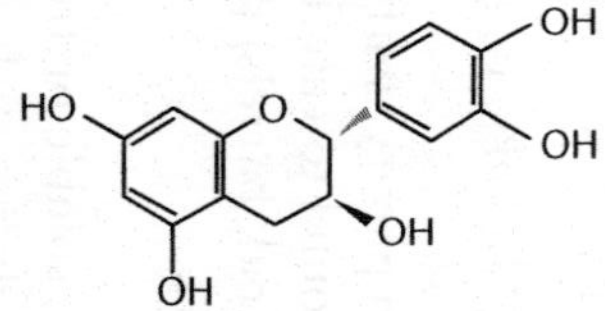 (-)-Epicatechin 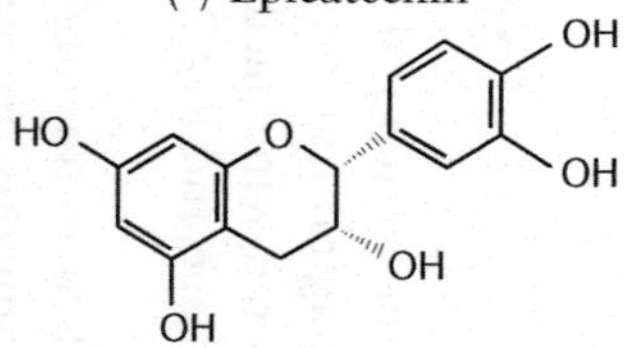	Cinnamon PA Type A 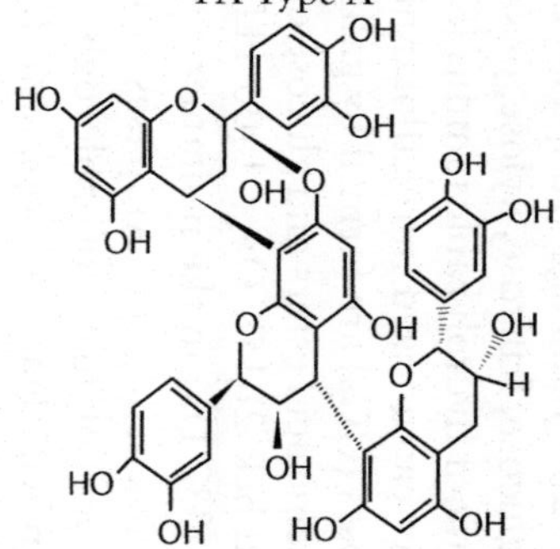	Sorghum PA Type B 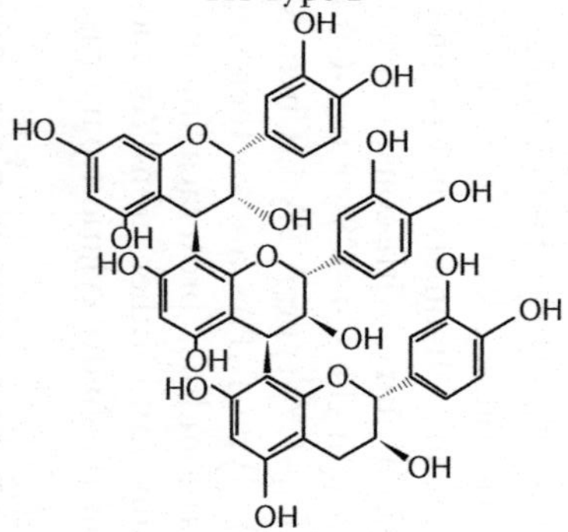	Killday et al. (2011); Amoako and Awika (2016)
Hydrolyzable tannins				
Gallotannins Gallic acid 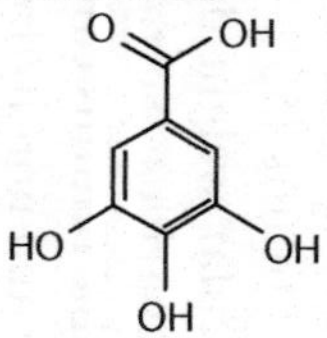	Ellagitannins Hexahydroxydiphenic acid 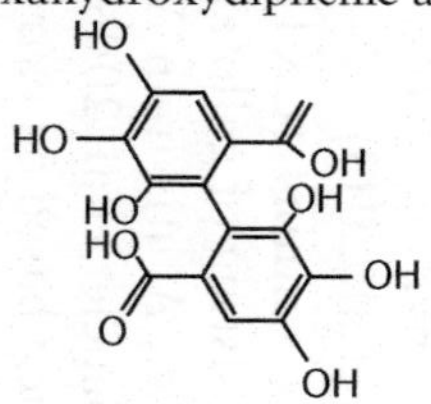	Mango (tetra-O-galloyl-glucose) Gallotannins 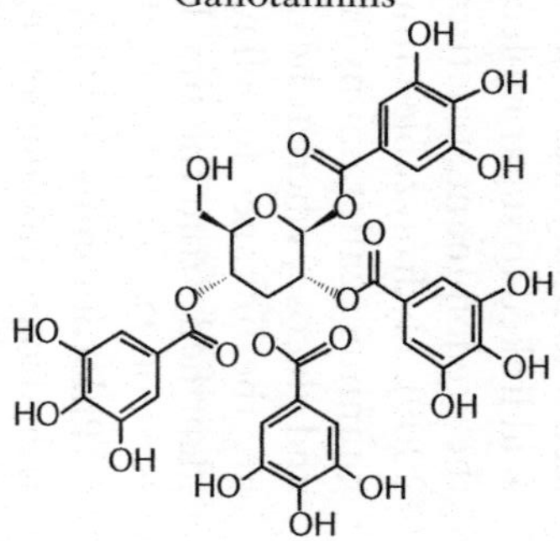	Walnut (pedunculagin) Ellagitannins 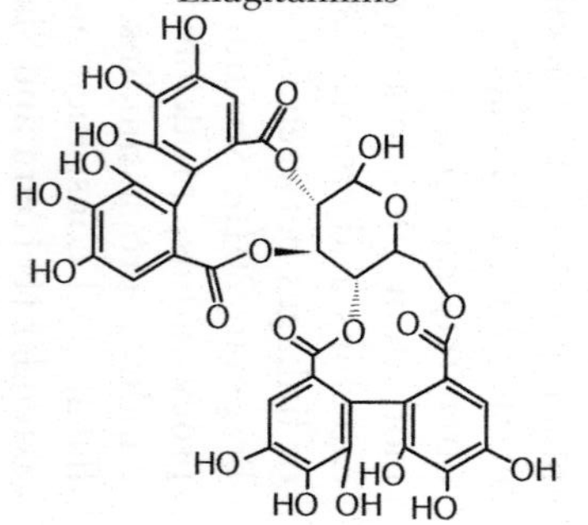	Berardini, Carle, and Schieber (2004); Larrosa et al. (2010)

(*Continued*)

TABLE 13.1 (CONTINUED) Tannins Classification Based on Their Chemical Structure

Basic Units	Source and Example	Reference
Phlorotannis Phloroglucinol (1,3,5-trihydroxybenzene) 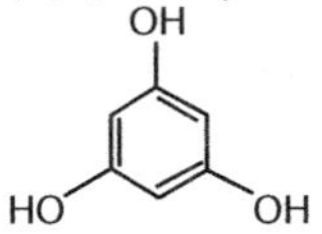 	*Eisenia bicyclis* (Phlorofucofuroeckol-A) 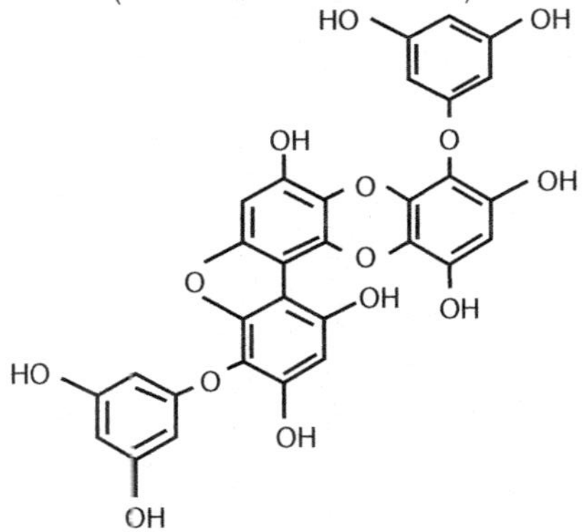 	Kim et al. (2013)
Complex tannins Example: Gallotannin or ellagitannin units bound to catechin or epicatechin	Red wine aged in oak barrels (Acutissimin A) 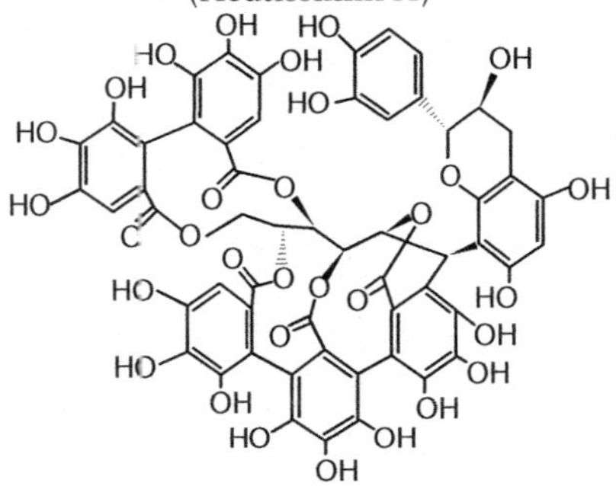	Saucier et al. (2006)

13.2.1 Association of Tannins with Other Compounds

The structural diversity of tannins enables strong interactions depending on their polyphenolic structures with different compounds, including but not limited to proteins, carbohydrates, polysaccharides, lipids, and metals (Jakobek 2005). The tannin–protein interaction is perhaps an essential phase in one of civilization's oldest processes, the transformation of hide into leather. Vegetable tannins were probably the earliest-used reagents (Romer et al. 2011). Condensed tannins (from mimosa and quebraco species) and hydrolyzable tannins (from chestnut and tara) display distinctive NMR spectroscopic signatures between resonance frequencies of ca. 71 ppm and 165 ppm, which contain some signals from leather collagen protein but numerous signals from tannins (Romer et al. 2011). This basic collagen crosslinking with tannic acid through hydrogen bonding and hydrophobic effects has recently been developed to prevent breast cancer recurrence (Jordan et al. 2016). For this purpose, an injectable matrix comprised of tannic acid crosslinked collagen beads and adipocytes serve as a tissue regeneration platform in patients post-lumpectomy (Jordan et al. 2016).

Tannin–protein interactions are specific and depend on the structure of both the protein and tannin. Protein characteristics that favor strong binding include large molecular size, open and flexible structures, and richness in proline. Similarly, tannins with high molecular weight and high conformational stability are prerequisite for strong bonding. Tannins characteristically interact strongly with protein forming complexes that affect taste, nutritional quality, and health; influencing many important aspects of ecology, agriculture, and plant and human biomedicine. For example, tannin complexes with protein at the rumen pH (6–7), thereafter the protected protein is released at the abomasum pH (<3.5), making it available for digestion (Engström et al. 2016). Tannins have therefore been proposed as chemical additives in ruminant feeds for protecting and decreasing protein fermentation and foaming to reduce symptoms of bloat. The size and flexibility of tannins dictate the formation of highly stabilized tannin–protein complexes and the oxidizability of tannins may determine the pH range that best expresses the bioactivity of the complexes (Engström et al. 2016). In wine, the astringency perception is due to the associations between salivary proteins (proline-rich proteins) and tannins through hydrophobic and hydrogen bond interactions (Watrelot et al. 2016). These salivary binding proteins (histatins or proline rich proteins [PRPs]) are endogenously increased to counteract high tannin intake in pigs (Cappai et al. 2014). The proline-rich sheath proteins of gastrointestinal nematode complexes with condensed tannins thereby attenuating their antiparasitic efficacy; tanniferous plant extracts have therefore been proposed as alternative control strategies instead of synthetic anthelmintics (Oliveira et al. 2011). The antimicrobial and antiviral activity of hydrolyzable tannins also depend on their strong interactions with proline-rich proteins occurring in the salivary pellicle or on the cell surfaces of many bacteria (Buzzini et al. 2008). Tannin toxicity in microorganisms is partly due to enzyme inhibition and substrate deprivation characteristics of protein–protein interactions (Reed 1995). Tannins from cranberries and pomegranates may slow starch digestion and offer specific approaches to control blood sugar levels by inhibiting the digestive enzymes α-amylase and glucoamylase (Barrett et al. 2013). Generally, larger and more complex tannins, such as those in pomegranate and cranberry, effectively inhibit these enzymes than the less-polymerized cocoa tannins. Tannin–protein interaction has been proposed as a strategy to remove or reduce peanut allergens in foods and drinks by forming stable allergen-bound insoluble complexes under gastric and intestinal conditions (Chung and Reed 2012).

Tannins have less affinity to carbohydrates than to proteins due to the strong hydrogen bond formation with protein's carboxyl group. Carbohydrates with high molecular weight, low solubility, and conformational flexibility increase the tannin–carbohydrate interaction. Both starch and cellulose complex with tannins, especially procyanidins; starch forms inclusion complexes with tannins, whereas cellulose has a direct surface interaction with tannins (Bachman and Hubbert 2004). Hydrophobic interactions increase carbohydrate–tannin binding efficiency. Thus, starch binds more efficiently to tannins than cellulose, but less efficiently than xyloglucose and gum arabic and amphihilic carbohydrate has lower affinity than pectin polysaccharides (Amoako and Awika 2016). Tannins also complex with pectin reducing astringency in Japanese persimmon fruit and favor juice extraction in ripe bananas (Kyamuhangire et al. 2006). Tannins have strong and specific interactions with starch, and the extent of the interaction is directly proportional to the amount of amylose in the starch (Amoako and Awika 2016). These amylose–tannin interactions can potentially modulate glucose metabolism and reduce the calorie density of starch-based foods. In vitro studies demonstrate that starch supplemented with tannins impede their hydrolysis and decrease their digestibility (Domínguez-Avila et al. 2017). The tannin–carbohydrate interaction is presumed to be partly responsible for the reduced glycemic response to carbohydrate foods and lower blood glucose response to legumes compared with cereal products (Thompson et al. 1984). Tannins can complex with carbohydrate substrates, especially cellulose, forming indigestible complexes with cell wall carbohydrates, reducing cell wall digestibility (Reed 1995). These insoluble tannin–carbohydrates as well as cell wall polysaccharide complexes are not completely extracted by solvents (Kunyanga et al. 2011). Tannin–carbohydrate interaction in addition to tannin–protein complex formation can have negative effects on fermentation, digestion, and methanogenesis of rumen (Gemeda and Hassen 2015).

Tannins are able to bind on lipid membrane surfaces due to gallic acid residues that interact with all lipid head group and cover the bilayer surface. The binding also favors electrostatic interaction between the π electrons in the phenol ring and $-N^{+}(CH_3)_3$ groups of phosphatidylcholines or sphingomyelin present in the outer lea et of plasma membrane (Tarahovsky 2008). The polyphenol–lipid interaction can change the fat absorption process (emulsification process and lipase activity), inhibit lipid oxidation, and delay polyphenol digestion (Jakobek 2015). The hydroxyl group from hydrolyzable tannins generally binds to the PO_2 groups from phospholipids (Sekowski et al. 2016). Tannins also exert antihypercholesterolimic effects in rats by modifying cholesterol transport, reducing intestinal cholesterol absorption and increasing bile acid excretion (Tebib, Besancon, and Rouanet 1994). For example, persimmon (*Diospyros kaki* L.) tannins interact with Niemann-Pick C1-Like 1 (NPC1L1), the main intestinal cholesterol transporter that directly modulate its expression and prevent dietary cholesterol uptake (Domínguez-Avila et al. 2017).

The interaction with metals is well known for the production of ferric tannate as a form of black ink, and for bismuth tannate as a pharmaceutical. Metal ion interaction with tannins depends on ion exchange, pH, and sometimes on selectivity for cooper, zinc, or cadmium. Some metal cations rapidly complex with tannins, specifically tannic acid, and even form visible precipitates, while others form complexes that remain soluble depending on conditions (Stephens and Bell 2016). Thus, tannins have been proposed as an immobilizing agent based on their binding with the important and clinically useful radioactive isotopes of Gallium for imaging or therapy (Stephens and Bell 2016). Black bean tannin chelate and sequester zinc, making it unavailable to metalloproteinases (metalloproteinase-2 and -9; MMP-2 and MMP-9) during the initial steps of the angiogenic process that may have beneficial implications in cancer prevention (Bawadi et al. 2005). Tannin fractions isolated from hazelnuts, walnuts, and almonds effectively chelated metals like copper, iron, and zinc at different levels. These metal–tannin interactions stabilize peroxidative activity and inhibit damage to different biological molecules (Karama 2009). The chelation capacity of tannins has been applied in removing metals from water (Anirudhan and Suchithra 2008; Tondi et al. 2008; Palma, Freer, and Baeza 2003).

13.3 ANALYTICAL METHODS IN FOODS

An earlier review on tannin analysis of food products (Deshpande et al. 1986) indicated the diverse and at times conflicting literature on tannin methodology. The difficulties encountered in tannin analysis include the lack of a common reference standard, quantitative extraction of tannins given their strong interactions with carbohydrate or protein matrices, and the biological diversity of tannin structures. Methods for tannin analysis range from simple colorimetric, UV spectrophotometric, chromatographic, and enzymatic to more sophisticated and expensive nuclear magnetic resonance (NMR) and near infrared (NIR) techniques. Some important characteristics of tannins should be considered for tannin quantifications; these include the structure of different tannin groups, particularly substituents in the aromatic ring responsible for antioxidant, protein precipitation, and metal interaction properties (Melone et al. 2013); degree of tannin polymerization (DP); differences in solubility and extraction methods; and the presence of compounds that can interfere in tannin analysis.

13.3.1 Short Historical Perspective of Tannin Analysis

The term *tannin* was first used in the late 18th century by the French chemist Armand Séguin to describe the chemical agents responsible for the fabrication of leather (Grasel and Ferrão 2016). Gravimetric method was used to extract and quantify tannins from different sources. Tannin was isolated and its effects on ferrous and ferric salts determined in 1798 (Trimble 1891). Almost a century later, tannin extracted from hops was described by its reaction with ferric chloride and potassium permanganate in the presence of an indigo solution (Heron 1896). This classical method, known as the "hide powder method," has now been discontinued and superseded by a more modern titration method, the Divergan method, which is based on tannin precipitation by absorption on polyvinylpyrrolidone (Antoine, Simon, and Pizzi 2004). A modified method based on potassium permanganate titration has been used to estimate tannin content in different food products (Tabasum et al. 2001). Tannin has also been obtained by alkaline hydrolysis and quantified gravimetrically (Perkin 1900). The gravimetric method formed part of the official method for analysis of tanning materials according to Circular 8 of the United States Department of Agriculture (AOAC 1901). This gravimetric method is often used to select plant species as tanning ingredients to recreate the past tanning process using vegetable tannins. For example, tannin was extracted from sweet thorn (*Acacia karroo*) by soaking 250 grams of ground bark in 1 liter of clean water for 12 hours (Rifkin 2011).

The most commonly used colorimetric methods for tannin analysis are the Folin–Dennis and Folin–Ciocalteau's Reagent (Folin and Ciocalteu 1927), the vanillin-HCl reaction (Broadhurst and Jones 1978), and the butanol–HCl reaction (Bate-Smith 1975) (Table 13.2). The Folin reaction, based on phosphomolybdic acid reduction by phenols in aqueous

TABLE 13.2 Methods for Analysis of Tannins

Method	Reaction	Used for	Important Points	Reference
Ferric or ferrous	Oxide-reduction	Phenolic compounds	Not specific for tannins. CT (green black products) and HT (blue-black products).	Falcão and Araújo (2011)
Folin-Ciocalteu	Oxide-reduction	Phenolic compounds	Not specific for tannins.	Folin and Ciocalteu (1927)
Prussian blue	Oxide-reduction	Phenolic hydroxyl groups	Not specific. Sensitive to ortho or para position phenolic group. Sensitive to T, pH and order of reagents applied. Less interferences than Folin–Ciocalteu.	Schofield, Mbugua, and Pell (2001); Price and Butler (1977)
Vanillin	Synthesis in presence of acid	CT	Not specific for CT, not sensitive to differences between procyanidins and prodelphinidins. Sensitive to T, vainillin concentration, reaction time and acid type.	Schofield, Mbugua, and Pell (2001)
Acid-butanol	Oxidative depolymerization	CT	Breakdown of the PA original structure and reduced polymer size. Sensitive to no. of phenolic groups in rings A and B, changes during the reaction such as T, proportion acid-butanol and water and time reaction. More sensitive presence of linkages C4–8 than C4–6.	Schofield, Mbugua, and Pell (2001); Hixson, Bindon, and Smith (2015)

(Continued)

TABLE 13.2 (CONTINUED) Methods for Analysis of Tannins

Method	Reaction	Used for	Important Points	Reference
Potassium iodate	Methanolysis followed by oxidation with KIO_3	HT (gallotannins and ellagitannins)	Not specific. Sensitive to reaction solvent, pH, and temperature.	Hartzfeld et al. (2002); Mueller-Harvey (2001); Willis, Allen, and Road (1998)
Rhodanine	Hydrolysis and after rhodamine reaction	HT (gallotannins)	Sensitive to number of gallic acid units.	Inoue and Hagerman (1988); Mueller-Harvey (2001)
Nitrous acid or sodium nitrite	Oxidation whit nitrous acid	HT (ellagitannins)	Selective for ellagic acid only.	Falcão and Araújo (2011); Mueller-Harvey (2001)
Amination	Regioselective amination with NH_3	CT and related polyphenols	Low amination on type B rings.	Hashida, Makino, and Ohara (2009)
Protein precipitation	Precipitation by interaction tannin-protein	CT	Sensitive to polymer size and pH.	Hagerman and Butler (1981)
Phloroglucinol degradation	Depolymerization (hydrolysis)	CT	Inability to remove. Bound tannins from cell wall.	Hixson, Bindon, and Smith (2015)
Acid hydrolysis with l and HPLC-MS analysis	Hydrolysis with benzylmercaptan or phloroglucino	CT	Lower yield than butanol–HCl method. Phloroglucinolysis produces lower yields than benzylmercaptan.	Pérez-Jiménez and Torres (2011)
Enzymatic treatments and HPLC analysis	Enzymatic	CT	Lower yield than butanol–HCl method.	Pérez-Jiménez and Torres (2011)

Note: Condensed tannins (CT), hydrolyzable tannins (HT), temperature (T).

alkali, has been recommended for tannin analysis in wines and distilled beverages (Reed 1995). The vanillin test is preferred for screening tannins in plants because of its sensitivity, specificity, and simplicity. The vanillin–HCl reagent is quite specific to a narrow range of flavonols and dihydrochalcones that have a single bond at the 2,3-position and free meta-hydroxyl groups on the B-ring, although it detects monomeric and polymeric flavonoids (Deshpande and Campbell 1992). The conventional HCl–vanillin assay forms the basis of a high-throughput 96-well microplate method for screening tannin content in sorghum grain (Herald et al. 2014). The protein precipitation method is used sometimes to evaluate the biological activity of tannins and the nutritional value of tannin containing foods. However, the method only measures polyphenolics that precipitate bovine serum albumin or other protein and does not distinguish between hydrolyzable and condensed tannins (Fuller 1989). The in vitro gas production technique, combined with a tannin binding agent (polyethylene glycol [PEG]), has been used as a bioassay to provide an index of biological effect of tannins on microbial utilization of feeds (Rodriguez, Frutos, and Fondevilla 2015).

Many high-performance liquid chromatography (HPLC) methods have become popular in differentiating polyphenolic compounds such as hydrolyzable tannins (Okuda et al. 1979). Several sophisticated specific methods of tannin analysis are available including Fourier transform infrared spectroscopy (FT-IR), x-ray, nuclear magnetic resonance (NMR), mass spectrometry (MS), matrix-assisted laser desorption/ionization time of flight (MALDI-TOF), HPLC, and coupling of these methods (Crestini, Lange, and Bianchetti 2016; Duval and Ave 2016; Ma et al. 2016; Santos et al. 2017).

13.3.2 Method Development/Enhancement for Current Use in Various Areas

Most industries require inexpensive and rapid methods of analysis. For example, the European Commission Regulation N2676/90, Community Methods for the Analysis of Wines requires determination of the total amount of all phenolic compounds where reagent is mixed with the wine sample and absorbance is measured at 765 nm. This method was improved to measure polyphenol, flavonoid, and anthocyanin concentrations in wine using a mobile phone camera for sample spot capture on a paper microzone and a remote computer with dedicated software for quantification (Vaher and Kaijurand 2012). The method yielded comparative results with conventional wine by oxidizing them using the Folin–Ciocalteau reagent (FCR). The amount can be expressed in gallic acid equivalents (GAE) if this particular polyphenolic compound is used for assay calibration. In this assay, the Folin–Ciocalteu spectrophotometry reduces reagent use and analysis time without using analytical instruments.

In wineries, spectral methods based on methyl cellulose precipitation are often used to measure condensed tannins in grapes and wines. For example, the Iland spectroscopic bovine serum albumin (BSA) precipitation method developed in Australia correlated with most of the South African wine data. Another spectrophotometric method, the Glories method developed in France, provided additional information, such as skin tannins and the contribution of seed tannins to the total phenolic profiles of these wines (Du Toit and Visgie 2012). Hemoglobin was proposed as a chromoprotein for spectrophotometric tannin analysis because of its water solubility and buffering ability compared to other proteins (Ferreira, Souza, and Nogueira 2003). However, colored compounds from blood interfered with the measurement of the flame atomic absorption spectrometry. Thus, a flow system, coupled with flame atomic absorption spectrometry (FIA-FAAS) was developed to evaluate tannin contents in pigeon pea (*Cajanus cajan*) based on the tannin–hemoglobin precipitation reaction. This method enabled determination of 30 samples per hour, relatively low standard deviation, low limit of detection (0.27 mg/L tannic acid), and applicable to different samples (Ferreira, Souza, and Nogueira 2003).

HPLC methods for tannin evaluation in grapes and wines are generally based on acid catalyzed depolymerisation of condensed tannins in the presence of strong nucleophile. Therefore, tannin contents of wines determined by HPLC are higher than those measured by the protein precipitation method (4.18 vs. 0.66 g/L; Watrelot et al. 2016). The HPLC–phloroglucinolysis is most appropriate for analytical studies where the amount and subunit composition of tannin are required (Seddon and Downey 2008). Tannin levels determined by the HPLC–phloroglucinolysis method were higher than those measured in 38 grape skin samples by protein and methylcellulose precipitation (Seddon and Downey 2008). Noncovalent interactions influence and modify tannin structure during wine production; this tannin–protein interaction in red wine has been examined with HPLC in conjunction with gas chromatography mass spectrometry (GC–MS; Watrelot, Schultz, and Kennedy 2017). The size and degree of tannin polymerization was also analyzed using centrifugal partition chromatography (CPC) followed by ultra high performance liquid chromatography coupled with electrospray ionization Quadrupole Time-of-Flight (UHPLC-ESI-Q-TOF) in relation to astringency perception of red wine (Ma et al. 2016).

Tannin–protein interactions have also been investigated using relevant enzyme assays and differential scanning calorimetry (DSC) to demonstrate enzyme inhibition by tannins from cocoa, pomegranates, cranberries, and grapes (Barrett et al. 2013). Similarly, microdifferential scanning calorimetry was used to examine the influence of tannin on the extent and structure of collagen arrangements in the thermal stability of leather during visible light exposure (Carsote et al. 2016). The protein–tannin interactions were recently investigated on a molecular level using a model system containing α-lactalbumin peptides and berry tannins (procyanidins) by applying both liquid chromatography–electrospray–tandem mass spectrometry (LC-ESI-MS) and size exclusion chromatography (SEC) to monitor oxidation, adduct formation, and binding reactions (Wang and Heinonen 2017).

Recent approaches combine high performance liquid chromatography (HPLC) with thyolysis and MALDI-TOF MS to obtain quantitative and qualitative information on the degree of tannin polymerization and procyanidin/prodelphinidin and cis/trans-ratios (Deng et al. 2016; Duval and Ave 2016). These methods are based on molecular mass and therefore do not require commercial tannin and/or internal standards. Nuclear magnetic resonance ^{13}C NMR has been used to characterize the structure of many hydrolyzable types of tannins and was used recently in demonstrating antimicrobial and cytotoxic activities of tannins against breast cancer cells (Zarin et al. 2016). A rapid and noninvasive method has been developed to classify natural tannin extracts by near-infrared spectroscopy (NIR) and multivariate analyses (Grasel and Ferrão 2016). The near-infrared (1200–2500 nm spectral range, 4 cm^{-1} resolution with 32 scans) was able to differentiate and precisely classify 49 samples of vegetable tannin extracts representing six commercially available types (Grasel and Ferrão 2016). NIR and multivariate calibration methods were used to directly determine tannin content in *Acacia mearnsii* bark to improve control during the tannin extraction process (Menezes et al. 2014). In another noninvasive method, Fourier transform infrared (FTIR), spectra is generally recorded in the 650–4000 cm^{-1} region to differentiate tannins from different plant sources extracted under different conditions (Chupin et al. 2013). FTIR both in absorbance and total reflectance (ATS) modes are used to evaluate tannin integrity; vegetable tannins present a fingerprint region between 1500 and 950 cm^{-1}, differentiating hydrolyzable from condensed tannins (Santos and dos Santos 2016). Condensed tannins in acacia (*Acacia mearnsii* wild) extracts were recently determined with a glassy carbon electrode chemically modified with gold nanoparticles stabilized in carboxymethylcellulose (Piovesan et al. 2017). However, low cost and simple quantitative methods are always preferred by the food industry, such as the spot test for soluble tannin in green tea using a portable diffuse reflectometer (Lima et al. 2012). The lab instrument, made of polytetrafluoroethylene (PTFE), uses a light emitting diode (LED) and a phototransistor as the detector; this diffuse reflectance spectroscopy enables in situ measurements at low cost (Lima et al. 2012). Methods of tannin analysis are continuously improving with new developments in techniques and approaches coupling existing and future technologies.

13.4 FOOD SOURCES: QUANTITY AND TYPE

Tannins, like other secondary metabolites, occur in many plants to protect them from environmental stress, particularly from insects, pathogens, and herbivores. Therefore, many plant foods and feeds and their products contain tannins at various levels. Good sources of proanthocyanidinds or condensed tannins are highly consumed beverages (coffee, tea, and wine), fruits (apples, blueberries, cranberries, grapes, peaches, and strawberries), dried fruits, mint, and basil. Hydrolyzable tannins are present in cereals (barley, oats, rice, and rye) and fruits (pomegranate, raspberries, and strawberries); red fruits are the main contributors of hydrolyzable tannins, such as ellagitannins (ET) consumed in Western diets (Lamy et al. 2016; Landete 2011).

13.4.1 Cereals

Tannins are generally believed to be limited to a few cereals (Table 13.3), particularly the so-called pseudocereals such as sorghum tannin, which has been extensively investigated. Generally, the pigmented and wild types of grains and cereals contain greater amounts of tannins. For example, proanthocyanidins (PAs) absent in common white rice are present in red and black rice genotypes (93.9–118.5 mg cyanidin equivalents/kg dry matter [DM]) (Finocchiaro, Ferrari, and Gianinetti 2010). Condensed tannin content of Tunisian six-row barley varieties were significantly higher than those of two-row barley (66 and 70 vs. 52 and 56 mg catechin equivalents (CE)/100 g fresh weight [FW]) determined by the vanillin–HCl method (Lahoura et al. 2014). Similarly, naked barley contains significantly lower amount of condensed tannins than regular barley (34.8 vs. 74.2 mg/100 g FW) (Bittner, Rzeppa, and Humpf 2013). PAs in barley are principally procyanidins composed entirely of catechin monomers and minor amounts of prodelpinidins (Naczk and Shahidi 2006; Gu et al. 2004). Barley tannin–protein interaction is critical in brewing; it commonly occurs during kettle boil or wort cooling resulting in the familiar hot and cold breaks, respectively. Beer brewed from malt containing high polyphenol (tannin) levels has a high tendency to produce nonbiological haze that is undesirable and can be avoided by using malt with low polyphenol content. In spelt (*Triticum spelta* L.) grain, tannin content ranged from 1.44 g to 2.35 g/kg DM and the average (1.96 g/kg) was similar to that of common wheat (2.02 g/kg) and considerably lower than in triticale (3.48 g/kg dm) (Grela 1996). Wheat from India had a total tannin content of 0.45 mg CE/g measured

TABLE 13.3 Food Sources

Food	Content		References
	Hydrolyzable Tannins	Condensed Tannins	
Cereals			
Sorghum (*Sorghum bicolor* L.)	–	10–68 mg CE/g DW	Awika and Rooney (2004)
Barley (*Hordeum vulgare*)	–	742 mg/kg FW	Bittner, Rzeppa, and Humpf (2013); Naczk and Shahidi (2006); Gu et al. (2004)
Naked barley	–	348 mg/kg FW	
Rice (*Oryza sativa*) (red and black)	–	93.9–118.5 mg cyanidin eq/kg DW	Finocchiaro, Ferrari, and Gianinetti (2010)
Corn (*Zea mays*)	–	60 mg CE/kg grain	Min et al. (2012).
Wheat (*Triticum aestivum*) (red and white)	–	60–90 mg CE/kg grain	Min et al. (2012).
Amaranth (*Amaranthus caudatus* L.)	–	2–3 g CE/kg db	Najdi Hejazi et al. (2016)
Pearl millet (*Pennisetum glaucum*)	–	170–220 mg CE/g DW	Eltayeb et al. (2007)
Legumes			
Pinto beans (*Phaseolus vulgaris*)	–	0.5–8 g/kg FW 1.5–3.3 mg/g DW	Bittner, Rzeppa, and Humpf (2013); Awika and Rooney (2004); Gu et al. (2004)
Small red beans	–	4.6 g/kg FW	Gu et al. (2004)
Red kidney beans	–	5.6 g/kg FW	Gu et al. (2004)
Black eye peas	–	0.3 g/kg FW	Gu et al. (2004)
Black beans	–	50–81 mg/kg FW	Bittner, Rzeppa, and Humpf (2013); Gu et al. (2004)
Lentils	–	3.2–10.4 g/kg DW 0.3–0.5 g/kg FW	Bittner, Rzeppa, and Humpf (2013); Awika and Rooney (2004)
Cowpea (*Vigna unguiculata*)	–	18–29 mg/kg DW	Awika and Rooney (2004)
Faba beans (*Vicia faba* L.)	–	700 mg/kg DW	Awika and Rooney (2004)
Kidney bean (*Phaseolus vulgaris*)	–	177 mg/kg FW	Bittner, Rzeppa, and Humpf (2013)
Green gram (*Vigna radiata* L.)	–	700 mg/kg FW 4.0 g CE/kg sample	Hithamani and Srinivasan (2014); Bittner, Rzeppa, and Humpf (2013)
Chickpea (*Cicer arietimum* L.)	–	700 mg CE/kg sample	Hithamani and Srinivasan (2014)
Azuki bean (*Phaseolus coccineus*)	–	240 mg/kg FW	Bittner, Rzeppa, and Humpf (2013)
Runner bean (*Phaseolus coccineus*)	–	202 mg/kg FW	Bittner, Rzeppa, and Humpf (2013)

(*Continued*)

TABLE 13.3 (CONTINUED) Food Sources

Food	Content		References
	Hydrolyzable Tannins	Condensed Tannins	
Calliandra (*Calliandra calothyrsus*)	–	263 g/kg DW	Stürm et al. (2007)
Flemingia (*Flemingia macrophylla*)	–	48 g/kg DW	Stürm et al. (2007)
Leucaena (*Leucaena leucocephala*)	–	46 g/kg DW	Stürm et al. (2007)
Fruits			
Strawberries (*Fragaria ananassa*)	650–850 mg/kg FW	0.5–1.4 g PAs/kg FW	Landete (2011); Hellström and Mattila (2008)
Rose hip (*Rosa rugosa*)	1.1 g/kg FW	7.0 g PAs/kg FW	Hellström and Mattila (2008); Koponen et al. (2007)
Raspberries (*Rubus idaeus*) (cultivated)	1.9 g/kg FW	302–788 mg PAs/kg FW	Landete (2011); Hellström and Mattila (2008); Gu et al. (2004)
Raspberries (*Rubus idaeus*) (wild)	2.7 g/kg FW	–	Landete (2011)
Cloudberries (*Rubus chamaemorus*)	3.1–3.6 g/kg FW	320 mg PAs/kg FW	Landete (2011); Hellström and Mattila (2008); Koponen et al. (2007)
Sea buckthorn (*Hipopophae rhamnoides*)	10 mg/kg FW	2.4 g PAs/kg FW	Hellström and Mattila (2008); Koponen et al. (2007)
Chokeberries (*Aronia mitschurinii*)	–	6.6–18.8 g PAs/kg FW	Hellström and Mattila (2008); Gu et al. (2004)
Sweet rowanberry (*Crataegosorbus mitschurinii*)	–	4.8 g PAs/kg FW	Hellström and Mattila (2008)
Rowanberry (*Sorbus aucuapria*)	–	4.3 g PAs/kg FW	Hellström and Mattila (2008)
Saskatoon berry (*Amelanchier ainifolia*)	–	2.8 g PAs/kg FW	Hellström and Mattila (2008)
Blueberry (*Vaccinium corymbosum*)	–	1.8–3.3 g PAs/kg FW	Gu et al. (2004)
Blackcurrant (*Ribes nigrum*)	–	1.5–4.0 g PAs/kg FW	Hellström and Mattila (2008); Gu et al. (2004)
European cranberry (*Vaccinium oxycoccus*)	–	4.1 g PAs/kg FW	Hellström and Mattila (2008)
Lingonberry (*Vaccinium vitis-idaea*)	–	3.9 g PAs/kg FW	Hellström and Mattila (2008)
Crowberry (*Empetrum nigrum*)	–	2.0 g PAs/kg FW	Hellström and Mattila (2008)

(*Continued*)

TABLE 13.3 (CONTINUED) Food Sources

Food	Content		References
	Hydrolyzable Tannins	Condensed Tannins	
Bilberry (*Vaccinium myrtillus*)	–	1.5 g PAs/kg FW	Hellström and Mattila (2008)
Plum (*Prunus domestica*)	–	1.0–2.2 g PAs/kg FW	Hellström and Mattila (2008); Gu et al. (2004)
Bog whortleberry (*Vaccinium iliginosum*)	–	1.0 g PAs/kg FW	Hellström and Mattila (2008)
Red currant (*Ribes rubrum*)	–	1.2 g PAs/kg FW	Hellström and Mattila (2008)
Gooseberry (*Ribes uva-crispa*)	–	1.0 g PAs/kg FW	Hellström and Mattila (2008)
Banana (*Musa sapientum*)	–	40–643 mg PAs/kg FW	Hellström and Mattila (2008); Gu et al. (2004)
Kiwi (*Actinidia chinensis*)	–	37–139 mg PAs/kg FW	Gu et al. (2004)
Blackberries (*Rubus laciniatus*)	–	270 mg PAs/kg FW	Gu et al. (2004)
Marion berries (*Rubus Marion*)	–	89 mg PAs/kg FE	Gu et al. (2004)
Apricots (*Prunus armeniaca*)	–	156 mg PAs/kg FW	Gu et al. (2004)
Avocado (*Persea americana*)	–	74–178 mg PAs/kg FW	Hellström and Mattila (2008); Gu et al. (2004)
Cherry (*Prunus avium*)	–	82–268 mg PAs/kg FW	Hellström and Mattila (2008); Gu et al. (2004)
Peach/nectarine (*Prunus persica*)	–	297–673 mg PAs/kg FW	Hellström and Mattila (2008); Gu et al. (2004)
Pear (*Pyrus communis*)	–	207–423 mg PAs/kg FW	Hellström and Mattila (2008); Gu et al. (2004)
Grape (*Vitis vinifera*)	–	430–815 mg PAs/kg FW	Hellström and Mattila (2008); Gu et al. (2004)
Apple (*Malus domestica*)	–	0.7–1.4 g PAs/kg FW	Gu et al. (2004)
Mango (*Mangifera indica* L.)	–	128 mg PAs/kg FW	Gu et al. (2004)
Vegetables			
Rhubarb (*Rheum rhaponticum*)	–	1.2 g PAs/kg FW	Hellström and Mattila (2008)
Indian squash	–	164 mg PAs/kg FW	Gu et al. (2004)
Beverages			
Pomegranate juices (200 mL/serving)	1 g of ET	–	Larrosa et al. (2010)

(*Continued*)

TABLE 13.3 (CONTINUED) Food Sources

Food	Content: Hydrolyzable Tannins	Content: Condensed Tannins	References
Pomegranate juices (100 percent) European commercial juices	415–1611 μmol ET/L	–	Borges, Mullen, and Crozier (2010)
Pomegranate juices blended with: grapes, aronia, berry, acai, lime, mandarine, orange, etc.	324–963 μmol ET/L	–	
Muscadine grapes juice	0.8 mg EAE/L	–	Lee and Talcott (2002)
Wine	2–9.1 mg/L	0.5–1.3 g CE/L	McRae et al. (2016); Gonzalo-Diago, Dizy, and Fernández-Zurbano (2013); Landete (2011); Saucier et al. (2006)
Bourbon whiskey	0.01–0.4 mg/L	–	Glabasnia and Hofmann (2006)
Apple juices	–	9 mg PAs/L	Gu et al. (2004)
Cranberry juice	–	231 mg PAs/L	Gu et al. (2004)
Grape juice	–	524 mg PAs/L	Gu et al. (2004)
Boysenberry (hybrid Rubus berry) commercial juice	–	11 mg PC/L	Furuuchi et al. (2011)
Beer	–	230 mg PAs/kg FW	Hellström and Mattila (2008)
Tea black or green (*Camellia sinlensis*)	–	33 and 122 mg PAs/kg FW	Hellström and Mattila (2008)
Nuts			
Walnut (*Juglans regia*)	400 mg of ET	125–670 mg PAs/kg FW	Bittner, Rzeppa, and Humpf (2013); Larrosa et al. (2010); Gu et al. (2004).
Peanuts (*Arachiis hypogaea*)	–	0.04–1.9 g PAs/kg FW	Bittner, Rzeppa, and Humpf (2013); Hellström and Mattila (2008)
Hazelnuts (*Corylus avellana*)	–	0.08–5.0 g PAs/kg FW	Bittner, Rzeppa, and Humpf (2013); Gu et al. (2004)
Pecans (*Carya illinoinensis*)	–	0.1–4.9 g PAs/kg FW	
Pistachios (*Pistacia vera*)	–	0.05–2.4 g PAs/kg FW	
Almonds (*Prunus dulcis*)	196–341 mg GAE/kg 0.5–0.6 g EAE/kg	0.07–1.8 g PAs/kg FW	Bittner, Rzeppa, and Humpf (2013); Xie, Roto, and Bolling (2012); Gu et al. (2004)

(*Continued*)

TABLE 13.3 (CONTINUED) Food Sources

Food	Content: Hydrolyzable Tannins	Content: Condensed Tannins	References
Cashews (*Anacardium occidentale*)	–	87–106 mg PAs/kg FW	Bittner, Rzeppa, and Humpf (2013); Gu et al. (2004)
Macadamia nut (*Macadamia ternifolia*)	–	1.5 mg PAs/kg FW	
Cocoa and coffee beans			
Cocoa beans (*Theobroma cacao*)	–	16.2–19.3 mg CE/g phenolic extract	Ortega et al. (2010)
Cocoa powder	–	14.6 g PAs/kg FW	Hellström and Mattila (2008)
Coffee beans (*C. arabica*) medium and dark roasted	–	26.9–38.3 mg CE/g sample	Campos-Vega et al. (2015)
Coffee beans (*C. arabica* and *C. canephora*) different roasting degrees	–	0–0.9 mg cyanidin chloride equivalents/g sample	Hečimović et al. (2011)
Food by-products			
Spent coffee (medium or dark roasted)	–	8–18.2 mg CE/g sample	Campos-Vega et al. (2015)
Spent coffee	–	0.2–2.9 percent GAE	Pujol et al. (2013)
Red coffee cherries pulp	–	15–20.1 g/kg DW PAs	Ramirez-Coronel et al. (2004)
Seeds of boysenberry	–	763 mg PC/kg sample	Furuuchi et al. (2011)
Apple pomace (seeds, stems, flesh and peels)	–	9.3 mg/kg PC B2	Kammerer et al. (2014)
Grape pomance (seeds, stems and skins	–	22.3 percent DW	Zhu et al. (2015)
Grape seeds	–	27–43.3 mg B1/g seed	Bozan, Tosun, and Özcan (2008)
Peels chestnuts	7.9–3542.6 μg/g DW	9–296.9 μg/g DW	Aires, Carvalho, and Saavedra (2016)
Peel and seed (kernel) of mango (*Mangifera indica* L.)	205 mg tannin/kg sample	–	Asif et al. (2016)
Stones and fruit pericarps of litchi fruits (*Litchi chinensis*)	–	172.9 and 210.2 mg/g DW	Zhou et al. (2011)
Hulls of faba beans	–	<0.1–3.5 percent CE	Jansman et al. (1995)
Seeds and husk of cacao	–	58.9 g CE/kg	Vargas-Magaña et al. (2014)
Pulp of cacao	–	14.8 g CE/kg	Vargas-Magaña et al. (2014)

Note: CE, catechin equivalents; DW, dry weight; EAE, ellagic acid equivalents; ET, ellagitanins; FW, fresh weight; GAE, gallic acid equivalent; PAs, proanthocyanidins; PC, procyanidins.

by the vanillin–HCl assay (Hithamani and Srinivasan 2014). The content of condensed tannins (182.7–440.7 mg/100 g seeds) varied among 30 maize genotypes with an average value of 300 mg/100 g (Kaur and Kaur 2016).

Minor amounts of tannins occur in the seeds of several amaranth species, particularly in dark-seeded types, for example, 0.3 percent in *Amarantus caudatus* (Pedersen, Kalinowski, and Eggum 1987) or 0.4–0.8 mg CE/100 g (Okoth et al. 2011) at levels typical of cereals. In finger millets, tannin content depends on environment with Indian cultivars containing 1 percent catechin equivalents, whereas some African cultivars had even higher values (3.4 percent CE measured by the vanillin–HCl method) (Hoseney, Varriano-Marston, and Dendy 1981). The high tannin millets altered the distribution of protein solubility fractions, thereby lowering their protein digestibility. Two pearl millet (*Pennisetum glaucum* L.) cultivars from Pakistan contained 0.17 and 0.22 percent CE tannins determined by the vanillin–HCl method (Eltayeb et al. 2007). Buckwheat (*Fagopyrum esculentum*) contain minor amounts of PAs (22.3 mg/100 g FW), whereas the content in buckwheat groats is lower (1.7 mg CE/g DM) (Awika and Rooney 2004). PAs in cereals such as corn and colored wheat are barely detectable (0.06–0.22 mg CE/g grain) and consist primarily of oligomers and small amount of polymers (Min et al. 2012).

Sorghum (*Sorghum bicolor* L.) contains condensed tannins that have significantly improved resistance to grain molds, birds, and major agronomic advantages, enabling their successful production in Africa. Sorghum genotypes with B1-B2-gene contain high tannin levels that can vary about two folds between genotypes. For example, sorghum genotypes with a pigmented testa can contain 10–15 times higher levels of tannins than those with unpigmented testa. However, tannin biosynthesis in sorghum grains may be inhibited under high growing temperature induced by climate change (Wu et al. 2016). Sorghum grains, particularly black sorghums, contain high levels of condensed tannins (68 mg/g, vanillin–HCl method) with high amount of polymers (>60 percent of polymers with degree of polymerization > 10) (Rooney and Awika 2005). The proline rich γ-kafirin sorghum prolamin protein with the most proline repeats binds the most condensed tannin (70–77 percent) and therefore decrease protein digestibility. Therefore, modification of γ-kafirin in breeding sorghums is being sought to improve digestibility and to lower calorie intake and reduce obesity incidence (Taylor et al. 2007). Even when complexed with proteins, sorghum tannins retained at least 50 percent of their antioxidant activity (Rooney and Awika 2005). Tannin sorghums are widely used in Africa in various food and food preparations, including beer, porridges, and unleavened breads. The tannin sorghum porridge is sought for its satiety, probably because of its low digestion rate and low glycemic index (Dykes and Rooney 2007). The tannin content of sorghum porridge ranged from 0.38 to 5.08 g CE/kg (Wu et al. 2016), whereas gruel made from fermented sorghum (0.2 percent DM) was not different from those produced from maize (Oboh and Amusan 2009).

13.4.2 Legumes

Highly pigmented legumes, particularly their seed coats or hulls, are rich sources of polyphenols, including tannins and natural antioxidants. The following tannin content (% DM) has been reported for *Phaseolus vulgaris* (0–0.7), *Lens esculenta* (1.0), *Cicer arietinum* (0.1–0.6), *Pisum sativum* (0.25), and *Vicia faba* (1.1) (Campos-Vega, Loarca-Piña, and Oomah 2010). Canadian grown legumes had tannin content (g CE/kg DM) for green lentils (5.4, range = 3–10.2), red lentils (6.1, range = 4.4–7.9), black turtle beans (4.9, range = 3.2–6.3), cranberry beans (8.4, range = 7.4–9.9), Dutch brown beans (9, range = 7.7–10.3), dark red kidney beans (7.5, range = 6.3–8.5), light red kidney bean (7.4, range = 4.9–8.6), pink beans (7.8, range = 7.4–8.5), pinto beans (11, range = 10.6–11.4), and small red beans (8.1, range = 5.1–10.7) (Wang and Daun 2004). This study also reported tannin contents (g CE/kg DM) of Australian grown field peas (0.2, range = 0.1–1.1), lentils (3.3, range = 0.6–6.3), kabuli chickpea (0.1, range = 0.1–0.4), desi chickpea (0.4, range = 0.1–0.9), and navy beans (4.9, range = 0.3–9.4) (Wang and Daun 2004). Condensed tannins were barely detectable in 17 field pea cultivars grown in western Canada, whereas the values varied widely for 9 grass pea breeding lines (0.89–5.18 g CE/kg) (Wang et al. 1998). Condensed tannin content ranged between 0–4.38 (1.17) g/kg for 100 lines of *Lathyrus sativus* (grass pea) germplasm; condensed tannins were positively correlated with seed coat pigmentation, with colored genotypes containing greater levels of tannins (Deshpande and Campbell 1992).

Single recessive genes (*tan tan*) for the absence of tannin have been identified in faba beans, dry beans, birdsfoot trefoil, and lentils. Generally, zero-tannin lines are characterized by the absence of anthocyanin pigments in various plant parts, including flower petals and dark seed coats, probably due to the blockage of the flavonoid biosynthetic pathway common to both anthocyanins and tannins. In lentils (*Lens culinaris* Medik.), condensed tannins are responsible for seed coat darkening during storage. Tannins are only present in the seed coat with concentration ranging from 35 to 93 g CE/kg (vanillin–HCl method) in seed coats of 87 lines from the USDA World Lentil Collection (Vaillancourt, Slinkard, and Reichert 1986).

Zero-tannin or tannin-free faba beans contain about 1 percent tannin, compared to 8–9 percent tannins in traditional or regular faba beans that are bitter to hogs, thereby restricting feed intake. Tannin-free faba beans are superior to tannin-containing faba beans as a protein source for monogastric animals. Tannin content varied (0.46–5.84 mg CE/g, determined by polyvinylpolypyrrolidone [PVPP] binding method) among 13 Canadian grown zero-tannin genotypes. Previous studies showed only small differences in condensed tannins concentrations (5.16, range = 4.8–5.64 g/kg) among different genotypes grown in replicated plots in six different locations in Manitoba and Saskatchewan in 1972 (Oomah et al. 2011). Traditional Algerian-grown faba bean (*Vicia faba*) subspecies *major* contained significantly higher tannin content than those of the subspecies *minor* (6.47 vs. 3.79 mg/g tannic acid equivalents [TAE]) by the BSA precipitation method. Tannin was concentrated in the hulls (35.5 and 36.1 mg/g TAE for *V. major* and *minor*, respectively), with only minor amounts in the cotyledons (1.75 and 2.75; 35.5 and 36.1 mg/g TAE for *V. major* and *minor*, respectively) (Boudjou et al. 2013). Tannin content of Algerian-grown lentil seed, hull, and cotyledon was 1.27, 46.27 and 0.40 mg/g CE, respectively (Boudjou et al. 2013) by polyvinylpolypyrrolidone (PVPP) binding/complexation. Pigeon pea (*Cajanus cajan* [L] Millsp) pod contained 26.9–90.8 and 31.9–49.7 g tannic acid/kg determined by the hemoglobin and BSA-binding methods, respectively (Ferreira, Souza, and Nogueira, 2003). Condensed tannin was also present in polysaccharides from beans (2.2, 1.3, and 0.8 mg CE/g of lyophilized polysaccharides from Mexican beans Negro 8025, Pinto Durango, and Bayo Madero seeds, respectively) (Campos-Vega et al. 2009). Moreover, cultivars with high condensed tannins displayed the highest short chain fatty acid production that can beneficially contribute to a healthy colon mucosa.

13.4.3 Fruits and Vegetables

Fruits, their botanical parts (skin, pulp, and seeds) and products (particularly beverages such as wine and juice) are the most commonly consumed sources of tannins. Condensed tannins (proanthocyanidins, PAs) occur naturally in many fruits and berries and vary in terms of degree of polymerization (DP). For example, procyanidins (PC) oligomers of DP > 10 have been detected in apples, whereas value >30 has been reported for grape seeds and skins (Kalili et al. 2013).

PAs particularly important in viticulture and enology have been extensively investigated due to their contribution to grape and wine quality. A total of 78 PAs have been identified in grape seeds (Kalili et al. 2013). Cranberry, blueberry, and strawberry are also rich sources of PAs (505, 332, and 145 mg/100 g of fresh fruits, respectively) (Das et al. 2016). Total PA content ranged between 18 and 92 g/kg dried fruit, quantified as procyanidin A2 by the dimethylaminocinnamaldehyde (DMAC) method for eight North American cranberry cultivars grown at four locations (Carpenter et al. 2014). Chokeberries (*Aronia mitschurinii*) and rose hip (*Rosa rugose*) are remarkably rich in PAs (1880 and 701 mg PAs/100g of fresh fruit, respectively) (Hellström and Mattila 2008). High proanthocyanidins are also present in northern berries such as sweet rowanberry (*Crataegosorbus mitschurinii*), rowanberry (*Sorbus aucuapria*), Saskatoon berry (*Amelanchier ainifolia*), and lingonberry (*Vaccinium vitis-idaea*) (276–485 mg PAs/100 g of fresh fruits) (Hellström and Mattila 2008). Both condensed tannins and hydrolyzable tannins (gallotannins and ellagitannins) occur in the edible parts of jambolan (*Syzygium cumini* [L.] Skeels) fruit, with greater amounts in the skin (Tavares et al. 2016). PAs in skin and pulp of jambolan fruit were 11.92 and 9.03 mg CE/kg FW, respectively. The hydrolyzable tannins of skin and pulp of jambolan fruit consisted of gallotannins (337 and 178 mg gallic acid equivalents/kg FW, respectively) and ellagitannins (286 and 106 mg castalagin acid equivalents/kg FW, respectively (Tavares et al. 2016).

Berries represent the richest dietary source of hydrolyzable tannins, generally ellagitannins (ET) at 2 g/kg in blackberries (Arapitsas 2012). Rubus berries are the primary sources of ET; native ellagitannin content in acetone extracts of cloudberry, raspberry and strawberry were 1600–2400, 2500 2600, and 80–180 mg/kg, respectively (Serrano et al. 2009). ET content varies in berries with the following values (mg PAs/100g) for cloudberries (360), cultivated and wild raspberries (190 and 270, respectively), strawberries (65–85), rosehip (107), and sea buckthorn (1) (Landete 2011; Koponen et al. 2007). ET identified in strawberries includes casuaricitin, sanguiin H-6 and HHDP; about 95 percent of these compounds are in the pulp, whereas only 4 percent are available in the seeds (Das et al. 2016). The peels and mesocarp of pomegranate (*Punica granatum* L.) are rich in hydrolyzable tannins (27–172 and 32–263 g/kg, respectively [Fischer et al. 2013]). These hydrolyzable tannins comprised of gallotannins and ellagitannins with gallotannin contents range from 0.1 to 1.7 g/kg and 0.1 to 0.6 g/kg in pomegranate peels and mesocarp, respectively. Punicalagin was the major ellagitanin (9.2–46 and 20.3–106.4 g/kg DM) in the peel and mesocarp, respectively (Fischer et al. 2013). In grapes, hydrolyzable tannins occur mainly in skins and seeds with sanguiin H-5, vescalagin, and castalagin as the major compounds present in high concentration in Muscadine grapes (*Vitis rotundifolia*), but at low levels in *Vitis vinifera* (Olejar, Vandermeer, and Kilmartin 2016; Landete 2011). Condensed tannins consisting mainly of procyanidins,

A-type linkages and prodelphinidin occur in crowberry (*Empetrum nigrum*), bilberry (*Vaccinium myrtillus*), and plum (*Prunus domestica*) (199, 148, and 105 mg PAs/100 g FW, respectively (Hellström and Mattila 2008). Lower amounts of PAC occur in other fruits (Table 13.3).

Vegetable tannins were probably the earliest used reagents in tanning; they leave a distinctive spectroscopic signature in tanned leather (Romer et al. 2011). For example, condensed tannins from mimosa (predominantly a prorobinetidin polymer) have a different spectra compared to those from quebraco (profistinidin) tannin (Romer et al. 2011). Only a few vegetables such as rhubarb (*Rheum rhaponticum*) and Indian squash contain low levels of PACs (120 and 16.4 mg PAs/100 g FW) consisting mainly of procyanidins and prodelphinidins (Hellström and Mattila 2008; Gu et al. 2004). Condensed tannins in vegetables (pumpkin [*Cucurbita maxima* L.], butternut (*Juglans cinerea* L.], and sweet potatoes [*Ipomoea batatas* L.]) ranged between 1.53 and 5.73 g/100 g DM, and from 3.15 to 5.73 g/100 g DM for leafy vegetables (drumstick [*Moringa oleifera* L.], pumpkin, and amaranth (*Amaranthus hybridus* L.] leaves) (Kunyanga et al. 2011). The high protein content (17.5–30.3 percent DM of which 5 percent is lysine) of amaranth leaves has been exploited as a potential leaf protein ingredient in feed production, particularly in the diet of Nile tilapia (*Oreochromis niloticus*). Tannin contents were 98.3 and 121.2 mg/100 g FW for *Amaranth hybridus* leaf and *A. hybridus* leaf protein concentrate, respectively (Ngugi et al. 2017).

13.4.4 Nuts

Betel nut (*Areca catechu* L.) is perhaps one of the richest sources of tannins (≈ 27 percent) used for human consumption in parts of Southeast Asia and South Pacific (Shahidi and Naczk 2004). Hydrolyzable tannins, particularly the ellagitannin pedunculagin, occur in walnuts with a portion of four walnuts providing 400 mg of ET (Landete 2011; Larrosa et al. 2010). Tannin content of pecan depends on cultivars ranging from 0.70 to 1.71 percent (Shahidi and Naczk 2004). Various nuts contain significant amount of PAs (mg PAs/100 g of fresh weight) (Table 13.3); for example, hazelnuts (*Corylus avellana*) and pecans (*Carya illinoinensis*) contain almost the same amount (500) of PAs. Similarly, peanuts (*Arachis hypogaea*) (4.4–186) with A-type linkage procyanidins and almonds (*Prunus dulcis*) contain between 4 to 186 mg PAs/100 g FW. Small amount of PAs (<10 mg PAs/100 g FW) are present in cashews (*Anacardium occidentale*) and macadamia nut (*Macadamia ternifolia*). The tannin content in nuts depends on the sample origin and the stage of ripeness. Almond hydrolyzable tannin content averaged 54.7 ± 2.3 mg ellagic acid and 27.4 ± 7.3 mg gallic acid per 100 g almond among California varieties (Xie, Roto, and Bolling 2012). Generally, proanthocyanidins in nuts consist of procyanidins, prodelphinidins and propelargonidins (Bittner, Rzeppa, and Humpf 2013; Hellström and Mattila 2008; Gu et al. 2004).

13.4.5 Cocoa and Coffee Beans

Cocoa (*Theobroma cacao*) consist primarily of PA dimers (B-type 26 percent), trimers, tetramers (trace amounts), and monomers (51 percent) (Barrett et al. 2013). The beans (Forastero variety) contain procyanidins (16.2–19.3 mg catechin/g of phenolic extract comprised of catechin, epicatechin, and different polymers (Ortega et al. 2010).

The procyanidin content decrease (from 60753 to 2974 μg/g) during fermentation to develop the characteristic cocoa flavors (Wollgast and Anklam 2000). However, the procyanidins (1460 mg PAs/100 g FW) are retained in cocoa powder (Hellström and Mattila 2008). Cocoa pod husk constitutes a major part of the whole matured cocoa fruit (700 g/kg) and contains 11.88 g/kg tannins (Murugan and Al-Sohaibani 2012).

Tannin contents of sundried coffee pulp vary depending on the method of analysis. Thus, gelatin and formaldehyde precipitation methods yielded 0.43 and 1.64 percent hydrolyzable tannins, respectively, whereas alkali solubilized 2.56 percent of condensed tannins. The content was ≈ 1 percent DW by a qualitative spot test (Clifford and Ramirez-Martinez 1991). Both coffee pulp and husk contain tannins (1.8–8.5 percent) (Murugan and Al-Sohaibani 2012). Condensed tannins in coffee beans depend on the extent of roasting with medium roast containing higher amount than dark roast (38.3 vs. 26.9 mg CE/g sample) (Campos-Vega et al. 2015).

13.4.6 Beverages

The tannin concentration measured by the protein precipitation method varied from 0.35 to 1.34 g/L and as lower than when measured by HPLC (2.75–6.16 g/L) for 34 wines (Watrelot et al. 2016). The beverage (brewing, wine, and soft drink) industry also uses food-grade tannins developed as stabilizing processing aids for antioxidant and taste improvement. For example, Ajinomoto OmniChem tannins (50–150 mg/L beverage) are designed to mask flavor, modify taste, stabilize natural colors, protect premature oxidation of vitamin C, and contribute to mouthfeel or astringency, thereby guaranteeing a constant end product. Proanthocyanidins in wine vary (499–1260 mg CE/L) depending on grape variety with procyanidins (80–91 percent) and prodelphinidins (8.8–19 percent) as the major components (Gonzalo-Diago, Dizy, and Fernández-Zurbano 2013). Different levels of ellagitannins are found in grape juice and wine; ET content was 0.8 mg ellagic equivalents/L in juice from cold-pressed white grapes, and 11.5–23 mg ellagic equivalents/L in juice and wine from hot-pressed red grapes (Lee and Talcott 2002). Hydrolyzable tannins increase in wine and spirits stored in oak barrels during aging due to the seepage of vescalagin and castalagin (2.2 and 8.1 mg/L, respectively) from oak wood giving rise to small amounts (2 mg/L) of acutissimin, epiacutissimin, and ethylvescalagin in wine (McRea et al. 2016; Landete 2011; Saucier et al. 2006). Similarly, bourbon whiskey contains castalagin and vescalagin (0.4 and 0.3 mg/L, respectively) and minor amounts of gandinin and roburin (0.01–0.1 mg/L) (Glabasnia and Hofmann 2006).

Pomegranate juice composition is strongly influenced by the extraction process and the source of the raw material. Hydrolyzable tannins comprised of gallotannins and ellagitannins and together with gallagyl esters account for 87 percent of the total phenolic compounds of pomegranate juice (Fischer et al. 2013). Gallotannins account for 0.2–24 percent (3.4–6.8 mg/L) in the juice, whereas the major component punicalagin represented 30–44 percent (0.4–3.6 g/L juice) of the hydrolyzable tannin of pomegranate juice (Fischer et al. 2013). A serving (200 mL) of pomegranate juice provides about 1 gram of ET (Larrosa et al. 2010). Good levels of ET are also found in European commercial juices; 415–1611 μmol ET/L in 100 percent pomegranate juice, and lesser amounts (324–963 μmol ET/L) in pomegranate juices blended with other fruits (grapes, aronia berry, acai, lime, mandarine, oranges, and others) (Borges, Mullen, and Crozier 2010). PAs content is high in cranberry and grape juices (231 and 524 mg of PAs/L, respectively), but low in currant and apple juices (0.15–9 mg of PAs/L) (Gu et al. 2004). Commercial juice concentrate of boysenberry contained 11 milligrams of procyanidin/L, consisting primarily of procyanidins and propelargonidin dimers (Furuuchi et al. 2011). The average PAs content of Ocean Spray cranberry juice cocktail was 30 mg/8 fl oz for 12 samples obtained from four different bottling locations quantified by the DMAC method (Cunningham et al. 2002). Other beverage sources of condensed tannins are beer (23 mg/100g FW), followed by black and green tea (3.3 and 12.2 mg/100 g FW), and apple cider (0.1–3.5 mg/L). PAs in these beverages consist of procyanidins, propelargonidins, and 3-O-gallates (Hellström and Mattila 2008).

13.4.7 Food By-Products

Rapeseed and canola meals, coproducts of the oil-processing industry, have been extensively investigated due to their use in animal feed. Condensed tannins are primarily concentrated in canola hulls (57–1556 mg CE/100 g hulls determined by the vanillin assay) and therefore occur in the meal after oil processing. However, the tannin content of canola meal varies (73–339 mg CE/100 g) depending on extraction solvent with acidified aqueous acetone (1 percent HCl in 70 percent acetone + four minutes boiling) and acidified methanol (1 percent HCl in 70 percent methanol), extracting the least and most tannins, respectively (Shahidi and Naczk 1995).

The economic disposal of residues from the wine industry has been a common concern since the early 1970s. Several attempts have been made in utilizing grape pomace (seeds, stems, and skins) for developing novel functional food products. Grape pomace is a rich source of condensed tannin (22.3 percent DM), whereas grape seed PAs (27–43.3 mg B1/g seed) consist primarily of proanthocyanidins B1 and B2, depending on grape variety (Zhu et al. 2015; Bozan, Tosun, and Özcan 2008). Separation of the crude tannin-rich grape seed extract led to the development of products such as MegaNatural BP and MegaNatural Gold with a specific array of molecular weights with demonstrated effects on supporting blood pressure in the normal range and oxidized cholesterol, respectively. MegaNatural-AZ grape seed-derived polyphenolic extracts (GSPE) is a highly purified, 100 percent water soluble polyphenolic extract containing less than 12 percent galloylated proanthocyanidins by weight of the total PAs. GSPE is designed as a potential therapeutic agent in neurodegenerative disorders including reduction (30–50 percent) in Alzheimer's disease–type cognitive deterioration demonstrated in mice by preventing amyloid formation in the brain (Pasinetti et al. 2010).

Seeds of boysenberry contain high amounts of proanthocyanidins (76.3 mg equivalent procyanidin/100 g), consisting of procyanidin and propelargonidin dimer and trimer (Furuuchi et al. 2011). Apple pomace obtained from the juice industry contains 9.3 mg/kg of procyanidin B2, composed of catechin dimers and other higher polymerized procyanidins (Kammerer et al. 2014). Peels obtained from industrial processing of chestnuts are good sources of condensed tannins consisting of (μg/g DW) (–)-epicatechin (9-66.8), (–)-epigallocatechin (13.6–213.4), (–)-epigallocatechin-O-gallate, and (+)-catechin (151-296.9); gallic acid (7.9–584.9), vescalagin (67.5-109.4), castalagin (49.6–100.4), acutissim (A and B), and ellagic acid (47.6–3542.6) comprised the hydrolyzable tannins (Aires, Carvalho, and Saavedra 2016). Peels and seeds are the principal wastes from mango (*Mangifera indica* L.) processing. Mango seed tannin (20.5 mg/100 g) consists mainly of gallotannins (Asif et al. 2016). A tannin-rich ethyl acetate fraction of mango gallotannins had a dry matter content of 24 g/L and exhibited an iron chelating capacity that was equivalent to a solution of 42.5 g/L (25 mM) tannic acid (Engels et al. 2009). Litchi (*Litchi chinensis*) stones and fruit pericarps are also good sources of condensed tannins (172.9 and 210.2 mg/g DW) (Zhou et al. 2011).

Some tannin-containing plant by-products have been under recent investigation for their anthelmintic (AH) activities; by-products from the nut industry in temperate areas, cabob pods in the Mediterranean region, and coffee by-products and cocoa fruit husks and leaves in Yucatan, Mexico (Hoste et al. 2015). Percolated by-products of *Coffea arabica* with tannin content (2.28 and 2.66 g epicatechin equivalents/100 g) exerted in vitro anthelmintic effect against exsheathment (Chan-Pérez et al. 2016). Plant by-products *Theobroma cacao* seed husks and pulp and percolated *Coffea arabica* contained 58.9, 14.84, and 2.02 g CE/kg, respectively, with the *T. cacao* extracts (seed husk and pulp) exerting mild ovicidal activity (Vargas-Magaña et al. 2014). The extent of roasting affects the condensed tannin content of spent ground coffee with higher content from dark to medium roast (18.2 vs. 8 mg CE/g) (Campos-Vega et al. 2015). The extraction method also influences the tannin content of spent ground coffee (0.17–2.93 percent gallic acid equivalent) (Pujol et al. 2013). Proanthocyanidins (15–20 g/kg DM) also occur in the pulp of red coffee berries (*C. arabica*) (Ramirez-Coronel et al. 2004).

13.5 TANNINS AND METABOLIC DISEASES

13.5.1 Obesity

The consumption of a diet containing tannins such as procyanidins have been reported to significantly reduce weight gain, decrease adipose tissue mass, and ameliorate insulin tolerance in various animal models and human studies (Table 13.4). Dietary tannins affect the expression of key genes involved in the regulation of antioxidative enzymes and of glucose and lipid metabolism. This effect was clearly demonstrated in *Cichorium intybus* methanolic extract reduction in blood glucose levels without inducing adipogenesis in 3T3-L1 adipocytes; tannins present in the extract conferred this effect since detannification nulled this effect. Both activities, glucose uptake and adipogenesis inhibition, were mediated by two independent factors insulin receptor (IR) and IGF1 but controlled by the only linkage PTP1B enzyme (Mutusami et al. 2008). Methanolic and aqueous extracts from chesnut by-products, containing 35.8 and 36.4 mg GAE/g (w/w) soluble and insoluble tannins, respectively, inhibited adipocyte differentiation by down-regulating the mRNA expression levels of CCAAT/enhancer binding protein (C/EBP)-β, C/EBPα, and peroxisome proliferator-activated receptor (PPAR)-γ in 3T3-L1 cells, inhibiting pre- and early stage adipogenesis in 3T3-L1 cells (Youn et al. 2016).

The pancreatic lipase (PL) enzyme is responsible for the digestion of 50–70 percent of dietary triglycerides. This enzyme hydrolyses triglycerides into monoacylglycerides and free fatty acids that can be absorbed by enterocytes; therefore, PL inhibition can reduce fat absorption, thereby decreasing energy uptake, a key target to mediate obesity (Chakrabarti et al. 2009). Different polypenolic-rich extracts from sources such as teas, fruits, and high tannin-containing plants inhibit PL during in vitro experiments with proanthocyanidins or ellagitannins, gallate esters, or stilbenoids considered to be the main active ingredients. The inhibitory mechanisms of these compounds are unknown, but it has been attributed to the ability of tannins to bind, complex, and precipitate proteins, although some studies reported noncompetitive or mixed inhibitions of PL (Sergent et al. 2012). The inhibitory effect of PL has also been reported in in vitro experiments.

TABLE 13.4 Health-Related Benefits

Target Disease	Source	Active Component	Preparation	Therapeutic Effect	Reference
Obesity	Chestnut (*Castanea crenata*) by-products	Hydrolysable tannins (Gallotannins, ellagitannins)	Water and methanolic extracts	Inhibit lipid accumulation and differentiation in adipocites	Youn et al. (2016)
	High tannin sorgum	Condensed tannins (High molecular weight proanthocyanidins)	Tannin extracts treated starch products	Increase resistant starch, alter starch digestion profile reducing the caloric intake	Amoako and Awika (2016)
	Red type III sorghum	Condensed tannins (proanthocyanidins, procyanidins anthocyanins)	Encapsulated in sorghum kafirin protein microparticles	Prevent blood glucose spike and decrease the maximum blood glucose level	Links et al. (2016)
	Cichorium intybus	Total tannins	Methanolic extract	Attenuate blood glucose level without adipogenesis induction	Muthusamy et al. (2008)
	Litchi flower	Condensed tannins, anthocyanins, and proanthocyanidins	Water extract	Suppress in vivo lipase activities	Wu et al. (2013)
	Cranberry and grape	Highly polymeric procyanidins	Extracts mixtures	Regulate microbiota	Masumoto et al. (2016)
	Pomegranate	Ellagitannins	Extract	Regulate microbiota, decrease inflammation markers	Yang et al. (2016)

(*Continued*)

TABLE 13.4 (CONTINUED) Health-Related Benefits

Target Disease	Source	Active Component	Preparation	Therapeutic Effect	Reference
Diabetes	Babul (*Acacia nilotica*)	Catechins, ellagic acid, gallic acid and tannic acid	Roasted and pulverized incorporated into biscuits	Decreases glycemic index	Kumari et al. (2014)
	Banaba (*Lagerstroemia speciosa* L.)	Ellagitannins	Aqueous and methanolic extracts	Hypoglycemic effect	Miura et al. (2012)
	Sorghum	Condensed tannins	Encapsulated in sorghum kafirin protein microparticles	Decrease α-amylase and α-glucosidase	Links et al. (2016)
	Strawberry raspberry, red grape, red wine, and green tea	Total tannins	Different extracts	Decrease α-amylase and α-glucosidase	McDougal et al. (2005)
	Cinnamon (*Cinnamomum verum*) bark	Total tannins		Regulate posprandial hypergycemia by α-glucosidase inhibition	Shihabudeen et al. (2011)
	Cinnamon Cinnamomum tamala and Cinnamomum cassia	Procyanidin B2 (cpd1), (−)-epicatechin (cpd2), cinnamtannin B1 (cpd3), procyanidin C1 (cpd4), parameritannin A1 (cpd5) and cinnamtannin D1 (cpd6) D1 (cpd6)	Cinnamon extracts and purified procyanidin oligomers	Increse β-cells viability and decrease ROS acumulation, increaseglucose-stimulated insulin secretion	Sun et al. (2016)

(*Continued*)

TABLE 13.4 (CONTINUED) Health-Related Benefits

Target Disease	Source	Active Component	Preparation	Therapeutic Effect	Reference
Cardiovascular disease	Pomegranate (*Punica granatum* L.)	Ellagitannins and anthocyanins (punicalagin)	(OXYLENT GR^{w}, Stiernon S.A.) Extract from pomegranate peel 30% polyphenol content and concentrations of 8% punicalagin and 5% ellagic acid	Reduce serum level of cholesterol (total and LDL), increase caecal content weight and caecal pool of bifidobacteria	Neyrink et al. (2013)
	Grape-wine and grape	Anthocyanins, procyanidins	Grape-wine extract and grape juice extract with 800 mg polyphenols	Decrease the 24-hour ambulatory systolic and diastolic blood pressure	Draijer et al. (2015)
	Pomegranate	Total tannins and athocyanins	Juice	Increase HDL- associated paraoxonase 1 (PON 1), breaking down atherosclerotic plaques	Aviram and Rosenblat, (2012)
	Ficus racemosa	Total tannins	Tannins extraction	Normalize total cholesterol and LDC levels in plasma	Velayutham, Sankaradoss, and Nazeer (2012)
Hepatic disease	Cinnamon (*Cinnamomum tamala* and *Cinnamomum cassia*)	Procyanidin B2 (cpd1), (–)-epicatechin (cpd2), cinnamtannin B1 (cpd3), procyanidin C1 (cpd4), parameritannin A1 (cpd5) and cinnamtanninD1 (cpd6) D1 (cpd6)	Cinnamon extracts and purified procyanidin oligomers	Increase β-cells viability and decrease ROS acumulation, increase glucose-stimulated insulin secretion	Sun et al. (2016)

Litchi flower–water extract (LFWE) containing a significant amount of phenolic acids, flavonoids, condensed tannins, anthocyanins, and proanthocyanidins effectively suppressed lipase activities in vivo in rats and decreased liver sizes, perirenal, and epididymal adipose tissues, and cell sizes of epididymal adipose tissues in hypercaloric diet–fed rats. Serum cholesterol and liver lipid levels also decreased in hypercaloric diet–fed LFWE-treated rats, resulting in higher fecal lipid concentrations (Wu et al. 2013).

Recently, the importance of gut microbiota has increased as it represents a major environmental factor that contributes to diseases such as obesity, diabetes, and metabolic syndrome. Most ingested tannins reach the colon where the gut microbiota plays a key role in tannin metabolism producing metabolites with health benefits (Marín et al. 2015). Information is limited on dysbiosis related to diseases like obesity that can affect the production of different tannin metabolites, but reduction in the obesity rates of germ-free mice has demonstrated that the gut microbiota can affect the host energy regulation and nutrient acquisition, as it improves insulin sensitivity and glucose tolerance (Masumoto et al. 2016). Non-absorbed highly polymeric procyanidins reach the colon and help prevent obesity, as well as impair lipid metabolism; these effects have been associated with changes in gut microbiota like the Firmicutes/Bacteroidetes ratio and the modulation of endogenous metabolites. Cranberry extracts and grape procyanidin mixtures with other polyphenols modulate the gut microbiota and improve obesity and diabetes in an animal model and have been associated with an increase in the proportion of *Akkermansia* in the gut microbiota of mice fed a high-fat diet. The administration of *Akkermansia* as a probiotic protects against obesity and prevents the high fat/high sucrose diet-induced increase in lipopolisacharides (LPS) release (Masumoto et al. 2016).

Ellagitannins are transformed by gut microbiota into ellagic acid; which is also metabolized by the gut bacteria producing urolithins A and B. Urolithin production varies among individuals, and about 30 percent of individuals cannot produce urolithin A when consuming ellagitannins-rich foods; this variability in microbial metabolism of ellagic acid has been attributed to differences in gut microbial ecology. Pomegranate extracts are an important source of ellagitannins, increase cecum bifidobacteria, decrease inflammation, and counteract the high fat–induced expression of inflammatory markers both in the colon and visceral adipose tissue (Yang et al. 2016). Ellagic acid, as well as urolithin A, C, and D triggers AMPK activation in cultured human adipocytes by inhibiting hASCs differentiation into adipocytes. This AMPK activation regulates energy homeostasis by inhibiting adipogenesis and de novo TG synthesis, and by augmenting fatty acids oxidation. Only a few studies have been conducted to determine whether AMPK is activated by urolithins in vivo. Therefore, further research is required to validate the potential AMPK activation by both elleagic acid and urolithins (Kang et al. 2016).

13.5.2 Diabetes

Tannins can positively affect diabetes by different mechanisms. These phenolic compounds reduce glycemia or blood glucose levels by decreasing the intestinal glucose absorption, an induction of β-cells regeneration and an insulin-like effect that can be exerted on insulin sensitive tissues such as adipose cells (Serrano et al. 2009).

The decrease in blood glucose has been studied by adding tannin-rich functional ingredients, such as the incorporation of Babul (*Acacia nilotica*) powder, in biscuits that positively regulated blood glucose levels; the effect was attributed to the high amount of tannins present in the tree powder (Kumari et al. 2014). Banaba (*Lagerstroemia speciosa L.*), a plant whose extracts have been used for many years in folk medicine to treat diabetes primarily in southeast Asia, also showed hypoglycemic effect in various animal models and human studies. Its antidiabetic effects have been attributed mainly to corosolic acid and ellagitannins, as well as a synergism between these compounds (Miura et al. 2012).

Tannins inhibit carbohydrate-hydrolysing enzymes, such as α-amylase and α-glucosidase in the small intestine. An in vitro analysis indicated that sorghum-condensed tannins can survive simulated gastric digestion and inhibit digestive amylases when encapsulated in sorghum kafirin protein microparticles; the microparticles decrease blood glucose levels similar to the synthetic inhibitor acarbose (Links et al. 2016).

The inhibition of these enzymes decreases the absorption of simple carbohydrates and can help regulate blood glucose levels. The reduction of these enzyme levels has also been studied using different food sources, such as soft fruits; the most effective extracts in inhibiting α-amylase were strawberry and raspberry, which contained appreciable amounts of soluble tannins. Other tannin-rich extracts from red grapes, red wine, and green tea were also effective inhibitors of α-amylase. On the contrary, α-glucosidase was more readily inhibited by blueberry and blackcurrant extracts and was related to the highest anthocyanin content; removing anthocyanins and tannins from the extracts also removed the inhibition of the enzymes (McDougal et al. 2005). Red kidney bean is recognized for its potent α-amylase inhibitory activity that is significantly increased by microwave treatment (Oomah et al. 2013). The postprandial hyperglycemia of diabetic rats showed a favorable effect by the administration of cinnamon (*Cinnamomum verum*) bark, containing an important amount of tannins that inhibited α-glucosidase activity (Shihabudeen et al. 2011).

Tannins exert antidiabetic effects in delaying the onset of insulin-dependent diabetes mellitus by regulating the antioxidant environment of pancreatic β-cells (Serrano et al. 2009). Extracts containing tannins can protect pancreatic β-cells and improve insulin secretion. This effect has been attributed to the protective effect of different procyanidin oligomers from cinnamon extracts on β-cells partly by reducing ROS (reactive oxygen species)–induced injury. The trimer procyanidins, cinnamtannin B1 and D1, inhibited cyclooxygenase-2, an enzyme associated with lipid peroxidation, whereas cinnamtannin B1 exerts antioxidative effects in pancreatic acinar cells; different studies from this group also showed that cinnamtannin D1 protects against PA-induced β-cell dysfunction by attenuating oxidative stress and reducing nuclear factor kappa B (NF- κB) activation (Sun, 2016). Recent evidence indicates that tannins can act on cells by modifying or interacting with specific proteins of important intracellular signaling pathways and improving hyperglycaemia (Serrano et al. 2009).

Hydrolyzable tannins are converted by microbial metabolism and gastric digestion into absorbable low molecular weight metabolites such as tannic acid. Commercially available tannic acids induce phosphorylation of the insulin receptor (IR) and act as translocation of glucose transporter 4 (GLUT 4), the protein factors involved in the signaling pathway of insulin-mediated glucose transport (Abdirahman et al. 2015). Protein glycation is a spontaneous in vivo reaction related to the degree and duration of hyperglycemia. The protein glycation can produce advanced glycation endproducts (AGEs) that can permanently alter protein structures and functions contributing to reduced artery, heart, and lung tissue elasticity and appear to have a significant role in the progression of general cardiovascular complications associated with diabetes. Catabolites derived from colonic microbial action against ellagitannins, urolithins, and pyrogallol, have shown a protective effect from glycation and reduced AGE formation by almost 50 percent. Pomegranate/raspberry ellagitannin-derived catabolites act as antiglycative agents, becoming good candidates in the control of cardiovascular hyperglycemia-related complications (Verzeloni et al. 2011).

13.5.3 Coronary Diseases

Coronary or cardiovascular disease (CD) represents a compilation of diseases and symptoms that affect the heart and blood vessels. Many studies have focused on the activity of food and food substances to prevent or mitigate CD incidence: fruits, vegetables, dark chocolate, red wine, and some plant-derived foods have shown CD risk reduction potential. Different studies have proposed that some dietary antioxidants such as proanthocyanidins prevent CD effectively. Some of these compounds reduce the oxidation of low-density lipoprotein (LDL), a key initiator of atherosclerosis, reduce overall oxidative stress and inflammatory markers, and have a positive impact on lipoprotein metabolism (Fava et al. 2006).

The consumption of tannin-rich foods and beverages such as tea, cocoa, grape juice, and red wine have shown improved endothelial function, measured by the changes in brachial artery flow-mediated dilation (Fava et al. 2006). Proanthocyanidins improve serum antioxidant status, decrease serum C-reactive protein and plasma homocysteine concentrations (Yang et al. 2013). Extracts from grape wine effectively decrease the 24-hour ambulatory systolic and diastolic blood pressure and the plasma concentration of the vasoconstrictor endothelin-1 in mildly hypertensive subjects. This effect was attributed to the catechins and proanthocyanidins contents from this type of grape (Draijer et al. 2015).

Pomegranate juice, rich in tannins and anthocyanins, protects against the oxidation of LDL and high-density lipoprotein (HDL), attenuating the development of atherosclerosis and related cardiovascular complications. Pomegranate antioxidants have also been reported to increase HDL-associated paraoxonase 1 (PON 1), which breaks down harmful oxidized lipids in lipoproteins, macrophages, and atherosclerotic plaques (Aviram and Rosenblat 2012). The cardioprotective effect of red wine has been attributed not only to resveratrol but to its synergic effect with proanthocyanidins, which exhibited strong antioxidant activity in the blood vessels by targeting the endothelium dependent relaxing cells, inhibiting the angiotensin-I converting enzyme and decreasing the tissue injury induced by ischemia (Chiarini et al. 2013).

Hyperlipidemia, a disease associated with alterations in the plasma lipid and lipoprotein profile, increases the risk of coronary heart diseases. Tannins from *Ficus racemosa*, used in Ayurvedic medicine in India, normalized total cholesterol and LDC levels in plasma in an animal model. It suggests that tannins modulate fatty acid catabolism in the liver, probably by regulating lipoprotein hydrolysis. The antioxidant status is restored by normalizing oxidative stress–related enzymes such as superoxide dismutase and catalase, and reducing the gluthathione peroxidase and gluthathione, which is important to prevent some cardiovascular diseases (Velayutham, Sankaradoss, and Nazeer 2012).

Amlamax™, a purified, standardized, dried extract of amla (*Emblica officinalis Gaertn.*), Indian gooseberry that contains about 35 percent galloellagitannins and other hydrolyzable tannins, reduced total and LDH cholesterols and enhanced the beneficial HDL cholesterol in a clinical study. These results compare with previous animal models with rabbits that showed similar tendencies (Antony, Benny, and Kaimal 2008). Persimmon tannins inhibit the expression of SREBP-2 and NPC1L1 genes, important genes for cholesterol synthesis and absorption, respectively. These compounds also promoted the cholesterol efflux by regulating mRNA and the protein levels of ABCA1, ABCG1, SR-BI, CYP7A1, and ABCG5/G8 (Zhenzhen et al. 2016). Rats with induced hypertension consuming freeze-dried fermented blueberries with probiotics showed significant blood pressure reduction and altered caecal microbiota (decreased Lachnospiraceare and increased Parabacteroides and Bacteroides types of bacteria). Similar observations have been reported as the Bacteroides species increased with a tannin diet (Ahrén et al. 2015), presumably related to the metabolites produced by the bacteria that can exert a protective effect against cardiovascular diseases.

13.5.4 Other Diseases

13.5.4.1 Cancer

The effect of tannins has been studied in the prevention of other diseases, such as cancer. Hydrolyzable and condensed tannins have shown anticancer activities in various studies; one of the mechanisms is attributed to the suppression of oxidative stress, an important target for cancer cells without damaging healthy cells. Tannins are transformed into epigallocatechin gallate and gallic acid (GA) by tannase, an enzyme produced by the gastrointestinal microbiota; GA is decarboxylated to form propyl gallate (PG) as the final product of the tannin metabolism (Hossain et al. 2013). Both GA and PG generate ROS that activate the tumor suppressor p53 in colorectal cancer cells. These compounds deplete cellular glutathione with selective cancer cell target killing action by modulating oxidative stress. GA also modulates NF-κB, Akt, and ATM kinase, signaling pathways to prevent carcinogenesis and it is a competitive inhibitor of the pro-inflammatory mediator COX-2 (Verma et al. 2013). Tea condensed tannins, such as (–)-epigallocatechin gallate (EGCG) favorably (but not significantly) changed serum prostate-specific antigen, serum insulin-like growth factor axis, and oxidative DNA damage in blood leukocytes in a randomized, double-blind placebo-controlled human study (Nguyen et al. 2011). Apple polyphenol extract rich in proanthocyanidins effectively suppressed the epidermal growth factor receptor phosphorylation inhibiting the growth of human colon carcinoma HT29 cells in vitro and antagonizing cancer promotion in vivo (Serrano et al. 2009). Condensed tannins isolated from black beans inhibited colon (Caco-2), breast (MCF-7 and Hs578T), and prostatic (DU 145) cancer cells by suppressing the fetal bovine serum stimulated cell migration, the secretion of matrix metalloproteinase-2 (MMP-2), matrix metalloproteinase-9 (MMP- 9), and vascular endothelial growth factor VEGF165 receptor expression (Bawadi et al. 2005). The same inhibitory effect was reported for different lung cell lines NCI-H1703, NCI-H460, A549 and human fibrosarcoma cell line HT1080 by the *Fructus phyllanthi* tannin fraction, a traditional Tibetan medicinal plant. The tannin fractions induced cell apoptosis and inhibited the migration and invasion of the cancer cells by decreasing MPPs expression through regulation of the MAPK pathway (Zhao et al. 2015). Apoptotic activity was increased in breast and prostate cancer cells when exposed to tannin extracts (Jordan et al. 2016). When these extracts are cross-linked with collagen and exposed to adipocytes, they induce caspase mediated apoptosis in ER+ and HER2+ breast cancer cells but reduce apoptotic inducing activity in normal breast epithelial cells.

13.5.4.2 Neuroprotective

Diverse tannin metabolites have demonstrated neuroprotective effects. Galloylated cyanogenic glucosides, gallotannins, ellagitannins, ellagic acid derivatives, and aromatic compounds isolated from the leaves of *P. Rotundifolia*, a Malaysian plant exerted neuroprotective activities against oxidative damage in NG108-15 cells challenged by H_2O_2 induction. Tannin-containing blueberry drink and dark chocolate also protect against H_2O_2-induced damage in peripheral blood mononuclear (PBMN) cells; fruits such as *Chrysobalanus icaco* or their parts such as pulp (Piquia and Açai) or peel (apple) protect against DNA damage in animal studies (Azqueta and Collins 2016). Several studies have reported that gallotannins increase the neuroblastoma–glioma hybrid cell viability proving tannins to have a preventive effect in neurodegenerative diseases such as Alzheimer's disease, Parkinson's disease, stroke, and dementia, reducing the oxidative stress-induced cell damage mediated by ROS, mainly superoxide anion and hydrogen peroxide (Tan et al. 2012).

13.5.4.3 Hepatoprotective

The hepatoprotective effect of tannin-rich sources has been reported in several in vitro and in vivo studies; most studies have evaluated extracts from some parts of plants containing different compounds and comparing them to isolated tannins. The mechanisms of action demonstrated by these compounds were the regulation of toxic metabolites as ROS and signal molecules on liver tissue, attributing the effect to the radical scavenging of the tannins. These outcomes were attained by chebulagic acid and chebulinic acid, two of the main bioactive hydrolyzable tannins of *Terminalia chebula Retz.* (Combretaceae), a tropical almond widely distributed in tropical and subtropical countries. The acids decreased extracellular matrix accumulation by inhibiting hepatic stellate cells proliferation, Smad2, Smad3, and Smad4 signal transduction, and procollagen I (α1) and III synthesis, as well as facilitating the resolution of fibrosis indirectly through PAI-1 inhibition, suggesting that both hydrolyzable tannins could be useful in controlling liver fibrosis (Chuang et al. 2011). Chebulagic acid present in *Triphala churna* inhibits tumor growth and VEGF-A mediated angiogenesis (Lu and Basu 2013). Different animal models showed that ellagitannins from walnut polyphenolic extracts and pomegranate suppressed carbon tetrachloride-induced liver injury in mice when orally administrated. Hydrolyzable tannin-containing pomegranate fruit extract (PFE) inhibits prosurvival signaling pathways in lung carcinoma A549 cells and tumor growth in nude mice (Khan et al. 2013). In A/J mice, PFE (0.2 percent, w/v) significantly reduced lung tumor multiplicity (by one-third), COX 2, RARβ, VEGF, CD31, and angiopoietin 2 expression and incidence of lung adenomas (Khan et al. 2013). Both hepatoprotective effects are attributed to peroxidate inhibition of hepatic lipids, reduced production of proinflammatory cytokines such as tumor necrosis factor (TNF)-R, interleukin (IL)-1β, and IL-6, and activation of nuclear factor (NF)- κB, which are involved in liver injury induction (Shimoda et al. 2008).

13.6 CONCLUDING REMARKS

The presence of tannins is important in plant physiology as a natural protection against pathogens, insects, herbivores, and other environmental stress. These environmental stressors are presumed to increase with climate change and global warming, thereby affecting tannin content in many plant and plant products. Tannins have evolved differently in each industry and sector, although it has long been considered an antinutrient in the food sector. This explains the overwhelming procedures developed to reduce or eliminate tannins in many plant and food products. For example, tannins are removed from green tea to obtain a product (PurTea) without the negative sensory attributes thereby reducing the overall catechin absorption within the body (Menayang, 2016). In contrast, tannin addition can remove/reduce peanut allergen in foods and drinks by forming stable allergen-bound insoluble complexes under gastric and intestinal conditions, and ultimately passing through without causing an allergic reaction (Chung and Reed 2012). Similarly, apple-condensed tannins can also inhibit the development of food allergies (Akiyama et al. 2005).

Numerous health benefits have been attributed to tannins in preventing or alleviating diseases and enhancing the effectiveness of treatments in some illnesses (Table 13.4). Ellagitannins, a hydrolyzable tannin from oak tree (*Quercus robur*) extract, Robuvit® significantly ($P < 0.05$) improved triathlon performance (swim, bike, and run) totalling to an average 10.6 percent less time to complete a triathlon. It is presumed that the physiologically inactive ellagitannins are transformed by intestinal microbiota into the active metabolites urolithins A–C that attenuates oxidative stress, particularly plasma free radicals (≈ 20 percent) (Vinciguerra, Belcaro, and Cacchio 2015). A similar mechanism has been invoked for the benefits of consuming cranberry juice containing high PA content. Furthermore, habitual high intake of PAs is associated with low fat mass independent of shared genetic and common environmental factors (Jennings et al. 2017). This may be related to the amylose–tannin interaction that can potentially be used to modulate glucose metabolism and reduce the calorie density of starch-based foods. Other beneficial attributes such as the antiviral activity against *Herpes simplex* virus type 1 are ascribed to the high amount of tannins (124 mg gallic acid equivalents/g) in extracts from *Schinus terebinthifolia* Raddi, a popular folk medicine (Nocchi et al. 2016).

Condensed tannins are one particular class of secondary products that are critically important in forage crops. Two types of intervention are required; reduction in tannin content of highly tanniferous species of tropical origin for increased palatability and nutritive value and tannin increase in some temperate forage to reduce bloat symptoms in livestock (Robbins and Morris 2000). Tannin decreases gas production, organic matter digestibility, total volatile fatty acid, and methane production in tanniferous browse plants. The low methanogenic potential and substantial ammonia generation of some of the browses may be potentially useful as rumen manipulating agents (Gemeda and Hassen 2015). Great strides have been made in developing zero/low-tannin variants of many crops including faba bean ever since the identification of single recessive genes (*tan tan*) responsible for the absence of tannins. In addition, hydrolyzable tannins at the right dosage are currently considered as a next-generation additive with many beneficial properties on gut health, particularly for their antimicrobial properties in broilers and piglets (Papanikou 2016). The evolution of tannins continues particularly in the developments and design of functional food ingredients with health benefits.

REFERENCES

Abdirahman, Y.A., K.K. Juma, M.J. Mukundi, S.M. Gitahi, and D.S. Agyirifo. 2015. In-vivo antidiabetic activity and safety of the aqueous stem bark extract of *Kleinia squarrosa*. *Journal of Diabetes and Metabolism* 6: 601.

Ahrén, I., J. Xu, G. Önning, C. Olsson, S. Ahrné, and G. Molin. 2015. Antihypertensive activity of blueberries fermented by *Lactobacillus plantarum* DSM 15313 and effects on the gut microbiota in healthy rats. *Clinical Nutrition* 34: 719–726.

Aires, A., R. Carvalho, and M.J. Saavedra. 2016. Valorization of solid wastes from chestnut industry processing: Extraction and optimization of polyphenols, tannins and ellagitannins and its potential for adhesives, cosmetic and pharmaceutical industry. *Waste Management* 48: 457–464.

Akiyama, H., Y. Sato, T. Watanabe, M.H. Nagaoka, Y. Yoshioka, T. Shoji, T. Kanda, K. Yamada, M. Totsuka, T. Teshima, J.-I. Sawada, Y. Goda, and T. Maitani. 2005. Dietary unripe apple polyphenol inhibits the development of food allergies in murine models. *FEBS Letters* 579: 4485–4491.

Amoako, D.B., and J.M. Awika. 2016. Polymeric tannins significantly alter properties and in vitro digestibility of partially gelatinized intact starch granule. *Food Chemistry* 208: 10–17.

Anirudhan, T.S., and P.S. Suchithra. 2008. Synthesis and characterization of tannin-immobilized hydrotalcite as a potential adsorbent of heavy metal ions in effluent treatments. *Applied Clay Science* 42: 214–223.

Antoine, M.-L., C. Simon, and A. Pizzi. 2004. UV spectrophotometric method for polyphenolic tannin analysis. *Journal of Applied Polymer Science* 91: 2729–2732.

Antony, B., Benny, M., and Kaimal, T.N.B. 2008. A Pilot clinical study to evaluate the effect of Emblica officinalis extract (Amlamax™) on markers of systemic inflammation and dyslipidemia. *Indian Journal of Clinical Biochemistry* 23(4): 378–381.

AOAC 1901. Official method for analysis of tanning materials: Adopted at the eighteenth convention of the Association of Official Agricultural Chemists. United States, Bureau of Chemistry: Association of Official Agricultural Chemists (U.S.), 8:6. Washington, DC.

Arapitsas, P. 2012. Hydrolyzable tannin analysis in food. *Food Chemistry* 135: 1708–1717.

Asif, A., U. Farooq, K. Akram, Z. Hayat, A. Shafi, F. Sarfraz, M.A.I. Sidhu, H. Rehman, and S. Aftab. 2016. Therapeutic potentials of bioactive compounds from mango fruit wastes. *Trends in Food Science & Technology* 53: 102–112.

Aviram, M., and M. Rosenblat. 2012. Pomegranate protection against cardiovascular diseases. *Evidence-Based Complementary and Alternative Medicine* 2012: e382763.

Awika, J.M., and L.W. Rooney. 2004. Sorghum phytochemicals and their potential impact on human health. *Phytochemistry* 65: 1199–1221.

Azqueta, A., and A. Collins. 2016. Polyphenols and DNA damage: A mixed blessing. *Nutrients* 8, 785.

Bachman, S.E., and M.E. Hubbert. 2004. Use of tannins and polymers to regulate digestion in animals. *US Patent* 2004/0047945 A1, issued March 11, 2004.

Barrett, A., T. Ndou, C.A. Hughey, C. Straut, A. Howell, Z. Dai, and G. Kaletunc. 2013. Inhibition of α-amylase and glucoamylase extracted from cocoa, pomegranates, cranberries, and grapes. *Journal of Agricultural and Food Chemistry* 61: 1477–1486.

Bate-Smith, E.C. 1975. Phytochemistry of proanthocyanidins. *Phytochemistry* 16: 1421–1572.

Bawadi, H.A., Bansode, R.R., Trappey, A., Truax, R.E., Losso, J.N. 2005. Inhibition of Caco-2 colon, MCF-7 and Hs578T breast, and DU 145 prostatic cancer cell proliferation by water-soluble black bean condensed tannins. *Cancer Letters* 218: 153–162.

Berardini, N., R. Carle, and A. Schieber. 2004. Characterization of gallotannins and benzophenone derivatives from mango (*Mangifera Indica* L. cv. Tommy Atkins) peels, pulp and kernels by high-performance liquid chromatography/electrospray ionization mass spectrometry. *Rapid Communications in Mass Spectrometry* 18: 2208–2216.

Bittner, K., S. Rzeppa, and H.U. Humpf. 2013. Distribution and quantification of flavan-3-ols and procyanidins with low degree of polymerization in nuts, cereals, and legumes. *Journal of Agricultural and Food Chemistry* 61: 9148–9154.

Borges G., W. Mullen, and A. Crozier. 2010. Comparison of the polyphenolic composition and antioxidant activity of European commercial fruit juices. *Food & Function* 1: 73–83.

Boudjou, S., B.D. Oomah, F. Zaidi, and F. Hosseinian. 2013. Phenolic content and antioxidant and anti-inflammatory activities of legume fractions. *Food Chemistry* 138: 1543–1550.

Bozan, B., G. Tosun, and D. Özcan. 2008. Study of polyphenol content in the seeds of red grape (*Vitis vinifera* L.) varieties cultivated in turkey and their antiradical activity. *Food Chemistry* 109: 426–430.

Broadhurst, R.B., and W.T. Jones. 1978. Analysis of condensed tannins using acidified vanillin. *Journal of the Science of Food and Agriculture* 29: 788–794.

Buzzini, P., P. Arapitsas, M. Goretti, E. Branda, B. Turchetti, P. Pinelli, F. Ieri, and A. Romani. 2008. Antimicrobial and antiviral activity of hydrolyzable tannins. *Mini-Reviews in Medical Chemistry* 8: 1179–1187.

Campos-Vega, R., G. Loarca-Piña, and B.D. Oomah. 2010. Minor components of pulses and their potential impact on human health. *Food Research International* 43: 461–482.

Campos-Vega, R., K. Vázquez-Sánchez, D. López-Barrera, G. Loarca-Piña, S. Mendoza-Díaz, and B.D. Oomah. 2015. Simulated gastrointestinal digestion and in vitro colonic fermentation of spent coffee (*Coffea arabica* L.): Bioaccessibility and intestinal permeability. *Food Research International* 77: 156–161.

Campos-Vega, R., R. Reynoso-Camacho, G. Pedraza-Aboytes, J.A. Acosta-Gallegos, S.H. Guzman-Maldonado, O. Paredes-Lopez, B.D. Oomah, and G. Loarca-Piña. 2009. Chemical composition and in vitro polysaccharide fermentation of different beans (*Phaseolus vulgaris* L.). *Journal of Food Science* 74: T59–T65.

Cappai, M.G., P. Wolf, W. Pinna, and J. Kamphues. 2014. Digestibility coefficients of crude nutrients in raw hulled acorns (*Quercus pubescens* Willd.) fed to growing pigs. *Animal Feed Science and Technology* 197: 148–154.

Carpenter, J.L., F.L. Caruso, A. Tata, N. Vorsa, and C.C. Neto. 2014. Variation in proanthocyanidin content and composition among commonly grown North American cranberry cultivars (*Vaccinium macrocarpon*). *Journal of the Science of Food and Agriculture* 94: 2738–2745.

Carsote, C., E. Badea, L. Miu, and G.D. Gatta. 2016. Study of the effect of tannins and animal species on the thermal stability of vegetable leather by differential scanning calorimetry. *Journal of Thermal Analysis Calorimetry* 124: 1255–1266.

Chakrabarti, R. 2009. Pharmacotherapy of obesity: Emerging drugs and targets. *Expert Opinion on Therapeutic Targets* 13: 195–207.

Chan-Pérez, J.I., J.F. Torres-Acosta, C.A. Sandoval-Castro, H. Hoste, G.S. Castañeda-Ramirez, G. Vilarem, and C. Mathieu. 2016. In vitro susceptibility of ten *Haemonchus contortus* isolates from different geographical origins towards acetone: Water extracts of two tannin rich plants. *Veterinary Parasitology* 217: 53–60.

Chen, M.H., A.M. Mcclung, and C.J. Bergman. 2016. Concentrations of oligomers and polymers of proanthocyanidins in red and purple rice bran and their relationships to total phenolics, flavonoids, antioxidant capacity and whole grain color. *Food Chemistry* 208: 279–287.

Chiarini, A., M. Micucci, M. Malaguti, R. Budriesi, P. Ioan, Monia Lenzi, C. Fimognari, T. Toschi, P. Comandini, and S. Hrelia. 2013. Sweet chestnut (*Castanea sativa* Mill.) bark extract: Cardiovascular activity and myocyte protection against oxidative damage. *Oxidative Medicine and Cellular Longevity*: 471790.

Chuang, H., L. Ng, L. Lin, J. Chang, J. Chen, T. Lin, and C. Lin. 2011. Hydrolyzable tannins of tropical almond show antifibrotic effects in TGF-β1-induced hepatic stellate cells. *Journal of the Science of Food and Agriculture* 91: 2777–2784.

Chung, S.-Y., and S. Reed. 2012. Removing peanut allergens by tannic acid. *Food Chemistry* 134: 1468–1473.

Chupin, L., C. Motillon, F.C.-E. Bouhtoury, A. Pizzi, and B. Charrier. 2013. Characterisation of maritime pine (*Pinus pinaster*) bark tannins extracted under different conditions by spectroscopic methods, FTIR and HPLC. *Industrial Crops and Products* 49: 897–903.

Clifford, M.N., and J.R. Ramirez-Martinez. 1991. Tannins in wet-processed coffee beans and coffee pulp. *Food Chemistry* 40: 191–200.

Corona, G., Y. Ji, P. Anegboonlap, S. Hotchkiss, C. Gill, P. Yaqoob, J.P.E. Spencer, and I. Rowland. 2016. Gastrointestinal modifications and bioavailability of brown seaweed phlorotannins and effects on inflammatory markers. *British Journal of Nutrition* 115: 1240–1253.

Crestini, C., H. Lange, and G. Bianchetti. 2016. Detailed chemical composition of condensed tannins via quantitative ^{31}P NMR and HSQC analyses: *Acacia catechu, Schinopsis balansae*, and *Acacia mearnsii. Journal of Natural Products* 79: 2287–2295.

Cunningham, D.G., S. Vannozzi, E. O'Shea, and R. Turk. 2002. Analysis and standardization of cranberry products. In *Quality Management of Nutraceuticals*, ed. Ho.C. et al., pp. 151–166. ACS Symposium Series, American Chemical Society, Washington, DC.

Dalzell, S.A., and G.L. Kerven. 2002. Accounting for co-extractable compounds (blank correction) in spectrophotometric measurement of extractable and total-bound proanthocyanidin in Leucaena Spp. *Journal of the Science of Food and Agriculture* 868: 860–868.

Das Q., M.R. Islam, M.F. Marcone, G.K. Warriner, and M.S. Diarra. 2016. Potential of berry extracts to control foodborne pathogens. *Food Control* 73: 650–662.

Deng, Y., G. Liang, Y. Shi, H. Li, J. Zhang, and X. Mao. 2016. Condensed tannins from *Ficus altissima* leaves: Structural, antioxidant, and antityrosinase properties. *Process Biochemistry* 51: 1092–1099.

Deshpande, S.S., and C.G. Campbell. 1992. Genotype variation in BOAA, condensed tannins, phenolics and enzyme inhibitors of grass pea (*Lathyrus sativus*). *Canadian Journal of Plant Science* 72: 1037–1047.

Deshpande, S.S., M. Cheryan, D.K. Salunkhe, and B.S. Luh. 1986. Tannin analysis of food products. *Critical Reviews in Food Science and Nutrition* 24: 401–449.

Domínguez-Avila, J.A., A. Wall-Medrano, G.R. Velderrain-Rodríguez, C.-Y.O. Chen, N.J. Salazar-López, M. Robles-Sánchez, and G.A. González-Aguilar. 2017. Gastrointestinal interactions, absorption, splanchnic metabolism and pharmacokinetics of orally ingested phenolic compounds. *Food & Function* 8: 15–38.

Draijer, R., Y. de Graaf, M. Slettenaar, E. de Groot, and C.I. Wright. 2015. Consumption of a polyphenol-rich grape-wine extract lowers ambulatory blood pressure in mildly hypertensive subjects. *Nutrients* 7: 3138–3153.

Du Toit, W.J., and M. Visagie. 2012. Correlations between South African red grape and wine colour and phenolic composition: Comparing the glories, iland and bovine serum albumin tannin precipitation methods. *South African Journal of Enology and Viticulture* 33: 33–41.

Duval, A., and L. Ave. 2016. Characterization and physicochemical properties of condensed tannins from *Acacia catechu*. *Journal of Agricultural and Food Chemistry* 64: 1751–1760.

Dykes, L., and L.W. Rooney. 2007. Phenolic compounds in cereal grains and their health benefits. *Cereal Foods World* 52: 105–111.

Eltayeb, M.M., A.B. Hassn, M.A. Sulieman, and E.E. Babiker. 2007. Effect of processing followed by fermentation on antinutritional factors content of pearl millet (*Pennisetum glaucum* L.) cultivars. *Pakistan Journal of Nutrition* 6: 463–467.

Engels, C., M. Knödler, Y.-Y. Zhao, R. Carle, M.G. Gänzle, and A. Schieber. 2009. Antimicrobial activity of gallotannins isolated from mango (*Magnifera indica* L.) kernels. *Journal of Agricultural and Food Chemistry* 57: 7712–7718.

Engström, M.T., X. Sun, M.P. Suber, M. Li, J.-P. Salminen, and A.E. Hagerman. 2016. The oxidative activity of ellagitannins dictates their tendency to form highly stable complexes with bovine serum albumin at increased pH. *Journal of Agricultural and Food Chemistry* 64: 8994–9003.

Falcão, L., and M.E.M. Araújo. 2011. Tannins characterisation in new and historic vegetable tanned leathers fibres by spot tests. *Journal of Cultural Heritage* 12: 149–156.

Fava, F., J.A. Lovegrove, R. Gitau, K.G. Jackson, and K.M. Tuohy. 2006. The gut microbiota and lipid metabolism: Implications for human health and coronary heart disease. *Current Medicinal Chemistry* 13: 3005–3021.

Ferreira, D., and D. Slade. 2002. Oligomeric proanthocyanidins: Naturally occurring O-heterocycles. *Natural Product Reports* 19: 517–541

Ferreira, E.C., G.B. Souza, and R.A. Nogueira. 2003. FIA-FAAS method for tannin determination based on a precipitation with hemoglobin. *Journal of the Brazilian Chemical Society* 14: 329–333.

Figueroa, M.C., A. Zafimahova, and P.G. Maldonado. 2015. Grape seed and apple tannins: Emulsifying and antioxidant properties. *Food Chemistry* 178: 38–44.

Finocchiaro, F., B. Ferrari, and A. Gianinetti. 2010. A study of biodiversity of flavonoid content in the rice caryopsis evidencing simultaneous accumulation of anthocyanins and proanthocyanidins in a black-grained genotype. *Journal of Cereal Science* 51: 28–34.

Fischer, U.A., A.V. Jaksch, R. Carle, and D.R. Kammerer. 2013. Influence of origin source, different fruit tissue and juice extraction methods on anthocyanin, phenolic acid, hydrolyzable tannin and isolariciresinol contents of pomegranate (*Punica granatum* L.) fruits and juices. *European Food Research and Technology* 237: 209–221.

Folin, O., and V. Ciocalteu. 1927. On tyrosine and tryptophane determination in proteins. *Journal of Biological Chemistry* 73: 627–650.

Fuller, S.J. 1989. Effect of tannins on the in vitro digestibility of jojoba meal proteins. MSc. thesis. San Jose State University: 63. http://scholarworks.sjsu.edu/etd_theses/3068.

Furuuchi, R., T. Yokoyama, Y. Watanabe, and M. Hirayama. 2011. Identification and quantification of short oligomeric proanthocyanidins and other polyphenols in boysenberry seeds and juice. *Journal of Agricultural and Food Chemistry* 59: 3738–3746.

Gemeda, B.S., and A. Hassen. 2015. Effect of tannin and species variation on *in vitro* digestibility, gas, and methane production of tropical browse plants. *Asian-Australasian Journal of Animal Sciences* 28: 188–199.

Glabasnia, A., and T. Hofmann. 2006. Sensory-directed identification of taste-active ellagitannins in american (*Quercus Alba* L.) and European oak wood (*Quercus Robur* L.) and quantitative analysis in bourbon whiskey and oak-matured red wines. *Journal of Agricultural and Food Chemistry* 54: 3380–3390.

Gonzalo-Diago, A., M. Dizy, and P. Fernández-Zurbano. 2013. Taste and mouthfeel properties of red wines proanthocyanidins and their relation to the chemical composition. *Journal of Agricultural and Food Chemistry* 61: 8861–8870.

Grasel, F.S., and M.F. Ferrão. 2016. A rapid and non-invasive method for the classification of natural tannin extracts by near-infrared spectroscopy and PLS-DA. *Analytical Methods* 8: 644–649.

Grela, E.R. 1996. Nutrient composition and content of antinutritional factors in spelt (*Triticum spelta* L) cultivars. *Journal of the Science of Food and Agriculture* 71: 399–404.

Gu, L., M.A. Kelm, J.F. Hammerstone, G. Beecher, J. Holden, D. Haytowitz, S. Gebhardt, and R.L. Prior. 2004. Concentrations of proanthocyanidins in common foods and estimations of normal consumption. *The Journal of Nutrition* 134: 613–617.

Hagerman, A.E, and L.G. Butler. 1981. The specificity of proanthocyanidin-protein interactions. *Journal of Biological Chemistry* 256: 4494–4497.

Hartzfeld, P.W., R. Forkner, M.D. Hunter, and A.E. Hagerman. 2002. Determination of hydrolyzable tannins (gallotannins and ellagitannins) after reaction with potassium iodate. *Journal of Agricultural and Food Chemistry* 50: 1785–1790.

Hashida, K., R. Makino, and S. Ohara. 2009. Amination of pyrogallol nucleus of condensed tannins and related polyphenols by ammonia water treatment. *Holzforschung* 63: 319–326.

Haslam, E. 2007. Vegetable tannins-lessons of a phytochemical lifetime. *Phytochemistry* 68: 2713–2721.

Hečimović, I., A. Belščak-Cvitanović, D. Horžić, and D. Komes. 2011. Comparative study of polyphenols and caffeine in different coffee varieties affected by the degree of roasting. *Food Chemistry* 129: 991–1000.

Hellström, J.K., and P.H. Mattila. 2008. HPLC determination of extractable and unextractable proanthocyanidins in plant materials. *Journal of Agricultural and Food Chemistry* 56: 7617–7624.

Herald, T.J., P. Gadgil, R. Perumal, S.R. Bean, and J.D. Wilson. 2014. High-throughput micro-plate HCl-vanillin assay for screening tannin content in sorghum grain. *Journal of the Science of Food and Agriculture* 94: 2133–2136.

Heron, J. 1896. Tannin of hops. *Journal of the Institute of Brewing* 3: 162–183.

Hithamani, G., and K. Srinivasan. 2014. Bioaccessibility of polyphenols from wheat (*Triticum aestivum*), sorghum (*Sorghum bicolor*), green gram (*vigna radiata*), and chickpea (*Cicer arietinum*) as influenced by domestic food processing. *Journal of Agricultural and Food Chemistry* 62: 11170–11179.

Hixson, J.L, K.A. Bindon, and P.A. Smith. 2015. Evaluation of direct phloroglucinolysis and colorimetric depolymerization assays and their applicability for determining condensed tannins in grape marc. *Journal of Agricultural and Food Chemistry* 63: 9954–9962.

Hoseney, R.C., E. Varriano-Marston, and D.A.V. Dendy. 1981. Sorghum and millets. In *Advances in Cereal Science and Technology*, Volume IV, ed. Y. Pomeranz, pp. 71–144. St. Paul, MN: American Association of Cereal Chemists.

Hossain, M.Z., S.F. Gilbert, K. Patel, S. Ghosh, A.K. Bhunia, and S.E. Kern. 2013. Biological clues to potent DNA-damaging activities in food and flavoring. *Food and Chemical Toxicology* 55: 557–567.

Hoste, H., J.F.J. Torres-Acosta, C.A. Sandoval-Castro, I. Mueller-Harvey, S. Sotiraki, H. Louvandini, S.M. Thamsborg, and T.H. Terrill. 2015. Tannin containing legumes as a model for nutraceuticals against digestive parasites in livestock. *Veterinary Parasitology* 212: 5–17.

Inoue, K.H., and A.E. Hagerman. 1988. Determination of gallotannin with rhodanine. *Analytical Biochemistry* 169: 363–369.

Jacobek, L. 2015. Interactions of polyphenols with carbohydrates, lipids and proteins. *Food Chemistry* 175: 556–567.

Jansman, A.J., M.W. Verstegen, J. Huisman, and J.W. Van Den Berg. 1995. Effects of hulls of faba beans (*Vicia faba* l.) with a low or high content of condensed tannins on the apparent ileal and fecal digestibility of nutrients and the excretion of endogenous protein in ileal digesta and feces of pigs. *Journal of Animal Science* 73: 118–127.

Jennings, A., A. MacGregor, T. Spector, and A. Cassidy. 2017. Higher dietary flavonoid intakes are associated with lower objectively measured body composition in women: Evidence from discordant monozygotic twins. *The American Journal of Clinical Nutrition* 1–9.

Jordan, L., B.-J. Walker, C. Moody, and B.W. Booth. 2016. Incorporation of natural compounds into biomaterials to prevent breast cancer recurrence. [abstract]. In *Proceedings of the 107th Annual Meeting of the American Association for Cancer Research*, April 16–20, 2016, New Orleans, LA. Philadelphia (PA): AACR; Cancer Research, 76 (14 Suppl): Abstract #2186.

Kalili, K.M., J. Vestner, M.A. Stander, and A. de Villiers. 2013. Toward unravelling grape tannin composition: Application of online hydrophilic interaction chromatography x reversed-phase liquid chromatography-time-of-flight mass spectrometry of grape seed analysis. *Analytical Chemistry* 85: 9107–9115.

Kammerer, D.R., J. Kammerer, R. Valet, and R. Carle. 2014. Recovery of polyphenols from the by-products of plant food processing and application as valuable food ingredients. *Food Research International* 65: 2–12.

Kang, I., T. Buckner, N.F. Shay, L. Gu, and S. Chung. 2016. Improvements in metabolic health with consumption of ellagic acid and subsequent conversion into urolithins: Evidence and mechanisms. *Advances in Nutrition: An International Review Journal* 7: 961–972.

Karama, M. 2009. Chelation of Cu (II), Zn(II), and Fe (II) by tannin constituents of selected edible nuts. *International Journal of Molecular Sciences* 10: 5485–5497.

Kaur, R., and K. Kaur. 2016. Appraisal of maize germplasm for identification of nutritionally rich nutraceutical genotypes. *Cereal Research Communications* 44: 501–512.

Khan, N., R. Sulliva, T. Hoang, T. Dittakavi, and H. Mukhtar. 2013. Delay in the progression of lung adenoma in adenocarcinoma in mice by oral consumption of pomegranate fruit extract. [abstract]. In *Proceedings of the 104th Annual Meeting of the American Association for Cancer Research*; April 6–10, 2013; Washington, DC. Philadelphia (PA): AACR; *Cancer Research*, 73 (8 Suppl): Abstract # 194.

Khanbabaee, K., and T. Ree. 2001. Tannins: Classification and definition. *Natural Product Reports* 18: 641–649.

Killday, K.B., M.H. Davey, J.A. Glinski, P. Duan, R. Veluri, G. Proni, F.J. Daugherty, and M.S. Tempesta. 2011. Bioactive A-type proanthocyanidins from *Cinnamomum cassia*. *Journal of Natural Products* 74: 1833–1841.

Kim, S.M., S.W. Kang, J.S. Jeon, Y.J. Jung, W.R. Kim, C.Y. Kim, and B.H. Um. 2013. Determination of major phlorotannins in *Eisenia bicyclis* using hydrophilic interaction chromatography: Seasonal variation and extraction characteristics. *Food Chemistry* 138: 2399–2406.

Koponen, J.M., A.M. Happonen, P.H. Mattila, and A.R. Törrönen. 2007. Contents of anthocyanins and ellagitannins in selected foods consumed in finland. *Journal of Agricultural and Food Chemistry* 55: 1612–1619.

Kumari, M., S. Jain, and R. Dave. 2014. Babul (*Acacia Nilotica*): A potential source of tannin and its suitability in management of type II diabetes. *Nutrition & Food Science* 44: 119–126.

Kunyanga, C.N., J.K. Imungi, M. Okoth, C. Momanyi, H.K. Biesalski, and V. Vadivel. 2011. Antioxidant and antidiabetic properties of condensed tannins in acetonic extract of selected raw and processed indigenous food ingredients from Kenya. *Journal of Food Science* 76(4): C560–C567.

Kyamuhangire, W., T. Krekling, E. Reed, and R. Pehrson. 2006. The microstructure and tannin content of banana fruit and their likely influence on juice extraction. *Journal of the Science of Food and Agriculture* 86: 1908–1915.

Lahoura, L., A. El Arem, F. Ghrairi, H. Chahdoura, H. Ben Salem, M. El Felah, and L. Achour. 2014. Phytochemical content and antioxidant properties of diverse varieties of whole barley (*Hordeum vulgare* L.) grown in Tunisia. *Food Chemistry* 145: 578–583.

Lamy, E., C. Pinheiro, L. Rodrigues, F. Capela e Silva, O. Silva, S. Tavares, and G. Rui. 2016. Determinants of tannin-rich food and beverage consumption: Oral perception vs. psychosocial aspects. In *Tannins: Bichemistry, Food Sources and Nutritional Properties*, ed. C. A. Combs. New York: Nova Science Publishers.

Landete, J. M. 2011. Ellagitannins, ellagic acid and their derived metabolites: A review about source, metabolism, functions and health. *Food Research International* 44: 1150–1160.

Larrosa, M., M.T. García, J.C. Espín, and F.A. Tomás. 2010. Ellagitannins, ellagic acid and vascular health. *Molecular Aspects of Medicine* 31: 513–539.

Lee, J.-H., and S.T. Talcott. 2002. Ellagic acid and ellagitannins affect on sedimentation in muscadine juice and wine. *Journal of Agricultural and Food Chemistry* 50: 3971–3976.

Lima, M.B., S.I.E. Andrade, I.S. Barreto, L.F. Almeida, E.C. Silva, and M.C.U. Araújo. 2012. Quantitative spot test analysis of soluble tannin in green tea using a portable diffuse reflectometer. *Analytical Methods* 4: 2329–2333.

Links, M.R., J. Taylor, M.C. Kruger, V. Naidoo, and J.R.N. Taylor. 2016. Kafirin microparticle encapsulated sorghum condensed tannins exhibit potential as an antihyperglycaemic agent in a small animal model. *Journal of Functional Foods* 20: 394–399.

Lu, K., and S. Basu. 2013. Chebulagic acid can inhibit vascular endothelial growth factor-A mediated angiogenesis. [abstract]. In *Proceedings of the 104th Annual Meeting of the American Association for Cancer Research*, April 6–10, 2013, Washington, DC. Philadelphia (PA): AACR; Cancer Research, 73 (8 Suppl): Abstract # 1617.

Ma, W., P. Waffo, M. Jourdes, and H. Li. 2016. Chemical affinity between tannin size and salivary protein binding abilities: Implications for wine astringency. *PloS One* 11: 1–14.

Marín, L., Miguélez, E.M., Villar, C.J., and Lombó, F. 2015. Bioavailability of dietary polyphenols and gut microbiota metabolism: Antimicrobial properties. *BioMed Research International* 1–18.

Masumoto, S., Terao, A., Yamamoto, Y., Mukai, T., Miura, T., and Shoji, T. 2016. Non-absorbable apple procyanidins prevent obesity associated with gut microbial and metabolomic changes. *Scientific Reports* 6.

McDougall, G.J., F. Shpiro, P. Dobson, P. Smith, A. Blake, and D. Stewart. 2005. Different polyphenolic components of soft fruits inhibit alpha-amylase and alpha-glucosidase. *Journal of Agricultural and Food Chemistry* 53: 2760–2766.

McRae, J.M., K.A. Bindon, J.L. Hixson, and P. Smith. 2016. Wine tannin structure and extraction: Review of current research. In *Tannins: Bichemistry, Food Sources and Nutritional Properties*, ed. C. A. Combs. New York: Nova Science Publishers.

Melone, F., R. Saladno, H. Lange, and C. Crestini. 2013. Tannin structural elucidation and quantitative ^{31}P NMR analysis. 2. Hydrolyzable tannins and proanthocyanidins. *Journal of Agricultural and Food Chemistry* 61: 9316–9324.

Menayang, A. (03-March 2016). Applied Food Sciences launches new green tea extract for functional beverages. *Nutra Ingredients-USA.com.*

Menezes, C.M., A.B. daCosta, R.R. Renner, L.F. Bastos, M.F. Ferrão, and V.L. Dressler. 2014. Direct determination of tannins in *Acacia mearnsii* bark using near-infrared spectroscopy. *Analytical Methods* 6: 8299–8305.

Min, B., L. Gu, A.M. McClung, C.J. Bergman, and M.-H. Chen. 2012. Free and bound total phenolic concentrations, antioxidant capacities, and profiles of proanthocyanidins and anthocyanins in whole grain rice (*Oryza Sativa* L.) of different bran colours. *Food Chemistry* 133: 715–22.

Miura, T., S. Takagi, and T. Ishida. 2012. Management of diabetes and its complications with banaba (*Lagerstroemia speciosa* L.) and corosolic acid. *Evidence-Based Complementary and Alternative Medicine* 2012: 871495.

Muccilli, V., N. Cardullo, C. Spatafora, V. Cunsolo, and C. Tringali. 2017. α-Glucosidase inhibition and antioxidant activity of an oenological commercial tannin. Extraction, fractionation and analysis by HPLC/ESI-MS/MS and ^{1}H NMR. *Food Chemistry* 215: 50–60.

Mueller-Harvey, I. 2001. Analysis of hydrolyzable tannins. *Animal Feed Science and Technology* 91: 3–20.

Murugan, K., and S. Al-Sohaibani. 2012. Coffee, tea, and cocoa. In *Valorization of Food Processing By-Products*, ed. M. Chandrasekaran, pp. 455–486. Boca Raton, FL: CRC Press.

Muthusamy, V.S., S. Anand, K.N. Sangeetha, S. Sujatha, B. Arun, and B.S. Lakshmi. 2008. Tannins present in *Cichorium intybus* enhance glucose uptake and inhibit adipogenesis in 3T3-L1 adipocytes through PTP1B inhibition. *Chemico-Biological Interactions* 174: 69–78.

Naczk, M., and F. Shahidi. 2006. Phenolics in cereals, fruits and vegetables: Occurrence, extraction and analysis. *Journal of Pharmaceutical and Biomedical Analysis, Nutraceuticals Analysis* 41: 1523–1542.

Najdi, S., V. Orsat, B. Azadi, and S. Kubow. 2016. Improvement of the in vitro protein digestibility of amaranth grain through optimization of the malting process. *Journal of Cereal Science* 68: 59–65.

Neyrinck, A.M., Van Hée, V.F., Bindels, L.B., De Backer, F., Cani, P.D., and Delzenne, N.M. 2013. Polyphenol-rich extract of pomegranate peel alleviates tissue inflammation and hypercholesterolaemia in high-fat diet-induced obese mice: Potential implication of the gut microbiota. *British Journal of Nutrition* 109(05): 802–809.

Ngugi, C.C., E. Oyoo-Okoth, J.O. Manyala, K. Fitzsimmons, and A. Kimotho. 2017. Characterization of nutritional quality of amaranth leaf protein concentrates and suitability of fish meal replacement in Nile tilapia feeds. *Aquaculture Reports* 5: 62–69.

Nguyen, M.M., F.R. Ahmann, R.B. Nagle, C.-H. Hsu, J.A. Tangrea, H.L. Parnes, M.H. Sokoloff, M.B. Gretzer, and H.-H.S. Chow. 2011. Randomized, double blind, placebo-controlled trial of polyphenol E in prostate cancer patients before prostatectomy: Evaluation of potential chemopreventive activities. *Cancer Prevention Research* 5: 290–298.

Nocchi, S.R., M.V.P. Companhoni, J.C. de Mello, B. Prado, D. Filho, C.V. Nakamura, C.A. Carollo, D.B. Silva, and T. Ueda-Nakamura. 2016. Antiviral activity of crude hydroethanolic extract from *Schinus terebinthifolia* against *Herpes simplex* virus type 1. *Planta Medica*, Oct 5.

Oboh, G., and T.V. Amusan. 2009. Nutritive value and antioxidant properties of cereal gruels produced from fermented maize and sorghum. *Food Biotechnology* 23: 17–31.

Okoth, J.K., S. Ochola, N.K. Gikonyo, and A. Makokha. 2011. Optimization of the period of steeping and germination of amaranth grain. *Journal of Agriculture and Food Technology* 1: 101–105.

Okuda, T., K. Mori, K. Seno, and T. Hatano. 1979. Constituents of Geranium thunbergII Sieb. et Zucc.: VII. High-Performance reversed-phase liquid chromatography of hydrolyzable tannins and related polyphenols. *Journal of Chromatography A* 171: 313–320.

Olejar, K.J., Ch.Vandermeer, and P.A. Kilmartin. 2016. Grape tannins: Structure, antioxidant and antimicrobial activity. In *Tannins: Bichemistry, Food Sources and Nutritional Properties*, ed. C. A. Combs. New York: Nova Science Publishers.

Oliveira, L.M.B., C.M.L. Bevilaqua, I.T.F. Macedo, S.M. Morais, M.V.B. Monteiro, C.C. Campello, W.L.C. Ribiro, and E.K.F. Batista. 2011. Effect of six tropical tanniferous plant extracts on larval exsheathment of *Haemonchus contortus*. *Revista Brasileira Parasitologia Veterinária* 20: 155–160.

Oomah, B.D., G. Luc, C. Leprelle, J.C.G. Drover, J.E. Harrison, and M. Olson. 2011. Phenolics, phytic acid, and phytase in Canadian-grown low-tannin faba bean (*Vicia faba* L.) genotypes. *Journal of Agricultural and Food Chemistry* 59: 3763–3771.

Oomah, B.D., L. Kotzeva, M. Allen, and P.Z. Bassinello. 2013. Microwave and micronization treatments affect dehulling characteristics and bioactive contents of dry beans (*Phaseolus vulgaris* L.). *Journal of the Science of Food and Agriculture* 94: 1349–1358.

Ortega, N., M.-P. Romero, A. Macià, J. Reguant, N. Anglès, J.-R. Morelló, and M.-J. Motilva. 2010. Comparative study of UPLC–MS/MS and HPLC–MS/MS to determine procyanidins and alkaloids in cocoa samples. *Journal of Food Composition and Analysis* 23: 298–305.

Palma, G., J. Freer, and J. Baeza. 2003. Removal of metal ions by modified pinus radiata bark and tannins from water solutions. *Water Research* 37: 4974–4980.

Papanikou, E. 2016. Phenolic compounds in animal feeds. *Pig International*. May/June 4–11.

Pasinetti, G.M., H. Ksiezak-Reding, I. Santa-Maria, J. Wang, and L. Ho. 2010. Development of a grape seed polyphenolic extract with anti-oligomeric activity as a novel treatment in progressive supranuclear palsy and other tauopathies. *Journal of Neurochemistry* 114: 1557–1568.

Pedan, V., N. Fischer, K. Bernath, T. Hühn, and S. Rohn. 2017. Determination of oligomeric proanthocyanidins and their antioxidant capacity from different chocolate manufacturing stages using the NP-HPLC-online-DPPH methodology. *Food Chemistry* 214: 523–532.

Pedersen, B., L.S. Kalinowski, and B.O. Eggum. 1987. The nutritive value of amaranth grain (*Amaranthus caudatus*). *Plant Foods for Human Nutrition* 36: 309–324.

Pérez-Jiménez, J., and J.L. Torres. 2011. Analysis of nonextractable phenolic compounds in foods: The current state of the art. *Journal of Agricultural and Food Chemistry* 59: 12713–12724.

Pérez, J., and F. Saura. 2015. Macromolecular antioxidants or non-extractable polyphenols in fruit and vegetables: Intake in four European countries. *Food Research International* 74: 315–323.

Perkin, A.G. 1900. Yellow colouring principles contained in various tannin matters. Part VII. *Arctostaphylos Uva Ursi, Hæmatoxylon Campeachianum, Rhus Metopium Myrica Gale, Coriaria Myrtifolia, and Robinia Pseudacacia*. *Journal of the Chemical Society, Transactions* 77: 423–432.

Ping, L., A. Pizzi, Z. Ding, and N. Brosse. 2012. Condensed tannins from grape pomace: Characterization by FTIR and MALDI TOF and production of environment friendly wood adhesive. *Industrial Crops & Products* 40: 13–20.

Piovesan, J.V., C.A. de Lima, E.R. Santana, and A. Spinelli. 2017. Voltammetric determination of condensed tannins with a glassy carbon electrode chemically modified with gold nanoparticles stabilized in carboxymethylcellulose. *Sensors and Actuators B: Chemical* 240: 838–847.

Price, M.L., and L.G. Butler. 1977. Rapid visual estimation and spectrophotometric determination of tannin content of sorghum grain. *Journal of Agricultural and Food Chemistry* 25: 1266–1273.

Pujol, D., C. Liu, J. Gominho, M.À. Olivella, N. Fiol, I. Villaescusa, and H. Pereira. 2013. The chemical composition of exhausted coffee waste. *Industrial Crops and Products* 50: 423–429.

Ramirez-Coronel, M.A., N. Marnet, V.S.K. Kolli, S. Roussos, S. Guyot, and C. Augur. 2004. Characterization and estimation of proanthocyanidins and other phenolics in coffee pulp (*Coffea arabica*) by thiolysis–high-performance liquid chromatography. *Journal of Agricultural and Food Chemistry* 52: 1344–1349.

Reed, J.S. 1995. Nutritional toxicology of tannins and related polyphenols in forage legumes. *Journal of Animal Science* 73: 1516–1528.

Rifkin, R.F. 2011. Assessing the efficacy of red ochre as a prehistoric hide tanning ingredient. *Journal African Archeology* 9: 131–158.

Robbins, M.P., and P. Morris. 2000. Metabolic engineering of condensed tannins and other phenolic pathways in forage and fodder crops. In *Metabolic Engineering of Plant Secondary Metabolism*, eds. R. Verpoorte and A.W. Alfermann, pp. 165 –177. Kluwer Academic Publishers.

Rodríguez, R., P. Frutos, and M. Fondevila. 2015. A new index to estimate reactivity and biological effect of tannins, using tropical browse legumes as a model. *Animal Feed Science and Technology* 205: 42–48.

Romer, F.H., A.P. Underwood, N.D. Senekal, S.L. Bonnet, M.J. Duer, D.G. Reid, and J.H. Van der Westhuizen. 2011. Tannin fingerprinting in vegetable tanned leather by solid state NMR spectroscopy and comparison with leathers tanned by other processes. *Molecules* 16: 1240–1252.

Rooney, L.W., and J.M. Awika. 2005. Speciality sorghums for healthful foods. In *Specialty Grains for Food and Feed*, eds. E. Abdel-Aal and P. Wood, pp. 283–312. St. Paul, MN: American Association of Cereal Chemists.

Santos, C., Á. Vargas, N. Fronza, and J. Henrique. 2017. Structural, textural and morphological characteristics of tannins from acacia mearnsii encapsulated using sol-gel methods: Applications as antimicrobial agents. *Colloids and Surfaces B: Biointerfaces* 151: 26–33.

Santos, C., and J.H.Z. dos Santos. 2016. Direct analysis of total tannins encapsulated in silica matrices by ultraviolet-visible diffuse reflectance spectroscopy. *Analytical Methods* 8: 3774–3779.

Saucier, C., M. Jourdes, Y. Glories, and S. Quideau. 2006. Extraction, detection, and quantification of flavano-ellagitannins and ethylvescalagin in a bordeaux red wine aged in oak barrels. *Journal of Agricultural and Food Chemistry* 54: 7349–7354.

Schofield, P., D.M. Mbugua, and A.N. Pell. 2001. Analysis of condensed tannins: A review. *Animal Feed Science and Technology* 91: 21–40.

Seddon, T.J., and M.O. Downey. 2008. Comparison of analytical methods for the determination of condensed tannins in grape skin. *Australian Journal of Grape and Wine Research* 14: 54–61.

Sekowski, S., M. Ionov, A. Dubis, and S. Mavlyanov. 2016. Biomolecular interactions of tannin isolated from oenothera gigas with liposomes. *The Journal of Membrane Biology* 249: 171–179.

Sergent, T., J. Vanderstraeten, J. Winand, P. Beguin, and Y. Schneider. 2012. Phenolic compounds and plant extracts as potential natural anti-obesity substances. *Food Chemistry* 135: 68–73.

Serrano, J., R. Puupponen, A. Dauer, and A.M. Aura. 2009. Review tannins: Current knowledge of food sources, intake, bioavailability and biological effects. *Molecular Nutrition & Food Research* 53: 310–329.

Shahidi, F., and M. Naczk. 1995. *Food Phenolics Sources Chemistry Effects Applications*. Lancaster, PA: Technomic Publishing Co. Inc.

Shahidi, F., and M. Naczk. 2004. *Phenolics in Foods and Nutraceuticals*. Boca Raton, FL: CRC Press.

Shihabudeen, H., D.H., and K. Thirumurugan. 2011. Cinnamon extract inhibits α-Glucosidase activity and dampens postprandial glucose excursion in diabetic rats. *Nutrition & Metabolism* 8: 46.

Shimoda, H., J. Tanaka, M. Kikuchi, T. Fukuda, H. Ito, T. Hatano, and T. Yoshida. 2008. Walnut polyphenols prevent liver damage induced by carbon tetrachloride and d-galactosamine: Hepatoprotective hydrolyzable tannins in the kernel pellicles of walnut. *Journal of Agricultural and Food Chemistry* 56: 4444–4449.

Stephens, R., and J.L. Bell. 2016. Radiolabelled material. *US Patent* 2016/0151518 A1, issued June 2, 2016.

Stürm, C.D., T.T. Tiemann, C.E. Lascano, M. Kreuzer, and H.D. Hess. 2007. Nutrient composition and in vitro ruminal fermentation of tropical legume mixtures with contrasting tannin contents. *Animal Feed Science and Technology* 138(1): 29–46.

Sun, P., T. Wang, L. Chen, B.W. Yu, Q. Jia, K.X. Chen, and H.Y. Wang. 2016. Trimer procyanidin oligomers contribute to the protective effects of cinnamon extracts on pancreatic β-cells *in vitro*. *Acta Pharmacologica Sinica* 37: 1083–1090.

Tabasum, S., S. Ahmad, N. Akhlaq, and K. Rahman. 2001. Estimation of tannins in different food products. *International Journal of Agriculture and Biology* 3: 529–530.

Tan, H., D.Z.H. Wong, S.K. Ling, Ch.H. Chuah, and H.A. Kadir. 2012. Neuroprotective activity of galloylated cyanogenic glucosides and hydrolyzable tannins isolated from leaves of *Phyllagathis rotundifolia*. *Fitoterapia* 83: 223–229.

Tarahovsky, Y.S. 2008. Plant polyphenols in cell-cell interaction and communication. *Plant Signalling & Behavior* 3: 609–11.

Tavares, I.M.C., E.S. Lago-Vanzela, L.P.G. Rebello, A.M. Ramos, S. Gómez-Alonso, E. García-Romero, R. Da-Silva, and I. Hermosín-Gutiérrez. 2016. Comprehensive study of the phenolic composition of the edible parts of jambolan fruit (*Syzygium cumini* (L.) Skeels). *Food Research International* 82: 1–13.

Taylor, J., S.R. Bean, B.P. Ioerger, and J.R.N. Taylor. 2007. Preferential binding of sorghum with γ-kafirin and the influence of tannin binding on kafirin digestibility and biodegradation. *Journal of Cereal Science* 46: 22–31.

Tebib, K., P. Besancon, and J.M. Rouanet. 1994. Dietary grape seed tannins affect lipoproteins, lipoprotein lipases and tissue lipids in rats fed. *The Journal of Nutrition* 124: 2451–2458.

Thompson, L.U., J.H. Yoon, D.J. Jenkins, T.M. Wolever, and A.L. Jenkins. 1984. Relationship between polyphenol intake and blood glucose response of normal and diabetic individuals. *The American Journal of Clinical Nutrition* 39: 745–751.

Tondi, G., C.W. Oo, A. Pizzi, A. Trosa, and M.F. Thevenon. 2008. Metal adsorption of tannin based rigid foams. *Industrial Crops & Products* 9: 336–340.

Trimble, H. 1891. The tannins a monograph on the history, preparation, properties, methods of estimation, and uses of the vegetable astringents, with an index to the literature. *Philadelphia: J. B. Lippincott Company* 2: 135–170.

Vaher, M., and M. Kaljurand. 2012. The development of paper microzone-based green analytical chemistry methods for determining the quality of wines. *Analytical and Bioanalytical Chemistry* 404: 627–633.

Vaillancourt, R., A.E. Slinkard, and R.D. Reichert. 1986. The inheritance of condensed tannin concentration in lentil. *Canadian Journal of Plant Science* 66: 241–246.

Vargas-Magaña, J.J., J.F.G. Torres-Acosta, A.J. Aguilar-Caballero, C.S. Sandoval-Castro, H. Hoste, and J.I. Chan-Pérez. 2014. Anthelmintic activity of acetone-water extracts against *Haemonchus contortus* eggs: Interactions between tannins and other plant secondary compounds. *Veterinary Parasitology* 206: 322–327.

Vazquez, A.A., J.E. Wong, M.A. Lerma, A.I. Martinez, F.J. Olivas, C.N. Aguilar, A. Wall, G.A. Gonzalez, E. Alvarez, and L.A. de la Rosa. 2017. Proanthocyanidins from the kernel and shell of pecan (*Carya illinoinensis*): Average degree of polymerization and effects on carbohydrate, lipid, and peptide hydrolysis in a simulated human digestive system. *Journal of Functional Foods* 28: 227–234.

Velayutham, R., N. Sankaradoss, and K.F.H. Nazeer Ahamed. 2012. Protective effect of tannins from *Ficus racemosa* in hypercholesterolemia and diabetes induced vascular tissue damage in rats. *Asian Pacific Journal of Tropical Medicine* 5: 367–373.

Verma, S., A. Singh, and A. Mishra. 2013. Gallic acid: Molecular rival of cancer. *Environmental Toxicology and Pharmacology* 35(3): 473–485.

Verzelloni, E., C. Pellacani, D. Tagliazucchi, S. Tagliaferri, L. Calani, L.G. Costa, and D. Del Rio. 2011. Antiglycative and neuroprotective activity of colon-derived polyphenol catabolites. *Molecular Nutrition and Food Research* 55(S1): S35–S43.

Vinciguerra, M.G., G. Belcaro, and M. Cacchio. 2015. Robuvit® and endurance in triathlon: Improvements in training performance, recovery and oxidative stress. *Minerva Cardioangiologica* 63: 403–409.

Wang, B. and M. Heinonen. 2017. Protein-tannin interactions of tryptic digests of α-lactalbumin and procyanidins. *Journal of Agricultural and Food Chemistry* 65: 148–155.

Wang, N., and J.K. Daun. 2004. The chemical composition and nutritive value of Canadian pulses. *Canadian Grain Commission Report*, Winnipeg, Manitoba: 19–29.

Wang, X., T.D. Warkentin, C.J. Briggs, B.D. Oomah, C.C. Campbell, and S. Woods. 1998. Total phenolics and condensed tannins in field pea (*Pisum sativum* L.) and grass pea (*Lathyrus sativus* L.). *Euphytica* 101: 97–102.

Watrelot, A.A., D.L. Schulz, and J.A. Kennedy. 2017. Wine polysaccharides influence tannin-protein interactions. *Food Hydrocolloids* 63: 571–579.

Watrelot, A.A., N.K. Byrnes, H. Heymann, and J.A. Kennedy. 2016. Understanding the relationship between red wine matrix, tannin activity, and sensory properties. *Journal of Agricultural and Food Chemistry* 64: 9116–9123.

Willis, R.B., P.R. Allen, and M.P. Road. 1998. Improved method for measuring hydrolyzable tannins using potassium iodate. *Analyst* 123: 435–439.

Wobber, V., B. Hare, and R. Wrangham. 2008. Great apes prefer cooked food. *Journal of Human Evolution* 55: 340–348.

Wollgast, J., and E. Anklam. 2000. Review on polyphenols in theobroma cacao: Changes in composition during the manufacture of chocolate and methodology for identification and quantification. *Food Research International* 33: 423–447.

Wu, G., S.K. Johnson, J.F. Bornman, S.J. Bennett, V. Singh, A. Simic, and Z. Fang. 2016. Effects of genotype and growth temperature on the contents of tannin, phytate and *in vitro* iron availability of sorghum grains. *PLoS ONE* 11: e0148712.

Wu. S., C. Chiu, D. Yang, Y. Lin, J. Tseng, and Y. Chen. 2013. Inhibitory effects of litchi (*Litchi chinensis* Sonn.): Flower-water extracts on lipase activity and diet-induced obesity. *Journal of Functional Foods* 5: 923–929.

Xie, L., A.V. Roto, and B.W. Bolling. 2012. Characterization of ellagitannins, gallotannins, and bound proanthocyanidins from California almond (*Prunus dulcis*) varieties. *Journal of Agricultural and Food Chemistry* 60: 12151–12156.

Yang, J., S.M. Henning, R. Lee, M. Hsu, E.M. Grojean, A. Ly, D. Heber, and Z. Li. 2016. Effects of pomegranate extract on high fat/high sucrose diet induced obesity are dependent on the intestinal formation of urolithin A. *The FASEB Journal* 30: 691.28–691.28.

Yoshihara, E., A.P. Minho, and M.H. Yamamura. 2013. Anthelmintic effect of condensed tannins in gastrointestinal nematodes of sheep (*Ovis aries*). *Semina: Ciências Agrárias, Landrina* 34: 3935–3950.

Youn, U.M. Shon, G. Kim, R. Katagiri, K. Harata, Y. Ishida, and S. Lee. 2016. Antioxidant and anti-adipogenic activities of chestnut (*Castanea Crenata*) by-products. *Food Science and Biotechnology* 25: 1169–1174.

Zarin, M.A., H.Y. Wan, A. Isha, and N. Armania. 2016. Antioxidant, antimicrobial and cytotoxic potential of condensed tannins from *Leucaena leucocephala* hybrid-Rendang. *Food Science and Human Wellness* 5: 65–75.

Zhao, H., T. Liu, X. Mao, S. Han, R. Liang, L. Hui, C. Cao, Y. You, and L. Zhang. 2015. Fructus phyllanthi tannin fraction induces apoptosis and inhibits migration and invasion of human lung squamous carcinoma cells in vitro via MAPK/MMP pathways. *Acta Pharmacologica Sinica* 36: 758–768.

Zhenzhen G., W. Zhu, J. Peng, X. Deng, and C. Li. 2016. Persimmon tannin regulates the expression of genes critical for cholesterol absorption and cholesterol efflux by LXRα independent pathway. *Journal of Functional Foods* 23: 283–293.

Zhou, H.-C., Y.-M. Lin, Y.-Y. Li, M. Li, S.-D. Wei, W.-M. Chai, and N.F.-Y., Tam. 2011. Antioxidant properties of polymeric proanthocyanidins from fruit stones and pericarps of litchi *Chinensis Sonn. Food Research International* 44: 613–620.
Zhou, Z., Y. Huang, J. Liang, M. Ou, J. Chen, and G. Li. 2016. Extraction, purification and anti-radiation activity of persimmon tannin from *Diospyros kaki* L.f. *Journal of Environmental Radioactivity* 162–163: 182–188.
Zhu, F., B. Du, L. Zheng, and J. Li. 2015. Advance on the bioactivity and potential applications of dietary fibre from grape pomace. *Food Chemistry* 186: 207–212.

SECTION IV

Antioxidant Power

CHAPTER 14

Antioxidant Power

J. Abraham Domínguez-Avila, Jacqueline Ruiz-Canizales, Ramón Pacheco-Ordaz, Mónica A. Villegas-Ochoa, and Gustavo A. González Aguilar

CONTENTS

14.1 INTRODUCTION

Several health benefits have been attributed to the consumption of phenolic compounds (PCs) from vegetable sources, and most are attributed to their high antioxidant power, which is quantified by various assays that are generally performed in vitro. PCs are involved in REDOX (*red*uction, *ox*idation) reactions, which involve the transfer of electrons either by themselves or together with a hydrogen atom. Molecules are reduced if they gain an electron or hydrogen atom, and they are oxidized if they lose an electron or hydrogen atom. These transfers take place in a myriad of reactions and are fundamental for life. A notable case is the electron transfer chain (ETC), which produces ATP within the mitochondria with the aid of molecular oxygen. Electrons that are being transported along the ETC can prematurely reach an oxygen atom, which gains an extra unpaired electron and is referred to as a reactive oxygen species (ROS), or free radical. The mitochondria, and specifically the ETC, are notorious producers of ROS, although ROS are also produced through other cellular processes.

Free radicals are a type of ROS that are extremely reactive and short-lived. Some have half-lives in the range of 10^{-9} seconds. Examples of free radicals include the hydroxyl (OH•), superoxide (O_2•-), peroxyl (ROO•), and lipoperoxyl (LOO•) radicals. They remove electrons from neighboring molecules to pair an unpaired electron and become stable. Stable ROS such as hydrogen peroxide (H_2O_2) can maintain their oxidizing properties and remain toxic to cells (Devasagayam et al. 2004). Biomolecules that are oxidized by free radicals or ROS may lose their function. For example, an oxidized enzyme may not exert catalytic activity, oxidized phospholipids may weaken the cell membrane, or oxidized nucleotides can change genetic information. Cells tolerate minuscule amounts of oxidative damage. In fact, ROS participate in the signaling pathways of healthy cells (Valko et al. 2007). ROS are kept within a tolerable range by the antioxidant system, which consists of a combination of exogenous antioxidants, where PCs are of high importance, and endogenous antioxidants. Both systems prevent ROS formation or neutralize the ones that are already present. When ROS and the antioxidant system are no longer in balance, cells enter a state known as oxidative stress through three distinct conditions: (1) An adequate antioxidant system with an overproduction of ROS, (2) a deficient antioxidant system with a normal production of ROS, or (3) a deficient antioxidant system with an overproduction of ROS.

Oxidative stress can be triggered by infection, genetic/chronic diseases, smoking, drinking, drug abuse, pharmacological treatment, exposure to UV or other radiation, dietary habits, sleep deprivation, exercise, contamination, metallic ions, and others. PCs are highly important in preventing or correcting oxidative stress. Interestingly, an overconsumption of antioxidants, mostly from supplements, can also induce oxidative stress (Gutteridge and Halliwell 2010).

Because oxidative stress can occur at any moment, the REDOX state can be quantitatively measured in cells or biological fluids. In addition, it has become customary to evaluate the antioxidant power of food, such as fruits and vegetables, or particular molecules that are contained within them. The following sections will detail how PCs interact with the antioxidant system, its biological importance, assays to measure antioxidant power, and factors that can alter it.

14.2 PHENOLIC COMPOUNDS AND THE ENDOGENOUS ANTIOXIDANT SYSTEM

ROS formation is ubiquitous, it occurs in healthy cells as undesirable by-products of normal processes, and can also be stimulated by different stressors. Aerobic organisms have evolved defensive systems to scavenge free radicals to maintain REDOX homeostasis and are classified as enzymatic (e.g., catalase, superoxide dismutase, glutathione peroxidase) and nonenzymatic (e.g., coenzyme Q10, α-lipoic acid, bilirubin) antioxidants. Figure 14.1 depicts the endogenous antioxidant system (Wu et al. 2017, Carocho and Ferreira 2013).

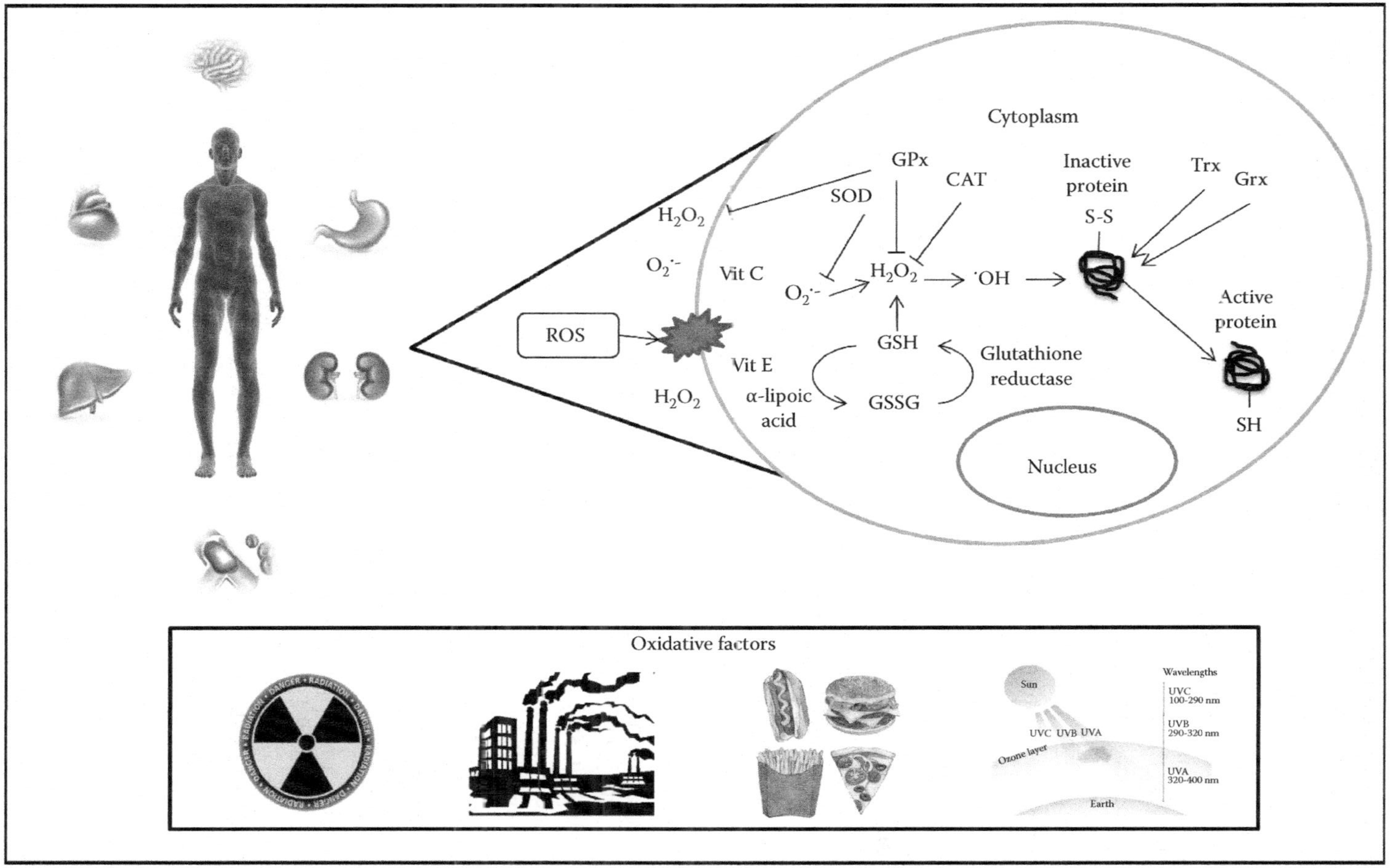

FIGURE 14.1 Endogenous antioxidant system. (Modified from Wu, Z., Liu, S., Zhao, J. et al., *Environmental and Experimental Botany*, 133, 1–11, 2017; Carocho, M., and Ferreira, I.C.F.R., *Food and Chemical Toxicology*, 51, 15–25, 2013.)

14.2.1 Enzymatic Antioxidants

The cell's first response against oxidative damage is mediated by enzymes, such as catalase, superoxide dismutase (SOD), and glutathione peroxidase (GPx). These enzymes are responsible for neutralizing oxidizing compounds such as H_2O_2, superoxide anion, and hydroxyl radicals (Carocho and Ferreira 2013). Several studies have found that different types of radiation (Megha et al. 2015), high-fat diets (Rindler et al. 2013), and some types of cancer (Liu et al. 2015) are stressful conditions that have a significant impact on the activity of said enzymes, which indicates that ROS production has increased.

Catalase (EC 1.11.1.6) is present in most organisms, especially in peroxisomes, but certain pathologies such as metabolic heart disease and cancer induce its mitochondrial, cytoplasmic, and membrane activity (Sverdlov et al. 2016). It reduces H_2O_2, which is produced in all cells, into water and molecular oxygen and can use H_2O_2 to oxidize toxins such as peroxynitrite and short-chain alcohols when the cell is under stressful conditions (Correa et al. 2005, Sahoo et al. 2009). It also protects some proteins associated with integrin pathways such as Grb2 and SHP2 (growth factor receptor-bound protein 2 and Src homology 2, respectively), and some transcription factors are involved in regulating its activity, such as Sp1 (specificity protein 1), NF-Y (nuclear factor Y), FoxO (forkhead box protein O), C/EBP-β (CCAAT-enhancer-binding protein β), PPARγ (peroxisome proliferator-activated receptor γ), and Oct-1 (octamer-binding transcription factor 1) (Nazıroğlu 2012, Glorieux et al. 2015).

In addition to catalase, GPx (EC 1.11.1.9) can also convert H_2O_2 or organic hydroperoxides to water using glutathione as a reducing agent. Known isoforms include GPx1 present in the cytosol and mitochondria, GPx2 found in the intestinal epithelium and GPx3 found in plasma (Brigelius-Flohé and Maiorino 2013). GPx1 deficiency has been implicated in endothelial and vascular dysfunction (Oelze et al. 2014), while GPx2 decreases inflammation, apoptosis, and colon carcinogenesis through Nrf2 modulation (Krehl et al. 2012). When GPx uses glutathione, two molecules are oxidized into glutathione disulfide (GSSG), which is no longer a reducing agent and must be reduced back to glutathione by glutathione reductase (EC 1.8.1.7), which does so with NADPH as an electron donor. GPx and glutathione reductase thus work in parallel.

SOD (EC 1.15.1.1) converts the superoxide anion into H_2O_2. Three isoforms are currently known: CuZnSOD or SOD1, MnSOD or SOD2, and ECSOD or SOD3, located in the cytosol, mitochondria, and extracellular space, respectively (Luangwattananun et al. 2016). CuZnSOD and MnSOD denote the ions present at the active site of each isoenzyme, copper–zinc, and manganese, respectively. SOD activity is associated with some transcription factors when cells are under oxidative stress, such as C/EBP, nuclear factor (NF)-κB, p53, and SP-1 (Che et al. 2016).

Thioredoxin reductase (EC 1.8.1.9) is an NADPH-dependent oxidoreductase that contains a REDOX disulfide catalytic active site. It is found in the cytosol (Trx1) and mitochondria (Trx2) (Zhang et al. 2016). This enzymatic complex protects cells from oxidation through reduction of peroxiredoxins and ribonucleotide reductase, and it inhibits apoptosis through signal regulating kinase 1 (Xu et al. 2016).

PCs from various sources have shown effects on enzymatic antioxidants in different models. When orange juices rich in PCs were administered to overweight or obese adults for 12 weeks, a decrease in catalase and glutathione reductase activity, in parallel to other markers of oxidative stress, resulted in increased protection against ROS (Rangel-Huerta et al. 2015). Fifty milliliters of olive oil administered to healthy adults for 30 days increased the activity of catalase and GPx, and it increased gene expression of

SOD, which were effects the authors attributed to the PCs contained in the oil (Oliveras-López et al. 2014). Postprandial gene expression of thioredoxin reductase decreased when elderly adults consumed a Mediterranean diet rich in PCs and omega-3 fatty acids compared to a diet rich in saturated fatty acids, which indicated that PCs exerted less oxidative burden on the individuals (Yubero-Serrano et al. 2013). Altogether, this information suggests that PCs have a strong influence on the enzymatic antioxidant system of adults.

14.2.2 Nonenzymatic Endogenous Antioxidants

Cells also produce nonenzymatic antioxidants such as glutathione, coenzyme Q (CoQ), lipoic acid, uric acid, and bilirubin, which synergize with enzymes or serve as their cofactors.

Glutathione (GSH) is a tripeptide (L-gamma-glutamyl-L-cysteinyl-glycine) that is present in all human cells and is often considered one of the main nonenzymatic antioxidants. It is synthesized in the cytoplasm and serves as a cofactor for GPx and glutathione-S-transferase (EC 2.5.1.18), among other enzymes (Carocho and Ferreira 2013, Marí et al. 2009). The sulfhydryl group of cysteine is easily oxidized, and is therefore essential due to its electron-donating properties that can neutralize ROS. It has been reported that GSH depletion is associated with the development and progression of chronic diseases, such as certain cancers and hypertension (Robaczewska et al. 2016, Wang et al. 2014). An eight-week supplementation of green tea has been shown to increase overall plasma antioxidant power and glutathione levels in adults with metabolic syndrome, which indicates reduced oxidative stress (Basu et al. 2013). Anthocyanin-rich red rice extracts favor reduction of the hepatic glutathione that protects against paracetamol-induced liver damage in a mouse model (Sinthorn et al. 2016).

The main cellular role of CoQ is as an electron transporter in the ETC, but it can also prevent lipid peroxyl radical formation (Jankowski et al. 2016). The polyisoprenylated benzoquinone structure of CoQ makes it highly apolar, and it is consequently localized in membranes. In addition to being produced endogenously, some foodstuffs such as argan oil are sources of CoQ, where it contributes to their antioxidant potential in conjunction with PCs (Venegas et al. 2011). Recent evidence demonstrated that resveratrol, a PC abundant in wine and other grape-derived products, can serve as a precursor to the aromatic ring of CoQ in bacterial, fungal, and mammalian cells (Xie et al. 2015), which highlights the intimate metabolic connection between PCs and endogenous antioxidants. Oral supplementation of CoQ to type 2 diabetic patients could reduce the risk of heart failure and improve glycemia (Mezawa et al. 2012, Fotino et al. 2013).

The α-lipoic acid (ALA) is an amphipathic organosulfur molecule that is distributed in the cytoplasm and cell membranes, where it is transformed into dihydrolipoic acid (DHLA). The main antioxidant moieties of ALA and DHLA are two sulfhydryl groups that can be easily oxidized and can therefore reduce ROS, such as hydroxyl radicals, hypochlorous acid, and singlet oxygen (Moini et al. 2002, Valko et al. 2006). Oral doses of ALA have been shown to significantly decrease diabetic symptomatic polyneuropathy, which presumably occurs through decreasing ROS and oxidative stress (Garcia-Alcala et al. 2015).

Uric acid is an end product of purine catabolism that is excreted by the kidneys and is therefore considered to be metabolic waste. Before being excreted, uric acid can perform antioxidant functions such as preventing an overproduction of oxo-heme oxidants, which are formed from the reaction of hemoglobin with peroxides (Simoyi et al. 2003). It also reduces lysis of erythrocytes and quenches singlet oxygen and hydroxyl radicals to form allantoin, which might be used as a marker of REDOX imbalance (Kand'ár et al. 2006). It has been recently shown that the tumor suppressor protein p53 stimulates the uric acid transporter SLC2A9 to increase intracellular uric acid concentrations and confer DNA protection by neutralizing ROS (Itahana et al. 2015).

Bilirubin is the end product of heme catabolism and is mainly excreted in bile, but similarly to uric acid, it can serve as an antioxidant before being excreted. Mild congenital hyperbilirubinemia, such as hyperbilirubinemia found in Gilbert syndrome patients, can have protective effects against diseases associated with oxidative stress, such as cardiovascular diseases and metabolic syndrome, presumably by modulating gene expression, inflammation, cellular adhesion, and several other processes in addition to antioxidant effects (Gazzin et al. 2016). A recent cohort study suggested that total serum bilirubin is inversely correlated with the severity of coronary atherosclerosis and inflammation (Akboga et al. 2015). Serum levels of bilirubin and uric acid were significantly lower in patients with polymyositis and dermatomyositis, which the authors suggest was related to increased oxidative stress (Chen et al. 2016). Fewer authors have focused on the antioxidant effects of increased uric acid or bilirubin, because any antioxidant benefit has to occur within physiological concentrations in healthy adults. Otherwise, conditions such as gout or jaundice take on a higher priority.

14.3 PHENOLIC COMPOUNDS AND THE EXOGENOUS ANTIOXIDANT SYSTEM

Cells prevent ROS damage through the previously discussed endogenous antioxidant system, but dietary components also exert an important role. The exogenous antioxidant system consists of molecules obtained from the diet, such as PCs and other molecules with antioxidant capacities.

14.3.1 Phenolic Compounds

PCs are secondary metabolites of vegetable origin that are involved in different physiological and biochemical processes of the plant, such as defense against pathogens, growth, pigmentation, and flavor (Naczk and Shahidi 2004). The PC content in vegetables and fruits depends on intrinsic and extrinsic factors, such as species, cultivar, ripening, characteristics of soil, climate, manipulation, storage, and others (Tomás-Barberán and Espin 2001). More than 8000 different PCs are currently known and are classified as flavonoids (flavanols, flavanones, flavones, isoflavones, and anthocyanidins) and non-flavonoids (phenolic acids such as hydroxycinnamic and hydroxybenzoic, stilbenes, lignans, condensed tannins, and hydrolyzable tannins). PCs have been an active area of study for several decades, partially because of the many beneficial effects associated with their consumption, such as antioxidant and anti-inflammatory effects and prevention of cardiovascular disease, obesity, neurodegenerative disorders, diabetes, and some types of cancer. In vitro antioxidant power has been the most analyzed property of PCs. Their molecular structure allows for resonance stabilization and allows them to stop the free radical chain reaction through electron or hydrogen atom donation from their hydroxyl groups (Bors et al. 1990, Fraga et al. 2010).

Flavonoids in particular are very effective antioxidants due to their metal-chelating potential, which is strongly dependent on the arrangement of the hydroxyl groups and carbonyl around the molecule. The presence of hydrogen- or electron-donating substituents can also allow ROS neutralization (Gülçin 2012). The proposed binding site for metals is the catechol moiety in ring B, the 3-hydroxyl, 4-oxo groups in the heterocyclic ring, and the 4-oxo, 5-hydroxyl groups between the heterocyclic and the A rings of the flavonoid (Gülçin 2012, Pietta 2000).

Resveratrol is a commonly studied PC present in red wine that has at least three different antioxidant mechanisms: (1) Competition with CoQ and decreasing the oxidative chain reaction of ROS, (2) scavenging superoxide anions formed in the mitochondria, and (3) inhibition of lipid peroxidation induced by products of the Fenton reaction (de la Lastra and Villegas 2007, Leonard et al. 2003). Considerable evidence continues to be generated about the particular health effects and mechanisms of action for other PCs such as quercetin, gallic acid, and catechins and the foods that contain these compounds.

14.3.2 Nonphenolic Compounds

In addition to PCs, vitamin A, vitamin C, E, minerals such as zinc and selenium, fatty acids, and dietary fiber exhibit varying degrees of antioxidant power. Thus, they comprise the exogenous antioxidant system.

The term *vitamin A* can be applied to retinoids (retinol, retinal, retinoic acid, and their esters) and some carotenoids (β-carotene, α-carotene, and β-cryptoxanthin), of which retinol and β-carotene are the most prominent. Retinyl palmitate (consumed from some animal sources, such as milk) is hydrolyzed to retinol by retinyl ester hydrolases in the intestinal lumen, and β-carotene (consumed mostly from vegetable sources) is metabolized to retinal and then finally to retinol by retinal reductase (Harrison 2005, Paik et al. 2001, Tang and Gudas 2011). Vitamin A is involved in cellular differentiation, embryonic development, hormone production, and vision (Weber and Grune 2012).

Retinol exerts its antioxidant activity by combining with peroxyl radicals before they start lipid peroxidation (Jee et al. 2006). Interestingly, retinoids can stimulate the activity of endogenous enzymes such as catalase. For example, Sertoli cells incubated for 24 hours with retinol and retinoic acid showed an increase in catalase activity but not in gene expression, which the authors suggest could be attributed to decreased catalase turnover or other mechanisms (Pasquali et al. 2008). Because of its antioxidant capacity, it was believed for several years that vitamin A could be used as a dietary supplement to treat or prevent disease, but subsequent in vitro and in vivo studies have shown that a pro-oxidant effect can occur under certain conditions (Penniston and Tanumihardjo 2006).

Eight different molecules can be referred to as vitamin E: α, β, γ, and δ-tocopherol and α, β, γ, and δ-tocotrienol. Their molecular structure has a 6-chromanol ring head and an isoprene side chain. The side chain is fully saturated in tocopherols and has three double bonds in tocotrienols. Vitamin E is obtained mostly from vegetable sources, such as nuts, seeds, grains, oils, and fruits, but some mushrooms and fishes can also contain vitamin E (Heleno et al. 2010, Jiang 2014). Because of its insoluble character, it is found in membranes, where it protects against phospholipid peroxidation (Descamps-Latscha et al. 2001). α-Tocopherol is the most powerful and most studied tocopherol, but others also show important antioxidant properties. For example, γ-tocopherol can trap reactive nitrogen species produced during inflammation (Jiang et al. 2001). Evidence also suggests that tocotrienols are better at scavenging peroxyl radicals due to a better distribution in the cell membrane (Wong and Radhakrishnan 2012). Vitamin E scavenges free radicals by transferring hydrogen atoms to yield a non-radical product and vitamin E radical (Niki 2014). Vitamin E has also gained much interest due its protective role in the prevention and treatment of cardiovascular diseases and cancer (Moya-Camarena and Jiang 2012).

Ascorbic acid or vitamin C is a water-soluble antioxidant with a hydroxylated structure similar to glucose or other monosaccharides. It can be easily oxidized to dehydroascorbic acid and then to carbon dioxide, oxalic acid, and threonic acid. Vitamin C can be found in animals and vegetables, but major dietary sources are citrus fruits, cherries, tomatoes, broccoli, cauliflower, and kiwifruit (Bursal and Gülçin 2011). The antioxidant power of vitamin C is high, and it is capable of scavenging ROS such as hydrogen peroxide, singlet oxygen, and superoxide anions by donating electrons, which yields a relatively unreactive ascorbate radical that can be reduced back to ascorbic acid by NADH reductases (Linster and Van Schaftingen 2007, Mamede et al. 2011). Vitamin C also synergizes with vitamin E against lipid peroxidation by reducing the α-tocopheryl radical formed when tocopherols are oxidized (Golumbic and Mattill 1941). It has been experimentally shown that free radical quenching by vitamin C can reduce the risk of cancer by decreasing in vivo damage to proto-oncogenes (Crott and Fenech 1999). It protects against plasma lipid and LDL oxidation, and is associated with the prevention of chronic diseases such as atherosclerosis and type 2 diabetes (El-Shafei and Saleh 2016, Rafighi et al. 2013). Vitamin C favors dietary iron absorption by reducing Fe^{3+} to Fe^{2+}, which is the form that can be transported by the enterocytes. Nevertheless, vitamin C can also have pro-oxidant effects in the presence of Fe^{2+} that result in ROS formation (Du et al. 2012).

Zinc is an essential trace element required for the function of various enzymes and other cellular proteins (Dock and Vahter 1999). Zinc modulates oxidative stress caused by free radicals through various mechanisms; it is present in the active site of SOD1 (CuZnSOD) and can stimulate the activity of SOD, catalase, and GPx and reduce the activity of NADPH oxidase (Bao et al. 2013, Prasad et al. 2001). It can also induce the production of metallothionein, which is a cysteine-rich protein that can scavenge free radicals. Zinc also decreases activation of the transcription factor NF-κB and interrupts the downstream inflammatory response caused by LDL oxidation.

Selenium is an essential trace mineral required for enzyme activity of approximately 25 selenoproteins, and 16 have an antioxidant role, such as GPx (Chen and Berry 2003, Mugesh and Singh 2000, Pappas et al. 2008). For example, some selenoenzymes protect DNA and other cellular components from oxidative damage and regulate synthesis of CoQ (Valko et al. 2006). Selenium can be obtained from foods such as garlic, mushrooms, grains, cabbage, carrots, and meat (Battin and Brumaghim 2009). Consumption of >400 μg of selenium per day results in selenosis (Johnson et al. 2003), while its deficiency is associated with a faster decline in cognitive function and poor performance in coordination assessment and motor speed (Steinbrenner and Sies 2013). In addition to elemental selenium, some seleno-compounds have been studied for their potential to reduce oxidative stress by scavenging free radicals. For example, seleno-carbamates possess a high in vitro scavenging activity of superoxide anion (Takahashi et al. 2005). Additionally, selenocysteine coordinates metal ions in enzymes, including molybdenum and tungsten in formate dehydrogenase, and nickel in NiFeSe hydrogenase (Garcin et al. 1999). Selenium can prevent ROS damage caused by heavy metal toxicity through its complexation with mercury (Raymond and Ralston 2009).

PCs can interact with non-PCs in the gastrointestinal tract, where PCs favor the bioaccessibility and bioactivity of α-tocopherol by preserving it in its reduced state. The presence of PCs in the gastrointestinal tract also prevents metal-catalyzed ROS formation by chelating iron, copper, or zinc ions, but without having negative consequences on their nutritional status (Domínguez-Avila et al. 2017).

14.4 BIOLOGICAL RELEVANCE OF OXIDATIVE STRESS

14.4.1 Biomarkers of Macromolecule Oxidation

Because ROS can nonspecifically oxidize any cellular component, oxidative stress can be estimated by quantifying known end products of oxidized macromolecules directly in cells, tissue samples, or biological fluids.

Unsaturated fatty acids, such as those of the cell membrane, are easily oxidized by ROS. Hydroxyl radicals remove a hydrogen atom from a double bond of a fatty acid that becomes a lipoperoxyl radical, and subsequent molecular rearrangements can lead to final products that either (1) have a conjugated double bond (-CH=CH-CH=CH-, named dienes), or (2) have a carbonyl group (-C=O, aldehydes or ketones). Although many different molecules can be generated by lipid oxidation, 4-hydroxynonenal and malondialdehyde (4-HNE and MDA, respectively) stand out as common lipid oxidation end products that are routinely produced. They have been linked to several pathologies such as cardiovascular diseases, Alzheimer's, and several other disorders (Ayala et al. 2014). Lipid oxidation is favored by free transition metal ions such as copper or iron, which catalyze ROS formation in vivo (Gutteridge and Halliwell 1990). Some PCs can decrease lipid peroxidation biomarkers; serum concentrations of 4-HNE and MDA are reduced by consumption of a pomegranate (*Punica granatum* L.) PC extract in type 2 diabetic patients (Basu et al. 2013). Flavonoid intake inversely correlates with serum MDA concentration in elderly adults, which indicates that flavonoid consumption mitigates lipid peroxidation (Gonzalez et al. 2013).

Carbohydrate and protein oxidation tend to occur alongside one another in so-called glycoxidation (glycation-oxidation) reactions. Oxidative conditions favor non-enzymatic protein glycation, which leads to the production of advanced glycation end products (AGEs) that can accumulate over time. N^{ε}-(carboxymethyl)-lysine (CML), fructoselysine, pentosidine, and glycated hemoglobin (HbA1c) are commonly studied AGEs (Ahmed et al. 1986, Fu et al. 1996, Gkogkolou and Bohm 2012). Proteins that are particularly susceptible to forming AGEs are hemoglobin and histones. Hemoglobin glycation is proportional to glycemia because entry of glucose into the erythrocyte is insulin-independent, which ultimately exposes hemoglobin to increased glucose whenever its serum concentration increases. Histones are rich in lysine, which favors a strong reaction with glucose at their ε-amino groups. AGEs can directly affect cells by hindering the activity of involved molecules or through crosslinking. They can also activate the receptor for AGEs (RAGE), which leads to positive feedback signaling that upregulates the inflammatory response (Goldin et al. 2006). Curcumin can prevent formation of AGEs by, at least in part, forming complexes with methylglyoxal, which is a precursor of AGEs (Sun et al. 2016). Furthermore, a recent systematic review concluded that consumption of PCs from various sources had no significant impact on healthy individuals or pre-diabetic patients, but they significantly decreased AGE formation in type 2 diabetic patients, which could hinder development of diabetic complications (Palma-Duran et al. 2015).

Oxidative damage to DNA yields several different end products from nucleotides. In addition, DNA-protein crosslinking, base-sugar crosslinking, inter-strand crosslinking, and strand breakage are also well known (Cadet and Wagner 2013). Due to its low reduction potential, guanine is the most easily oxidized nitrogenous base (Steenken and Jovanovic 1997), and its major oxidation product 8-hydroxy-2′-deoxyguanosine (8OHdG) is commonly found in urine. Because 8OHdG is produced when DNA is damaged, it has been used as a biomarker after exposure to carcinogenic substances (Valavanidis et al. 2009). However, other researchers have used it as a biomarker for several unrelated conditions, such as atherosclerosis, diabetes, arthritis, dermatitis, and strenuous exercise, among other conditions. The widespread quantification of 8OHdG showed that ROS-derived DNA damage was comorbid with several pathologies. Consumption of a chokeberry (*Aronia melanocarpa* L.) extract in a beverage was found to significantly decrease urinary 8OHdG in metabolic syndrome patients and healthy adults, which suggested that there was mitigation of oxidative stress and protection against DNA damage (Bernabe et al. 2013).

14.4.2 Oxidative Stress in Healthy Individuals

Basal ROS production is evident in the cumulative damage of skin collagen, which leads to the appearance of wrinkles as part of normal aging. Collagen can be damaged by ROS that attack its amino acid sidechains, thus leading to increased strand breaks (Berlett and Stadtman 1997). Collagen fragments produced from oxidative stress favor expression of collagen-degrading matrix metalloproteinases (MMPs), which further degrade collagen in a positive feedback loop that reduces the mechanical properties of skin (Fisher et al. 2009). Oxidative stress in the skin can be caused by, among several other factors, UV radiation and dietary habits. Thus, reduced UV exposure and a diet that includes antioxidant-rich foods can delay the process. For example, Bae et al. (2008) suggested that epigallocatechin gallate, which is a PC present in green tea and other sources, can inhibit UV-induced photoaging in mice due to its ability to prevent ROS production, and can regulate the signaling pathways involved in cell survival. Similarly, skin fragility can be mitigated by oral administration of α-tocopherol in obese diabetic mice by modulating the expression of MMP genes (Ibuki et al. 2012).

Physical exercise can acutely affect ROS production in healthy organisms because it requires an increase in mitochondrial ATP production. Evidence of exercise-induced ROS has been accepted since the 1970s when Dillard et al. (1978) reported molecular evidence of lipid peroxidation in response to exercise, which could be mitigated by vitamin E. In the following decades, more studies have been published, but they have shown inconsistent results regarding the specific oxidant effects induced by exercise in healthy adults (Urso and Clarkson 2003). This lack of consistency occurs because the correlation between exercise and oxidative stress is multifactorial (intensity, duration, physical condition, etc.). Thus, there is no single *best* approach to address it. Some patients might benefit from PC supplementation, while an adequate diet that supplies the necessary amount of antioxidants might suffice for others. Consumption of PCs from several sources, such as chokeberries, can have a positive impact in athletes by preventing oxidative DNA damage due to chronic exercise, as measured by decreased urinary biomarkers (Garcia-Flores et al. 2016). Additionally, personalized regiments have been suggested for managing the oxidative consequences of exercise (Pingitore et al. 2015).

14.4.3 Oxidative Stress in Disease

Oxidative stress is exacerbated in sick individuals, and illness can easily overwhelm the antioxidant system. In fact, oxidative stress has been demonstrated alongside endocrine, cardiovascular, neurodegenerative, cancerous, and other diseases.

Diabetes is a well-documented condition in which oxidative stress has profound multi-organ implications. For example, the retina and renal glomerular cells are extremely susceptible to chronic hyperglycemia because these cells are insulin-independent. Hyperglycemia in the retina leads to increased ROS and AGE production, which can contribute to loss of vision. A recent study suggested that adults with diabetic retinopathy can significantly benefit from antioxidant supplementation (pycnogenol, vitamin E, and CoQ) to reduce their oxidant burden and macular damage, which could potentially preserve their vision (Domanico et al. 2015). Hyperglycemia favors ROS and apoptosis of renal glomerular cells, which further stresses the remaining cells and accelerates the loss of renal function (Singh et al. 2011).

Atherosclerosis is a significant component of cardiovascular disease. It is characterized by increased total and LDL cholesterol (among others). ROS can oxidize LDL to produce oxidized LDL (oxLDL), which is considered a highly atherogenic particle, as compared to LDL (Steinberg 1997). LDL or oxLDL can build up and form a plaque in the tunica intima of major arteries, which can increase in size, accrue dead cells, and calcify over decades. Eventually, this plaque occludes blood flow and can culminate in an acute coronary event (Badimon et al. 2012). Although the development of atherosclerosis is multifactorial, oxidative stress is intimately linked to hypertension, hypercholesterolemia, smoking, and diabetes, which are known risk factors for atherosclerosis (Li et al. 2013). Dietary habits can also play a significant part in atherosclerosis. For example, diets rich in PCs can reduce oxidative stress and prevent LDL oxidation (Annuzzi et al. 2014). Consumption of a Mediterranean diet has been linked to reduced atherogenesis, which could be attributed in part to bioactive PCs that can normalize serum cholesterol, mitigate inflammation, and modulate signaling pathways and enzyme activities relevant to atherogenesis (Massaro et al. 2010).

14.5 FACTORS RELATED TO THE ANTIOXIDANT POWER OF PCs

The antioxidant power of any compound can be attributed to the number and position of hydroxyl groups, sulfhydryl groups, double bonds, and so on (Agati et al. 2012, Hix et al. 2004). The antioxidant power of PCs is often quantified in vitro using high purity compounds, and although several health-promoting effects related to PC consumption are attributed to their high antioxidant power, there is no *absolute* value for it. When analyzing their antioxidant contribution in vegetables or their extracts, values will vary due to the presence of other compounds, the food matrix, and food processing, among other factors, which could result in significantly higher or lower values.

The antioxidant power of plants is mainly due to the diversity and concentration of PCs, which can differ between species, cultivar, and agricultural practices, among other factors. Abiotic factors can strongly influence the production of PCs and other secondary metabolites. For example, hydric restriction has been shown to promote an increase in total phenolic acids and flavonoids, as well as antioxidant activity, compared to standard irrigation during sorghum production (Wu et al. 2017), while a similar behavior was reported in UV-C irradiated tomatoes (Maharaj et al. 2014). It is well documented that a PC increase is mediated by phenylalanine ammonia-lyase through the phenylpropanoid pathway, which has increased activity under stressful conditions (Wu et al. 2017, Martinelli et al. 2012).

Food processing and storage alter bioactivity through exposure to oxygen, light, other compounds, variations in pH, and other factors, which ultimately changes the molecular structure of PCs. Rodriguez-Mateos et al. (2014) demonstrated a significant reduction of 27 to 42 percent in anthocyanins and phenolic acids from wild blueberries after baking them. Martins and de Rosso (2016) found a second-order rate of degradation of tomato carotenoids after a thermal treatment. Corrales-Bañuelos et al. (2016) also mentioned that alkaline and thermal processing during production of tortillas significantly reduce carotenoid content and the antioxidant activity of the flour. The concentration and antioxidant activity changes can be attributed to the oxidation of functional groups (such as hydroxyl or sulfhydryl) or unsaturated bonds in the carbon skeleton of these molecules, which essentially renders them ineffective as antioxidants (El-Agamey et al. 2004).

Recent studies mentioned that addition of natural antioxidants to food products can prevent oxidation of some components and improve their functional properties (Rashidinejad et al. 2016). For example, added antioxidants such as myricetin, rosmarinic acid, and carnosic acid can protect vegetal oils from oxidation during storage and frying (Guo et al. 2016, Guitard et al. 2016). PCs can donate electrons or hydrogen atoms during lipid peroxidation, which reduces the formation of saturated fatty acids. Therefore, these molecules are a suitable option to use as food additives to minimize the utilization of synthetic antioxidants that might have undesired effects in humans (Guo et al. 2016, Ayala-Zavala et al. 2011).

Other studies suggested that PCs can have additive, synergistic, or antagonist effects that can modify their bioactivity. Palafox-Carlos et al. (2012) found that the antioxidant activity of mango pulp (*Mangifera indica* L.) was due to the additive effect of major phenolic acids (gallic acid, chlorogenic acid, vanillic acid, and protocatechuic acid). Antagonism was observed between β-carotene and α-tocopherol to prevent chlorophyll oxidation (Smyk 2015). Antioxidant compounds of vegetable origin could be used to design food systems that increase their health benefits, such as the modifications made by Velderrain-Rodríguez et al. (2015) when they applied a synergistic combination of antioxidants from mango peels (gallic acid and protocatechuic acid) in edible coatings to enhance the functional properties of fresh-cut papaya (*Carica papaya* L.).

The food matrix can also have significant effects on antioxidant values. It has been demonstrated that macromolecules such as carbohydrates, lipids, and proteins can interact with PCs through hydrogen bonds and electrostatic and van der Waals interactions, which reduces their antioxidant power as well as bioaccessibility and bioavailability (Quirós-Sauceda et al. 2014, Palafox-Carlos et al. 2011). This reduction has been experimentally corroborated for pineapple (*Ananas comosus* L.), mango, and papaya during in vitro gastric digestion (Velderrain-Rodríguez et al. 2016). Furthermore, during the intestinal stage of in vitro digestion, a decrease in PC content from papaya was observed, which could be attributed to oxidation of caffeic acid due to alkaline conditions (Friedman and Jürgens 2000). Orally ingested PCs that remain bound to food matrix components, such as dietary fiber, can resist digestion by human enzymes and will only be released by colonic microbiota in the large intestine, where their antioxidant power can change due to bacterial metabolism (Palafox-Carlos et al. 2011).

Because antioxidant power can be lower when PCs are bound to other macromolecules, improved release through food processing is desirable and has been attempted by research groups. Rodríguez-Roque et al. (2015) found that nonthermal treatments such

TABLE 14.1 General Overview and Examples of Available Systems to Encapsulate Antioxidants

Type of Particle	Technique	Wall Material	Core Compound	Purpose	Reference
Micro-particles	Spray drying	Maltodextrin	Anthocyanins	Color stability	de Souza et al. 2015
	Freeze drying	Arabic gum/ maltodextrin	Anthocyanins	Storage stability	Rocha-Parra et al. 2016
	Molecular inclusion	β-cyclodextrin	Essential oils	Enhance antimicrobial activity	Del Toro-Sánchez et al. 2010
Nano-particles	Precipitation	Zein-pectin/alginate	Curcumin	Enhance water-solubility	Huang et al. 2016
	Electrospray	Zein	Curcumin	Storage stability	Gomez-Estaca et al. 2012
Vesicles	Double emulsion	Oat polar lipid fraction	Betalain	Intestinal delivery	Kaimainen et al. 2015
	Liposomes	Soy lecithin/alginate	Grape pomace extract	Oral delivery	Manconi et al. 2016
	Niosomes	Gelot 64	Resveratrol	Dermic delivery	Pando et al. 2015

as high-intensity pulsed electric fields and high-pressure processing increased the release of PCs from fruit juice–based beverages due to breakdown of the cell wall structure. Similar behavior was observed by van het Hof et al. (2000), who saw an increase in lycopene release from homogenized and heated tomatoes.

Other more specific strategies have been developed to preserve the antioxidant properties of PCs by increasing their stability and minimizing undesired interactions within the environment or with components of the food matrix. Encapsulation protects compounds by packaging labile molecules within inert carrier materials, which have been shown to increase the shelf life of various PCs (Zokti et al. 2016, Tumbas Šaponjac et al. 2016). Different parameters of the encapsulation process can be optimized to obtain the best results. For example, the choice of encapsulating material can affect PCs, but ultimately these choices are dictated by the final application of the encapsulated product. Table 14.1 briefly lists various encapsulation systems that are currently in use.

14.6 COMMONLY USED ANTIOXIDANT POWER ASSAYS

Consumption of PCs from fruits, vegetables, and drinks such as red wine, juices, and teas has been associated with health benefits, such as the prevention of several chronic degenerative diseases and slowing the aging process. These effects are related to the antioxidant power of the molecules present in these food and beverages, and as previously discussed, the effects are dependent on several factors, including the assay used to quantify them. Although various assays are available, two general mechanisms of action are generally recognized: hydrogen atom transfer (HAT) and single electron transfer (SET). An assay is HAT-based when the compound quenches free radicals by donating a hydrogen atom, which typically comes from a hydroxyl group. SET-based assays rely on the transfer a single electron from an antioxidant to a free radical or oxidized molecule. In addition, some assays rely on a mixed HAT/SET mechanism.

14.6.1 DPPH

The DPPH (2,2-diphenyl-1-picrylhydrazyl) assay uses the stable DPPH free radical. It has an extra electron that is delocalized on the whole molecule and does not dimerize as other free radicals do (Molyneux 2004). This method is measured at ambient temperature to avoid thermal degradation. Electron delocalization generates an intense purple color with maximum absorption (λ_{max}) at 515 nm. If the radical is reduced, the purple color and absorbance at λ_{max} decrease, which depends on the concentration of the antioxidant. The DPPH assay involves mixing the sample of interest with a DPPH radical solution (50 µM) at a 40:1 proportion (0.1 mL of sample with 3.9 mL of radical solution) using methanol as solvent (Brand-Williams, Cuvelier, and Berset 1995). The mixtures are incubated for 5–30 minutes, and the absorbance is read at 515–520 nm. Radical scavenging percentage is calculated as follows:

$$\%\ \text{inhibition of DPPH radical} = \left[(\text{Abs}_{\text{initial}} - \text{Abs}_{\text{final}})/\text{Abs}_{\text{initial}}\right] \times 100$$

where $\text{Abs}_{\text{initial}}$ is the absorbance before the reaction and $\text{Abs}_{\text{final}}$ is the absorbance after the reaction. The mechanism of action of the DPPH assay is mixed HAT and SET (Apak et al. 2016).

14.6.2 TEAC

The Trolox equivalent antioxidant capacity (TEAC) assay uses the ABTS radical [2,2′-azino-bis(3-ethylbenzothiazoline-6-sulfonic acid)], which is stable but does not occur naturally. It has to be produced by reacting the molecule with an oxidant, which is typically sodium persulfate or manganese dioxide. This assay measures a decrease in blue/green color and absorbance at 743 nm (645, 734, 750, and 815 nm have also been used) once the ABTS radical reacts with antioxidants in the sample of interest (Re et al. 1999, Seeram et al. 2006, Su et al. 2007). TEAC typically uses Trolox as a standard, which is the water-soluble analog of vitamin E that the assay is named for. A standard curve of Trolox typically uses 0–350 μM concentrations, from which Trolox equivalents in a sample are calculated (Seeram et al. 2006). The TEAC assay has been used to determine the antioxidant capacity of fruit and vegetable extracts, soft drinks, alcoholic drinks, tea, and coffee (Pellegrini et al. 2003). The mechanism of action of the TEAC assay is mixed HAT and SET (Apak et al. 2016).

14.6.3 FRAP

The ferric reducing antioxidant power (FRAP) assay measures the reduction of iron from ferric (Fe^{3+}) to ferrous (Fe^{2+}) when complexed with TPTZ [2,4,6-tri(2-pyridyl)-s-triazine] under acid conditions (pH 3.6). The FRAP reagent is prepared by mixing 1 part TPTZ (10 mM in 40 mM HCl) with 10 parts of acetate buffer (300 mM) and 1 part of $FeCl_3$ (20 mM). The ratio of TPTZ:iron is 1:2 (Fe^{3+} 1.67 mM, TPTZ 0.83 mM). The solution also contains other oxidants besides Fe^{3+}:TPTZ, such as other Fe^{3+} species capable of reacting with antioxidants (Benzie and Strain 1996, Huang et al. 2005). Reduction of Fe^{3+}: TPTZ to Fe^{2+}:TPTZ generates an intense blue color that can be measured at 593 nm. The FRAP assay is performed by mixing the sample with the FRAP reagent at a 1:10 ratio, and the absorbance at 593 nm is read after five seconds and then every 15 seconds for four minutes to calculate the change in absorbance ($\Delta A = A_{4min} - A_{0min}$), which is linearly proportional to the concentration of antioxidants. FRAP values are calculated by using a standard curve prepared with Trolox, ascorbic acid, or other standards. Other studies recommend reading the absorbance value after 30 minutes of incubation at 37°C (Benzie and Strain 1999) or even incubating the samples for several hours at 37°C before recording the absorbance (Pulido et al. 2000). The mechanism of action of the FRAP assay is SET (Apak et al. 2016).

14.6.4 ORAC

The oxygen radical absorbance capacity (ORAC) method measures the antioxidant scavenging activity against the peroxyl radical induced by AAPH [2,2′-azobis(2-amidinopropane) dihydrochloride] at 37°C (Pisoschi and Negulescu 2012, Thaipong et al. 2006). Initially, β-phycoerythrin (β-PE), a fluorescent protein used as a probe, can provide an indication of the damage from its reaction with the peroxyl radical through loss of fluorescence (Cao et al. 1993). β-PE was eventually replaced due to its loss of fluorescence by interacting with some polyphenols even without adding the radical generator. Additionally, β-PE is photobleached under plate reader conditions, among other conditions (Cao and Prior 1999, Ou et al. 2001). Fluorescein is now used as a probe because it overcomes the limitations of β-PE, and the reaction products of fluorescein with peroxyl radical have been characterized. The product patterns were consistent with a classic HAT mechanism (Naguib 2000). The ORAC method provides a direct measure of antioxidant capacity for breaking the hydrophilic and lipophilic chains in the presence of peroxyl radicals. Peroxyl radicals are the most prevalent free radicals, which is why the ORAC assay is relevant to in vivo conditions. Additionally, the method can be adapted to numerous sample matrices in addition to fruits and vegetables, such as plasma and tissue (Prior 2015).

14.6.5 CUPRAC

The chromogenic oxidizing reagent of the cupric ion reducing antioxidant capacity (CUPRAC) method is bis(neocuproine)copper(II) chloride [Cu(II)-Nc], which reacts with n-electron antioxidants:

$$n\text{Cu(Nc)}_2^{2+} + n\text{-electron reductant} \leftrightarrow n\text{Cu(Nc)}_2^{1+} + n\text{-electron oxidized product} + n\text{H}^{1+}$$

where the liberated protons may be buffered with a relatively concentrated ammonium acetate buffer solution. In this reaction, the reactive hydroxyl groups of PCs (or other antioxidants) are oxidized to the corresponding quinones and Cu(II)-Nc is reduced. Then, absorbance of the highly colored Cu(I)-Nc complex is read at 450 nm (Alam et al. 2013, Apak et al. 2007, 2016).

14.6.6 Electrochemical Techniques

The previously mentioned assays are colorimetric and are the most common, but electrochemical methods are an alternative for determining the antioxidant power of bioactive compounds. They do not require time-consuming sample preparation and have low costs (Sochor et al. 2013). Most methods are voltammetric and amperometric, and they measure current intensity (generated by the oxidation or reduction of an electroactive analyte) that flows between a working electrode and a reference electrode (Pisoschi and Negulescu 2012).

Cyclic voltammetry is used to evaluate the total antioxidant power of low molecular weight antioxidants in fruits, vegetables, plasma, and tissue (Chevion et al. 1997, Chevion et al. 1999). The working electrode potential is ramped linearly over time and is linearly scanned from an initial value, to an intermediate value, to a final value, while recording the respective current intensity (Pisoschi and Negulescu 2012). Cyclic voltammetry parameters are the intensities of the cathodic and anodic peaks (Ic and Ia), the anodic oxidation potential (Ea), and the cathodic oxidation potential (Ec). Analysis of the tracing yields the values of the biological oxidation potential, including E and $E_{1/2}$, which relate to the nature of the specific molecules, the intensity Ia of the anodic current and the area of anodic wave.

Because there is no gold standard method, it is common not to rely on a single assay when quantifying antioxidant power but to report values obtained from at least two of the aforementioned methods.

14.7 CONCLUSIONS

Antioxidants consumed in the diet are part of the exogenous antioxidant system and defend cells against oxidative stress. Phenolic compounds are a significant component of the exogenous antioxidant system and can regulate the activity of the endogenous antioxidant system or synergize with it. Phenolic compounds have been known to decrease the concentration of several biomarkers of oxidative stress in human, animal, and in vitro models. This important bioactivity of phenolic compounds is often related to their high antioxidant power for which various methods are available to quantify it, although no gold standard method is available for this purpose.

ACKNOWLEDGMENTS

This work is part of CONACYT Fronteras de la Ciencia project 563: "Un enfoque multidisciplinario de la farmacocinética de polifenoles de mango Ataulfo. Interacciones moleculares, estudios preclínicos y clínicos."

REFERENCES

Agati, G., Azzarello, E., Pollastri, S., and Tattini, M. 2012. Flavonoids as antioxidants in plants: Location and functional significance. *Plant Science* 196:67–76.

Ahmed, M.U., Thorpe, S.R., and Baynes, J.W. 1986. Identification of n-epsilon-carboxymethyllysine as a degradation product of fructoselysine in glycated protein. *Journal of Biological Chemistry* 261:4889–4894.

Akboga, M.K., Canpolat, U., Sahinarslan, A. et al. 2015. Association of serum total bilirubin level with severity of coronary atherosclerosis is linked to systemic inflammation: Promising results from a large-scale cohort study. *Journal of Biological Chemistry* 65:A1683–A1683.

Alam, M.N., Bristi, N.J., and Rafiquzzaman, M. 2013. Review on in vivo and in vitro methods evaluation of antioxidant activity. *Saudi Pharmaceutical Journal* 21:143–152.

Annuzzi, G., Bozzetto, L., Costabile, G. et al. 2014. Diets naturally rich in polyphenols improve fasting and postprandial dyslipidemia and reduce oxidative stress: A randomized controlled trial. *The American Journal of Clinical Nutrition* 99:463–471.

Apak, R., Güçlü, K., Demirata, B. et al. 2007. Comparative evaluation of various total antioxidant capacity assays applied to phenolic compounds with the CUPRAC assay. *Molecules* 12:1496–1547.

Apak, R., Ozyurek, M., Guclu, K., and Capanoglu, E. 2016. Antioxidant activity/capacity measurement. 1. classification, physicochemical principles, mechanisms, and electron transfer (et)-based assays. *Journal of Agricultural and Food Chemistry* 64:997–1027.

Ayala, A., Munoz, M.F., and Arguelles, S. 2014. Lipid peroxidation: Production, metabolism, and signaling mechanisms of malondialdehyde and 4-hydroxy-2-nonenal. *Oxidative Medicine and Cellular Longevity* 2014:1–31.

Ayala-Zavala, J.F., Vega-Vega, V., Rosas-Domínguez, C. et al. 2011. Agro-industrial potential of exotic fruit byproducts as a source of food additives. *Food Research International* 44:1866–1874.

Badimon, L., Padro, T., and Vilahur, G. 2012. Atherosclerosis, platelets and thrombosis in acute ischaemic heart disease. *European Heart Journal: Acute Cardiovascular Care* 1:60–74.

Bae, J.Y., Choi, J.S., Choi, Y.J. et al. 2008. Epigallocatechin gallate hampers collagen destruction and collagenase activation in ultraviolet-B-irradiated human dermal fibroblasts: Involvement of mitogen-activated protein kinase. *Food and Chemical Toxicology* 46:1298–1307.

Bao, B., Ahmad, A., Azmi, A. et al. 2013. The biological significance of zinc in inflammation and aging. In *Inflammation Advancing and Nutrition*, eds. I. Rahman and D. Bagchi, pp. 15–27. New York: Elsevier, Inc.

Basu, A., Betts, N.M., Mulugeta, A. et al. 2013. Green tea supplementation increases glutathione and plasma antioxidant capacity in adults with the metabolic syndrome. *Nutrition Research* 33:180–187.

Basu, A., Newman, E.D., Bryant, A.L., Lyons, T.J., and Betts, N.M. 2013. Pomegranate polyphenols lower lipid peroxidation in adults with type 2 diabetes but have no effects in healthy volunteers: A pilot study. *Journal of Nutrition and Metabolism* 2013:1–7.

Battin, E.E., and Brumaghim, J.L. 2009. Antioxidant activity of sulfur and selenium: A review of reactive oxygen species scavenging, glutathione peroxidase, and metal-binding antioxidant mechanisms. *Cell Biochemistry and Biophysics* 55:1–23.

Benzie, I.F., and Strain, J. 1996. The ferric reducing ability of plasma (FRAP) as a measure of "antioxidant power": The FRAP assay. *Analytical Biochemistry* 239:70–76.

Benzie, I.F., and Strain, J. 1999. Ferric reducing/antioxidant power assay: Direct measure of total antioxidant activity of biological fluids and modified version for simultaneous measurement of total antioxidant power and ascorbic acid concentration. *Methods in Enzymology* 299:15–27.

Berlett, B.S., and Stadtman, E.R. 1997. Protein oxidation in aging, disease, and oxidative stress. *Journal of Biological Chemistry* 272:20313–20316.

Bernabe, J., Mulero, J., Cerde, B. et al. 2013. Effects of a citrus based juice on biomarkers of oxidative stress in metabolic syndrome patients. *Journal of Functional Foods* 5:1031–1038.

Bors, W., Heller, W., Michel, C., and Saran, M. 1990. Flavonoids as antioxidants: Determination of radical-scavenging efficiencies. *Methods in Enzymology* 186:343–355.

Brand-Williams, W., Cuvelier, M.E., and Berset, C. 1995. Use of a free-radical method to evaluate antioxidant activity. *LWT-Food Science and Technology* 28:25–30.

Brigelius-Flohé, R., and Maiorino, M. 2013. Glutathione peroxidases. *Biochimica et Biophysica Acta (BBA)-General Subjects* 1830:3289–3303.

Bursal, E., and Gülçin, İ. 2011. Polyphenol contents and in vitro antioxidant activities of lyophilised aqueous extract of kiwifruit (*Actinidia deliciosa*). *Food Research International* 44:1482–1489.

Cadet, J., and Wagner, J.R. 2013. DNA base damage by reactive oxygen species, oxidizing agents, and UV radiation. *Cold Spring Harbor Perspectives in Biology* 5: a012559 1–19.

Cao, G., Alessio, H.M., and Cutler, R.G. 1993. Oxygen-radical absorbance capacity assay for antioxidants. *Free Radical Biology and Medicine* 14:303–311.

Cao, G., and Prior, R.L. 1999. Measurement of oxygen radical absorbance capacity in biological samples. *Methods in Enzymology* 299:50–62.

Carocho, M., and Ferreira, I.C.F.R. 2013. A review on antioxidants, prooxidants and related controversy: Natural and synthetic compounds, screening and analysis methodologies and future perspectives. *Food and Chemical Toxicology* 51:15–25.

Che, M., Wang, R., Li, X., Wang, H.Y., and Zheng, X.F.S. 2016. Expanding roles of superoxide dismutases in cell regulation and cancer. *Drug Discovery Today* 21:143–149.

Chen, J., and Berry, M.J. 2003. Selenium and selenoproteins in the brain and brain diseases. *Journal of Neurochemistry* 86:1–12.

Chen, Z., Su, Z., Pang, W. et al. 2016. Antioxidant status of serum bilirubin and uric acid in patients with polymyositis and dermatomyositis. *International Journal of Neuroscience* 127:1–7.

Chevion, S., Berry, E.M., Kitrossky, N., and Kohen, R. 1997. Evaluation of plasma low molecular weight Antioxidant capacity by cyclic voltammetry. *Free Radical Biology and Medicine* 22:411–421.

Chevion, S., Chevion, M., Chock, P.B., and Beecher, G.R. 1999. Antioxidant capacity of edible plants: Extraction protocol and direct evaluation by cyclic voltammetry. *Journal of Medicinal Food* 2:1–10.

Corrales-Bañuelos, A.B., Cuevas-Rodríguez, E.O., Gutiérrez-Uribe, J.A. et al. 2016. Carotenoid composition and antioxidant activity of tortillas elaborated from pigmented maize landrace by traditional nixtamalization or lime cooking extrusion process. *Journal of Cereal Science* 69:64–70.

Correa, M., Pascual, M., Sanchis-Segura, C., Guerri, C., and Aragon, C.M.G. 2005. Lead-induced catalase activity differentially modulates behaviors induced by short-chain alcohols. *Pharmacology Biochemistry and Behavior* 82:443–452.

Crott, J.W., and Fenech, M. 1999. Effect of vitamin C supplementation on chromosome damage, apoptosis and necrosis ex vivo. *Carcinogenesis* 20:1035–1041.

De la Lastra, C.A., and Villegas, I. 2007. Resveratrol as an antioxidant and pro-oxidant agent: Mechanisms and clinical implications. *Biochemical Society Transactions* 35:1156–1160.

De Souza, V.B., Thomazini, M., de Carvalho Balieiro, J.C., and Fávaro-Trindade, C.S. 2015. Effect of spray drying on the physicochemical properties and color stability of the powdered pigment obtained from vinification byproducts of the Bordo grape (*Vitis labrusca*). *Food and Bioproducts Processing* 93:39–50.

Del Toro-Sánchez, C.L., Ayala-Zavala, J.F., Machi, L. et al. 2010. Controlled release of antifungal volatiles of thyme essential oil from β-cyclodextrin capsules. *Journal of Inclusion Phenomena and Macrocyclic Chemistry* 67:431–441.

Descamps-Latscha, B., Drüeke, T., and Witko-Sarsat, V. 2001. Dialysis-induced oxidative stress: Biological aspects, clinical consequences, and therapy. *Seminars in Dialysis* 14:193–199.

Devasagayam, T.P., Tilak, J.C., Boloor, K.K. et al. 2004. Free radicals and antioxidants in human health: Current status and future prospects. *Journal of the Association of Physicians of India* 52:794–804.

Dillard, C.J., Litov, R.E., Savin, W.M., Dumelin, E.E., and Tappel, A.L. 1978. Effects of exercise, vitamin E, and ozone on pulmonary function and lipid peroxidation. *Journal of Applied Physiology* 45:927–932.

Dock, L., and Vahter, M. 1999. Metal toxicology. *General and Applied Toxicology* 2:2057–2060.

Domanico, D., Fragiotta, S., Cutini, A. et al. 2015. Circulating levels of reactive oxygen species in patients with nonproliferative diabetic retinopathy and the influence of antioxidant supplementation: 6-month follow-up. *Indian Journal of Ophthalmology* 63:9–14.

Domínguez-Avila, J.A., Wall-Medrano, A., Velderrain-Rodríguez, G.R. et al. 2017. Gastrointestinal interactions, absorption, splanchnic metabolism and pharmacokinetics of orally ingested phenolic compounds. *Food & Function* 8:15–38.

Du, J., Cullen, J.J., and Buettner, G.R. 2012. Ascorbic acid: Chemistry, biology and the treatment of cancer. *Biochimica et Biophysica Acta (BBA)-Reviews on Cancer* 1826:443–457.

El-Agamey, A., Lowe, G.M., McGarvey, D.J. et al. 2004. Carotenoid radical chemistry and antioxidant/pro-oxidant properties. *Archives of Biochemistry and Biophysics* 430:37–48.

El-Shafei, R.A., and Saleh, R.M. 2016. Pharmacological effects of vitamin C & E on diclofenac sodium intoxicated rats. *Biomedicine & Pharmacotherapy* 84:314–322.

Fisher, G.J., Quan, T., Purohit, T. et al. 2009. Collagen fragmentation promotes oxidative stress and elevates matrix metalloproteinase-1 in fibroblasts in aged human skin. *The American Journal of Pathology* 174:101–114.

Fotino, A.D., Thompson-Paul, A.M., and Bazzano, L.A. 2013. Effect of coenzyme Q(10) supplementation on heart failure: A meta-analysis. *The American Journal of Clinical Nutrition* 97:268–275.

Fraga, C.G., Galleano, M., Verstraeten, S.V., and Oteiza, P.I. 2010. Basic biochemical mechanisms behind the health benefits of polyphenols. *Molecular Aspects of Medicine* 31:435–445.

Friedman, M., and Jürgens, H.S. 2000. Effect of pH on the stability of plant phenolic compounds. *Journal of Agricultural and Food Chemistry* 48:2101–2110.

Fu, M.X., Requena, J.R., Jenkins, A.J. et al. 1996. The advanced glycation end product, n-(carboxymethyl)lysine, is a product of both lipid peroxidation and glycoxidation reactions. *Journal of Biological Chemistry* 271:9982–9986.

Garcia-Alcala, H., Vichido, C.I.S., Macedo, S.I. et al. 2015. Treatment with alpha-lipoic acid over 16 weeks in type 2 diabetic patients with symptomatic polyneuropathy who responded to initial 4-week high-dose loading. *Journal of Diabetes Research* 2015:1–8.

Garcia-Flores, L.A., Medina, S., Cejuela-Anta, R. et al. 2016. DNA catabolites in triathletes: Effects of supplementation with an aronia-citrus juice (polyphenols-rich juice). *Food & Function* 7:2084–2093.

Garcin, E., Vernede, X., Hatchikian, E. et al. 1999. The crystal structure of a reduced [NiFeSe] hydrogenase provides an image of the activated catalytic center. *Structure* 7:557–566.

Gazzin, S., Vitek, L., Watchko, J., Shapiro, S.M., and Tiribellil, C. 2016. A novel perspective on the biology of bilirubin in health and disease. *Trends in Molecular Medicine* 22:758–768.

Gkogkolou, P., and Bohm, M. 2012. Advanced glycation end products: Key players in skin aging? *Dermato-endocrinology* 4:259–270.

Glorieux, C., Zamocky, M., Sandoval, J.M., Verrax, J., and Calderon, P.B. 2015. Regulation of catalase expression in healthy and cancerous cells. *Free Radical Biology and Medicine* 87:84–97.

Goldin, A., Beckman, J.A., Schmidt, A.M., and Creager, M.A. 2006. Advanced glycation end products: Sparking the development of diabetic vascular injury. *Circulation* 114:597–605.

Golumbic, C., and Mattill, H. 1941. Antioxidants and the autoxidation of fats. XIII. The antioxygenic action of ascorbic acid in association with tocopherols, hydroquinones and related compounds. *Journal of the American Chemical Society* 63 (5):1279–1280.

Gomez-Estaca, J., Balaguer, M.P., Gavara, R., and Hernandez-Munoz, P. 2012. Formation of zein nanoparticles by electrohydrodynamic atomization: Effect of the main processing variables and suitability for encapsulating the food coloring and active ingredient curcumin. *Food Hydrocolloids* 28:82–91.

Gonzalez, S., Cuervo, A., and Lasheras, C. 2013. Polyphenol intake in elderly people is associated with lipid oxidative damage. *Journal of the American College of Nutrition* 32:384–390.

Guitard, R., Paul, J.F., Nardello-Rataj, V., and Aubry, J.M. 2016. Myricetin, rosmarinic and carnosic acids as superior natural antioxidant alternatives to α-tocopherol for the preservation of omega-3 oils. *Food Chemistry* 213:284–295.

Gülçin, I. 2012. Antioxidant activity of food constituents: An overview. *Archives of Toxicology* 86:345–391.

Guo, Q., Gao, S., Sun, Y. et al. 2016. Antioxidant efficacy of rosemary ethanol extract in palm oil during frying and accelerated storage. *Industrial Crops and Products* 94:82–88.

Gutteridge, J.M.C., and Halliwell, B. 1990. Measurement and mechanism of lipid-peroxidation in biological-systems. *Trends in Biochemical Sciences* 15:129–135.

Gutteridge, J.M.C., and Halliwell, B. 2010. Antioxidants: Molecules, medicines, and myths. *Biochemical and Biophysical Research Communications* 393:561–564.

Harrison, E.H. 2005. Mechanisms of digestion and absorption of dietary vitamin A. *Annual Review of Nutrition* 25:87–103.

Heleno, S.A., Barros, L., Sousa, M.J., Martins, A., and Ferreira, I.C. 2010. Tocopherols composition of Portuguese wild mushrooms with antioxidant capacity. *Food Chemistry* 119:1443–1450.

Hix, L.M., Lockwood, S.F., and Bertram, J.S. 2004. Bioactive carotenoids: Potent antioxidants and regulators of gene expression. *Redox Report* 9:181–191.

Huang, D., Ou, B., and Prior, R.L. 2005. The chemistry behind antioxidant capacity assays. *Journal of Agricultural and Food Chemistry* 53:1841–1856.

Huang, X., Huang, X., Gong, Y. et al. 2016. Enhancement of curcumin water dispersibility and antioxidant activity using core–shell protein–polysaccharide nanoparticles. *Food Research International* 87:1–9.

Ibuki, A., Akase, T., Nagase, T. et al. 2012. Skin fragility in obese diabetic mice: Possible involvement of elevated oxidative stress and upregulation of matrix metalloproteinases. *Experimental Dermatology* 21:178–183.

Itahana, Y., Han, R., Barbier, S. et al. 2015. The uric acid transporter SLC2A9 is a direct target gene of the tumor suppressor p53 contributing to antioxidant defense. *Oncogene* 34:1799–1810.

Jankowski, J., Korzeniowska, K., Cieślewicz, A., and Jabłecka, A. 2016. Coenzyme Q10 – A new player in the treatment of heart failure? *Pharmacological Reports* 68:1015–1019.

Jee, J.P., Lim, S.J., Park, J.S., and Kim, C.K. 2006. Stabilization of all-trans retinol by loading lipophilic antioxidants in solid lipid nanoparticles. *European Journal of Pharmaceutics and Biopharmaceutics* 63:134–139.

Jiang, Q. 2014. Natural forms of vitamin E: Metabolism, antioxidant, and anti-inflammatory activities and their role in disease prevention and therapy. *Free Radical Biology and Medicine* 72:76–90.

Jiang, Q., Christen, S., Shigenaga, M.K., and Ames, B.N. 2001. γ-Tocopherol, the major form of vitamin E in the US diet, deserves more attention. *The American Journal of Clinical Nutrition* 74:714–722.

Johnson, L.J., Meacham, S.L., and Kruskall, L.J. 2003. The antioxidants-vitamin C, vitamin E, selenium, and carotenoids. *Journal of Agromedicine* 9:65–82.

Kaimainen, M., Marze, S., Järvenpää, E., Anton, M., and Huopalahti, R. 2015. Encapsulation of betalain into w/o/w double emulsion and release during in vitro intestinal lipid digestion. *LWT-Food Science and Technology* 60(2, Part 1):899–904.

Kand'ár, R., Žáková, P., and Mužáková, V. 2006. Monitoring of antioxidant properties of uric acid in humans for a consideration measuring of levels of allantoin in plasma by liquid chromatography. *Clinica Chimica Acta* 365:249–256.

Krehl, S., Loewinger, M., Florian, S. et al. 2012. Glutathione peroxidase-2 and selenium decreased inflammation and tumors in a mouse model of inflammation-associated carcinogenesis whereas sulforaphane effects differed with selenium supply. *Carcinogenesis* 33:620–628.

Leonard, S.S., Xia, C., Jiang, B.H. et al. 2003. Resveratrol scavenges reactive oxygen species and effects radical-induced cellular responses. *Biochemical and Biophysical Research Communications* 309:1017–1026.

Li, H.G., Horke, S., and Forstermann, U. 2013. Oxidative stress in vascular disease and its pharmacological prevention. *Trends in Pharmacological Sciences* 34:313–319.

Linster, C.L., and Van Schaftingen, E. 2007. Vitamin C. *The FEBS Journal* 274:1–22.

Liu, Z., Shen, C., Tao, Y. et al. 2015. Chemopreventive efficacy of menthol on carcinogen-induced cutaneous carcinoma through inhibition of inflammation and oxidative stress in mice. *Food and Chemical Toxicology* 82:12–18.

Luangwattananun, P., Yainoy, S., Eiamphungporn, W. et al. 2016. Engineering of a novel tri-functional enzyme with MnSOD, catalase and cell-permeable activities. *International Journal of Biological Macromolecules* 85:451–459.

Maharaj, R., Arul, J., and Nadeau, P. 2014. UV-C irradiation effects on levels of enzymic and non-enzymic phytochemicals in tomato. *Innovative Food Science & Emerging Technologies* 21:99–106.

Mamede, A.C., Tavares, S.D., Abrantes, A.M. et al. 2011. The role of vitamins in cancer: A review. *Nutrition and Cancer* 63:479–494.

Manconi, M., Marongiu, F., Castangia, I. et al. 2016. Polymer-associated liposomes for the oral delivery of grape pomace extract. *Colloids and Surfaces B: Biointerfaces* 146:910–917.

Marí, M., Morales, A., Colell A., García-Ruiz, C., and Fernández-Checa, J.C. 2009. Mitochondrial glutathione, a key survival antioxidant. *Antioxidants & Redox Signaling* 11:2685–2700.

Martinelli, F., Basile, B., Morelli, G., d'Andria, R., and Tonutti, P. 2012. Effects of irrigation on fruit ripening behavior and metabolic changes in olive. *Scientia Horticulturae* 144:201–207.

Martins, P.L.G., and de Rosso, V.V. 2016. Thermal and light stabilities and antioxidant activity of carotenoids from tomatoes extracted using an ultrasound-assisted completely solvent-free method. *Food Research International* 82:156–164.

Massaro, M., Scoditti, E., Carluccio, M.A., and De Caterina, R. 2010. Nutraceuticals and prevention of atherosclerosis: Focus on omega-3 polyunsaturated fatty acids and Mediterranean diet polyphenols. *Cardiovascular Therapeutics* 28:e13–e19.

Megha, K., Deshmukh, P.S., Banerjee, B.D. et al. 2015. Low intensity microwave radiation induced oxidative stress, inflammatory response and DNA damage in rat brain. *Neurotoxicology* 51:158–165.

Mezawa, M., Takemoto, M., Onishi, S. et al. 2012. The reduced form of coenzyme Q10 improves glycemic control in patients with type 2 diabetes: An open label pilot study. *Biofactors* 38:416–421.

Moini, H., Packer, L., and Saris, N.E.L. 2002. Antioxidant and prooxidant activities of α-lipoic acid and dihydrolipoic acid. *Toxicology and Applied Pharmacology* 182:84–90.

Molyneux, P. 2004. The use of the stable free radical diphenylpicrylhydrazyl (DPPH) for estimating antioxidant activity. *Songklanakarin Journal of Science and Technology* 26:211–219.

Moya-Camarena, S.Y., and Jiang, Q. 2012. The role of vitamin E forms in cancer prevention and therapy—Studies in human intervention trials and animal models. In *Nutraceuticals and Cancer*, ed. F.H. Sarkar, pp. 323–354. New York: Springer.

Mugesh, G., and Singh, H.B. 2000. Synthetic organoselenium compounds as antioxidants: Glutathione peroxidase activity. *Chemical Society Reviews* 29:347–357.

Naczk, M., and Shahidi, F. 2004. Extraction and analysis of phenolics in food. *Journal of Chromatography A* 1054:95–111.

Naguib, Y.M. 2000. A fluorometric method for measurement of oxygen radical-scavenging activity of water-soluble antioxidants. *Analytical Biochemistry* 284:93–98.

Nazıroğlu, M. 2012. Molecular role of catalase on oxidative stress-induced Ca_{2+} signaling and TRP cation channel activation in nervous system. *Journal of Receptors and Signal Transduction* 32:134–141.

Niki, E. 2014. Role of vitamin E as a lipid-soluble peroxyl radical scavenger: In vitro and in vivo evidence. *Free Radical Biology and Medicine* 66:3–12.

Oelze, M., Kröller-Schön, S., Steven, S. et al. 2014. Glutathione peroxidase-1 deficiency potentiates dysregulatory modifications of endothelial nitric oxide synthase and vascular dysfunction in aging. *Hypertension* 63:390–396.

Oliveras-López, M.J., Berna, G., Jurado-Ruiz, E., de la Serrana, H.L.G., and Martin, F. 2014. Consumption of extra-virgin olive oil rich in phenolic compounds has beneficial antioxidant effects in healthy human adults. *Journal of Functional Foods* 10:475–484.

Ou, B., Hampsch-Woodill, M., and Prior, R.L. 2001. Development and validation of an improved oxygen radical absorbance capacity assay using fluorescein as the fluorescent probe. *Journal of Agricultural and Food Chemistry* 49:4619–4626.

Paik, J., During, A., Harrison, E.H. et al. 2001. Expression and characterization of a murine enzyme able to cleave β-carotene. The formation of retinoids. *Journal of Biological Chemistry* 276:32160–32168.

Palafox-Carlos, H., Ayala-Zavala, J.F., and González-Aguilar, G.A. 2011. The role of dietary fiber in the bioaccessibility and bioavailability of fruit and vegetable antioxidants. *Journal of Food Science* 76:R6–R15.

Palafox-Carlos, H., Yahia, E.M., and González-Aguilar, G.A. 2012. Identification and quantification of major phenolic compounds from mango (*Mangifera indica*, cv. Ataulfo) fruit by HPLC–DAD–MS/MS-ESI and their individual contribution to the antioxidant activity during ripening. *Food Chemistry* 135:105–111.

Palma-Duran, S.A., Vlassopoulos, A., Lean, M., Govan, L., and Combet, E. 2015. Nutritional intervention and impact of polyphenol on glycohaemoglobin (HbA1c) in non-diabetic and type 2 diabetic subjects: Systematic review and meta-analysis. *Critical Reviews in Food Science and Nutrition* 57:975–986.

Pando, D., Matos, M., Gutiérrez, G., and Pazos, C. 2015. Formulation of resveratrol entrapped niosomes for topical use. *Colloids and Surfaces B: Biointerfaces* 128:398–404.

Pappas, A., Zoidis, E., Surai, P., and Zervas, G. 2008. Selenoproteins and maternal nutrition. *Comparative Biochemistry and Physiology Part B: Biochemistry and Molecular Biology* 151:361–372.

Pasquali, M.A., Gelain, D.P., Zanotto-Filho, A. et al. 2008. Retinol and retinoic acid modulate catalase activity in Sertoli cells by distinct and gene expression-independent mechanisms. *Toxicology in Vitro* 22:1177–1183.

Pellegrini, N., Serafini, M., Colombi, B. et al. 2003. Total antioxidant capacity of plant foods, beverages and oils consumed in Italy assessed by three different in vitro assays. *The Journal of Nutrition* 133:2812–2819.

Penniston, K.L., and Tanumihardjo, S.A. 2006. The acute and chronic toxic effects of vitamin A. *The American Journal of Clinical Nutrition* 83:191–201.

Pietta, P.G. 2000. Flavonoids as antioxidants. *Journal of Natural Products* 63:1035–1042.

Pingitore, A., Lima, G.P.P., Mastorci, F. et al. 2015. Exercise and oxidative stress: Potential effects of antioxidant dietary strategies in sports. *Nutrition* 31:916–922.

Pisoschi, A.M., and Negulescu, G.P. 2012. Methods for total antioxidant activity determination: A review. *Biochemistry and Analytical Biochemistry* 1:1–10.

Prasad, A.S., Bao, B., Beck, F.W., and Sarkar, F.H. 2001. Zinc activates NF-κB in HUT-78 cells. *Journal of Laboratory and Clinical Medicine* 138:250–256.

Prior, R.L. 2015. Oxygen radical absorbance capacity (ORAC): New horizons in relating dietary antioxidants/bioactives and health benefits. *Journal of Functional Foods* 18:797–810.

Pulido, R., Bravo, L., and Saura-Calixto, F. 2000. Antioxidant activity of dietary polyphenols as determined by a modified ferric reducing/antioxidant power assay. *Journal of Agricultural and Food Chemistry* 48:3396–3402.

Quirós-Sauceda, A.E., Palafox-Carlos, H., Sayago-Ayerdi, S.G. et al. 2014. Dietary fiber and phenolic compounds as functional ingredients: Interaction and possible effect after ingestion. *Food & Function* 5:1063–1072.

Rafighi, Z., Shiva A., Arab S., and Yusuf R.M. 2013. Association of dietary vitamin c and e intake and antioxidant enzymes in type 2 diabetes mellitus patients. *Global Journal of Health Science* 5:183–187.

Rangel-Huerta, O.D., Aguilera, C.M., Martin, M.V. et al. 2015. Normal or high polyphenol concentration in orange juice affects antioxidant activity, blood pressure, and body weight in obese or overweight adults. *Journal of Nutrition* 145:1808–1816.

Rashidinejad, A., Birch, E.J., Sun-Waterhouse, D., and Everett, D.W. 2016. Effect of liposomal encapsulation on the recovery and antioxidant properties of green tea catechins incorporated into a hard low-fat cheese following in vitro simulated gastrointestinal digestion. *Food and Bioproducts Processing* 100:238–245.

Raymond, L.J., and Ralston, N.V. 2009. Selenium's importance in regulatory issues regarding mercury. *Fuel Processing Technology* 90:1333–1338.

Re, R., Pellegrini, N., Proteggente, A. et al. 1999. Antioxidant activity applying an improved ABTS radical cation decolorization assay. *Free Radical Biology and Medicine* 26:1231–1237.

Rindler, P.M., Plafker, S.M., Szweda, L.I., and Kinter, M. 2013. High dietary fat selectively increases catalase expression within cardiac mitochondria. *Journal of Biological Chemistry* 288:1979–1990.

Robaczewska, J., Kedziora-Kornatowska, K., Kozakiewicz, M. et al. 2016. Role of glutathione metabolism and glutathione-related antioxidant defense systems in hypertension. *Journal of Physiology and Pharmacology* 67:331–337.

Rocha-Parra, D.F., Lanari, M.C., Zamora, M.C., and Chirife, J. 2016. Influence of storage conditions on phenolic compounds stability, antioxidant capacity and colour of freeze-dried encapsulated red wine. *LWT-Food Science and Technology* 70:162–170.

Rodriguez-Mateos, A., Cifuentes-Gomez, T., George, T.W., and Spencer, J.P.E. 2014. Impact of cooking, proving, and baking on the (poly)phenol content of wild blueberry. *Journal of Agricultural and Food Chemistry* 62:3979–3986.

Rodríguez-Roque, M.J., de Ancos, B., Sánchez-Moreno, C. et al. 2015. Impact of food matrix and processing on the in vitro bioaccessibility of vitamin C, phenolic compounds, and hydrophilic antioxidant activity from fruit juice-based beverages. *Journal of Functional Foods* 14:33–43.

Sahoo, R., Bhattacharjee, A., Majumdar, U. et al. 2009. A novel role of catalase in detoxification of peroxynitrite in *S. cerevisiae*. *Biochemical and Biophysical Research Communications* 385:507–511.

Seeram, N.P., Henning, S.M., Niu, Y. et al. 2006. Catechin and caffeine content of green tea dietary supplements and correlation with antioxidant capacity. *Journal of Agricultural and Food Chemistry* 54:1599–1603.

Simoyi, M.F., Falkenstein, E., Van Dyke, K., Blemings, K.P., and Klandorf, H. 2003. Allantoin, the oxidation product of uric acid is present in chicken and turkey plasma. *Comparative Biochemistry and Physiology Part B: Biochemistry and Molecular Biology* 135:325–335.

Singh, D.K., Winocour, P., and Farrington, K. 2011. Oxidative stress in early diabetic nephropathy: Fueling the fire. *Nature Reviews Endocrinology* 7:176–184.

Sinthorn, W., Chatuphonprasert, W., Chulasiri, M., and Jarukamjorn, K. 2016. Thai red rice extract provides liver protection in paracetamol-treated mice by restoring the glutathione system. *Pharmaceutical Biology* 54:770–779.

Smyk, B. 2015. Singlet oxygen autoxidation of vegetable oils: Evidences for lack of synergy between β-carotene and tocopherols. *Food Chemistry* 182:209–216.

Sochor, J., Dobes, J., Krystofova, O. et al. 2013. Electrochemistry as a tool for studying antioxidant properties. *International Journal of Electrochemical Science* 8:8464–8489.

Steenken, S., and Jovanovic, S.V. 1997. How easily oxidizable is DNA? One-electron reduction potentials of adenosine and guanosine radicals in aqueous solution. *Journal of the American Chemical Society* 119:617–618.

Steinberg, D. 1997. Low density lipoprotein oxidation and its pathobiological significance. *Journal of Biological Chemistry* 272:20963–20966.

Steinbrenner, H., and Sies, H. 2013. Selenium homeostasis and antioxidant selenoproteins in brain: Implications for disorders in the central nervous system. *Archives of Biochemistry and Biophysics* 536:152–157.

Su, L., Yin, J.J., Charles, D. et al. 2007. Total phenolic contents, chelating capacities, and radical-scavenging properties of black peppercorn, nutmeg, rosehip, cinnamon and oregano leaf. *Food Chemistry* 100:990–997.

Sun, Y.P., Gu, J.F., Tan, X.B. et al. 2016. Curcumin inhibits advanced glycation end product-induced oxidative stress and inflammatory responses in endothelial cell damage via trapping methylglyoxal. *Molecular Medicine Reports* 13:1475–86.

Sverdlov, A.L., Elezaby, A., Qin, F. et al. 2016. Mitochondrial reactive oxygen species mediate cardiac structural, functional, and mitochondrial consequences of diet-induced metabolic heart disease. *Journal of the American Heart Association* 5: e002555 1–13.

Takahashi, H., Nishina, A., Fukumoto, R.H. et al. 2005. Selenocarbamates are effective superoxide anion scavengers in vitro. *European Journal of Pharmaceutical Sciences* 24:291–295.

Tang, X.H., and Gudas, L.J. 2011. Retinoids, retinoic acid receptors, and cancer. *Annual Review of Pathology: Mechanisms of Disease* 6:345–364.

Thaipong, K., Boonprakob, U., Crosby, K., Cisneros-Zevallos, L., and Byrne, D.H. 2006. Comparison of ABTS, DPPH, FRAP, and ORAC assays for estimating antioxidant activity from guava fruit extracts. *Journal of Food Composition and Analysis* 19:669–675.

Tomás-Barberán, F.A., and Espin, J.C. 2001. Phenolic compounds and related enzymes as determinants of quality in fruits and vegetables. *Journal of the Science of Food and Agriculture* 81:853–876.

Tumbas Šaponjac, V., Čanadanović-Brunet, J., Ćetković, G. et al. 2016. Encapsulation of beetroot pomace extract: RMS optimization, storage and gastrointestinal stability. *Molecules* 21:1–16.

Urso, M.L., and Clarkson, P.M. 2003. Oxidative stress, exercise, and antioxidant supplementation. *Toxicology* 189:41–54.

Valavanidis, A., Vlachogianni, T., and Fiotakis, C. 2009. 8-hydroxy-2′-deoxyguanosine (8-OHdG): A critical biomarker of oxidative stress and carcinogenesis. *Journal of Environmental Science and Health Part C* 27:120–139.

Valko, M., Leibfritz, D., Moncol, J. et al. 2007. Free radicals and antioxidants in normal physiological functions and human disease. *The International Journal of Biochemistry & Cell Biology* 39:44–84.

Valko, M., Rhodes C., Moncol J., Izakovic M., and Mazur M. 2006. Free radicals, metals and antioxidants in oxidative stress-induced cancer. *Chemico-Biological Interactions* 16:1–40.

Van het Hof, K.H., de Boer, B.C.J., Tijburg, L.B.M. et al. 2000. Carotenoid bioavailability in humans from tomatoes processed in different ways determined from the carotenoid response in the triglyceride-rich lipoprotein fraction of plasma after a single consumption and in plasma after four days of consumption. *Journal of Nutrition* 130:1189–1196.

Velderrain-Rodríguez, G.R., Ovando-Martínez, M., Villegas-Ochoa, M. et al. 2015. Antioxidant capacity and bioaccessibility of synergic mango (cv. Ataulfo) peel phenolic compounds in edible coatings applied to fresh-cut papaya. *Food and Nutrition Sciences* 6:365–373.

Velderrain-Rodríguez, G., Quirós-Sauceda, A., Mercado-Mercado, G. et al. 2016. Effect of dietary fiber on the bioaccessibility of phenolic compounds of mango, papaya and pineapple fruits by an in vitro digestion model. *Food Science and Technology (Campinas)* 36:188–194.

Venegas, C., Cabrera-Vique, C., Garcia-Corzo, L. et al. 2011. Determination of coenzyme Q10, coenzyme Q9, and melatonin contents in virgin argan oils: Comparison with other edible vegetable oils. *Journal of Agricultural and Food Chemistry* 59:12102–12108.

Wang, J., Wang, K., Wang, Y. et al. 2014. A novel application of pulsed electric field (PEF) processing for improving glutathione (GSH) antioxidant activity. *Food Chemistry* 161:361–366.

Weber, D., and Grune, T. 2012. The contribution of β-carotene to vitamin A supply of humans. *Molecular Nutrition & Food Research* 56:251–258.

Wong, R.S., and Radhakrishnan, A.K. 2012. Tocotrienol research: Past into present. *Nutrition Reviews* 70:483–490.

Wu, Z., Liu, S., Zhao, J. et al. 2017. Comparative responses to silicon and selenium in relation to antioxidant enzyme system and the glutathione-ascorbate cycle in flowering Chinese cabbage (Brassica campestris L. ssp. chinensis var. utilis) under cadmium stress. *Environmental and Experimental Botany* 133:1–11.

Xie, L.T.X., Williams, K.J., He, C.W.H. et al. 2015. Resveratrol and para-coumarate serve as ring precursors for coenzyme Q biosynthesis. *Journal of Lipid Research* 56:909–919.

Xu, J., Cheng, Q., and Arnér, E.S.J. 2016. Details in the catalytic mechanism of mammalian thioredoxin reductase 1 revealed using point mutations and juglone-coupled enzyme activities. *Free Radical Biology and Medicine* 94:110–120.

Yubero-Serrano, E.M., Gonzalez-Guardia, L., Rangel-Zuñiga, O. et al. 2013. Postprandial antioxidant gene expression is modified by Mediterranean diet supplemented with coenzyme Q10 in elderly men and women. *Age* 35:159–170.

Zhang, L., Cheng, Q., Zhang, L. et al. 2016. Serum thioredoxin reductase is highly increased in mice with hepatocellular carcinoma and its activity is restrained by several mechanisms. *Free Radical Biology and Medicine* 99:426–435.

Zokti, J., Sham Baharin, B., Mohammed, A., and Abas, F. 2016. Green tea leaves extract: Microencapsulation, physicochemical and storage stability study. *Molecules* 21:1–24.

SECTION V

Phenolic Compounds in Different Foodstuffs

CHAPTER 15

Phenolic Compounds in Wines

Angelita Gambuti and Luigi Moio

CONTENTS

15.1 INTRODUCTION

Phenolic compounds are responsible for color, mouthfeel sensation, astringency, bitterness, and for the numerous beneficial effects linked to a moderate consumption of wine. Some of them can even contribute to wine aroma. These grape secondary metabolites, derivatives of the pentose phosphate, shikimate, and phenylpropanoid pathways, are phytochemicals that play an important role in grape growth and reproduction, contributing to color, sensory characteristics, and providing protection against pathogens and predators. In grapes there are simple phenols and more complex polyphenolic compounds that have multiple phenol rings within the structure. This large group of compounds includes numerous phenolic classes: Phenolic acids, stilbenoids, flavonols, anthocyanins, flavanols, and condensed tannins. Derived compounds, such as esters, methyl ethers, glycosides, and so on, are also included. They are located in different parts of the grape cluster. The skin of a grape berry contains anthocyanins (in black grapes), flavanols and condensed tannins, phenolic acids and stilbenoids; the pulp mainly contains phenolic acids; the seeds and stems contain flavanols and condensed tannins. They are extracted from each part of a grape bunch during winemaking and, their concentration and localization in grapes, as well as factors affecting their extraction, are of fundamental importance for final wine quality. Dissimilarities in grape phenolic composition give way to different winemaking procedures to produce red and white wines. Usually red wines are richer in phenolic compounds and much more enduring.

Among many environmental and cultural factors that influence the content and quality of phenolics within the grape berry, the grape variety and genetic pattern are the most important. Differences among phenolic patterns of wines obtained from different grape varieties are evident for red wines: Full-bodied red wines generally contain higher levels of polyphenols and are deep red in color. This kind of wine is obtained from grape varieties such as cabernet sauvignon and franc, sangiovese, nebbiolo, and aglianico. Varieties that are deep in color but less rich in tannins include merlot, zinfandel, and syrah, while pinot noir grapes feature lower red-colored wines.

During wine production and aging, grape native phenols are further transformed into complex molecules, which result in several modifications in the color and sensory attributes of wines. The knowledge of their reactivity and evolution is thus fundamental to understand their role in final wine characteristics. The chemical properties of phenolic compounds important in wine are essentially three: Their reactivity towards oxygen, their susceptibility to electrophilic aromatic substitution, and their ability to form hydrogen bonds and to participate in hydrophobic interactions. Despite the fact that phenols are weak acids they don't show an acid behavior at wine pH 3–4 because their ionization is too small. Several sensory and healthy properties of wine originate from these phenol reactions. To be more specific, the oxidation of some phenols to quinones results in the stabilization of color and in the decrease of red wine astringency, but when the oxygen exposure is excessive, they determine the appearance of oxidation off-flavor in white and red wines. Electrophilic aromatic substitution reactions determine important changes, such as the stabilization of red wine color. Hydrophobic interactions and hydrogen bonds are the main interactions that occur between tannins and other macromolecules as proteins and polysaccharides and are responsible for astringency and colloidal behavior of wines.

15.2 CHEMICAL CLASSIFICATION OF GRAPE AND WINE PHENOLICS

Grape and wine phenolics constitute a wide group of compounds showing a great diversity of structures, ranging from simple molecules (monomers and oligomers) to polymers. Although numerous scientists have focused on wine phenolics, the overall wine phenol composition is still not well understood. Low molecular weight molecules that can be separated and assayed with HPLC are well known, but the molecular structure of condensed tannins and new compounds formed during wine aging is still not completely known because of the great structural complexity of these compounds. Though a complete determination of wine phenolic composition remains a major challenge of wine research, a differentiation of them for chemical classes with same structural features that confer specific sensory and enological properties is commonly considered.

They are grouped into two categories, flavonoids and nonflavonoids. Each group is further divided into several families, with structural features that confer specific properties such as color, mouthfeel sensations, and aroma. The most abundant and important for wine quality belong to the flavonoid class.

Flavonoids can be divided into various classes on the basis of their molecular structure. They have a common skeleton (C6–C3–C6) consisting of two benzene rings (A and B) linked by an oxygen containing a pyran ring (C). The three main groups of flavonoids and the molecular structure of each group of flavonoids are listed in Figure 15.1. Differences in the oxidation state and substitution of ring C defines the different classes of flavonoids. The substitution pattern of ring B defines the member of the class. The three classes of flavonoid compounds are flavan-3-ols and condensed tannins, anthocyanins, and flavonols. These compounds possess many important properties influencing wine sensory quality, stability, and longevity. Flavan-3-ol monomers and their oligomers and polymers, called condensed tannins or proanthocyanidins, are the most abundant class. The most important property of proanthocyanidins is their ability to bind proteins rendering plant material unpalatable to both animals and microbes (Swain and Bate-Smith 1962). People have benefited from them since ancient times to tan animal skins to form leather; this is why these compounds are called tannins. The word *tannins* is not exclusive to this class of compounds, but it includes each vegetable-derived compound that can be used to tan animal skins. In wine, two kinds of tannins can be found: Condensed tannins derived from grape berry skins and seeds and hydrolyzable tannins derived from wood. Anthocyanins are pigmented compounds responsible for the red wine color and longevity; they are essentially located in black grape skins and are extracted during the first phases of winemaking. They vary among grape varieties and are proposed as a tool to differentiate grape cultivars. The third group, flavonols, are less abundant than condensed tannins and anthocyanins. They are important for reactions related to wine color stabilization and for some healthy properties of wines. Wine flavonoid patterns change over time, increasing storage time and becoming more complex. These compounds can combine with numerous substances including polysaccharides, proteins, and other polyphenols to influence wine color, stability, and flavor over time.

Four classes of phenolic compounds belong to the nonflavonoids group, phenolic acids (hydroxycinnamic and benzoic acids), stilbenes, hydrolyzable tannins, and volatile phenols. Hydroxycinnamic acids and benzoic acids are located in the pulp and skin (Adams 2006). They are the main phenolics in white wines. The former are found as tartrate esters in grapes and are responsible for browning of white musts and wines. Benzoic acids are present in free form or as esters of more complex structures in grapes and wood used to store wines. They can be released in wine by hydrolysis during aging. Stilbenes are found as free (e.g., *trans*-resveratrol) and glycosides in grapes. They are phytoalexin produced in response to fungal attack and UV irradiation, well known for their healthy properties. The hydrolyzable tannins, which are ester-linked oligomers of gallic acid or ellagic acid with glucose or other sugars, are not of grape origin but can be found in wines aged or treated with wood. Volatile phenols are minor constituents that can have different origin and impact wine flavor. Some of them arise from the alcoholysis of lignin and could positively affect wine aroma; others result from microbiological pathways or external contamination and, if present in high concentrations, they cause wine off-flavors that negatively affect wine quality such as the cork and smoke taint and the “Brett” character.

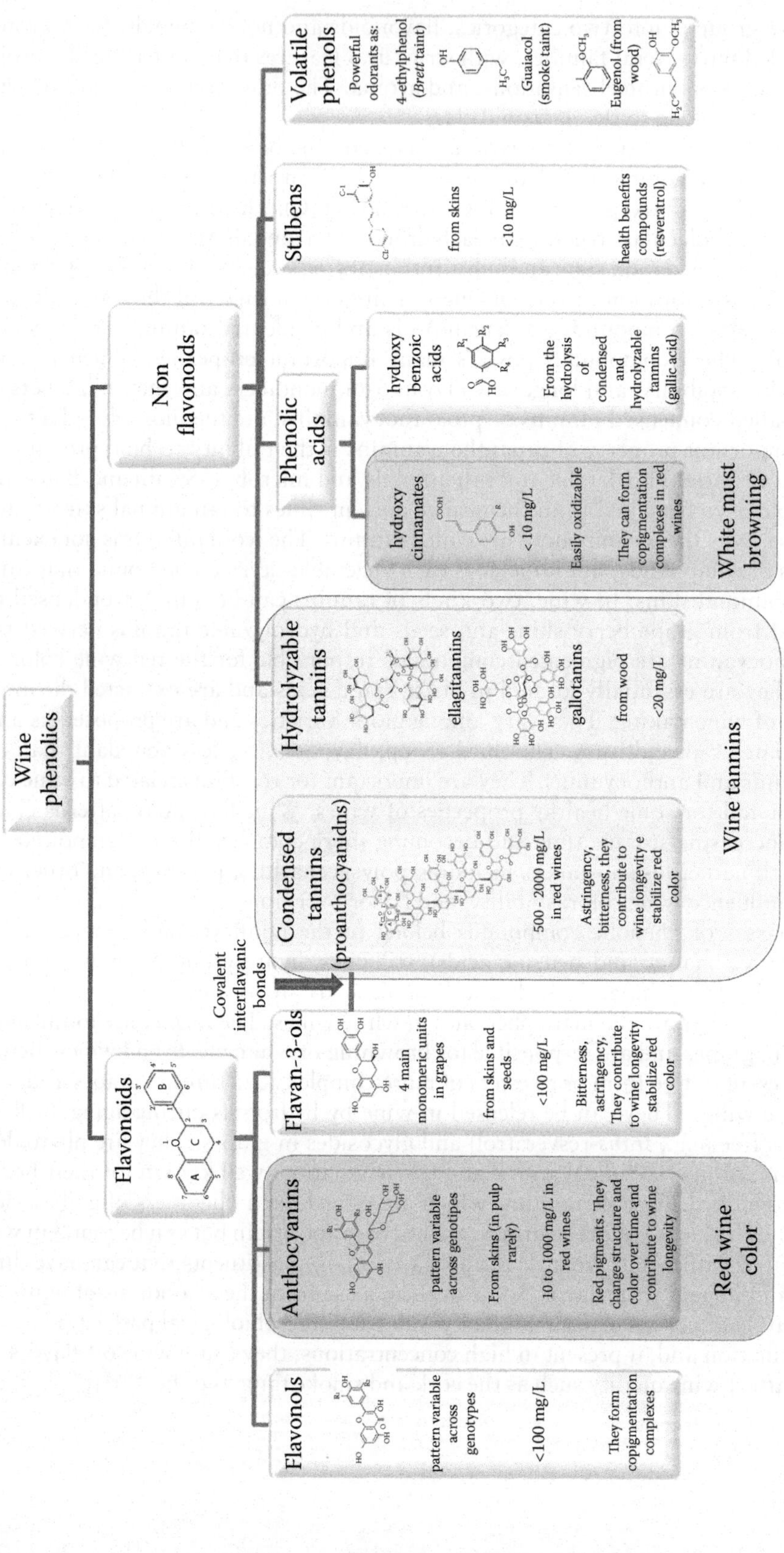

FIGURE 15.1 Flavonoid and nonflavonoid phenolic compounds in wines. Schematic showing phenolic compounds, along with their role for wine quality and associated chemical structure.

15.3 ANTHOCYANINS

Anthocyanins are pigments that give black and red grapes their color. Apart from a few *Tenturier* grape varieties, where they are located also in the pulp, anthocyanins are stored as colored inclusions in grape skin vacuoles and are transferred into the must during the first phases of the red winemaking process. The whole pool of native anthocyanins and compounds derived from their reaction with other wine components is responsible for color and color stability of red wines.

Grapes synthesize anthocyanins via the flavonoid pathway for a multitude of biological roles. Most of them are linked to their ability to act as antioxidants because they quickly reduce oxidizing species (Kahkonen and Heinonen 2003) and, in some cases (anthocyanins having a catechol nucleus), chelate transition metal ions (Goto et al. 1986); thus, they are potentially involved in protection against solar exposure and ultraviolet radiation, defense against free radicals, resistance against many different pathogens and, due to their color, attraction of predators for seed dispersal (Chalker-Scott 1999; Schaefer et al. 2004). As widely showed for anthocyanins derived from other vegetable sources, grape anthocyanins possess numerous therapeutic properties for human health, which include free radical scavenging, antioxidant, anticancer, antimutagenic, antimicrobial, and antiviral activity and prevention of cardiovascular disease (Xia et al. 2010). In contrast to the mounting evidence supporting their health benefits, their bioavailability is still questioned (Lila et al. 2016).

The content of these pigments in black grape varieties usually ranges between 500 and 900 mg/kg. Some varieties, such as pinot noir, differentiate for lower content of total native anthocyanins (330 mg/kg of grapes; Mazza et al. 1999); others, such as zinfandel, differentiate for higher content (1528 mg/kg of grapes; Nelson 2015). As they are extracted during the first phases of fermentation, their concentration in wines depends on the winemaking practice adopted. Generally, in young red wines, they range from 200 to 1000 mg/L; in rosé wines, the concentration of total anthocyanins is below 50 mg/L. In aged wines, the concentration of native anthocyanins decreases quickly; after one year a mean decrease of 30–40 percent occurs, while it increases the content of stable pigments derived from them.

Grape native anthocyanins constitute a large group and the distribution of anthocyanins in the grapes belonging to *V. vinifera* cultivars, which accounts for over 99 percent of the recorded grape wine production in the world, varies mainly according to variety. This is why they have been postulated as chemical markers and employed for chemotaxonomic purposes (Mattivi et al. 2006; Ortega-Regules et al. 2006; González-Neves et al. 2007; Ferrandino et al. 2012; Figueiredo-González et al. 2012; Muccillo et al. 2014). However, their use as a molecular marker is limited by the fact that during wine aging they are involved in numerous reactions changing their original distribution and concentration in grape skin.

15.3.1 Chemical Structure

Structurally, the core of anthocyanins, called anthocyanidin, is the flavylium (2-phenylbenzopyrilium), which has the typical C6–C3–C6 flavonoid skeleton (Figure 15.2). It contains one heterocyclic benzopyran ring (as the C ring), one fused aromatic ring (as the A ring) and one phenyl constituent (as the B ring). The *V. vinifera* common native anthocyanidins, delphinidin, cyanidin, peonidin, petunidin, and malvidin, differ in the

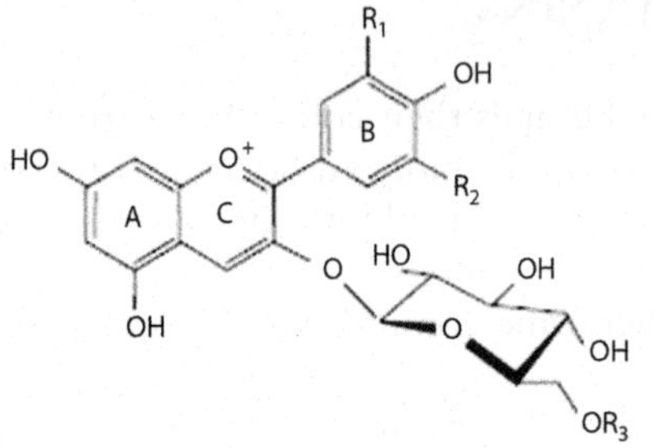

Antocyanidin moiety	R_1	R_2
Cyanidin	OH	H
Delphinidin	OH	OH
Malvidin	OCH_3	OCH_3
Peonidin	OCH_3	H
Petunidin	OCH_3	OH

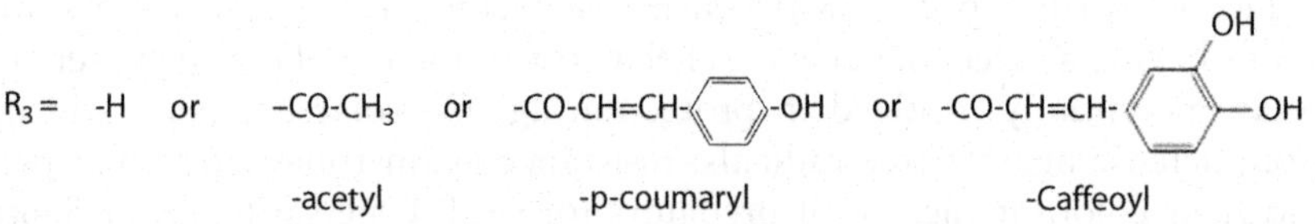

FIGURE 15.2 Main anthocyanins found in *Vitis vinifera*.

hydroxyl or methoxyl substituents in position 3′, 5′ on the B ring. Grape anthocyanins are glycosides and acylglycosides of anthocyanidins, with the proportion of 3′, 5′ forms showing variation among cultivars as well as the contribution of acetylated forms. In *V. vinifera* grapevines, glucose molecules are linked to the anthocyanidin through glycosidic bonds at the C3 position to form 3-O-monoglucoside anthocyanins; on the contrary, in the American *Vitis* species, the 3,5-O-diglucoside anthocyanins are common (Jánváry et al. 2009). Although occasionally found in trace amounts in red wines derived from *V. vinifera*, diglucosylated anthocyanins are considered markers of non–*V. vinifera* grapes. In several countries some of these varieties cannot be used to produce wines; diglucosylated anthocyanins in a wine is an indication of fraud. Malvidin is usually the predominant anthocyanidin in most red grapes (Mattivi et al. 2006; Figueiredo-González et al. 2012). Most cultivars possess acylated forms of anthocyanins, but some have a much simpler profile, with the cultivar pinot noir being a notable example, possessing only the five basic anthocyanidin 3-glucosides (Mazza et al. 1999). Traces of pelargonidin-3-O-glucoside and its acetyl and p-coumaroyl derivatives have been also found in *V. vinifera* species (Wang et al. 2003; He et al. 2010).

15.3.2 Color, Reactions, and Mouthfeel Properties in Wine

Anthocyanins are water-soluble pigments usually represented in the red flavylium cation form, the color of which shifts from orange to purple depending on the hydroxylation and methylation pattern of the B ring of the anthocyanidins. Blueness is enhanced with the increasing of free hydroxyl groups, whereas redness intensifies with the raising of the methylation of the hydroxyl groups. Thus, malvidin is the reddest individual anthocyanidin (Mazza and Francis 1995) while delphinidin and derivatives are associated with blueness; cyanidin and derivatives are reddish. B ring substituent patterns regulate also the stability of these pigments in wine. Anthocyanin-O-methylation results in a higher stability, while the existence of adjacent hydroxyl groups of o-diphenols makes the molecule more sensitive to oxidation. Therefore, cyanidin, delphinidin, and petunidin are less stable compared to malvidin and peonidin (Mazza and Francis 1995).

When speaking of anthocyanins chemistry in wine, three factors need to be considered: (1) They are pigments changing color with pH of the aqueous solution owing to proton transfer and hydration reactions; (2) they react with sulfur dioxide present in wine (sulfite bleaching); (3) their contribution to wine color changes depending on reactions among them and other wine components. First, an equilibrium among four anthocyanin species, namely the red flavylium cations (AH'), the blue quinoidal anhydrobase (A), the colorless hemiketal carbinol (B), and the yellow chalcone is established (Figure 15.3). The equilibrium among these species depends on pH. At wine pH 3–4, anthocyanins occur mostly as the colorless, hydrated hemiketal form (Brouillard and Delaporte 1977). Second, sulfur dioxide, the most common preservative used in winemaking due to its antioxidant and antimicrobial properties, reacts as nucleophile with the flavylium cation on the C ring at the C4 position of anthocyanidin molecule. The compound formed is a colorless pseudobase chromen-4-sulfonic acid (Figure 15.3). Third, during winemaking and aging native anthocyanins can be degraded because they undergo acid-catalyzed reaction, intramolecular cyclization, or, in an excess of oxygen, they can form insoluble complexes or brown compounds (Waterhouse et al. 2016). Despite all of this, red color in wine is still ensured. This is because two stabilization processes have occurred: An association mechanism called copigmentation, and the conversion of grape anthocyanins into other new pigments. Copigmentation is a noncovalent interaction that determines a stabilization of red color (Boulton 2001) by auto-association (Asen et al. 1972; Gonzalez-Manzano et al. 2008) and intermolecular mechanisms (Asen et al. 1972; Gonzalez-Manzano et al. 2008). Essentially, copigmentation consists of a decrease of the percentage of hydrated colorless hemiketal form of anthocyanins, particularly important at wine pH due to the fact that colored anthocyanins are planar structures that can interact with other planar species (copigments) to form molecular stacks (π–π hydrophobic interaction) from which water is excluded (Oliveira et al. 2014). The flavylium ion is thus trapped and protected from hydration. Copigmentation leads to the exhibition of a

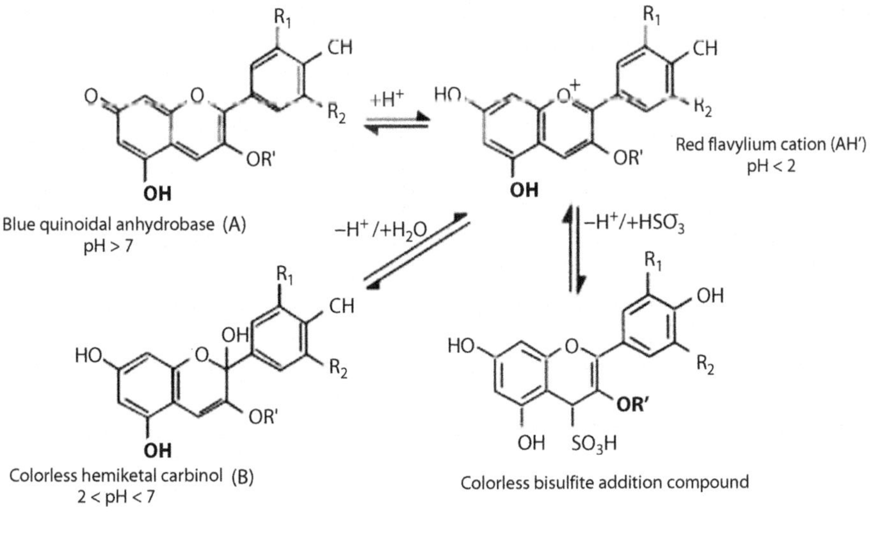

FIGURE 15.3 Reactions of proton transfer, hydration, and sulfite bleaching of anthocyanins in wine (R_1, R_2 = H, OH, OCH_3, R' = glucose). (Adapted from Cheynier, V., Dueñas-Paton, M., Salas, E. et al., *American Journal of Enology and Viticulture*, 57(3), 298–305, 2006.)

greater color than would be expected by the pigment concentration (Boulton 2001). The formation of new anthocyanin-derived pigments via covalent binding with other wine compounds is another important phenomenon occurring during wine aging. These new pigments are more stable and less bleachable by sulfur dioxide than native anthocyanins (Sarni-Manchado et al. 1996; Bakker and Timberlake 1997; Asenstorfer et al. 2001). They show great structural diversity, ranging from the low molecular weight orange pyranoanthocyanins (Sarni-Manchado et al. 1996), the violet ethylene bridged species (Escribano-Bailon et al. 2001; Timberlake and Bridle 1976; Atanasova et al. 2002) and the blue flavanyl-vinylpyranoanthocyanins (Mateus et al. 2003), to the large polymers such as the red tannin–anthocyanin adducts (Remy et al. 2000).

The main sensory attribute of anthocyanins is color and, for the concentration found in wine, typically 0.2 to 1.2 g/L, they should not contribute significantly to astringency and bitterness (Vidal et al. 2004). Nevertheless, recent results (Ferrer-Galego et al. 2015) have suggested a really slight sensation of astringency elicited by glucosylated anthocyanins due to the formation of soluble aggregates between anthocyanins and salivary proteins. Future experiments may better elucidate the contribution of native pigments to this important wine sensory attribute. Conversely, it is now accepted that during wine aging, the incorporation of anthocyanins into tannin structures, which are referred to as polymeric pigments, may instead lead to a decrease in astringency (Weber et al. 2012). Astringency is a tactile sensation mainly due to the hydrophobic interactions between saliva proteins and polyphenols (McRae and Kennedy 2011). As pigments derived from the incorporation of anthocyanins are more polar than consistently structured proanthocyanidins, they interact less with proteins resulting in an attenuated sensory effect. The sensory role of other anthocyanins derived pigments remain to be investigated.

15.4 FLAVANOLS AND CONDENSED TANNINS

Flavan-3-ols and their polymeric condensation products, namely proanthocyanidins, are the most abundant polyphenolic compounds in grapes mainly accumulated in berry seeds, skins, and in stems. They include a wide group of compounds with a great heterogeneity of structure characterized by a flavan-3-ol unit. Oligomeric and polymeric proanthocyanidins, the latter also called *condensed tannins* (usually when mean polymerization degree is ≥ 5), are polymers that differ for the kind of constitutive units, the type of interflavan linkages, and the degree of polymerization. They are end products of the flavonoid biosynthetic pathway and, like other flavonoids, they participate in plant protection against biotic (herbivores, pathogens) and abiotic stresses (UV radiation, heat). The concentration of total flavan-3-ols and proanthocyanidin in grape berry ranges between 500 and 2500 mg/kg in fresh berries. In almost all varieties these occurred in higher amounts in the seeds than in the skin (Mattivi et al. 2009; Downey et al. 2003a). Rarely, a higher amount of tannins in skins has also been found (Cerpa-Calderón and Kennedy 2008). A comparison among data reported in literature is not easy due to the great differences among procedures used to extract them from the solid parts of grape and methods of analysis used.

15.4.1 Chemical Structure

Grape flavan-3-ols are flavonoid compounds that present a hydroxyl group at carbon C-3 of the central pyranic ring C (Figure 15.4). Because of the two different substituents on a

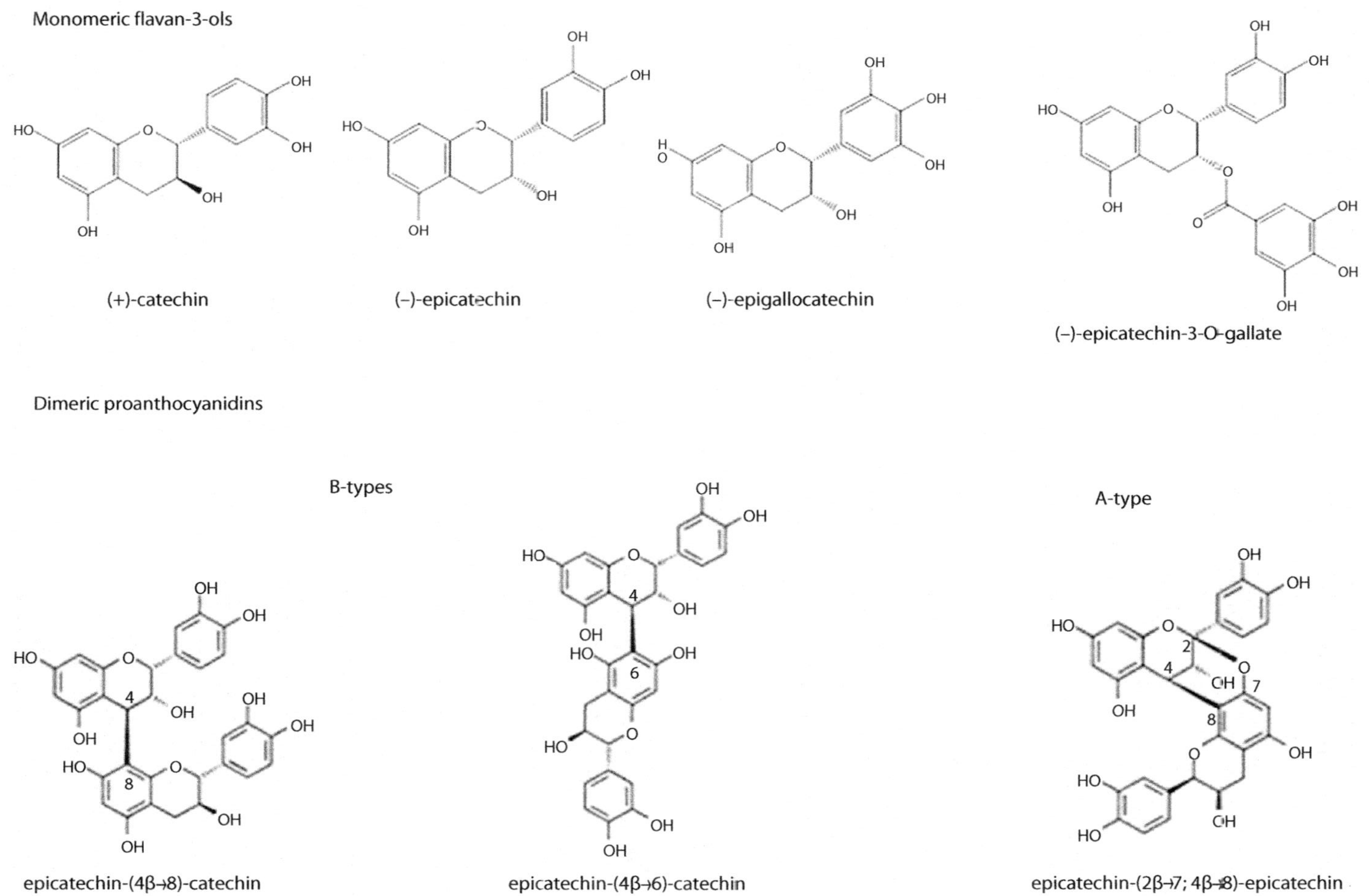

FIGURE 15.4 Grape-derived monomeric flavan-3-ols and dimeric B- and A-types proanthocyanidins.

double bond on ring C, the hydroxyl group at C-3 and the attached B ring at carbon C–2, flavan-3-ols can exist in *cis* and *trans* forms. Grape flavan-3-ols always show a carbon C-2 in the 2R configuration. The carbon C-3 on ring C can be instead found in the 3S or 3R configuration. Therefore the two stereoisomers present in grape and wine are (+)-catechin with 2R,3S configuration (*trans* conformation) and (−)-epicatechin with a 2R,3R configuration (*cis* conformation). Ring B can be monohydroxylated as in (+)-catechin and (−)-epicatechin or trihydroxylated to give rise to (+)-gallocatechin or (−)-epigallocatechin, respectively. In grape (−)-epicatechin can also be found esterified with gallic acid to give the (−)-epicatechin gallate. Low levels of flavan-3-ols glucosylated in the hydroxyl group of carbon C-3 of the pyranic ring C have also been detected in grapes and wine (Delcambre and Saucier 2012). Condensed tannins are oligomers and polymers of monomeric flavan-3-ols; they are also called *proanthocyanidins* because they release anthocyanidins when heated under acidic conditions. They derive from the biochemical condensation of flavan-3-ol monomers with the formation of covalent interflavanic bonds as the C4–C6 or C4–C8 in the B-type series; additional C2–O–C7 or C2–O–C5 bonds are instead characteristic of A-type structures (Figure 15.4). When the constitutive units are (+)-catechin and (−)-epicatechin, they can be also called *procyanidins* as they release cyanidin after acid-catalyzed cleavage. This is the case of seed tannins. When the polymer release gallocatechin units after acid cleavage, it is called *prodelphinidin*. Because of after acid cleavage of skin tannins, both cyanidin and delphinidin are released; those tannins are known as *proanthocyanidins*. All constitutive units and linkages can be distributed at random within a polymer; therefore, there are a great number of possible isomers that increase the chain length. The chemical nature of grape proanthocyanidins varies depending on grape source. Seed proanthocyanidins are constitute of (+)-catechin, (−)-epicatechin, and (−)-epicatechin gallate linked by C4–C8 and/or C4–C6 bonds (B type) and show a high level of galloylation, a large proportion of epicatechin, and a low mDP corresponding to a majority of oligomeric tannins (Vivas et al. 2004). Skin tannins also contain (−)-epigallocatechin and trace amounts of (+)-gallocatechin (Boido et al. 2011) and (−)-epigallocatechin gallate (Souquet et al. 1996), they have higher polymerization degree and a low level of galloylation compared to seed proanthocyanidins (Souquet et al. 1996; Prieur et al. 1994). Stem tannins are made up of the four monomers (+)-catechin, (−)-epicatechin, (−)-epicatechin-3-gallate, and (−)-epigallocatechin, with a low level of galloylation and a medium value for mDP (>5) (Del Llaudy et al. 2008; Vivas et al. 2004).

15.4.2 Reactions with Proteins and Their Role for Wine Mouthfeel Properties

The main property of flavan-3-ols and their oligomers and polymers is their ability to precipitate proteins. An excess of grape proteins in wine can cause haziness but, due to the presence of tannins during the fermentation–maceration of black grapes, most of the proteins are precipitated out. Therefore, the presence of a residual amount of unstable protein is of great concern for white wines but not for red ones. However, an excessive presence of these molecules may be not positive because they elicit two main sensations, bitterness and astringency. Both these attributes are typical of red wine but, if present at high intensity, they determine negative consumer reactions. Despite the importance of these mouthfeel sensations, the chemical structure/sensory activity relationship is still not well known. The first difficulty is linked to the fact that phenolic compounds can often elicit both sensations simultaneously. In addition, astringency is such a complex sensation that the study of this sensation is complicated and the physiological mechanisms involved are still not well understood. It is generally accepted that astringency is due to the tannin-induced interaction or precipitation of salivary proteins in the oral cavity. These interactions are thought to be governed by hydrophobic interactions and the formed complexes are then stabilized by hydrogen bonds (Jöbstl et al. 2004). Both aromatic nuclei and hydroxyl groups of the aromatic ring of phenolic compounds provide the main binding sites for tannin–protein interaction. However, mechanisms other than tannin–salivary protein interaction or precipitation could be involved. In addition, other factors such as temperature, pH of the solution, ionic strength, and the presence of carbohydrates have been shown to affect the tannin–protein interaction. The relationship between tannin structure and astringency has been evaluated in many studies (Robichaud and Noble 1990; Sun et al. 2013; de Freitas and Mateus 2012). Monomers and oligomers are more bitter than astringent while galloylation and the B-ring trihydroxylation of polymers enhance astringency (Gawel 1998). In agreement with Quijada-Morín et al. (2012), astringency is more affected by the subunit composition of procyanidins than by the total concentration or the average degree of polymerization. However, flavan-3-ol structures present in wine are numerous and complex, so all these results are not conclusive and more studies are necessary to deepen our knowledge on this important subject. To make matters more complicated, there is evidence that astringency involves several sensations that are simultaneously perceived (Green 1993) and that are very different from each other; hence, a wide range of subqualities have been traditionally used to describe wine astringency, including "drying," "puckering," "rough," "sappy," "harsh," "woody," and "green" (Gawel et al. 2000). In a recent study, Ferrer-Gallego et al. (2015) evaluated trihydroxylated B-ring and dihydroxylated B-ring flavanols. They found that dihydroxylated B-ring flavanols (catechins and procyanidins) were more astringent, bitter, dry, rough, unripe, and persistent than trihydroxylated B-ring flavanols (epigallocatechin and prodelphinidins). These latter compounds proved to be smoother, more velvety, and viscous.

The origin and mechanism of bitterness perception is well known. This taste is perceived by the activation of the human bitter taste receptors, TAS2Rs. Several studies report data showing structural characteristics that enhance bitterness. More than 15 years ago, Peleg et al. (1999) found that (–)-epicatechin was more bitter than the stereoisomer (+)-catechin and that these both were more bitter than the procyanidin trimers, catechin-(4-8)-catechin-(4-8)-catechin and catechin-(4-8)-catechin-(4-8)-epicatechin. In contrast, other reports have shown that bitterness of polyphenols increases with molecular weight. Hufnagel and Hofmann (2008a) found that procyanidin dimers and a trimer were more bitter than (–)-epicatechin. More recently, Soares et al. (2013), studying the interaction with human taste receptors, suggested that the catechol or galloyl group (which has only one more hydroxyl group than catechol) are critical features (although not essential) for the interaction of polyphenol compounds with the bitter receptor TAS2R5. This is in agreement with studies showing that the higher the hydrophobicity of a bitter substance is, the higher its interaction with the bitter receptor (Kumazawa et al. 1986). The hydrophobicity is increased by the galloyl group. As a result, the interaction with the receptor is expected to be stronger. Interesting results have been reported by Ferrer-Gallego et al. (2016) evaluating synergisms on bitterness and on astringent subqualities when the phenolic compounds were tested as mixtures in comparison to individual compounds. When catechin and epicatechin were tasted together, the perception of the mouthfeel characteristics of the mixture solution were quite different to that elicited for the solutions containing the components alone. A shift towards more bitterness was detected.

Flavan-3-ols and condensed tannins contribution to wine properties is not limited only to astringency and bitterness: They can contribute to the copigmentation or the formation of new pigments stabilizing the color of red wine (Francia-Aricha et al. 1997). Furthermore, depending on their structure, they consume oxygen or react with products of oxidation; thus, they are key compounds for wine stability and longevity (Da Silva et al. 1991). Based on their powerful antioxidant properties and their ability to bind proteins and to form stable complexes with metal ions, they show numerous health effects when introduced into the human diet and numerous reviews on grape flavanols and proanthocyanidins and their importance in human health and disease are reported in literature (Bagchi et al. 2000; Schroeter et al. 2010; Nunes et al. 2016).

15.4.3 Flavonols

Flavonols are flavonoids in which the two benzene rings A and B are joined by a three-carbon chain that is part of a heterocyclic C ring with a 3-hydroxyflavone backbone and a double bond (Figure 15.5). Flavonols of *Vitis vinifera* red grape cultivars occur as glycosylated forms (glucosides, galactosides, rhamnosides, rutinosides, and glucuronides) of the six flavonoid structures, kaempferol, quercetin, isorhamnetin, myricetin, laricitrin, and syringetin. The major flavonols of the *Vitis vinifera* grape cultivars occur as 3-O-glycosylated. Aglycones differ in the B-ring substitution pattern: Kaempferol is monohydroxylated in position 4′; quercetin is dihydroxylated in positions 3′ and 4′; and myricetin is trihydroxylated in positions 3′, 4′ and 5′; isorhamnetin is the methylated form of quercetin; and laricitrin and syringetin are the methylated forms of myricetin. The total amount and pattern of flavonols is highly variable across genotypes. Generally the total content of flavonols in *Vitis vinifera* grapes vary between black and white cultivars with the first richer, ranging between 3 and 80 mg/kg and having a mean value of 35 mg/kg. White varieties can reach maximum values of 30 mg/kg of grapes (Mattivi et al. 2006; Figueiredo-González et al. 2012). In almost all grape varieties quercetin and its glycosides are the main flavonols but, in few cases, such as cabernet sauvignon, sagrantino, and teroldego, myricetin is the major flavonol. In white wine, only quercetin, kaempferol, and isorhamnetin have been detected while myricetin and methylated forms laricitrin and syringetin are missing (Jeffery et al. 2008; Mattivi et al. 2006). Given the differences in flavonols pattern among cultivars, these compounds, together with anthocyanins, have been investigated for chemo-taxonomical purposes (Mattivi et al. 2006; Figueiredo-González et al. 2012).

Flavonols act in grape as UV- and photo-protectors (Price et al. 1995) and their synthesis is strongly affected by sunlight. Likely this is because they are mainly located in

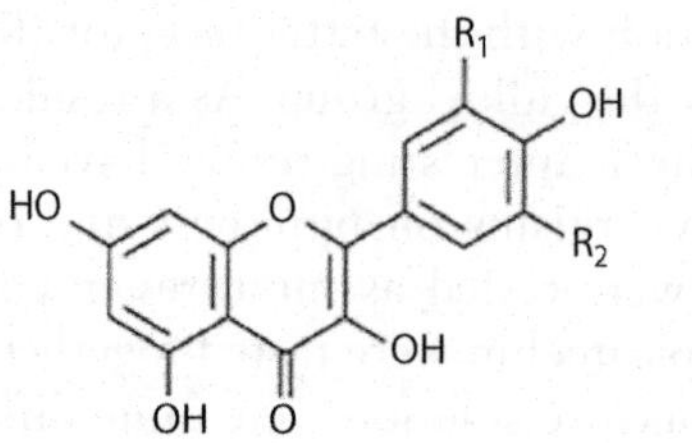

	R_1	R_2
Kaempferol	H	H
Quercetin	OH	H
Isorhamnetin	OCH_3	H
Myricetin	OH	OH
Laricitrin	OCH_3	OH
Syringetin	OCH_3	OCH_3

FIGURE 15.5 Flavonol aglycones of grape and wine.

the outer epidermis of the skin, only in teinturier grape cultivars they have been detected also in flesh of berries (Castillo-Munoz et al. 2009). Their biosynthesis in plant tissues is so greatly influenced by exposure to sunlight that quercetin 3-glucoside concentrations were four to eight times greater in skins from sun-exposed clusters than in those from shaded clusters (Spayd et al. 2002). Other factors such as temperature are less influent (Spayd et al. 2002). The concentration of flavonols in flowers was high and decreased between flowering and berry set. It then stabilizes during berry development (Downey et al. 2003b). Their concentration in wines depends on vinification procedure applied and on the degree of hydrolysis that flavonol glycosides undergo during winemaking and wine aging. These are important limiting factors for their use as a molecular marker for recognizing cultivar used to produce wines.

15.4.3.1 The Role in Wine

One of most important roles played by flavonols in wine is to act as copigments, enhancing color response of anthocyanins. They are yellow flavonoid pigments but more than 40 years ago Asen et al. (1972) found that quercetin glycosides (rhamnose and glucose) caused shifts of 15 to 20 nm and enhancements of 150 to 200 percent in absorbance of aqueous solutions containing cyanidin 3,5-diglucoside. Numerous other studies confirmed that flavonols are among the best copigmentation cofactors of red wine (Baranac et al. 1996; Teixeira et al. 2013; Lambert et al. 2011). The larger extended π conjugation of these compounds that facilitates the π–π interactions to form stable copigmentation complexes is the cause of their ability to enhance the color response of anthocyanins in red wine (Boulton 2001). Their action as copigments is even important during vinification because the formation of copigmentation complexes between anthocyanins and flavonols causes an augmentation of the extraction of anthocyanins during winemaking (Darias-Martìn et al. 2001; Rustioni et al. 2012).

Flavonols are bitter and astringent but their contribution to wine mouthfeel is probably negligible when compared to that of flavan-3-ols and their derivatives—this is why few studies deal with their contribution to wine mouthfeel. Trying to relate polyphenolic composition to sensory data for different commercial wines, Preys et al. (2006) hypothesized a relationship between flavonol aglycones (myricetin and quercetin) and bitterness. In contrast, after their purification and isolation from red wine, flavan-3-ols were found to not be of major importance for astringency and bitter taste but flavonols glycosides are reported to be a velvety astringent with syringetin-3-O-D-glucopyranoside and quercitin-3-O-D-galactopyranoside showing the lowest recognition threshold concentrations (Hufnagel and Hofmann 2008a, 2008b). Recently, Ferrer-Gallego et al. (2016) completed a study aimed at evaluating the effect of the addition of quercetin 3-O-glucoside on the astringency and bitterness of red and white wines, observing that red wines became more astringent, rough, green, dry, bitter, and persistent in its presence. The addition of the flavonol to the white wine resulted in greenness, persistence, and bitterness qualities close to red wines' characteristics.

15.4.4 Phenolic Acids

Phenolic acids constitute the most important group of non-flavonoid phenols in wine. There are two main groups of compounds with different properties belonging to this chemical class: The hydroxyl derivatives of benzoic acids, containing seven carbon atoms, C6–C1, and the hydroxyl derivatives of cinnamic acids, having a C6–C3 skeleton in

Hydroxycinnamates

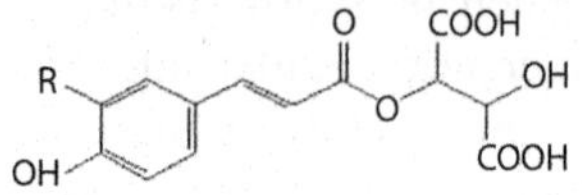

R = H *p*-coumaric acid, tartrate ester (coutaric acid)
R = OH caffeic acid, tartrate ester (caftaric acid)
R = OCH_3 ferulic acid, tartrate ester (fertaric acid)

Hydroxybenzoic acids

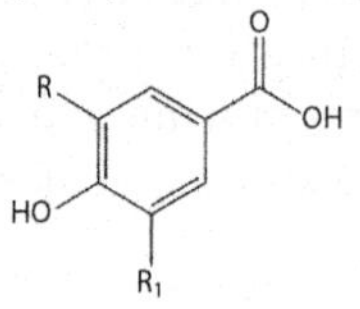

R = OH, R_1 = H protocatechuic acid
R = OCH_3, R_1 = H vanillic acid
R = OH, R_1 = OH gallic acid
R = OCH_3, R_1 = OCH_3 syringic acid

FIGURE 15.6 Hydroxycinnamates and hydroxybenzoic acids found in grape.

which a conjugated double bond exists between the phenolic ring and the carboxylic group. Benzoic acids in grapes are conjugated as esters or glycosides whereas cinnamic acids mainly exist in grape berry as tartaric acid esters.

Hydroxycinnamic acids reside in the vacuoles of whole berry cells and hence are present in both white and red wines, with concentrations 10 times higher in red wines (100–300 mg/L) (Ribéreau-Gayon et al. 2006). As they are the major phenols in grape juice, they constitute the major class of phenolics in white wines. The main hydroxycinnamic acids found in grapes and wines are the tartaric esters of caffeic, coumaric, and ferulic acids (Figure 15.6). These acids differ in the number of hydroxyl and methoxyl substituents. The esters have the usual name of caftaric acid (caffeoyltartaric acid), p-coutaric acid (coumaroyltartaric acid), and fertaric acid (feruloyltartaric acid).

Cinnamic acids display *cis* or *trans* isomeric forms, the latter being the most abundant in nature. These isomers are convertible either enzymatically or through the action of light. They are transferred from grape pulp and skin (the latter in the case of black grapes) to must and, during fermentation process, the partial hydrolysis of esters may take place. Free hydroxycinnamic acids produced may in turn be transformed into corresponding ethyl esters (Pozo-Bayón et al. 2003).

In grape and wines hydroxybenzoic acids are less abundant, seven of them have been found in wine where they can origin from grape but also wood used during wine aging. The most abundant is gallic acid and its content is higher in aged wines because it is released by the hydrolysis of gallate esters of both condensed and hydrolyzable tannins. Besides gallic acid, of which concentrations can reach values of 70 mg/L (Teissedre et al. 1996), hydroxybenzoic acid concentrations are below 10 mg/L in wines.

15.4.4.1 The Role in Wine

Hydroxycinnamates are very important for white wine quality because they are responsible for browning. When grapes are crushed, polyphenol oxidases (tyrosinase from all grapes and laccase from botritized ones) are released and, in the presence of oxygen, oxidize the hydroxycinnamates to benzoquinones. Quinones are then polymerized in a series of reactions, resulting in the formation of brown pigments (Cheynier et al. 1986). However, in grapes there is another compound important for must browning—a tripeptide, glutathione (GSH) that quickly react with hydroxycinnamate-derived quinone, forming a colorless product called a *grape reaction product* (GRP). This latter is no longer a substrate for further oxidation by polyphenol oxidases (Salgues et al. 1986), practically limiting must browning. Therefore the browning potential of musts of white grapes is more properly related to the ratio between hydroxycinnamic acids and GSH present (Cheynier et al. 1989) than to the presence of the oxidation substrates. Hydroxycinnamates are also involved in a serious issue for some wines, the appearance of 4-ethylphenol taint, a compromising spoilage product of the enzymatic decarboxylation, and reduction of hydroxycinnamic acids (especially *p*-coumaric acid) by *Dekkera/Brettanomyces* sp. yeasts (Chatonnet et al. 1992).

These acids also act as wine copigments because they protect the anthocyanidin from hydration due to intramolecular copigmentation complex that enhances color and a bathochromic shift of solutions containing anthocyanins (Dangles et al. 1993; Darias-Martín et al. 2002; Eiro and Heinonen 2002). Some of them have been also evaluated as external polyphenolic cofactors to increase the copigmentation reactions in wines (Aleixandre-Tudó et al. 2013). Recently Zhang et al. (2015) studied the copigmentations of malvidin-3-O-glucoside with five hydroxybenzoic cofactors (*p*-hydroxybenzoic acid, protocatechuic acid, gallic acid, vanillic acid, and syringic acid) and detected that syringic acid had a stronger copigmentation effect than the other four phenolic acids investigated. Most purified phenolic acids of wine are astringent while the ethyl esters are more bitter than astringent (Hufnagel and Hofmann 2008a). However, their concentration is so low in comparison to their threshold that they do not make any direct contribution to the taste of wine (Vèrette et al. 1988).

15.5 MINOR NONFLAVONOIDS CONSTITUENTS OF WINE AND THEIR IMPACT ON THE CHEMICAL AND SENSORY PROPERTIES OF WINES

15.5.1 Hydrolyzable Tannins

Hydrolyzable tannins are a wide group of phenolic compounds that mainly originate from wood used to mature wines, or from enotannins added during vinification, usually to precipitate proteins. These compounds are called *hydrolyzable* because they are esters of phenolic acids with a sugar core (Jourdes et al. 2011) and are easier to hydrolyze than condensed tannins. As well as for the proanthocyanidins they are called tannin for their ability to tan proteins. Two main groups belong to this class: Gallotannins, which are polygalloyl esters, and ellagitannins, which give ellagic acid (the dilactone formed by the association of two molecules of gallic acid) after acid hydrolysis. Gallic acid dimers and polymers and complexes of gallotannins and ellagitannins have also been observed in wine (Haslam 1998; Okuda et al. 1990). The two most abundant ellagitannins extracted from oak heartwood are vescalagin and castalagin, being isomers of five gallic acid moieties bonded via a central glucose. Roburins are dimers of castalagin and vescalagin, with or without links to xylose or lyxose. Molecules deriving from the association of flavanols or anthocyanins with ellagitannin form flavano-ellagitannin or anthocyano-ellagitannins, respectively (Saucier et al. 2006; Chassaing et al. 2010). The concentrations in wines vary widely from 0 to 20 mg/L (González-Centeno et al. 2016) because the amount of hydrolyzable tannins released into the wine depends on several factors: The species of wood used (oak being the most used in wineries), botanical origin of the wood, type of seasoning and toasting of the wood, kind of wood used (barrel, staves, chips), and the length of contact between wood and wine. Concerning their effect on wine quality, ellagitannins can take part in oxidation reactions that may originate from acetaldehyde and can favor the reactions between flavanols and between flavanols and anthocyanins (Vivas and Glories 1996). As a consequence, ellagitannins affect wine color and astringency. However, oak ellagitannins, grandinin, roburin E, castalagin, and vescalagin can slightly contribute to the astringency and bitterness of oak-matured red wines (Puech et al. 1999; Glabasnia and Hofmann 2006).

15.5.2 Stilbenes

Stilbenes are nonflavonoid compounds consisting of an essential structural skeleton of two aromatic rings joined by an ethylene bridge (C6–C2–C6). They are phytoalexins, plant chemicals inhibitory to microorganisms, which are synthesized as a response to the interaction with a pathogen. Even abiotic stress, such as irradiation with ultraviolet light, stimulate their biosynthesis in plants (Langcake and Pryce 1977). Stilbenes exist in monomeric, oligomeric, polymeric forms, and are substituted with sugars, methyl, methoxy, and other residues to form a large group of compounds called *stilbenoids*. In a recent review, 19 stilbene compounds were found in grapes (Flamini et al. 2013).

Resveratrol (3,5,4′-trihydroxy-*trans*-stilbene) is the most known stilbene. There are several comprehensive reviews focusing on resveratrol because of the extensive scientific attention it has attracted over the past few decades due to its potential involvement as a preventive agent for numerous diseases. Although low bioavailability, poor solubility, and a high rate of metabolic breakdown strongly limit the health benefits attributed to stilbenes, modern technologies, such as microencapsulation (Penalva et al. 2015) or nanoparticles (da Rocha Lindner et al. 2015), are promising tools to better deliver these diet-derived chemopreventive molecules to the target tissues. In grapes and wines, resveratrol is the major stilbene. Both isomer forms, *cis*- and *trans*-resveratrol, were detected, with the latter predominant (Lamuela-Raventos et al. 1995). Stilbenes are mainly located in berry skins and mostly in glucosylated form (Creasy and Coffee 1988). This is why their concentrations in red wines are higher compared with rosé and white wines. However, their total level in wines rarely exceed 10 mg/L (Lamuela-Raventos et al. 1995; Bavaresco et al. 2015). Significant differences between grape varieties have been observed, primarily attributed to the variety itself, growing conditions, fungal pressure, and climatic variables (Bavaresco et al. 2016). In wines, all factors affecting extraction from skins during winemaking, maceration time, temperature, pressing, fining, and aging (Gambuti et al. 2004, 2007) can determine changes in their levels. Significant antimicrobial activity of resveratrol against yeasts, acetic acid bacteria (Pastorkova et al. 2013), and lactic acid bacteria (García-Ruiz et al. 2011) has been reported, which suggest that this substance could be a suitable agent to control growth of wine spoilage microorganisms. Concerning sensory properties of resveratrol, Koga et al. (2015) found a bitter taste detection threshold of 90 mg resveratrol/L in an aqueous solution containing 1.2 percent of ethanol. In an experiment where higher concentrations were used (200 mg/L) to fortify wines, bitterness of white wine was increased (Gaudette and Pickering 2011), while no effect on taste was observed for red wine.

15.5.3 Volatile Phenols

In the wine field, the term *volatile phenols* usually means the group of compounds made up of two molecules, 4-ethylphenol and 4-ethylguaiacol, produced by *Brettanomyces bruxellensis*. Depending on their concentration levels, they are considered detrimental to wine flavor; their off-flavor is called the "Brett" character. The "Brett" character has been a cause of great concern in enology. It is commonly described as an off-flavor evoking stables, animals, horse sweat, leather, Band-Aid, spicy, smoky, and medicinal (Chatonnet et al. 1990, 1992). *Brettanomyces/Dekkera* produces 4-ethylphenol and 4-ethylguaiacol from hydroxycinnamic acids, *p*-coumaric and ferulic acids, respectively, that are converted into 4-vinylphenol and 4-vinylguaiacol by the enzyme hydroxycinnamate decarboxylase and then reduced by the enzyme vinylphenol reductase (Chatonnet et al. 1992). When the levels of these volatile phenols exceed the perception threshold (440 μg L–1 for 4-ethylphenol and 135 μg L–1 for 4-ethylguaiacol in a model solution; Chatonnet et al. 1992) the wine is compromised. An optimal substrate for the proliferation of *Brettanomyces* are wood casks, but this taint can even appear at different stages during the production and ageing processes (Malfeito-Ferreira 2011). Owing to the higher presence of hydroxycinnamic acids, red wines are usually more contaminated than white wines. Although numerous preventive and curative strategies (Lisanti et al. 2017) to find a solution to this serious contamination have been proposed, none of them has been totally efficient thus far.

Other volatile phenols, such as 2,4,6-trichlorophenol TCP (Alvarez-Rodríguez et al. 2002) and 2,4,6-tribromophenol TBP (Chatonnet et al. 2004), are important for wine flavor because they can be methylated by microorganisms present in cork or in the winery to give the powerful odorants 2,4,6-trichloroanisole TCA and 2,4,6-tribromoanisole TBA, which confer musty and mold off-flavor taint at very low concentrations. They originate from the process of disinfecting cork bark and winery equipment, which transforms the phenol into TCP and TBP. Another detrimental off-flavor of wine due to volatile phenols is the "smoke taint," due to grapevine smoke exposure (Kennison et al. 2007, 2008). A range of volatile phenols, including guaiacol, 4-methylguaiacol, 4-ethylguaiacol, and 4-ethylphenol, have been identified in juice, unwooded wine, acid, and enzyme hydrolyates prepared from smoke-affected grapes (Kennison et al. 2007, 2008). These compounds accumulate in the leaves and fruit of smoke-affected grapevines in glycoconjugate forms (Hayasaka et al. 2010) that are released during alcoholic and malolactic fermentation, causing the appearance of unpleasant tones including "pharmaceutical," "dirty," "ash tray," "medicinal," "camp fire," or "burnt," and reduce the perception of varietal fruit aroma. Agreeable volatile phenols originate instead from wood and confer a distinctive character to wood-aged wine. The most important are guaiacol (*o*-methoxyphenol), syringaldehyde (hydroxy-3,4-dimethoxybenzaldehyde), and eugenol (2-methoxy-4-(2-propenyl) phenol), which are produced by the lignin breakdown during wood toasting and are responsible for the odor of burn (Aiken and Noble 1984), vanilla, and cloves (Aiken and Noble 1984; Chatonnet et al. 1990; Feuillat et al. 1999), respectively. The vanillin (4-hydroxy-3-methoxybenzaldehyde) originates directly from lignin degradation. It directly influences wine aroma by attributing a pleasant character of vanilla (Puech 1987).

15.6 ENOLOGICAL ASPECTS AND CONCLUDING REMARKS

The palatability of wine as well as most winery practices are governed by the presence of phenolics in grapes. The difference between white and red winemaking is the first notable consequence. White wines are produced by minimizing the extraction of flavan-3-ols and condensed tannins from grapes which, in the absence of anthocyanins, may confer excessive astringency and bitterness to wine. Prior to fermentation, the juice for white wine productions is usually separated from skins and seeds by a sequence of processes including destemming, crushing, draining, and pressing. All these phases are conducted taking care to avoid the oxidation of main phenolics extracted from pulp, the hydroxycinnamates. When alternative practices such as a quick maceration of grapes (skin-contact) is applied, a higher extraction of flavonoids (mainly flavan-3-ols) from skins and seeds occur. During winemaking, the formation of quinones from the oxidation of phenolics has important implications for the wine aroma and for the efficacy of preservative agents used. During the past few decades, the studies of the relationships among quinones, tannin phloroglucinol group, natural and exogenous antioxidants (e.g., glutathione, sulfur dioxide, and ascorbic acid), and aromatic varietal thiols have clarified the oxidation phenomenon (Waterhouse and Nikolantonaki 2015). However, the impact of prefermentative techniques, such as skin maceration and grape pressing, on the balance between extraction of phenolic and aroma compounds from grapes and their oxidation by-products still need to be studied in depth and investigated.

In red wines, the extraction of anthocyanins from skins along with flavan-3-ols and condensed tannins, confers to must and successively to wine, color, and agreeable mouthfeel properties. This is why the extraction of phenolics from grapes is the crucial step of winemaking. This consists of two simultaneous processes: Fermentation and maceration of the solid part of the grape berries. Usually it happens after the destemming and crushing of grape bunches. The extractability of the different phenolics from grape to must-wine is strongly linked to the characteristics of different grape varieties. The three main groups of phenolics involved in this step are anthocyanins located in the upper cellular layers of the hypodermis of skins, skin tannins located in the vacuoles of thick-walled hypodermal cells and associated with cell wall polysaccharides (Adams 2006; Gagné et al. 2006), and seed tannins compartmentalized in thin-walled parenchyma cells located between the cuticle and the inner lignified layers of the seeds (Thorngate and Singleton 1994). The final phenolic composition of red wine is also governed by the following: (1) Variables affecting their diffusion into must-wine, (2) their reactivity with other phenolic compounds and with metabolites of alcoholic fermentation, and (3) their sorption on grape pomaces. Numerous studies and reviews have dealt with these simultaneous phenomena and on factors affecting them, such as temperature, time of contact, alcohol level, sulfur dioxide, cap management, and so on. Whereas some trends are clear (e.g., the early peak and subsequent decline in anthocyanins during fermentation), some phenomena still are not well known (e.g., reactions that give copigmentation and the formation of new pigments stabilizing color), and other behaviors only recently have been investigated by scientific community (Bindon et al. 2016; Springer et al. 2016), such as the role of cell wall composition and structure on the final extraction of astringent tannins from skins and seeds.

Once produced, the aging behavior of both white and red wines is linked to their phenolic pattern. Nowadays it is clear that the shelf life of wine depends on their reactions under slow oxidative conditions. However, the role of each class of phenolics into the resistance to oxygen of wine and the effect of these oxidative reactions on the evolution of wine aroma and mouthfeel need more investigation.

These studies, together with a deeper knowledge of intrinsic factors of grape variety affecting the extraction of phenolics from grapes to must-wine, still present a challenge worth exploring in an industry as competitive as that of wine production.

REFERENCES

Adams, D.O. 2006. Phenolics and ripening in grape berries. *American Journal of Enology and Viticulture.* 57 (3): 249–56.

Aiken, J.W., and Noble, A.C. 1984. Comparison of the aromas of oak- and glass-aged wines. *American Journal of Enology and Viticulture.* 35 (4): 196–99.

Aleixandre-Tudó, J.L., Alvarez, I., Lizama, V. et al. 2013. Impact of caffeic acid addition on phenolic composition of tempranillo wines from different winemaking techniques. *Journal of Agricultural and Food Chemistry.* 61 (49): 11900–912.

Alvarez-Rodríguez, M.L., Laura López-Ocaña, J.M., López-Coronado, E.R. et al. 2002. Cork taint of wines: Role of the filamentous fungi isolated from cork in the formation of 2,4,6-trichloroanisole by o methylation of 2,4,6-trichlorophenol. *Applied and Environmental Microbiology.* 68 (12): 5860–69.

Asen, S., Stewart, R.N., and Norris, K.H. 1972. Co-pigmentation of anthocyanins in plant tissues and its effect on color. *Phytochemistry.* 11 (3): 1139–44.

Asenstorfer, R.E., Hayasaka, Y., and Jones, G.P. 2001. Isolation and structures of oligomeric wine pigments by bisulfite-mediated ion-exchange chromatography. *Journal of Agricultural and Food Chemistry.* 49 (12): 5957–63.

Atanasova, V., Fulcrand, H., Cheynier, V., and Moutounet, M. 2002. Effect of oxygenation on polyphenol changes occurring in the course of wine-making. *Analytica Chimica Acta.* 458: 15–27.

Bagchi, D., Bagchi, M., Stohs, S.J. et al. 2000. Free radicals and grape seed proanthocyanidin extract: Importance in human health and disease prevention. *Toxicology.* 148 (2): 187–97.

Bakker, J., and Timberlake, C.F. 1997. Isolation, identification, and characterization of new color-stable anthocyanins occurring in some red wines. *Journal of Agricultural and Food Chemistry.* 8561 (96): 35–43.

Baranac, J.M., Petranović, N.A., and Dimitrić-Marković, J.M. 1996. Spectrophotometric study of anthocyan copigmentation reactions. *Journal of Agricultural and Food Chemistry.* 44 (5): 1333–36.

Bavaresco, L., Morreale, G., and Flamini, R. 2015. Genetic and physiological bases of healthy attributes of grapes and wine. *Acta Horticulturae.* No. 1082(April): 263–74.

Bavaresco, L., Lucini, L., Busconi, M. et al. 2016. Wine resveratrol: From the ground up. *Nutrients.* 8 (4): 222.

Bindon, K.A., Li, S., Kassara, S. et al. 2016. Retention of proanthocyanidin in wine-like solution is conferred by a dynamic interaction between soluble and insoluble grape cell wall components. *Journal of Agricultural and Food Chemistry.* 64 (44): 8406–19.

Boido, E., García-Marino, M., Dellacassa, E. et al. 2011. Characterisation and evolution of grape polyphenol profiles of *Vitis vinifera* L. Cv. Tannat during ripening and vinification. *Australian Journal of Grape and Wine Research.* 17 (3): 383–93.

Boulton, R. 2001. The copigmentation of anthocyanins and its role in the color of red wine: A critical review. *American Journal of Enology and Viticulture.*

Brouillard, R., and Delaporte, B. 1977. Chemistry of anthocyanin pigments. 2. kinetic and thermodynamic study of proton transfer, hydration, and tautomeric reactions of malvidin 3-glucoside. *Journal of the American Chemical Society.* 99 (26): 8461–68.

Castillo-Muñoz, N., Fernández-González, M., Gómez-Alonso, S. et al. 2009. Red-color related phenolic composition of garnacha tintorera (*Vitis vinifera* L.) grapes and red wines. *Journal of Agricultural and Food Chemistry*. 57 (17): 7883–91. doi:10.1021/jf9002736.

Cerpa-Calderón, F.K., and Kennedy, J.A. 2008. Berry integrity and extraction of skin and seed proanthocyanidins during red wine fermentation. *Journal of Agricultural and Food Chemistry*. 56 (19): 9006–14.

Chalker-Scott, L. 1999. Environmental significance of anthocyanins in plant stress responses. *Photochemistry and Photobiology*. 70 (1): 1–9.

Chassaing, S., Lefeuvre, D., Jacquet, R. et al. 2010. Physicochemical studies of new anthocyano-ellagitannin hybrid pigments: About the origin of the influence of oak c-glycosidic ellagitannins on wine color. *European Journal of Organic Chemistry*. 2010 (1): 55–63.

Chatonnet, P., Boidron, J.N., Pons, M. 1990. Elevage des vins rouges en fûts de chêne: Évolution de certains composés volatils et de leur Impact arômatique. *Sciences Des Aliments*. 10 (3): 565–87.

Chatonnet, P., Bonnet, S., and Boutou, S. et al. 2004. Identification and responsibility of 2, 4, 6-tribromoanisole in musty, corked odors in wine. *Journal of Agricultural and Food Chemistry*. 52 (5): 1255–62.

Chatonnet, P., Dubourdie, D., Boidron, J.-B. et al. 1992. The origin of ethylphenols in wines. *Journal of the Science of Food and Agriculture*. 60 (2): 165–78.

Cheynier, V.F., Trousdale, E.K., Singleton, V.L. et al. 1986. Characterization of 2-s-glutathionyl caftaric acid and its hydrolysis in relation to grape wines. *Journal of Agricultural and Food Chemistry*. 34 (2): 217–21.

Cheynier, V., Dueñas-Paton, M., Salas, E. et al. 2006. Structure and properties of wine pigments and tannins. *American Journal of Enology and Viticulture*. 57 (3): 298–305.

Cheynier, V., Souquet, J.M., and Moutounet, M. 1989. Glutathione content and glutathione to hydroxycinnamic acid ratio in *Vitis vinifera* grapes and musts. *American Journal of Enology and Viticulture*. 40 (4): 320–24.

Creasy, L.L., and Coffee, M. 1988. Phytoalexin Production Potential of Grape Berries. *Journal of the American Society for Horticultural Science*. 113 (2): 230–34.

Da Rocha Lindner, G., Bonfanti Santos, D., Colle, D. et al. 2015. Improved neuroprotective effects of resveratrol-loaded polysorbate 80-coated poly(lactide) nanoparticles in MPTP-induced parkinsonism. *Nanomedicine*. 10 (7): 1127–38.

Da Silva, J.M.R., Darmon, N., Fernandez, Y. et al. 1991. Oxygen free radical scavenger capacity in aqueous models of different procyanidins from grape seeds. *Journal of Agricultural and Food Chemistry*. 39 (9): 1549–52.

Dangles, O., Saito, N., and Brouillard, R. 1993. Anthocyanin intramolecular copigment effect. *Phytochemistry*. 34 (1): 119–24.

Darias-Martín, J., Martín-Luis, B., Carrillo-López, M. et al. 2002. Effect of caffeic acid on the color of red wine. *Journal of Agricultural and Food Chemistry*. 50 (7): 2062–67.

Darias-Martín, J., Carrillo, M., Díaz, E. et al. 2001. Enhancement of red wine colour by pre-fermentation addition of copigments. *Food Chemistry*. 73 (2): 217–20.

De Freitas, V., and Mateus, N. 2012. Protein/polyphenol interactions: Past and present contributions. Mechanisms of astringency perception. *Current Organic Chemistry*. 16 (6): 724–46.

Del Llaudy, M.C., Canals, R., Canals J.M. et al. 2008. Influence of ripening stage and maceration length on the contribution of grape skins, seeds and stems to phenolic composition and astringency in wine-simulated macerations. *European Food Research and Technology*. 226 (3): 337–44.

Delcambre, A., and Saucier, C. 2012. Identification of new flavan-3-ol monoglycosides by UHPLC-ESI-Q-TOF in grapes and wine. *Journal of Mass Spectrometry*. 47 (6): 727–36.

Downey, M.O., Harvey, J.S., and Robinson, S.P. 2003a. Analysis of tannins in seeds and skins of Shiraz grapes throughout berry development. *Australian Journal of Grape and Wine Research*. 9 (1): 15–27.

Downey, M.O., Harvey, J.S., and Robinson, S.P. 2003b. Synthesis of flavonols and expression of flavonol synthase genes in the developing grape berries of Shiraz and Chardonnay (*Vitis vinifera* L.). *Australian Journal of Grape and Wine Research*. 9 (2): 110–21.

Eiro, M.J., and Heinonen, M. 2002. Anthocyanin color behavior and stability during storage: Effect of intermolecular copigmentation. *Journal of Agricultural and Food Chemistry*. 50 (25): 7461–66.

Escribano-Bailón, T., Álvarez-García, M., Rivas-Gonzalo, J.C. et al. 2001. Color and stability of pigments derived from the acetaldehyde-mediated condensation between malvidin 3-O-glucoside and (+)-catechin. *Journal of Agricultural and Food Chemistry*. 49 (3): 1213–17.

Ferrandino, A., Carra, A., Rolle, L. et al. 2012. Profiling of hydroxycinnamoyl tartrates and acylated anthocyanins in the skin of 34 *Vitis vinifera* genotypes. *Journal of Agricultural and Food Chemistry*. 60 (19): 4931–45.

Ferrer-Gallego, R., Quijada-Morín, N., Brás, N.F., Gomes, P., de Freitas, V., Rivas-Gonzalo, J.C., and Escribano-Bailón, M.T. 2015. Characterization of sensory properties of flavanols—A molecular dynamic approach. *Chemical Senses*. 40(6), 381–390.

Ferrer-Gallego, R., Brás, N.F., García-Estévez, I. et al. 2016. Effect of flavonols on wine astringency and their interaction with human saliva. *Food Chemistry*. 209: 358–364.

Feuillat, F., Keller, R., Sauvageot, F. et al. 1999. Characterization of French oak cooperage (*Quercus robur* L., *Quercus petraea* Liebl.). Research of the study group on barrel-aging Burgundy wines. *American Journal of Enology and Viticulture*. 50 (4): 513–18.

Figueiredo-González, M., Martínez-Carballo, E., Cancho-Grande, B. et al. 2012. Pattern recognition of three *Vitis vinifera* L. red grapes varieties based on anthocyanin and flavonol profiles, with correlations between their biosynthesis pathways. *Food Chemistry*. 130 (1): 9–19.

Flamini, R., Mattivi, F., De Rosso, M. et al. 2013. Advanced knowledge of three important classes of grape phenolics: Anthocyanins, stilbenes and flavonols. *International Journal of Molecular Sciences*. 14 (10): 19651–69.

Francia-Aricha, E.M., Guerra, M.T., Rivas-Gonzalo, J.C. et al. 1997. New anthocyanin pigments formed after condensation with flavanols. *Journal of Agricultural and Food Chemistry*. 45 (6): 2262–66.

Gagné, S., Saucier, C., and Gény, L. 2006. Composition and cellular localization of tannins in cabernet sauvignon skins during rowth. *Journal of Agricultural and Food Chemistry*. 54: 9465–71.

Gambuti, A., Strollo, D., Erbaggio, A. et al. 2007. Effect of winemaking practices on color indexes and selected bioactive phenolics of Aglianico wine. *Journal of Food Science*. 72 (9): S623–28.

Gambuti, A., Strollo, D., Ugliano, M. et al. 2004. Trans-resveratrol, quercetin, (+)-catechin, and (–)-epicatechin content in South Italian monovarietal wines: Relationship with maceration time and marc pressing during winemaking. *Journal of Agricultural and Food Chemistry*. 52: 5747–51.

García-Ruiz, A., Moreno-Arribas, M.V., Martín-Álvarez, P.J. et al. 2011. Comparative study of the inhibitory effects of wine polyphenols on the growth of enological lactic acid bacteria. *International Journal of Food Microbiology*. 145 (2): 426–31.

Gaudette, N.J. and Pickering, G.J. 2011. Sensory and chemical characteristics of trans-resveratrol-fortified wine. *Australian Journal of Grape and Wine Research*. 17 (2): 249–57.

Gawel, R. 1998. Red wine astringency: A review. *Australian Journal of Grape and Wine Research*. 4 (2): 74–95.

Gawel, R., Oberholster, A., and Francis, I.L. 2000. A 'mouth-feel wheel': Terminology for communicating the mouth-feel characteristics of red wine. *Australian Journal of Grape and Wine Research*. 6 (3): 203–7.

Glabasnia, A., and Hofmann, T. 2006. Sensory-directed identification of taste-active ellagitannins in American (*Quercus alba* L.) and European oak wood (*Quercus robur* L.) and quantitative analysis in bourbon whiskey and oak-matured red wines. *Journal of Agricultural and Food Chemistry*. 54 (9): 3380–90.

González-Centeno, M.R., Chira, K., and Teissedre, P.-L. 2016. Ellagitannin content, volatile composition and sensory profile of wines from different countries matured in oak barrels subjected to different toasting methods. *Food Chemistry*. 210: 500–11.

Gonzalez-Manzano, S., Santos-Buelga, C., Dueñas, M. et al. 2008. Colour implications of self-association processes of wine anthocyanins. *European Food Research and Technology*. 226 (3): 483–90. doi:10.1007/s00217-007-0560-9.

González-Neves, G., Franco, J., Barreiro, L., Gil, G., Moutounet, M., and Carbonneau, A. 2007. Varietal differentiation of Tannat, Cabernet-Sauvignon and Merlot grapes and wines according to their anthocyanic composition. *European Food Research and Technology*. 225 (1): 111–17.

Goto, T., Tamura, H., Kawai, T., Hoshino, T., Harada, N., and T. Kondo. 1986. Chemistry of metalloanthocyanins. *Annals of the New York Academy of Sciences*. 471 (1): 155–73.

Green, B.G. 1993. Oral astringency: A tactile component of flavor. *Acta Psychologica*. 84 (1): 119–25.

Haslam, E. 1998. Practical polyphenolics: From structure to molecular recognition and physiological action. Cambridge University Press.

Hayasaka, Y., Baldock, G.A., Pardon, K.H. et al. 2010. Investigation into the formation of guaiacol conjugates in berries and leaves of grapevine *Vitis vinifera* L. Cv. cabernet sauvignon using stable isotope tracers combined with HPLC-MS and MS/MS analysis. *Journal of Agricultural and Food Chemistry*. 58 (4): 2076–81.

He, F., He, J.J., Pan, Q.H. et al. 2010. Mass-spectrometry evidence confirming the presence of pelargonidin-3-O-glucoside in the berry skins of Cabernet Sauvignon and Pinot Noir (*Vitis vinifera* L.). *Australian Journal of Grape and Wine Research*. 16 (3): 464–68. doi:10.1111/j.1755-0238.2010.00107.x.

Hufnagel, J.C. and Hofmann, T. 2008a. Orosensory-directed identification of astringent mouthfeel and bitter-tasting compounds in red wine. *Journal of Agricultural and Food Chemistry*. 56 (4): 1376–86.

Hufnagel, J.C., and Hofmann, T. 2008b. Quantitative reconstruction of the nonvolatile sensometabolome of a red wine. *Journal of Agricultural and Food Chemistry*. 56 (19): 9190–99.

Jánváry, L., Hoffmann, T., Pfeiffer, J. et al. 2009. A double mutation in the anthocyanin 5-O-glucosyltransferase gene disrupts enzymatic activity in *Vitis vinifera* L. *Journal of Agricultural and Food Chemistry*. 57 (9): 3512–18.

Jeffery, D.W., Parker, M., and Smith, P.A. 2008. Flavonol composition of Australian red and white wines determined by high-performance liquid chromatography. *Australian Journal of Grape and Wine Research*. 14 (3): 153–61.

Jöbstl, E., O'Connell, J., Fairclough, J.P.A. et al. 2004. Molecular model for astringency produced by polyphenol/protein interactions. *Biomacromolecules*. 5 (3): 942–49.

Jourdes, M., Michel, J., Saucier, C. et al. 2011. Identification, amounts, and kinetics of extraction of c-glucosidic ellagitannins during wine aging in oak barrels or in stainless steel tanks with oak chips. *Analytical and Bioanalytical Chemistry*. 401 (5): 1531–39.

Kahkonen, M.P., and Heinonen, M. 2003. Antioxidant activity of anthocyanins and their aglycons. *Journal of Agricultural and Food Chemistry*. 628–33.

Kennison, K.R., Gibberd, M.R., Pollnitz A.P. et al. 2008. Smoke-derived taint in wine: The release of smoke-derived volatile phenols during fermentation of Merlot juice following grapevine exposure to smoke. *Journal of Agricultural and Food Chemistry*. 56 (16): 7379–83.

Kennison, K.R., Kerry, L., Wilkinson, H.G. et al. 2007. Smoke-derived taint in wine: Effect of postharvest smoke exposure of grapes on the chemical composition and sensory characteristics of wine. *Journal of Agricultural and Food Chemistry*. 55 (26): 10897–901.

Koga, C.C., Becraft, A.R., Lee, Y. et al. 2015. Taste detection thresholds of resveratrol. *Journal of Food Science*. 80 (9): S2064–70.

Kumazawa, T., Kashiwayanagi, M., and Kurihara, K. 1986. Contribution of electrostatic and hydrophobic interactions of bitter substances with taste receptor membranes to generation of receptor potentials. *Biochimica et Biophysica Acta*. 888 (1): 62–69.

Lambert, S.G., Asenstorfer, R.E., Williamson, N.M. et al. 2011. Copigmentation between malvidin-3-glucoside and some wine constituents and its importance to colour expression in red wine. *Food Chemistry*. 125 (1): 106–15.

Lamuela-Raventos, R.M., Romero-Perez, A.I., Waterhouse, A.L., and de la Torre-Boronat, M.C. 1995. Direct HPLC analysis of cis- and trans-resveratrol and piceid isomers in spanish red *Vitis vinifera* wines. *Journal of Agricultural and Food Chemistry*. 43 (2): 281–83.

Langcake, P., and Pryce, R.J. 1977. The production of resveratrol and the viniferins by grapevines in response to ultraviolet irradiation. *Phytochemistry*. 16 (8): 1193–96.

Lila, M.A., Burton-Freeman, B., Grace, M. et al. 2016. Unraveling anthocyanin bioavailability for human health. *Annual Review of Food Science and Technology*. 7: 375–93.

Lisanti, M.T., Gambuti, A., Genovese, A., Piombino, P., and Moio, L. 2017. Treatment by fining agents of red wine affected by phenolic off-odour. *European Food Research and Technology*. 243(3): 501–510.

Malfeito-Ferreira, M. 2011. Yeasts and wine off-flavours: A technological perspective. *Annals of Microbiology*. 61 (1): 95–102.
Mateus, N., Silva, A.M., Rivas-Gonzalo, J.C. et al. 2003. A new class of blue anthocyanin-derived pigments isolated from red wines. *Journal of Agricultural and Food Chemistry*. 51 (7): 1919–23.
Mattivi, F., Guzzon, R., Vrhovsek, U., Stefanini, M., and Velasco, R. 2006. Metabolite profiling of grape: Flavonols and anthocyanins. *Journal of Agricultural and Food Chemistry*. 54 (20): 7692–7702.
Mattivi, F., Vrhovsek, U., Masuero, D., and Trainotti, D. 2009. Differences in the amount and structure of extractable skin and seed tannins amongst red grape varieties. *Australian Journal of Grape and Wine Research*. 15 (1): 27–35.
Mazza, G., and Francis, F.J. 1995. Anthocyanins in grapes and grape products. *Critical Reviews in Food Science and Nutrition*. 35 (4): 341–71.
Mazza, G., Fukumoto, L., Delaquis, P., Girard, B., and Ewert, B. 1999. Anthocyanins, phenolics, and color of Cabernet franc, Merlot, and Pinot noir wines from British Columbia. *Journal of Agricultural and Food Chemistry*. 47 (10): 4009–17.
McRae, J.M., and Kennedy, J.A. 2011. Wine and grape tannin interactions with salivary proteins and their impact on astringency: A review of current research. *Molecules*. 16 (3): 2348–2364.
Muccillo, L., Gambuti, A., Frusciante, L. et al. 2014. Biochemical features of native red wines and genetic diversity of the corresponding grape varieties from Campania region. *Food Chemistry*. 143: 506–13.
Nelson, C.C. 2015. Zinfandel productivity is managed by pruning systems, applied water amounts, and rootstocks in a hot climate. Doctoral dissertation.
Nunes, M.A., Pimentel, F., Costa, A.S. et al. 2016. Cardioprotective properties of grape seed proanthocyanidins: An update. *Trends in Food Science and Technology*. 57: 31–39.
Okuda, T., Yoshida, T., and Hatano, T. 1990. Oligomeric hydrolyzable tannins. A new class of plant polyphenols. *Heterocycles*. 30: 30–37.
Oliveira, J., Brás, N.F., da Silva, M.A. et al. 2014. Grape anthocyanin oligomerization: A putative mechanism for red color stabilization? *Phytochemistry*. 105: 178–85.
Ortega-Regules, A., Romero-Cascales, I., López-Roca, J.M. et al. 2006. Anthocyanin fingerprint of grapes: Environmental and genetic variations. *Journal of the Science of Food and Agriculture*. 86 (10): 1460–67.
Pastorkova, E., Zakova, T., Landa, P. et al. 2013. Growth inhibitory effect of grape phenolics against wine spoilage yeasts and acetic acid bacteria. *International Journal of Food Microbiology*. 161 (3): 209–13.
Peleg, H., Gacon, K., Schlich, P. et al. 1999. Bitterness and astringency of flavan-3-ol monomers, dimers and trimers. *Journal of the Science of Food and Agriculture*. 79 (8): 1123–28.
Penalva, R., Esparza, I., Larraneta, E. et al. 2015. Zein-based nanoparticles improve the oral bioavailability of resveratrol and its anti-inflammatory effects in a mouse model of endotoxic shock. *Journal of Agricultural and Food Chemistry*. 63 (23): 5603–11.
Pozo-Bayón, M.Á., Hernández, M.T., Martín-Álvarez, P.J. et al. 2003. Study of low molecular weight phenolic compounds during the aging of sparkling wines manufactured with red and white grape varieties. *Journal of Agricultural and Food Chemistry*. 51 (7): 2089–95.

Preys, S., Mazerolles, G., Courcoux, P. et al. 2006. Relationship between polyphenolic composition and some sensory properties in red wines using multiway analyses. *Analytica Chimica Acta*. 563 (1): 126–36.

Price, S.F., Breen, P.J., Valladao, M. et al. 1995. Cluster sun exposure and quercetin in pinot noir grapes and wine. *American Journal of Enology and Viticulture*. 46 (2): 187–94.

Prieur, C., Rigaud, J., Cheynier, V. et al. 1994. Oligomeric and polymeric procyanidins from grape seeds. *Phytochemistry*. 36 (3): 781–84.

Puech, J.L. 1987. Extraction of phenolic compounds from oak wood in model solutions and evolution of aromatic aldehydes in wines aged in oak barrels. *American Journal of Enology and Viticulture*. 38 (3): 236–38.

Puech, J.-L., Feuillat, F., and Mosedale, J.R. 1999. The tannins of oak heartwood: Structure, properties, and their influence on wine flavor. *American Journal of Enology and Viticulture*. 50 (4): 469–78.

Quijada-Morín, N., Regueiro, J., Simal-Gándara, J. et al. 2012. Relationship between the sensory-determined astringency and the flavanolic composition of red wines. *Journal of Agricultural and Food Chemistry*. 60 (50): 12355–61.

Remy, S., Fulcrand, H., Labarbe, B. et al. 2000. First confirmation in red wine of products resulting from direct anthocyanin-tannin reactions. *Journal of the Science of Food and Agriculture*. 80 (6): 745–51.

Ribéreau-Gayon, P., Glories, Y., Maujean, A. et al. 2006. Phenolic compounds. In *Handbook of Enology*, pp. 141–203. Chichester, UK: John Wiley & Sons, Ltd.

Robichaud, J.L., and Noble, A.C. 1990. Astringency and bitterness of selected phenolics in wine. *Journal of the Science of Food and Agriculture*. 53 (3): 343–53.

Rustioni, L., Bedgood, D.R., Failla, O. et al. 2012. Copigmentation and anti-copigmentation in grape extracts studied by spectrophotometry and post-column-reaction HPLC. *Food Chemistry*. 132 (4): 2194–2201.

Salgues, M., Cheynier, V., Gunata, Z. et al. 1986. Oxidation of grape juice 2-s-glutathionyl caffeoyl tartaric acid by *Botrytis cinerea* Laccase and characterization of a new substance: 2,5-Di-s-glutathionyl caffeoyl tartaric acid. *Journal of Food Science*. 51 (5): 1191–94.

Sarni-Manchado, P., Fulcrand, H., Souquet, J.M. et al. 1996. Stability and color of unreported wine anthocyanin-derived pigments. *Journal of Food Science*. 61 (5): 938–41.

Saucier, C., Jourdes, M., Glories, Y. et al. 2006. Extraction, detection, and quantification of flavano-ellagitannins and ethylvescalagin in a Bordeaux red wine aged in oak barrels. *Journal of Agricultural and Food Chemistry*. 54 (19): 7349–54.

Schaefer, H.M., Schaefer, V., and Levey, D.J. 2004. How plant–animal interactions signal new insights in communication. *Trends in Ecology & Evolution*. 19 (11): 577–84.

Schroeter, H., Heiss, C., Spencer, J.P. et al. 2010. Recommending flavanols and procyanidins for cardiovascular health: Current knowledge and future needs. *Molecular Aspects of Medicine*. 31 (6): 546–57.

Soares, S., Kohl, S., Thalmann, S., Mateus, N., Meyerhof, W., and De Freitas, V. 2013. Different phenolic compounds activate distinct human bitter taste receptors. *Journal of Agricultural and Food Chemistry*. 61 (7): 1525–33.

Souquet, J.-M., Cheynier, V., Brossaud, F. et al. 1996. Polymeric proanthocyanidins from grape skins. *Phytochemistry*. 43 (2): 509–12.

Spayd, S.E., Tarara, J.M., Mee, D.L. et al. 2002. Separation of sunlight and temperature effects on the composition of *Vitis vinifera* cv. Merlot berries. *American Journal of Enology and Viticulture.* 53 (3): 171–82.

Springer, L.F., Chen, L.A., Stahlecker, A.C., Cousins, P., and Sacks, G.L. 2016. Relationship of soluble grape-derived proteins to condensed tannin extractability during red wine fermentation. *Journal of Agricultural and Food Chemistry.* 64 (43): 8191–99.

Sun, B., Sá, M.D., Leandro, C., Caldeira, I., Duarte, F.L., and Spranger, I. 2013. Reactivity of polymeric proanthocyanidins toward salivary proteins and their contribution to young red wine astringency. *Journal of Agricultural and Food Chemistry.* 61 (4): 939–46.

Swain, T. and Bate-Smith, E.C. 1962. Flavonoid compounds. In *Comparative Biochemistry*, pp. 55–61. London: Academic Press. doi:10.1007/978-1-4899-2909-9_18.

Teissedre, P.L., Frankel, E.N., Waterhouse, A.L. et al. 1996. Inhibition of *in vitro* human LDL oxidation by phenolic antioxidants from grapes and wines. *Journal of the Sciene of Food and Agriculture.* 70: 55–61.

Teixeira, N., Cruz, L., Brás, N.F. et al. 2013. Structural features of copigmentation of oenin with different polyphenol copigments. *Journal of Agricultural and Food Chemistry.* 61 (28): 6942–48.

Thorngate, J.H., and Singleton, V.L. 1994. Localization of procyanidins in grape seeds. *American Journal of Enology and Viticulture.* 45 (2): 259–62.

Timberlake, C.F., and Bridle, P. 1976. Interactions between anthocyanins, phenolic compounds, and acetaldehyde and their significance in red wines. *American Journal of Enology and Viticulture.* 27 (3).

Vèrette, E., Noble, A.C., and Somers, T.C. 1988. Hydroxycinnamates of *Vitis vinifera*: Sensory assessment in relation to bitterness in white wines. *Journal of the Science of Food and Agriculture.* 45 (3): 267–72.

Vidal, S., Francis, L., Noble, A., Kwiatkowski, M., Cheynier, V., and Waters, E. 2004. Taste and mouth-feel properties of different types of tannin-like polyphenolic compounds and anthocyanins in wine. *Analytica Chimica Acta.* 513 (1): 57–65.

Vivas, N., and Glories, Y. 1996. Role of oak wood ellagitannins in the oxidation process of red wines during aging. *American Journal of Enology and Viticulture.* 47: 103–7.

Vivas, N., Nonier, M.F., de Gaulejac, N.V. et al. 2004. Differentiation of proanthocyanidin tannins from seeds, skins and stems of grapes (*Vitis vinifera*) and heartwood of Quebracho (*Schinopsis balansae*) by matrix-assisted laser desorption/ionization time-of-flight mass spectrometry and thioacidolysis/liquid chromatography/electrospray ionization mass spectrometry. *Analytica Chimica Acta.* 513 (1): 247–56.

Wang, H., Race, E.J., and Shrikhande, A.J. 2003. Characterization of anthocyanins in grape juices by ion trap liquid chromatography–mass spectrometry. *Journal of Agricultural and Food Chemistry.* 51 (7): 1839–44.

Waterhouse, A.L., and Nikolantonaki, M. 2015. Quinone reactions in wine oxidation. Chapter 18, pp. 291–301, eds. S.B. Ebeler, G. Sacks, S. Vidal, and P. Winterhalter, Advances in Wine Research. American Chemical Society.

Waterhouse, A., Sacks, G., and Jeffery, D. 2016. Topics related to aging. In *Understanding Wine Chemistry*, eds. A.L. Waterhouse, G.L. Sacks, and D.W. Jeffery, pp. 294–317. John Wiley & Sons.

Weber, F., Greve, K., and Durner, D. et al. 2012. Sensory and chemical characterization of phenolic polymers from red wine obtained by gel permeation chromatography. *American Journal of Enology and Viticulture.*

Xia, E.-Q., Deng, G.-F., Guo, Y.-J. et al. 2010. Biological activities of polyphenols from grapes. *International Journal of Molecular Sciences.* 11 (2): 622–46.

Zhang, B., Liu, R., He, F. et al. 2015. Copigmentation of malvidin-3-O-glucoside with five hydroxybenzoic acids in red wine model solutions: Experimental and theoretical investigations. *Food Chemistry.* 170: 226–233.

CHAPTER 16

Phenolic Compounds in Cereals and Legumes

María Teresa Espino Sevilla, Cristian Jiménez Martínez, and Anaberta Cardador-Martínez

CONTENTS

16.1 PHENOLIC COMPOUND IN LEGUMES

Legumes are phanerogams, belonging to the Fabaceas (Papilonaceas) family. These dicotyledonous plants typically produce fruits in the form of pods called legumes. *Legume* is a word from Latin origin (Legumen), which means leguminous plant. Another term for the edible seeds of these plants is *pulse*, from Latin *puls*, which means "pottage."

Legumes are the third-largest family of flowering plants, comprising 751 genera and 19,500 species (Gutiérrez-Uribe et al. 2016). The best-known species are the common bean (*Phaseolus vulgaris*), chickpea (*Cicer arietum*), lentil (*Lens esculenta*), Ayocote bean (*Phaseolus coccineus*), alfalfa (*Medicago sativa*), soybean (*Glycine max*), peanut (*Arachis hypogaea*), and in some regions of the American and European continents, *Lupinus* genera. Depending of the species of legume in question, immature pods, mature or immature seeds, and leaves and roots are used (Olmedilla-Alonso et al. 2010).

Legumes constitute an important part of the diet for large sectors of the world population, as they are a good source of proteins, carbohydrates, minerals, and B vitamins. However, this food group contains several compounds of nonnutritive character that hinder or inhibit the uptake of nutrients, producing physiological and biochemical adverse effects in humans and animals—and in some cases, they can be toxic. These compounds traditionally have been called antinutritional but recently it was found that, in an acceptable proportion, they are beneficial to health.

Many phenolic compounds and mixtures thereof are frequent in a wide variety of fruits, vegetables, legumes, and cereals (Adom and Liu 2002; Stratil et al. 2006). Among functions developed by phenolic compounds in plant tissues are protection against UV radiation, pigmentation, and stimulation of nodules' N2 fixation (Denny and Buttriss 2007). Also, some classes of phenolic compounds, such as flavonoids, have antimicrobial activity due to their ability to form complexes between soluble extracellular proteins and the microbial cell wall (Lin and Weng 2006). For instance, phytoalexins are groups of phenolic compounds produced by plants to defend themselves when they have been injured and invaded (Dixon 2001). In response to an invader fungus, resveratrol (Figure 16.1) is synthesized from p-coumaroyl CoA and malonyl CoA (Soleas et al. 1997). In legumes, the content of phenolic compounds (Table 16.1) varies and depends on the type, variety, and solvent used for extraction (water,

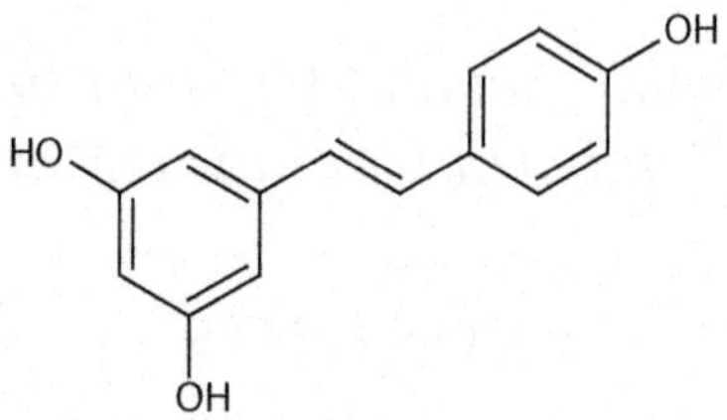

FIGURE 16.1 Resveratrol structure.

TABLE 16.1 Total Phenolics (TPC), Flavonoids, and Tannin Content of Legumes of Common and Unconventional Use

Legume	TPC (mg GAE[a]/g)	Flavonoids (mg RUE[b]/g)	Tannins (mg CAE[c]/g)	Reference
Soybean	1.57–5.57	1.06–4.04	0.37–1.96	Gan et al. 2016
Beans				
Navy beans	0.42 ± 0.04			Anton et al. 2008
Pinto beans	1.92 ± 0.12			Anton et al. 2008
Common beans	0.57–6.99	0.92–4.24	0.47–5.16	Xu and Chang 2010
Peanuts	18.21 ± 0.15			Anton et al. 2008
Lentil	4.86–9.60	3.04–4.54	3.73–10.2	Xu and Chang 2010
Faba bean	36.69	15.81	0.545–20	Boudjou et al. 2013
Chickpea	19.42 ± 2.72	10.65 ± 1.43	0.78–2.72	Nithiyanantham et al. 2012
Pisum sativum	12.88 ± 0.68	7.93 ± 0.37	0.203–10.5	Nithiyanantham et al. 2012
Lima beans	4.3	7.6	4.1	Agostini-Costa et al. 2015
Black beans	11.74 ± 0.07		0.37–6.74	Amarowicz and Pegg 2008
Mung beans	2.14 ± 0.10	1.95 ± 0.21		Guo et al. 2012
Unconventional Legumes				
Cowpea	3.30 ± 0.11		1.75–5.90	Aguilera et al. 2013
Jack bean	2.30 ± 0.05			Aguilera et al. 2013
Doilchos	0.72 ± 0.02			Aguilera et al. 2013
Mucuna	37.40 ± 0.42			Aguilera et al. 2013
P. lunatus	4.3 ± 2.8	7.5 ± 5.0	4.1 ± 3.1	Agostini-Costa et al. 2015

[a] GAE: Gallic acid equivalents.
[b] RUE: Rutin equivalents.
[c] CAE: Catechin equivalents.

methanol, ethanol, acetone, or mixtures), extraction time, and quantification method. However, the most-used method to quantify phenolics is the colorimetric reaction with the Folin–Ciocalteu reagent in alkaline conditions. The content of phenolic compounds (Table 16.1) in legumes can range from 1–50 milligrams equivalent gallic acid (GAE)/gram. For example, Chukwumah et al. (2009) indicated that the total phenol content of the main cultivars of peanut (seeds) consumed in the United States varies from 0.94 to 2.28 milligrams GAE/gram of fresh peanut with an average content of 1.43 milligrams GAE/gram. The phenolic compounds concentration in *Lupinus* was 1.16–6.92 milligrams GAE/gram; faba bean, 0.2–2.38 milligrams GAE/gram; pea, 0.96–2.26 milligrams GAE/gram; and chickpea, 0.52–1.27 milligrams GAE/gram (Magalhaes et al. 2017). Among the phenolic compounds found in methanol extracts of chickpea are p-hydroxybenzoic acid, gentisic acid, syringic acid, luteolin, myricetin, quercetin, pinocembrin, kaempferol, and biochanin (Aguilera et al. 2013). Their concentrations depend on the variety of legume (Magalhaes et al. 2017).

16.1.1 Phenolic Acids

Phenolic acids of the family of benzoic acid and trans-cinnamic acid are synthesized through the phenylpropanoid metabolic pathway. Five phenolic acids of the benzoic type and their derivatives (gallic acid, protocatechuic acid, p-hydroxybenzoic acid, protocatechualdehyde, and 2,3,4-trihydroxybenzoic acid), vanillic acid, and four phenolic acids of the cinnamic type (chlorogenic acid, p-coumaric acid, m-coumaric acid, and sinapic acid) were detected in lentil cultivars (Xu and Chang 2010). In some varieties of dark bean, phenolic acids like p-hydroxybenzoic, genistic, syringic acid have been quantified, the last mainly in dark varieties (Dueñas et al. 2016). Aguilera et al. (2011b) reported that these compounds play an important role in antioxidant properties, antimicrobial activity, and anti-inflammatory activities of the legumes.

16.1.2 Flavonoids

Flavonoids are the most common and widely distributed group of phenolic compounds in plants. Their basic structure consists of a diphenylpropane core formed by two aromatic rings external and a bridge of three carbons, which can be closed (as flavons, flavonols, and anthocyanidins) or open (as chalcones). Flavonols such as quercetin, kaempferol, and myricetin have antimutagenic, anticancer, and antihypertensive activities. These compounds play a key role in plants, such as UV protection, pigmentation, stimulation of N_2 fixation nodules, and resistance against some diseases (Denny and Buttriss 2007). In legumes, different flavonoids have been identified, for example, in pea varieties have been found exclusively prodelphinidin (gallocatechin and epigallocatechin). The subunits procyanidin- and prodelphinidin-type flavan-3-ol are the most abundant in seeds of faba bean and lentil with a concentration ranging from 2.7 to 6.5 milligrams/gram on wet basis (Jin et al. 2012).

Procyanidin B2, B3, and a procyanidin tetramer (0.1 to 0.5 mg/g) have been identified in lentil (López-Amorós et al. 2006). Procyanidin B2, C1, and C2 were found in husk extract of red bean (*Phaseolus vulgaris*), brown, and black (Madhujith et al. 2004). Flavonoids are usually found as glycosides. Among flavonoids epicatechin (flavan-3-ol), apigenin, and luteolin glycosides (flavones), and myricetin-3-O-rhamnoside (flavonol), have been identified in several varieties of beans with differences in the profile of these compounds. In *Vicia faba*, myricetin-3-O-rhamnoside and derivatives have been detected (El-Mergawi and Taie 2014). In some varieties of mature faba bean luteolin-8-C-glucoside and luteolin-6-C-glucoside have been detected; however, these compounds have not been found in green stage (broad beans) (Abu-Reidah et al. 2014; Baginsky et al. 2013). The presence of glucosides of luteolin, myricetin, and quercetin were detected in varieties of chickpea (Aguilera et al. 2011a). In peas, glycosylated flavones have been detected; for instance, derivatives of luteolin and apigenin, such as apigenin-8-C-glucoside apigenin-6-C-glucoside, luteolin-6-C-glucoside, and luteolin-3,7-di-O-glucoside, and also quercetin and kaempferol glucosides (Dueñas et al. 2004; Troszyńska et al. 2002).

Besides flavonols and anthocyanins, isoflavones such as formononetin, isoformononetin, and biochanin A that, similarly to other flavonoids, are naturally found as glycosides and acyl glycosides. The concentration of these compounds is also related to physiological changes that occur during germination and as a response to different elicitors (Guardado-Félix et al. 2017).

16.1.3 Lignans

Lignans are formed by two phenylpropane units (Figure 16.2), which are linked by oxidative dimerization (Ignat et al. 2011). These compounds normally exist in nature in free form, and glycosylated derivatives are produced only in small quantities. Besides serving as antioxidants, it has been shown that lignans are capable of inhibiting the

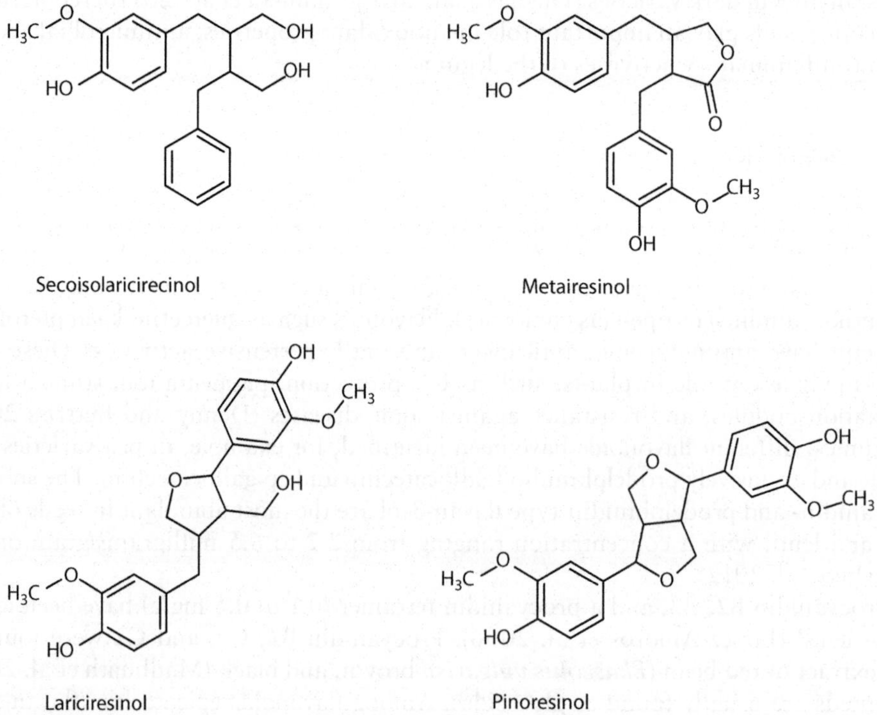

FIGURE 16.2 Representative structures of lignanes.

promoting effects on breast cancer by binding to estrogen receptors (Pianjing et al. 2010). Fourteen lignans have been found in soy-derived products, among these are secoisolariciresinol, matairesinol, syringaresinol, lariciresinol, isolariciresinol, and pinoresinol (Peñalvo et al. 2005).

16.1.4 Stilbenes

Stilbenes consists of two phenyl groups linked by double bond; the pterostilbene, piceatannol, and resveratrol are the representative main compounds of this family (Leopoldini et al. 2011). Resveratrol has been the subject of special attention because of their health-promoting effects, such as cardioprotective, neuroprotective, anticancer, antidiabetic, and prevention of premature aging (Pandey and Rizvi 2011). Resveratrol has been found in legumes such as peanuts, *Cassia* spp., and *Pterolobium hexapetallum* (Chukwumah et al. 2009).

16.1.5 Tannins

Tannins are a group of phenolic polymers with astringent properties and are synthesized by polymerization of flavan derivatives or polymerization of quinone units (Chung et al. 1998); they can be subdivided into classes according to their chemical composition: Hydrolyzable or condensed.

Hydrolyzable tannins can be divided into gallotannins and ellagitannins. Gallotannins are composed of subunits of esterified gallic acid to glucose. Ellagitannins are simple polymers of ellagic acid and gallic acid. Hydrolyzable tannins are so called because they are readily hydrolyzed in acidic or weak alkalis to obtain monomeric units. Condensed tannins, named proanthocyanidins (PACs), generate monomers of anthocyanidins when they are heated with acid.

In foods, PACs are usually classified as procyanidins or prodelfinidins according to the chemistry of their subunits' flavan-3-ol. Procyanidins are composed of (–)-epicatechin, while prodelfidins are composed of epigallocatechin subunits. These compounds have been located mainly on the husk or testa of seeds. The tannin content in legumes varies between 0.1 and 5 percent of the total weight (expressed as milliequivalent of catechin or tannic acid) (Reddy et al. 1985).

Condensed tannins ranged from 9.49 to 35.70 milliequivalent of (+)-catechin (CAE)/gram of bean flour in 62 wild and weedy Mexican bean collections from diverse origins (Espinosa-Alonso et al. 2006), the wide variation observed was more related to their genotype than to the color factor. In varieties of chickpea and pea, Nithiyanantham et al. (2012) found 10.85 and 6.85 milligrams/gram of extract.

Anthocyanins such as delphinidin-3-glucoside, petunidin-3-glucoside, malvidin-3-glucoside, malvidin-3,5-diglucoside and proanthocyanidins (monomers, dimers, trimers, tetramers, pentamers, hexamers) have been isolated from black beans (Aparicio-Fernandez et al. 2005). Both procyanidin- and prodelphinidin-type flavan-3-ol subunits were abundant in faba bean and lentil seeds (Jin et al. 2012). The biological activity of crude methanol extracts is due to the presence of flavonoid monomers and polymers (condensed tannins), hydrolyzable tannins, and phenolic compounds. Tannins isolated from adzuki bean (*Vigna angularis*) have antioxidant properties (Ariga et al. 1988; Han et al. 2015) and extracts containing tannins from a variety of dry beans (*P. vulgaris*) have been shown to inhibit iron-catalyzed oxidation of soybean (*Glycine max* [L.] Merr.) oil (Ganthavorn and Hughes 1997).

16.2 PHENOLIC COMPOUNDS IN CEREALS

Cereal grains have played a key role in the traditional diets of human beings throughout the world because they are an excellent source of protein, dietary fiber, starch, micronutrients, and bioactive compounds with low levels of fat. The whole grain offers many health-promoting components, which include phenolic compounds (Dykes and Rooney 2007).

Structurally the seeds are composed of three main parts, including the endosperm, embryo, and seed coat. The endosperm is the primary starch storage portion but also contains some protein. The embryo is the oil storage portion, high in protein and minerals. The seed coat, also called the pericarp or bran, consists mainly of cellulose and hemicellulose with some protein and lignin.

Epidemiological studies have consistently shown that the consumption of whole cereals may contribute to the prevention of many chronic diseases, such as obesity, cardiovascular disease, type 2 diabetes, and some cancers (Liu 2007). As the most common phenolic compounds in cereals, phenolics acids have strong antioxidant activity and may modulate cellular oxidative status and prevent oxidation of biologically important molecules such as DNA, proteins, and membrane lipids from oxidative damage (Yu et al. 2005).

Phenolic compounds in cereals fall into soluble and insoluble or linked. In the first group the free phenols, glycosylated, and esterified are in greater amounts in the peripheral layers of grain (pericarp, testa, and aleurone cells), while their concentration is lower in the endosperm (Yu et al. 2001). These compounds include phenolic acids (benzoic and cinnamic acids), tannins, coumarins, flavonoids, and alkyl resorcinol, which are responsible for the flavor, texture (such as mouthfeel of beer), color, taste, and oxidative stability of plant foods (Naczk and Shahidi 2006). The phenolic content in cereals depends on many factors such as climatic and agro technical conditions in cultivation and harvesting, ripeness of the material, harvest time, storage conditions, effect of genetic factors, and variety-dependent variability (Naczk and Shahidi 2006).

In cereal grains, since phenolic compounds are located mainly in the pericarp, they can be concentrated by decorticating of grain to produce bran, which can be incorporated into a food product (i.e., breads, cookies, and tortillas) with increased dietary fiber levels and nutraceutical properties (Dykes and Rooney 2007). In corn, particularly when it is lime-cooked to produce tortillas, most of the phenolics are lost into the wastewater (Gutiérrez-Uribe et al. 2010).

Cereals contain a wide range of phenolic compounds. A significant amount of phenolic acids such as ferulic, caffeic, p-hydroxybenzoic, protocatechuic, p-coumaric, vanillic, and syringic acids are typical to cereals. These compounds occur in the grain primarily in the bound form as conjugates with sugars, fatty acids, or protein. Research indicates that cereal grains contain special phenolic compounds, such as ferulic acid and diferulates, which are not present in significant quantities in fruit and vegetables (Bunzel et al. 2001). Other compounds present in cereals are flavonoids, condensed tannins, coumarins, and alkyl-resorcinol (Naczk and Shahidi 2004).

16.2.1 Phenolic Acids in Cereals

Phenolic acids are the most important and the largest group of antioxidants in terms of incidence in cereal grains (Table 16.2). Phenolic acids can be classified as free or bound phenolic acids (Renger and Steinhart 2000). The extraction methods are important to recover these compounds. Free phenolic acids are extractable by a polar solvent, such as water, methanol, ethanol, and acetone; bound phenolic acids, which are covalently bound to structural components of the cell wall such as cellulose, hemicellulose, lignin, pectin, and rod-shaped structural proteins are commonly extracted by adding a strong alkali to the residue obtained after the extraction of free phenolic (Acosta-Estrada et al. 2014). The solvent solution and the number of extraction steps adopted for the extraction of phenolic compounds are an important factor in their recovery; moreover, the literature data about the effects of the solvent type and number of extraction steps for the recovery of free phenolics is limited. Regarding bound phenolics, a great trouble arises from the alkaline hydrolysis treatment. The increase in matrix viscosity promoted by the alkali addition may hinder the extraction of these phenolics (Zhou et al. 2004).

Flavonoids in cereals include anthocyanins, flavanols, flavones, flavonoids, and flavanones. Cereal grains contain a variety of flavonoids, located generally in the pericarp. The six common anthocyanidins found in cereals are cyanidin, malvinidin, pelragonidin, delphinidin, petunidin, and peonidin (Kaur et al. 2014). Table 16.2 shows some phenolics found in cereals.

Maize (*Zea mays*) contains more total phenols and a higher antioxidant power than cereals like wheat, rice, and oats (Adom and Liu 2002). The main phenolic is ferulic acid, which represents about 85 percent of total phenols and is concentrated in the grain pericarp as free or esterified to the hemicellulose of the cell wall (de la Parra et al. 2007).

Phenolics in maize germ are associated with tolerance to *Fusarium* spp. by decreasing its toxin production (Bakan et al. 2003), while those found in pericarp are related with tolerance to storage pests. These phenolics may be bound with polysaccharides and, therefore, when arabinoxylans were extracted from resistant corn, they contained a higher amount of hydroxycinnamic acids (Ayala-Soto et al. 2014).

On other hand, purple corn is an important source of anthocyanins and other polyphenols that are distributed throughout the plant (Pedreschi and Cisneros-Zevallos 2007). Various phenolic phytochemicals have been characterized in purple corn, including total polyphenols, flavonoids, flavonols, flavanols, and anthocyanins (Ramos-Escudero et al. 2012).

Wheat is the main source of cereal products, but they are mostly produced from refined white flour from which peripheral tissues (germ and envelopes) were removed. However, these tissues, which are eliminated in the so-called bran fraction and serve mainly for animal feeding, contain most of the micronutrients, fiber, and phytochemicals of the grain that could contribute greatly to increasing the nutritional quality of human food if included in flours or used as food ingredients. There is increasing agreement that sufficient whole grain consumption has a protective effect against the development of diet-related disorders such as cardiovascular disease and type 2 diabetes, and also against certain cancers (Jones 2006). One of the most abundant phenolic compounds in wheat grain is ferulic acid, accounting for 90 percent of the total polyphenols in wheat grain (Manach et al. 2004). As in corn, in bran, feruic acid is largely located as a structural

TABLE 16.2 Phenolic Acids Reported in Cereals Grains

Phenolic Acid	Grain	Reference
Hydroxybenzoic acids		
Gallic	Rice, sorghum	Walter et al. 2013
Protocatechuic	Maize, oat, rice, sorghum, wheat	Mazza and Gao 2005; Tian et al. 2005; Walter et al. 2013
p-Hydroxybenzoic	Maize, oat, rice, sorghum, wheat	Mazza and Gao 2005; Tian et al. 2005; Walter et al. 2013
Gentisic	Milled, sorghum	McDonough et al. 2000
Salycilic	Barley, sorghum, wheat	Kim, Tsao, Yang, and Cui 2006; Mazza and Gao 2005
Vanillic	Barley, maize, oat, rice, sorghum, wheat	Kim, Tsao, Yang, and Cui 2006; Mazza and Gao 2005; McDonough et al. 2000; Zhou et al. 2004
Syringic		
Hydroxycinnamic acids:		
Ferulic	Barley, maize, oat, rice, sorghum, wheat	Kim, Do, and Lee 2006; Mattila et al. 2005; Mazza and Gao 2005; McDonough et al. 2000; Zhou et al. 2004
Caffeic	Maize, oat, rice, sorghum and wheat	Kim, Tsao, Yang, and Cui 2006; Mattila et al. 2005; Zhou et al. 2004
Cinnamic	Sorghum, wheat, rice	McDonough et al. 2000; Tian et al. 2004
Sinapic	Barley, oat, rice, sorghum	Kim et al. 2006; Mattila et al. 2005; Mazza and Gao 2005; McDonough et al. 2000; Tian et al. 2004
Flavonoids		
Anthocyanidins	Rice, wheat, maize, sorghum	Escribano-Bailón et al. 2004; Masisi et al. 2016; Reddy et al. 2007

component of the cell walls of aleurone and pericarp (Hemery et al. 2007). Most of the ferulic acid is covalently bound to complex polysaccharides in the cell walls, mainly arabinoxylans (Bunzel et al. 2001).

Phenolic content varies significantly among wheat cultivars. Work on wheat phenolics has largely focused on the effects exerted on wheat flour quality, particularly involvement in pigmentation of both flour and bread. However, with the increasing popularity of functional foods, it has become important to focus on cereal fractions with potential health benefits. In addition to its value as dietary fiber, wheat bran can be a valuable source of phenolic compounds. Fractionation by traditional roller milling of wheat separates bran layers enriched in phenolic compounds.

Oats are a cereal commonly consumed as whole grains and known to provide healthy nutrients to humans. The phenolic antioxidants in oats have been studied extensively since the 1930s (Hole et al. 2013). A wide spectrum of active phytochemicals can be found in oats, which may act as antioxidants in different ways and combinations and even synergistically. The main sources of antioxidants found in oats are phenolic compounds of various classes such as hydroxycinnamic acids and avenanthramides and, to lesser extent, flavonoids (Peterson et al. 2002).

Research has shown the presence of protocatechuic, p-hydroxybenzoic acid, vanillic, syringic, ferulic, caffeic, p-coumaric, and sinapic in oats (Wu and Prior 2005). In their free form they can be found in both oat groats and hull, with caffeic acid found to a larger extent in the oat groats while p-coumaric, ferulic, and sinapic acid are more concentrated in the hull fraction (Emmons and Peterson 1999; Xing and White 1997).

Rice (*Oryza sativa* L.) is a staple food in many countries of Asia. It is considered an important source of energy for populations from developed and developing countries. More than just energy, rice is a source of proteins, minerals, vitamins, and bioactive compounds (Monks et al. 2013). The most consumed rice is polished rice, which is industrially prepared by removing the pericarp and aleurone layers of rice caryopsis and the germ, causing the loss of important bioactive compounds.

Ferulic acid and p-coumaric acid are the major phenolic compounds in rice and exist in the form of free, soluble conjugated, and in soluble bound. Nevertheless, rice also contains anthocyanins, proanthocyanidins, tocopherols, and oryzanol (Deng et al. 2013; Kim, Do, and Lee 2006).

Black and wild rice are important sources of anthocyanins and phenolic acids, while red rice is an important source of proanthocyanidins and phenolic acids. On other hand, according to the results presented by Zaupa et al. (2015) and Zhang et al. (2015), the main phenolic acids found in both pigmented and nonpigmented rice are protocatechuic acid, synaptic acid, vanillic acid, p-coumaric acid, and ferulic acid. The pigmented black, red, and wild rice grains are rich sources of phenolic compounds while just a small amount of phenolic compounds are found in brown rice and polished rice (Paiva et al. 2014; Walter et al. 2013). In particular for rice, an alternative process to reduce the loss of free phenolics due to polishing is parboiling (Paiva et al. 2016).

Buckwheat, amaranth, and quinoa are three of the most widely used pseudo-cereals, but their production is dwarfed by the true cereals (Gorinstein et al. 2007). The cereals and pseudocereals are essentially a starchy crop. However, they may contain significant quantities of protein and oil, and it is frequently these constituents that determine their suitability for a specific end use. Based on high contents of polyphenols, anthocyanins, flavonoids, and their antioxidant activities pseudocereals such as buckwheat, quinoa, and amaranth can be a substitute for cereals for common and atherosclerotic diets and sometimes in allergic cases (Gorinstein et al. 2007).

REFERENCES

Abu-Reidah, I. M., Contreras, M., Arráez-Román, D. et al. 2014. UHPLC-ESI-QTOF-MS-based metabolic profiling of *Vicia faba* L.(Fabaceae) seeds as a key strategy for characterization in foodomics. *Electrophoresis*. 35(11): 1571–1581.

Acosta-Estrada, B. A., Gutiérrez-Uribe, J. A., and Serna-Saldívar, S. O. 2014. Bound phenolics in foods, a review. *Food Chemistry*. 152: 46–55.

Adom, K. K., and Liu, R. H. 2002. Antioxidant activity of grains. *Journal of Agricultural and Food Chemistry*. 50(21): 6182–6187.

Agostini-Costa, T. D. S., Teodoro, A. F. P., Alves, R. D. B. D. N. et al. 2015. Total phenolics, flavonoids, tannins and antioxidant activity of lima beans conserved in a Brazilian Genebank. *Ciência Rural*. 45(2): 335–341.

Aguilera, Y., Diaz, M. F., Jimenez, T. et al. 2013. Changes in nonnutritional factors and antioxidant activity during germination of nonconventional legumes. *Journal of Agricultural and Food Chemistry*. 61(34): 8120–8125.

Aguilera, Y., Dueñas, M., Estrella, I. et al. 2011a. Phenolic profile and antioxidant capacity of chickpeas (*Cicer arietinum* L.) as affected by a dehydration process. *Plant Foods for Human Nutrition.* 66(2): 187–195.

Aguilera, Y., Estrella, I., Benitez, V. et al. 2011b. Bioactive phenolic compounds and functional properties of dehydrated bean flours. *Food Research International.* 44 (3): 774–780.

Amarowicz, R., and Pegg, R. B. 2008. Legumes as a source of natural antioxidants. *European Journal of Lipid Science and Technology.* 110(10): 865–878.

Anton, A. A., Ross, K. A., Beta, T., Fulcher, R. G., and Arntfield, S. D. 2008. Effect of predehulling treatments on some nutritional and physical properties of navy and pinto beans (*Phaseolus vulgaris* L.). *LWT-Food Science and Technology.* 41(5): 771–778.

Aparicio-Fernandez, X., Yousef, G. G., Loarca-Pina, G. et al. 2005. Characterization of polyphenolics in the seed coat of black Jamapa bean (*Phaseolus vulgaris* L.). *Journal of Agricultural and Food Chemistry.* 53(11): 4615–4622.

Ariga, T., Koshiyama, I. and Fukushima, D. 1988. Antioxidative properties of procyanidins B-1 and B-3 from azuki beans in aqueous systems. *Agricultural and Biological Chemistry.* 52(11): 2717–2722.

Ayala-Soto, F. E., Serna-Saldívar, S. O., García-Lara, S., and Pérez-Carrillo, E. 2014. Hydroxycinnamic acids, sugar composition and antioxidant capacity of arabinoxylans extracted from different maize fiber sources. *Food Hydrocolloids.* 35: 471–475.

Baginsky, C., Peña-Neira, Á., Cáceres, A. et al. 2013. Phenolic compound composition in immature seeds of fava bean (*Vicia faba* L.) varieties cultivated in Chile. *Journal of Food Composition and Analysis.* 31(1): 1–6.

Bakan, B., Bily, A. C., Melcion, D. et al. 2003. Possible role of plant phenolics in the production of trichothecenes by *Fusarium graminearum* strains on different fractions of maize kernels. *Journal of Agricultural and Food Chemistry.* 51(9): 2826–2831.

Boudjou, S., Oomah, B. D., Zaidi, F., and Hosseinian, F. 2013. Phenolics content and antioxidant and anti-inflammatory activities of legume fractions. *Food Chem.* 138(2–3): 1543–1550.

Bunzel, M., Ralph, J., Marita, J. M. et al. 2001. Diferulates as structural components in soluble and insoluble cereal dietary fibre. *Journal of the Science of Food and Agriculture.* 81(7): 653–660.

Chukwumah, Y. C., Walker, L. T., Verghese, M. et al. 2009. Effect of frequency and duration of ultrasonication on the extraction efficiency of selected isoflavones and trans-resveratrol from peanuts (*Arachis hypogaea*). *Ultrasonics Sonochemistry.* 16(2): 293–299.

Chung, K.-T., Wong, T. Y., Wei, C.-I. et al. 1998. Tannins and human health: A review. *Critical Reviews in Food Science and Nutrition.* 38(6): 421–464.

De la Parra, C., Serna Saldivar, S. O., and Liu, R. H. 2007. Effect of processing on the phytochemical profiles and antioxidant activity of corn for production of masa, tortillas, and tortilla chips. *Journal of Agricultural and Food Chemistry.* 55(10): 4177–4183.

Deng, G.-F., Xu, X.-R., Zhang, Y., Li, D., Gan, R.-Y., and Li, H.-B. 2013. Phenolic compounds and bioactivities of pigmented rice. *Critical Reviews in Food Science and Nutrition.* 53(3): 296–306.

Denny, A., and Buttriss, J. 2007. Plant foods and health: Focus on plant bioactives. *Synthesis Report.* (4): 1–64.

Dixon, R. A. 2001. Natural products and plant disease resistance. *Nature*. 411(6839): 843–847.

Dueñas, M., Estrella, I., and Hernández, T. 2004. Occurrence of phenolic compounds in the seed coat and the cotyledon of peas (*Pisum sativum* L.). *European Food Research and Technology*. 219(2): 116–123.

Dueñas, M., Sarmento, T., Aguilera, Y. et al. 2016. Impact of cooking and germination on phenolic composition and dietary fibre fractions in dark beans (*Phaseolus vulgaris* L.) and lentils (*Lens culinaris* L.). *LWT-Food Science and Technology*. 66: 72–78.

Dykes, L., and Rooney, L. W. 2007. Phenolic compounds in cereal grains and their health benefit. *Cereal Foods World*. 52(3): 105–111.

El-Mergawi, R., and Taie, H. A. 2014. Phenolic composition and antioxidant activity of raw seeds, green seeds and sprouts of ten faba bean (*Vicia faba*) cultivars consumed in Egypt. *International Journal of Pharma and Bio Sciences*. 5(2).

Emmons, C. L., and Peterson, D. M. 1999. Antioxidant activity and phenolic contents of oat groats and hulls. *Cereal Chemistry*. 76(6): 902–906.

Escribano-Bailón, M. T., Santos-Buelga, C., and Rivas-Gonzalo, J. C. 2004. Anthocyanins in cereals. *Journal of Chromatography A*. 1054(1–2): 129–141.

Espinosa-Alonso, L. G., Lygin, A., Widholm, J. M. et al. 2006. Polyphenols in wild and weedy Mexican common beans (*Phaseolus vulgaris* L.). *Journal of Agricultural and Food Chemistry*. 54(12): 4436–4444.

Gan, R.-Y., Lui, W.-Y., Wang, M.-F., Corke, H., and Sui, Z.-Q. 2016. Accumulation of solvent-soluble and solvent-insoluble antioxidant phenolics in edible bean sprouts: Implication of germination. *Functional Foods in Health and Disease*. 6(8): 519–535.

Ganthavorn, C., and Hughes, J. S. 1997. Inhibition of soybean oil oxidation by extracts of dry beans (*Phaseolus vulgaris*). *Journal of the American Oil Chemists' Society*. 74(8): 1025–1030.

Gorinstein, S., Medina-Vargas, O. J., Jaramillo, N. O. et al. 2007. The total polyphenols and the antioxidant potentials of some selected cereals and pseudocereals. *European Food Research and Technology*. 225: 321–328.

Guardado-Félix, D., Serna-Saldivar, S. O., Cuevas-Rodríguez, E. O., Jacobo-Velázquez, D. A., and Gutiérrez-Uribe, J. A. 2017. Effect of sodium selenite on isoflavonoid contents and antioxidant capacity of chickpea (*Cicer arietinum* L.) sprouts. *Food Chemistry*. 226: 69–74.

Guo, X., Li, T., Tang, K., and Liu, R. H. 2012. Effect of germination on phytochemical profiles and antioxidant activity of mung bean sprouts (*Vigna radiata*). *Journal of Agricultural and Food Chemistry*. 60(44): 11050–11055.

Gutiérrez-Uribe, J. A., Guajardo-Flores, D., and López-Barrios, L. 2016. Legumes in the diet. In *The Encyclopedia of Food and Health*, vol. 3, eds. B. Caballero, P. Finglas, F. Toldrá, pp. 539–543. Oxford: Academic Press.

Gutiérrez-Uribe, J. A., Rojas-García, C., García-Lara, S., and Serna-Saldivar, S. O. 2010. Phytochemical analysis of wastewater (nejayote) obtained after lime-cooking of different types of maize kernels processed into masa for tortillas. *Journal of Cereal Science*. 52(3): 410–416.

Han, K.-H., Kitano-Okada, T., Seo, J.-M. et al. 2015. Characterisation of anthocyanins and proanthocyanidins of adzuki bean extracts and their antioxidant activity. *Journal of Functional Foods*. 14: 692–701.

Hemery, Y., Rouau, X., Lullien-Pellerin, V. et al. 2007. Dry processes to develop wheat fractions and products with enhanced nutritional quality. *Journal of Cereal Science*. 46(3): 327–347.

Hole, A. S., Kjos, N. P., Grimmer, S. et al. 2013. Extrusion of barley and oat improves the bioaccessibility of dietary phenolic acids in growing pigs. *Journal of Agricultural and Food Chemistry*. 61(11): 2739–2747.

Ignat, I., Volf, I., and Popa, V. I. 2011. A critical review of methods for characterisation of polyphenolic compounds in fruits and vegetables. *Food Chemistry*. 26(4): 1821–1835.

Jin, A., Ozga, J. A., Lopes-Lutz, D. et al. 2012. Characterization of proanthocyanidins in pea (*Pisum sativum* L.), lentil (*Lens culinaris* L.), and faba bean (*Vicia faba* L.) seeds. *Food Research International*. 46(2): 528–535.

Jones, J. 2006. Grain-based foods and health. *Cereal Foods World*. 51(3): 108–113.

Kaur, K. D., Jha, A., Sabikhi, L. et al. 2014. Significance of coarse cereals in health and nutrition: A review. *Journal of Food Science and Technology*. 51(8): 1429–1441.

Kim, J. Y., Do, M. H., and Lee, S. S. 2006. The effects of a mixture of brown and black rice on lipid profiles and antioxidant status in rats. *Annals of Nutrition and Metabolism*. 50(4): 347–353.

Kim, K.-H., Tsao, R., Yang, R., and Cui, S. W. 2006. Phenolic acid profiles and antioxidant activities of wheat bran extracts and the effect of hydrolysis conditions. *Food Chem*. 95(3): 466–473.

Leopoldini, M., Russo, N., and Toscano, M. 2011. The molecular basis of working mechanism of natural polyphenolic antioxidants. *Food Chemistry*. 125(2): 288–306.

Lin, J.-K., and Weng, M.-S. 2006. Flavonoids as nutraceuticals. In *The Science of Flavonoids*, ed. E. Grotewold, pp. 213–238. New York: Springer.

Liu, R. H. 2007. Whole grain phytochemicals and health. *Journal of Cereal Science*. 46(3): 207–219.

López-Amorós, M. L., Hernández, T., and Estrella, I. 2006. Effect of germination on legume phenolic compounds and their antioxidant activity. *Journal of Food Composition and Analysis*. 19(4): 277–283.

Madhujith, T., Amarowicz, R. and Shahidi, F. 2004. Phenolic antioxidants in beans and their effects on inhibition of radical-induced DNA damage. *Journal of the American Oil Chemists' Society*. 81(7): 691–696.

Magalhaes, S. C., Taveira, M., Cabrita, A. R. et al. 2017. European marketable grain legume seeds: Further insight into phenolic compounds profiles. *Food Chemistry*. 215: 177–184.

Manach, C., Scalbert, A., Morand, C. et al. 2004. Polyphenols: Food sources and bioavailability. *The American Journal of Clinical Nutrition*. 79(5): 727–747.

Masisi, K., Beta, T., and Moghadasian, M. H. 2016. Antioxidant properties of diverse cereal grains: A review on *in vitro* and *in vivo* studies. *Food Chem*. 196: 90–97.

Mattila, P., Pihlava, J.-M., and Hellström, J. 2005. Contents of phenolic acids, alkyl- and alkenylresorcinols, and avenanthramides in commercial grain products. *Journal of Agricultural and Food Chemistry*. 53(21): 8290–8295.

Mazza, G., and Gao, L. 2005. Blue and purple grains. In *Speciality Grains for Food and Feed*, ed. A. International, pp. 313–350. St. Paul, MN: American Association of Cereal Chemists.

McDonough, C. M., Rooney, L. W., and Serna-Saldivar, S. O. 2000. The millets. In *Handbook of Cereal Science and Technology*, eds. K. Kulp and J. G. Ponte, pp. 177–201. New York: CRC Press.

Monks, J. L. F., Vanier, N. L., Casaril, J. et al. 2013. Effects of milling on proximate composition, folic acid, fatty acids and technological properties of rice. *Journal of Food Composition and Analysis*. 30(2): 73–79.

Naczk, M., and Shahidi, F. 2004. Extraction and analysis of phenolics in food. *Journal of Chromatography A*. 1054(1): 95–111.
Naczk, M., and Shahidi, F. 2006. Phenolics in cereals, fruits and vegetables: Occurrence, extraction and analysis. *Journal of Pharmaceutical and Biomedical Analysis*. 41(5): 1523–1542.
Nithiyanantham, S., Selvakumar, S., and Siddhuraju, P. 2012. Total phenolic content and antioxidant activity of two different solvent extracts from raw and processed legumes, *Cicer arietinum* L. and *Pisum sativum* L. *Journal of Food Composition and Analysis*. 27(1): 52–60.
Olmedilla-Alonso, B., Farré-Rovira, R., Asensio-Vegas, C. et al. 2010. Papel de las leguminosas en la alimentación actual. *Revista Española de Nutrición Humana y Dietética*. 14(2): 72–76.
Paiva, F., Vanier, N., Berrios, J. D. J. et al. 2014. Physicochemical and nutritional properties of pigmented rice subjected to different degrees of milling. *Journal of Food Composition and Analysis*. 35(1): 10–17.
Paiva, F. F., Vanier, N. L., Berrios, J. D. J. et al. 2016. Polishing and parboiling effect on the nutritional and technological properties of pigmented rice. *Food Chemistry*. 191: 105–112. doi: https://doi.org/10.1016/j.foodchem.2015.02.047
Pandey, K. B., and Rizvi, S. I. 2011. Anti-oxidative action of resveratrol: Implications for human health. *Arabian Journal of Chemistry*. 4(3): 293–298.
Pedreschi, R., and Cisneros-Zevallos, L. 2007. Phenolic profiles of Andean purple corn (*Zea mays* L.). *Food Chemistry*. 100(3): 956–963.
Peñalvo, J. L., Haajanen, K. M., Botting, N. et al. 2005. Quantification of lignans in food using isotope dilution gas chromatography/mass spectrometry. *Journal of Agricultural and Food Chemistry*. 53(24): 9342–9347.
Peterson, D. M., Hahn, M. J., and Emmons, C. L. 2002. Oat avenanthramides exhibit antioxidant activities in vitro. *Food Chemistry*. 79(4): 473–478.
Pianjing, P., Thiantanawat, A., Rangkadilok, N. et al. 2010. Estrogenic activities of sesame lignans and their metabolites on human breast cancer cells. *Journal of Agricultural and Food Chemistry*. 59(1): 212–221.
Ramos-Escudero, F., Muñoz, A. M., Alvarado-Ortíz, C., Alvarado, A., and Yánez, J. A. 2012. Purple corn (*Zea mays* L.) phenolic compounds profile and its assessment as an agent against oxidative stress in isolated mouse organs. *Journal of Medicinal Food*. 15(2): 206–215.
Reddy, A. M., Reddy, V. S., Scheffler, B. E. et al. 2007. Novel transgenic rice overexpressing anthocyanidin synthase accumulates a mixture of flavonoids leading to an increased antioxidant potential. *Metabolic Engineering*. 9(1): 95–111.
Reddy, N., Pierson, M., Sathe, S. et al. 1985. Dry bean tannins: A review of nutritional implications. *Journal of the American Oil Chemists Society*. 62(3): 541–549.
Renger, A., and Steinhart, H. 2000. Ferulic acid dehydrodimers as structural elements in cereal dietary fibre. *European Food Research and Technology*. 211(6): 422–428.
Soleas, G. J., Diamandis, E. P., and Goldberg, D. M. 1997. Resveratrol: A molecule whose time has come? And gone? *Clinical Biochemistry*. 30(2): 91–113.
Stratil, P., Klejdus, B., and Kubáň, V. 2006. Determination of total content of phenolic compounds and their antioxidant activity in vegetables evaluation of spectrophotometric methods. *Journal of Agricultural and Food Chemistry*. 54(3): 607–616.

Tian, S., Nakamura, K., Cui, T., and Kayahara, H. 2005. High-performance liquid chromatographic determination of phenolic compounds in rice. *Journal of Chromatography A*. 1063(1–2): 121–128.

Tian, S., Nakamura, K., and Kayahara, H. 2004. Analysis of phenolic compounds in white rice, brown rice, and germinated brown rice. *Journal of Agricultural and Food Chemistry*. 52(15): 4808–4813.

Troszyńska, A., Estrella, I., López-Amóres, M. L. et al. 2002. Antioxidant activity of pea (*Pisum sativum* L.) seed coat acetone extract. *LWT-Food Science and Technology*. 35(2): 158–164.

Walter, M., Marchesan, E., Massoni, P., da Silva, L., Sartori, G., and Ferreira, R. 2013. Antioxidant properties of rice grains with light brown, red and black pericarp colors and the effect of processing. *Food Research International*. 50(2): 698–703.

Wu, X., and Prior, R. 2005. Systematic identification and characterization of anthocyanins by HPLC-ESI-MS/MS in common foods in the United States: Fruits and berries. *Journal of Agricultural and Food Chemistry*. 53(7): 2589–2599.

Xing, Y., and White, P. 1997. Identification and function of antioxidants from oat groats and hulls. *Journal of the American Oil Chemists' Society*. 74(3): 303–307.

Xu, B., and Chang, S. 2010. Phenolic substance characterization and chemical and cell-based antioxidant activities of 11 lentils grown in the Northern United States. *Journal of Agricultural and Food Chemistry*. 58(3): 1509–1517.

Yu, J., Vasanthan, T., and Temelli, F. 2001. Analysis of phenolic acids in barley by high-performance liquid chromatography. *Journal of Agricultural and Food Chemistry*. 49(9): 4352–4358.

Yu, L. L., Zhou, K., and Parry, J. W. 2005. Inhibitory effects of wheat bran extracts on human LDL oxidation and free radicals. *LWT-Food Science and Technology*. 38(5): 463–470.

Zaupa, M., Calani, L., Del Rio, D., Brighenti, F., and Pellegrini, N. 2015. Characterization of total antioxidant capacity and (poly) phenolic compounds of differently pigmented rice varieties and their changes during domestic cooking. *Food Chemistry*. 187: 338–347.

Zhang, H., Shao, Y., Bao, J., and Beta, T. 2015. Phenolic compounds and antioxidant properties of breeding lines between the white and black rice. *Food Chemistry*. 172: 630–639.

Zhou, Z., Robards, K., Helliwell, S. et al. 2004. The distribution of phenolic acids in rice. *Food Chemistry*. 87(3): 401–406.

CHAPTER 17

Phenolic Compounds in Herbs and Spices

Laura A. de la Rosa, Nina del Rocío Martínez-Ruíz, J. Abraham Domínguez-Avila, and Emilio Alvarez-Parrilla

CONTENTS

17.1 INTRODUCTION

Spice (from Latin, *specĭes*) is often used to refer to certain aromatic vegetable products used whole, sliced, or powdered for seasoning food to give them special flavor and scent. The concept of seasoning includes spices, but this term can also refer to products that do not have a direct vegetable origin such as salt, vinegar, beer, and others. The hard parts of certain aromatic plants, such as seeds or bark, are technically considered a spice. Aromatic herbs are sometimes confused with spices; their separation is not very clear, although flavors and scents in aromatic herbs are less concentrated and softer than in spices. Herbs and spices added to food can be fresh, dried, or industrially manufactured. Small quantities provide odors and flavors, generating a tasty and palatable food (Fálder-Rivero 2005).

The use of aromatic herbs and spices (H&S) is practically as old as humanity. Primitive humans collected all kinds of fruits and herbs. There were harmless and poisonous herbs, and with this, their recollection became selective. These people used many plants for medicinal purposes and to accelerate the death of hunted animals. The majority of spices can be considered native from tropical regions of Asia, and the Moluccan Islands in Indonesia, also known as the "Spice Islands." Some spices were grown in the Mediterranean (anise, mustard) and others were taken to Europe by the pathfinder of

America (e.g., vanilla, chili peppers, cocoa, and annatto). In China, Emperor Shen Nung (3000 BC), compiled the knowledge of aromatic and medicinal plants. The Sumerians used many aromatic plants in the preparation of food and medicines. This knowledge was taken to Mesopotamia and Egypt. Aromatic herbs such as laurel and thyme were cultivated in Babylon. Some exotic spices, such as saffron and cinnamon, migrated to Egypt and Greece from the East. Sesame, poppy seeds, coriander, and cumin entered Greece from Minor Asia. The Phoenicians spread spices and aromatic herbs such as onion, garlic, mint, peppermint, thyme, rosemary, oregano, coriander, dill, and others in the Mediterranean. They used some plants to prevent the spoilage of food. In the Middle Ages, many spices and herbs from the East (e.g., cinnamon, clove, pepper, ginger, saffron, and cardamom) were highly sought due to the difficult transport of those spices during the fall of the Roman Empire. The Arabs of North Africa carried spices and herbs from the Far East to Europe. Meanwhile, Marco Polo, through the East, and Columbus, through the West, tried to obtain spices from Asian producers. In the seventeenth century, the Netherlands, Portugal, Italy, and England monopolized the spice commerce, while the growing of different aromatic plants such as mustard began to spread in Central Europe (Civitello 2008; Fálder-Rivero 2005). Finally, in the eighteenth and nineteenth centuries, the active ingredients contained in many herbs and spices began to be synthesized in laboratories (Civitello 2008). Today, spices like oregano and rosemary are used as preservatives and antioxidants in industrially manufactured foods such as sausages and other meat products. Herbs and spices are an important group because they contain active principles like alkaloids, essences, flavonoid glycosides, tannins, minerals, polyphenolic compounds (PCs), and others that may bring benefits to human health (Cameroni 2012).

Culinary use of herbal extracts and essential oils have increased in popularity as natural sources of preservative agents, mainly due to the extensive growth of these plants and their effective and safe consumption. Spices such as garlic, onion, cinnamon, clove, and mustard have been effective in inhibiting the microbial spoilage of food (Stankevičius et al. 2011). Parsley, fresh or dried, inhibits the growth of *Listeria monocytogenes* L. and *Escherichia coli* (Manderfeld et al. 1997). However, the functional activity of PCs contained in herbs and spices is influenced by factors like food culture and sensory preference of individuals (Mercado-Mercado et al. 2013). An average PC consumption of 1 g/day was suggested 25 years ago (Scalbert and Williamson 2000). However, there is limited data that reveals with certainty the amount and type of PC consumed today. The profiles and identification of PCs in herbs and spices are still limited due mainly to the wide variety of structures of the natural compounds, the lack of availability of commercial standards (Vallverdú-Queralt et al. 2014), and the different methods used to identify and quantify these compounds (Stankevičius et al. 2011). The aim of this chapter is to present relevant information about the main PCs, their biological activity, and the proximate analysis and vitamin content found in the herbs and spices of greatest use in the world. In order to systematize this analysis, the USDA National Nutrient Database for Standard Reference was used.

17.2 NUTRITIONAL CHARACTERISTICS

One of the most complete available compositional sets of information on the nutritional properties of H&S is that published by the USDA (2014). Table 17.1 summarizes the compositional and nutritional characteristics of the main H&S consumed in the United States and other Western countries (USDA 2014). As previously stated, H&S are a wide

TABLE 17.1 Macronutrient Contents (g/100 g) in Herbs and Spices

Herbs and Spices	Water	Protein	Lipids	Carbohydrates	Fiber
Annatto paste[a]	60.60	3.10	2.80	22.60	
Anise seed	9.54	17.60	15.90	50.02	14.60
Basil (fresh)	92.06	3.15	0.64	2.65	1.60
Basil dried	10.35	22.98	4.07	47.75	37.70
Bay leaf	5.44	7.61	8.36	74.97	26.30
Cardamom	8.28	10.76	6.7	68.47	28.00
Celery seed	6.04	18.07	25.27	41.35	11.80
Cinnamon (ground)	10.58	3.99	1.24	80.59	53.10
Clove (ground)	9.87	5.97	13.00	65.53	33.90
Coriander leaf (dried)	7.30	21.93	4.78	52.1	10.40
Coriander seed	8.86	12.37	17.77	54.99	41.90
Cumin seed	8.06	17.81	22.27	44.24	10.50
Dill seed	7.70	15.98	14.54	55.17	21.10
Dill weed (dried)	7.30	19.96	4.36	55.82	13.60
Dill weed (fresh)	85.95	3.46	1.12	7.02	2.10
Epazote (raw)	89.21	0.33	0.52	7.44	3.80
Fennel seed	8.81	15.8	14.87	52.29	39.8
Garlic powder	6.45	16.55	0.73	72.73	9.00
Ginger (ground)	9.94	8.98	4.24	71.62	14.10
Mustard seed (ground)	5.27	26.08	36.24	28.09	12.20
Nutmeg (ground)	6.23	5.84	36.31	49.29	20.80
Onion powder	5.39	10.41	1.04	79.12	15.20
Oregano (dried)	9.93	9.00	4.28	68.92	42.50
Paprika	11.24	14.14	12.89	53.99	34.90
Parsley (dried)	5.89	26.63	5.48	50.64	26.70
Parsley (fresh)	87.71	2.97	0.79	6.33	3.30
Dry pepper (Pasilla)	14.84	12.35	15.85	51.13	26.80
Pepper	12.46	10.39	3.26	63.95	25.30
Pepper (red)	8.05	12.01	17.27	56.63	27.20
Peppermint (fresh)	78.65	3.75	0.94	14.89	8.00
Rosemary (dried)	9.31	4.88	15.22	64.06	42.60
Rosemary (fresh)	67.77	3.31	5.86	20.7	14.10
Saffron	11.90	11.43	5.85	65.37	3.90
Sage (ground)	7.96	10.63	12.75	60.73	40.30
Star anise[b]		0.12			
Tarragon (dried)	7.74	22.77	7.24	50.22	7.40
Thyme (dried)	7.79	9.11	7.43	63.94	37.00
Thyme (fresh)	65.11	5.56	1.68	24.45	14.00
Turmeric (ground)	12.85	9.68	3.25	67.14	22.70
Vanilla extract	52.58	0.06	0.06	12.65	

Source: All data from the U.S. Department of Agriculture, USDA National Nutrient Database for Standard Reference, Release 27, 2014, except where indicated:

[a] Alvarez-Parrilla, E. et al., *Food Science and Technology Campinas*, 34, 371–378, 2014.

[b] Dinesha, R. et al., *Journal of Pharmacology and Phytochemistry*, 2, 98–103, 2014.

range of fresh and dry seeds, leafs, flowers, and other plant parts. According to water content, H&S can be classified as fresh (water content higher than 65%), dry (water content lower than 23%), and seeds (water content lower than 10%). Annatto paste (60.6% water content) is an exception to this behavior, and can be explained considering that annatto paste is a complex mixture that includes annatto seeds mixed with corn starch, ground pepper, salt, and other condiments. Protein content varies from 0.06 g/100 g in vanilla extract to 26.63 g/100 g in dried parsley. In general terms, seeds and dry H&S show higher protein content compared with fresh samples. Lipid content varies from 0.06 g/100 g in vanilla extract to 36.31 g/100 g in nutmeg. Interestingly, vanilla extract showed the lowest protein and lipid content, mainly because this sample is an ethanolic extract. As in the case of proteins, seed showed the higher lipid content, followed by dry H&S and finally fresh samples. Carbohydrate content ranged from 2.65 g/100 g in fresh basil to 80.59 g/100 g in cinnamon, and showed the same pattern in lipids and proteins. Of particular interest is the fact that most herbs and spices contain at least a modest fiber amount and a low sugar content, which makes them an adequate option to include as part of a healthy diet.

17.3 MAIN PHENOLIC COMPOUNDS FOUND IN H&S

H&S are, in general terms, good sources of phenolic compounds and other bioactive phytochemicals. For this reason, many of them have been traditionally used not only as food ingredients but also as natural remedies. Table 17.2 summarizes the information reported by several authors on the content of total phenolic compounds, total flavonoids, and the major phenolic compounds found in some of the most popular herbs and spices around the world. All data are reported in dry weight basis and, when possible, in the same units, in order to make comparison possible.

Phenolic compounds are a diverse group of phytochemicals containing at least one hydroxyl group in an aromatic ring. Classes of phenolic compounds include the phenolic acids (hydroxyphenolic and hydroxycinnamic acids) and their derivatives, flavonoids (flavones, isoflavones, flavonols, flavanones, flavan-3-ols, anthocyanidins, and others) and their derivatives, and other compounds with diverse structures including coumarins, stilbenes, volatile compounds, and others (Naczk and Shahidi 2004; Andrés-Lacueva et al. 2010). This structural complexity makes their identification and quantification a difficult task; moreover, since they are secondary metabolites, their contents in the plant tissues are dependent on many factors including genetics, developmental stage, environment, and growing conditions (Beato et al. 2011; Lv et al. 2012; Bae et al. 2014). Many authors studying phenolic compounds in H&S and other foods have used the Folin–Ciocalteu assay (Singleton and Rossi 1965) to quantify total phenolic compounds. This is a simple, sensitive, and reproducible spectroscopic method based on the generic reducing property of all polyphenols, which makes it quite unspecific (Pérez-Jiménez et al. 2010). This low specificity has the advantage that the method may account for almost all classes of phenolic compounds, but the disadvantage that other nonphenolic reducing compounds (ascorbic acid or sugars, for example) can interfere and cause overestimation of the content of total phenolic compounds (Naczk and Shahidi 2004). Nevertheless, the Folin–Ciocalteu assay is very useful as a first approach to compare the content of phenolic compounds in different plant samples. When assayed with the Folin–Ciocalteu method, total phenolic compounds are expressed as equivalent to a simple phenolic standard (mainly gallic acid) or to the main phenolic compound found in the sample. Table 17.2 shows that the

TABLE 17.2 Content and Profile of Phenolic Compounds Present in Herbs and Spices

Common Name	Scientific Name	Family	Total Phenolic Compounds (% of Dry Material)	Total Flavonoids (mg/g of Dry Material)	Main Polyphenolic Compounds	References
Annatto (achiote)	*Orellana americana* Kuntze	Bixaceae	0.09–0.18 (1)		*Phenolic acids*: caffeoyl derivatives. *Flavonoids*: hypolaetin.	Cardarelli et al. 2008
Anise seed	*Pimpinella anisum* L.	Apiaceae	0.05–5.39 (1)	52.6 (6)	*Phenolic acids*: protocatechuic, caffeic, gallic, ferulic and rosmarinic acids, caffeoyl derivatives. *Flavonoids*: catechin. *Other*: anethole, eugenol, epirosmanol, carvacrol.	Liu et al. 2008; Shan et al. 2005; Christova-Bagdassarian et al. 2014; Rothwell et al. 2013
Basil	*Ocimum basilicum* L.	Lamiaceae	3.64–4.32 (1)		*Phenolic acids*: rosmarinic, vanillic, gentisic, caffeic, ferulic and chlorogenic acids, caffeoyl derivatives. *Flavonoids*: catechin, kaempferol. *Other*: eugenol, epirosmanol, carvacrol, linalool, estragole, lignans.	Shan et al. 2005; Rothwell et al. 2013
Bay leaf	*Laurus nobilis* L.	Lauraceae	4.17 (1)		Phenolic acids and derivatives, flavonoids, and volatile phenols.	Shan et al. 2005; Rothwell et al. 2013
Cardamom	*Elettaria cardamomum* (L.) *Maton*	Zingiberaceae	0.46 (1)		*Phenolic acids*: caffeic, protocatechuic, and *p*-coumaric acids.	Shan et al. 2005; Rothwell et al. 2013
Celery seed	*Apium graveolens* var. *dulce* Pers.	Apiaceae		4.2(?)	*Flavonoids*: luteolin, apigenin, chrysoeriol, and their glycosides.	Lin et al. 2007; Bhagwat et al. 2014; Rothwell et al. 2013

(*Continued*)

TABLE 17.2 (CONTINUED) Content and Profile of Phenolic Compounds Present in Herbs and Spices

Common Name	Scientific Name	Family	Total Phenolic Compounds (% of Dry Material)	Total Flavonoids (mg/g of Dry Material)	Main Polyphenolic Compounds	References
Chili pepper (dry spice)	*Capsicum annuum* var. *Annuum* L.	Solanaceae	0.86–1.03 (1)	5.5 (4)	*Phenolic acids*: *p*-coumaric and ferulic acids. *Flavonoids*.	Lu et al. 2011; Alvarez-Parrilla et al. 2014; Shan et al. 2005
Cinnamon	*Cinnamomum versum* J. Presl	Lauraceae	6.34–11.90 (1)		*Phenolic acids*: 2-hydroxybenzoic, protocatechuic, syringic, caffeic, and *p*-coumaric acids. *Flavonoids*: catechin. *Other*: cinnamyl aldehydes.	Shan et al. 2005; Rothwell et al. 2013
Clove	*Syzygium aromaticum* (L.) Merr. & L.M. Perry	Myrtaceae	14.38–19.44 (1)	46.3 (6)	*Phenolic acids*: gallic, protocatechuic, syringic, and *p*-coumaric acids. *Flavonoids*: quercetin, kaempferol. *Other*: eugenol, hydrolizable tannins.	Liu et al. 2008; Shan et al. 2005; Pérez-Jiménez et al. 2010; Rothwell et al. 2013
Coriander (leaf)	*Coriandrum sativum* L.	Apiaceae	0.36–2.22 (1)		*Phenolic acids*: vanillic, *p*-coumaric, ferulic acids. *Flavonoids*: kaempferol, quercetin.	Vanisha et al. 2010; Pérez-Jiménez et al. 2010; Wangensteen et al. 2004; Rothwell et al. 2013
Coriander seed	*Coriandrum sativum* L.	Apiaceae	0.02–1.89 (1)	0.11 (?)		Wangensteen et al. 2004; Christova-Bagdassarian et al. 2014

(*Continued*)

TABLE 17.2 (CONTINUED) Content and Profile of Phenolic Compounds Present in Herbs and Spices

Common Name	Scientific Name	Family	Total Phenolic Compounds (% of Dry Material)	Total Flavonoids (mg/g of Dry Material)	Main Polyphenolic Compounds	References
Cumin seed	*Cuminum cyminum* L.	Apiaceae	0.11–2.04 (1)	0.5–5.9 (4)	*Phenolic acids*: gallic, caffeic, chlorogenic, syrigic, vanillic, *p*-coumaric, ferulic, rosmarinic, cinnamic, and other acids. *Flavonoids*: luteolin, eriodictyol, catechin, quercetin, apigenin, kaempferol.	Ani et al. 2006; Mariod et al. 2009; Rebey et al. 2012; Pérez-Jiménez et al. 2010; Rothwell et al. 2013
Dill (leaf)	*Anethum graveolens* L.	Apiaceae	1.40–13.65 (1)	37.2 (4)	*Phenolic acids*: vanillic and protocatechuic acids. *Flavonoids*: isorhamnetin, kaempferol, myricetin, quercetin, catechin.	Shyu et al. 2009; Shan et al. 2005; Stankevičius et al. 2011; Bhagwat et al. 2014; Rothwell et al. 2013
Dill seed	*Anethum graveolens* L.	Apiaceae	13.05 (1)	33.6 (4)		Shyu et al. 2009
Epazote (wormseed)	*Chenopodium ambrosioides* L.	Chenopodiaceae	0.8 (1)	2.9 (4)	*Phenolic acids*: *p*-coumaric acid. *Flavonoids*: quercetin and kaempferol glycosides.	Barros et al. 2013; Mercado-Mercado et al. 2013
Fennel seed	*Foeniculum vulgare Mill.*	Apiaceae	0.63–9.00 (1)	1.1–6.8 (4)	*Phenolic acids*: gallic, caffeic and ferulic acids. *Flavonoids*: quercetin.	Anwar et al. 2009; Przygodzka et al. 2014; Oktay et al. 2003; Rothwell et al. 2013
Garlic powder	*Allium sativum* L.	Amaryllidaceae	0.01–10.80 (1)	0.15–0.60 (6)	*Phenolic acids*: gallic, caffeic, ferulic, vanillic, *p*-coumaric, *m*-coumaric. *Flavonoids*: quercetin, apigenin, catechin, epicatechin.	Kim et al. 2013; Chen et al. 2013; Beato et al. 2011; Alarcón-Flores et al. 2014; Wongsa et al. 2012

(*Continued*)

TABLE 17.2 (CONTINUED) Content and Profile of Phenolic Compounds Present in Herbs and Spices

Common Name	Scientific Name	Family	Total Phenolic Compounds (% of Dry Material)	Total Flavonoids (mg/g of Dry Material)	Main Polyphenolic Compounds	References
Ginger	*Zingiber officinale* Roscoe	Zingiberaceae	0.06–1.08 (1); 0.70–1.58 (2)	0.34–3.8 (7); 8.0 (4)	*Phenolic acids*: caffeic acid. *Other*: Gingerols and shogaols (0.13–8.33 g/100g of dry material).	Oboh et al. 2012; Przygodzka et al. 2014; Pawar et al. 2011; Ranilla et al. 2010; Cheng et al. 2011; Schwertner and Rios 2007; Rothwell et al. 2013
Mustard seed	*Brassica juncea* (L.) Czern.	Brassicaceae	0.78–1.37 (3)		*Phenolic acids*: sinapic acid. *Other*: sinapine.	Dubie et al. 2013; Engels et al. 2012
Nutmeg	*Myristica fragrans* Houtt.	Myristicaceae	0.26–1.08 (1)	10.0 (4)	*Phenolic acids*: protocatechuic, syringic, caffeic, and *p*-coumaric acids. *Flavonoids*: catechin. *Other*: volatile phenolics.	Su et al. 2007; Shan et al. 2005; Przygodzka et al. 2014; Rothwell et al. 2013
Onion powder	*Allium cepa* L.	Amaryllidaceae	0.46–7.41 (1)	2.2–17.7 (6)	*Phenolic acids*: gallic and ferulic acids. *Flavonoids*: quercetin and kaempferol (and their glycosides).	Prakash et al. 2007; Cheng et al. 2013; Stankevičius et al. 2011
Oregano	*Origanum vulgare* L.	Lamiaceae	0.22–23.50 (1)	57.1–132.0 (6)	*Phenolic acids*: rosmarinic, caffeic, chlorogenic, ferulic, protocatechuic, *p*-coumaric, *p*-hydroxybenzoic, syringic, gallic, vanillic. *Flavonoids*: luteolin, apigenin, quercetin, kaempferol, myricetin. *Other*: lignans.	Vallverdú-Queralt et al. 2014; Teixeira et al. 2013; Aranha and Jorge 2012; Licina et al. 2013; Martins et al. 2014; Danila et al. 2011; Rababah et al. 2010; Bhagwat et al. 2014; Rothwell et al. 2013

(*Continued*)

TABLE 17.2 (CONTINUED) Content and Profile of Phenolic Compounds Present in Herbs and Spices

Common Name	Scientific Name	Family	Total Phenolic Compounds (% of Dry Material)	Total Flavonoids (mg/g of Dry Material)	Main Polyphenolic Compounds	References
Paprika	*Capsicum annuum* L.	Solanaceae	1.36 (1); 0.24 (4)		*Flavonoids*: quercetin, luteolin, kaempferol, apigenin, and myricetin.	Vega-Gálvez et al. 2009; Park et al. 2011; Bae et al. 2014
Parsley	*Petroselinum crispum* (Mill.) Fuss	Apiaceae	0.97–2.92 (1)		*Phenolic acids*: 4-hydroxybenzoic, gallic, gentisic, vanillic, *p*-coumaric, and caffeic acid. *Flavonoids*: apigenin, luteolin, isorhamnetin. *Other*: furanocoumarins.	Pérez-Jiménez et al. 2010; Shan et al. 2005; Bhagwat et al. 2014; Rothwell et al. 2013
Pepper, black, green, or white	*Piper nigrum* L.	Piperaceae	0.30–2.80 (1)	3.0–23.6 (7)	Phenolic amides	Pérez-Jiménez et al. 2010; Shan et al. 2005
Pepper, red or cayenne	*Capsicum annuum* L. var Cayenne	Solanaceae	0.88–1.70 (4)		*Flavonoids*: quercetin, luteolin, kaempferol, apigenin, and myricetin.	Bae et al. 2012, 2014
Peppermint	*Mentha x piperita* L.	Lamiaceae	2.58–19.12 (1)	8.6–43.3 (4)	*Phenolic acids*: rosmarinic, caffeic, syringic, *p*-coumaric, *o*-coumaric, ferulic, chlorogenic. *Flavonoids*: hesperidin, hesperetin, quercetin, luteolin, apigenin, other flavones, catechin, gallocatechin-gallate, epigallocatechin-gallate, eriocitrin, eriodictyol.	Pérez et al. 2014; Capecka et al. 2005; Lv et al. 2012; Riachi et al. 2015; Bhagwat et al. 2014; Rothwell et al. 2013

(*Continued*)

TABLE 17.2 (CONTINUED) Content and Profile of Phenolic Compounds Present in Herbs and Spices

Common Name	Scientific Name	Family	Total Phenolic Compounds (% of Dry Material)	Total Flavonoids (mg/g of Dry Material)	Main Polyphenolic Compounds	References
Rosemary	*Rosmarinus officinalis* L.	Lamiaceae	2.52–9.4 (1)	15.0 (7)	*Phenolic acids*: syringnic, vanillic, chlorogenic, caffeic, ferulic, *p*-coumaric, and rosmarinic acids. *Flavonoids*: naringenin, luteolin and catechin. *Other*: carnosic acid, carnosol, carvacrol.	Pérez-Jiménez et al. 2010; Shan et al. 2005; Bhagwat et al. 2014; Rothwell et al. 2013
Saffron	*Crocus sativus* L.	Iridaceae	0.65–0.83 (1)	3.5–3.6 (6)	*Flavonoids*: naringenin, kaempferol, and taxifolin	Baba et al. 2015; Rothwell et al. 2013
Sage	*Salvia officinalis* L.	Labitae or Lamiaceae	2.92–5.32 (1)		*Phenolic acids*: gallic, syringic, vanillic, 5-caffeoylquinic, caffeic, ferulic, *p*-coumaric, and rosmarinic acids. *Flavonoids*: apigenin, luteolin. *Other*: carnosic acid, phenolic volatiles.	Pérez-Jiménez et al. 2010; Shan et al. 2005; Bhagwat et al. 2014; Rothwell et al. 2013
Star anise	*Illicium verum* Hook.f.	Illiciaceae	1.49–17.00 (1); 23.77 (4)	14.00–115.84 (4)	*Phenolic acids*: protocatechuic, caffeic and *p*-coumaric acids. *Flavonoids*: rutin, kaempferol, quercetin. *Other*: anethole.	Lu et al. 2011; Kannatt et al. 2014; Hoque et al. 2011; Chung 2009; Padmashree et al. 2007; Przygodzka et al. 2014; Ohira et al. 2009; Rothwell et al. 2013

(*Continued*)

TABLE 17.2 (CONTINUED) Content and Profile of Phenolic Compounds Present in Herbs and Spices

Common Name	Scientific Name	Family	Total Phenolic Compounds (% of Dry Material)	Total Flavonoids (mg/g of Dry Material)	Main Polyphenolic Compounds	References
Tarragon	*Artemisia dracunculus* L.	Asteraceae	0.57–10.2 (1)	22.0 (7)	*Flavonoids*: luteolin, isorhamnetin, kaempferol, quercetin. *Other*: coumarins, estragole.	Pérez-Jiménez et al. 2010; Eisenman et al. 2011; Bhagwat et al. 2014
Thyme	*Thymus vulgaris* L.	Lamiaceae	1.82–15.7 (1)	34.0 (7)	*Phenolic acids*: gallic, syringic, vanillic, caffeic, ferulic, *p*-coumaric, and rosmarinic acids. *Flavonoids*: apigenin, luteolin. *Other*: thymol, phenolic diterpenes.	Pérez-Jiménez et al. 2010; Shan et al. 2005; Bhagwat et al. 2014; Rothwell et al. 2013
Turmeric	*Curcuma longa* L.	Zingiberaceae	2.00–9.18 (5)		*Phenolic acids*: caffeic, *p*-coumaric, and protocatechuic acids. *Other*: curcuminoids (curcumin, demethoxycurcumin and bisdemethoxycurcumin).	Jayaprakasha et al. 2002; Suhaj 2006
Vanilla extract	*Vanilla planifolia* Andrews	Orchidaceae	0.50 (1)	16.5 (4)	*Phenolic acids*: vanillic acid. Other: vanillin, 4-hydroxybenzyl alcohol, 4-hydroxybenzaldehyde.	Maruenda et al. 2013; Salazar-Rojas et al. 2012; Przygodzka et al. 2014; Sinha et al. 2007; Pérez-Silva et al. 2011

Note: (1) GAE (gallic acid equivalents); (2) TAE (tannic acid equivalents); (3) SAE (sinapic acid equivalents); (4) CE (catechin equivalents); (5) Total curcuminoids; (6) RE (rutin equivalents); (7) QE (quercetin equivalents).

content of total phenolic compounds varies widely among commonly consumed H&S. Clove (*Syzygium aromaticum* L.), oregano (*Origanum vulgare* L.), peppermint (*Mentha x piperita* L.), and star anise (*Illicium verum* Hook f.) are among the spices with the highest contents of total phenolic compounds (close to 20% gallic acid equivalents [GAE] in dry weight); however, many other spices are also good sources of these compounds, containing around 5–15% in dry weight (anise, cinnamon, dill, fennel, garlic powder, onion powder, rosemary, sage, tarragon, thyme, and turmeric). Actually, clove, peppermint, and star anise are the three foods richest in polyphenolic compounds according to the Phenol-Explorer database (Rothwell et al. 2013), and many other spices are among the 100 richest dietary sources of polyphenols (Pérez-Jiménez et al. 2010). Important variation can be observed also in the values of total phenolic compounds found for the same herb or spice by different authors or even in a single study. For example, fennel seeds have been reported to contain between 0.63 and 9.00 g GAE/g of dry weight, depending on the sample (whole seeds or commercial spices) and solvent used for extraction (water, methanol, ethanol, or water/alcohol combinations). The same is true for almost all H&S listed in Table 17.2. The most common sources of natural variation are plant variety or type of sample, and the most common experimental factor of variation is the type of solvent used for extraction. In general, all studies show that the best solvents for the extraction of phenolic compounds from H&S samples are water or alcohol/water (50/50 or 80/20) combinations.

Among the different classes of phenolic compounds, flavonoids are remarkable for their bioactivity and are widely distributed among the plant kingdom. They comprise several structural classes and in nature are found mostly glycosylated in different positions and with different types and numbers of sugar moieties. Identification and quantification of individual flavonoids is usually carried out by HPLC, and many times only the aglycone component of the molecule can be identified, since commercial standards for the glycosides are difficult to obtain. However, a simple spectroscopic method has been developed, and is sometimes used to quantify total flavonoids by their reaction with sodium nitrate followed by the formation of a flavonoid–aluminum complex, which is detected at 510 nm (Zhishen et al. 1999). Total flavonoids determined by the aluminum complexation method in spices are reported as equivalents of catechin (a flavan-3-ol), quercetin (a flavonol), and rutin (a quercetin glycoside), which are among the most widely distributed flavonoid compounds. Their contents range from under 1 milligram/gram of dry sample (in coriander and cumin seed, and garlic powder) to over 100 milligrams/gram in oregano and star anise, which are also high in total phenolic compounds.

Individual phenolic compounds (flavonoids and other classes) have been identified in herbs and spices by means of HPLC and, more recently, HPLC/MS. As mentioned before, HPLC analysis is hampered by a lack of standards due to the great diversity of individual phenolic compounds in the plant kingdom. Some phenolic compounds are fairly common in almost all herbs and spices, and actually in almost all plant foods, like the flavonoid quercetin and the hydroxycinammic acid caffeic acid; however, they are almost always present in the form of derivatives. Caffeic acid is usually sterified with quinic acid forming chlorogenic acid or other caffeoyl-quinic derivatives; quercetin is always glycosylated (Andrés-Lacueva et al. 2010). In H&S, the flavones luteolin and apigenin are some of the most common flavonoids (see Table 17.2) and the presence of volatile phenols, such as eugenol, anethole, and carvacrol, is also frequent. In addition to to these ubiquitous phenolic compounds, some families of H&S contain some unique phenolic components, many of them with strong biological activity.

Rosmarinic acid is a hydroxycinnamic acid abundant in spices from the Lamiaceae family (basil, oregano, peppermint, rosemary, sage, thyme) and some of the Apiaceae family (anise and cumin). Other characteristic compounds of some species of the family Lamiaceae are the phenolic terpenoids carnosol, carnosic acid (rosemary and sage), and thymol (thyme). Peppermint also contains the uncommon flavonoid eriodictyol.

Sinapic acid and its derivative sinapine are the major phenolic compounds in mustard, from the Brassicaceae family, but cannot be found in other spices although they are also present in other foods, but are not common (Rothwell et al. 2013). Vanilla extract (*Vanilla planifolia* Andrews, family Orchidaceae) is recognized for its unique component vanillin, derived from vanillic acid (which is, in contrast, a quite common hydroxybenzoic acid) and phenolic alcohols and aldehydes.

Finally, spices from the Zingiberaceae family possess some of the most remarkable phenolic compounds. Ginger (*Zingiber officinale* Roscoe) contains gingerols and shogaols, while turmeric, the dried rhizome of *Curcuma longa L.*, contains high amounts (up to 9.18%) of curcuminoids, unique polyphenolic compounds, with strong biological activity.

17.4 HEALTH EFFECTS OF H&S

Recently, H&S have been considered an important source of bioactive compounds like polyphenols with benefits to human health (i.e., stimulants digestive, anti-inflammatory, antimicrobial, and anticarcinogenic activity, among others). As formerly reported, H&S present a high content of PCs such as flavonoids and hydroxycinnamic acids, which have shown biological effects in the chelation for heavy metals, scavenging radicals, inhibition of cell proliferation, enzyme modulation, and signal transduction (Vallverdú-Queralt et al. 2014). The radicals scavenging activity of PCs (antioxidant capacity) has been associated with vasodilator, vascular protective, antithrombotic, antilipemic, antiatherosclerotic, and antiapoptotic effects (Mercado-Mercado et al. 2013). An indirect protective effect of PCs has been proposed, in which PCs reduce food lipid oxidation (conjugated dienes, hydroperoxides, hexanal) during cooking and storage, preventing the formation of free radicals and cytotoxic and genotoxic compounds in the gastrointestinal tract (Mercado-Mercado et al. 2013). In this context, annatto and dry hot peppers showed protection against chopped cooked pork meat stored at 4 °C for 16 days, indicating that the use of spices or spice extracts may protect meat against lipid oxidation (Alvarez-Parrilla et al. 2014). Even though several H&S have been used in traditional medicine, the number of clinical studies to prove their activity is scarce. Among H&S, garlic and turmeric may be two of the most studied spices. In the following section the effect of H&S on several diseases will be described.

17.4.1 Cardiovascular Diseases (CD)

The beneficial effects of H&S on CD have been related to the decrease on plasma triglycerides, cholesterol, and LDL-cholesterol, as well as platelet aggregation and LDL-oxidation inhibition (Viuda-Martos et al. 2011). Lee et al. (2006) reported that carnasol, a phenolic diterpene present in H&S from the Laminaceae family (basil, oregano, peppermint, rosemary, sage, thyme) inhibited platelet aggregation by inhibition of the thromboxane A2 receptor. Other studies have associated the presence of flavonoids (quercetin, luteolin, and kaempherol) and chlorogenic acid–impeded platelet aggregation (Viuda-Martos et al. 2011). Srinivasan (2005) reported that the consumption of large amounts of garlic and onion reduced total cholesterol in humans. Meta-analysis and clinical trials showed that supplementation with curcumin, the main component of turmeric, significantly reduced the risk factor of cardiovascular disease and has been effective against atherosclerosis, myocardial infarction, platelet aggregation, and LDL-oxidation (Prasad et al. 2014).

17.4.2 Cancer

Even though clinical studies have not conclusively correlated the consumption of H&S with protection against cancer, several in vitro and animal studies have demonstrated the protective effect of H&S extracts against several cancers (Wargovich et al. 2001; Viuda-Martos et al. 2011). This protective effect of H&S phenolic compounds against cancer has been related to different mechanisms which include: (1) scavenging of ROS species, (2) scavenging of carcinogenic species, (3) inhibition of phase I enzymes, (4) induction of phase II enzymes, and (5) induction of apoptosis, among others (Viuda-Martos et al. 2011; Mercado-Mercado et al. 2013). Several authors have suggested that capsaicin, the main phytochemical of peppers, promotes apoptosis of carcinogenic cells, probably through a stress-related mechanism (Sánchez et al. 2008). Carnosol, a phenolic diterpene found in rosemary, showed anticarcinogenic activity in leukemia cells due to cell apoptosis caused by mitochondrial membrane depolarization (Dörrie et al. 2001). Probably one of the most studied H&S phenolic compounds as anticarcinogenic is curcumin. It has been demonstrated its action against leukemia, lymphoma, melanoma, sarcoma, as well as gastrointestinal, breast, ovarian, lung, and neurological cancers, among others, by modulating multiple cell signaling pathways (Prasad et al. 2014).

17.4.3 Inflammation

Probably one of the most-studied effects of H&S PCs is their anti-inflammatory activity. There are several in vitro, in vivo, and clinical trials that demonstrate the anti-inflammatory effect of extracts or pure PCs (curcumin, eugenol, capsaicin) (Srinivasan 2005). Currently, there are several topical creams and tablets used to treat inflammatory processes associated with rheumatism, backache, skin rashes, and as odontological analgesics, in which extracts or pure PC such as capsaicin are used. The administration of a single dose of curcumin, eugenol, or capsaicin reduced by about 50 percent the carrageenan-induced inflammation in rats. Several H&S extracts (bay leaf, anise, lemon myrtle, pepper) and pure PCs (curcumin, galangin, quercetin) have shown anti-inflammatory activity in cell and rat models through the inhibition of COX-2, COX-1, and iNOS enzymes (Viuda-Martos et al. 2011; Guo et al. 2014; Prasad et al. 2014). Srinvansan et al. (2005) proposed that the anti-inflammatory activity of H&S is due to the phenolic compounds that reduce the synthesis of prostaglandin E_2 and other proinflammatory compounds.

17.4.4 Obesity

In the last few years there has been an increasing interest in studying the effect of H&S in obesity and metabolic syndrome control, because these two diseases are becoming prime health issues in both developed and developing countries. According to the World Health Organization, every year 2.8 million people died from overweight- and obesity-related diseases (WHO 2013). Several mechanisms of action have been proposed to explain the antiobesity effect of H&S. They are known to activate the gastric system, increasing salivation and release of gastric and bile juices, which favor the digestion process and exert a thermogenic effect (Viuda-Martos et al. 2011). Another mechanism of action of polyphenolic compounds present in H&S in weight control, is that these compounds are able to inhibit pancreatic lipase through a noncompetitive pattern, decreasing lipid absorption (Wu et al. 2013). In vitro and in vivo studies have shown that flavonoids, stilbenes, phenolic acids, capsaicin, and procyanidins control obesity through downregulating adipogenesis as well as some obesity-related inflammatory markers (Mukherjee et al. 2015). Capsaicin inhibited adipogenesis and promoted adipocyte apoptosis through the activation of AMKP in pre-adipocyte 3T3-L1 cells (Alcalá Hernández et al. 2015). The antiobesity and anti-inflammatory activity of this compound has been also attributed to the reduction of adipose TNFα, MCP-1, and IL-6, and the increase of adiponectine levels when measured in obese mice fed with a capsaicin rich diet (Kang et al. 2007).

17.5 CONCLUSION

H&S are, in general terms, good sources of phenolic compounds and other bioactive phytochemicals. The major PCs present in H&S are flavones such as luteolin and apigenin, and volatile phenolics such as eugenol, carbacrol, and anetole. They also contain common phenolic acids and unique compounds such as rosmarinic acid, gingeriols, and curcuminoids. Their large diversity makes it necessary to carry out systematic characterization of compounds in some spices that have not been well-studied. In vitro and in vivo studies, and some clinical trials, suggest that the polyphenolics compounds present in H&S may be associated with beneficial effects against several diseases, including cardiovascular diseases, cancer, inflammation, and obesity.

ACKNOWLEDGMENTS

Financial support provided by the Consejo Nacional de Ciencia y Tecnología, CONACYT, Mexico (CB-2011-01-167932) is acknowledged. The authors wish to thank Dr. Helvia Rosa Pelayo Benavides for her assistance with the taxonomical classification of the H&S.

REFERENCES

Alarcón-Flores, M.I., Romero-González, R., Martínez-Vidal, J.L., and Garrido-Frenich, A. 2014. Determination of phenolic compounds in artichoke, garlic and spinach by ultra-high-performance liquid chromatography coupled to tandem mass spectrometry. *Food Analytical Methods* 7:2095–2106.

Alcalá Hernández, C.F., Ardilla de la Rosa, L.A., Wall-Medrano, A., López Díaz, J.A., and Álvarez Parrilla, E. 2015. Mecanismos de diferenciación de adipocitos. *Nutrición Hospitalaria* 32:545–555.

Alvarez-Parrilla, E., Mercado-Mercado, G., de la Rosa, L.A., Diaz, J.A.L., Wall-Medrano, A., and González-Aguilar, G.A. 2014. Antioxidant activity and prevention of pork meat lipid oxidation using traditional Mexican condiments (pasilla dry pepper, achiote, and mole sauce). *Food Science and Technology Campinas* 34:371–378.

Andrés-Lacueva, C., Medina-Remon, A., Llorach, R., Urpi-Sarda, M., Khan, N., Chiva-Blanch, G., Zamora-Ros, R., Rotches-Ribalta, M., and Lamuela-Raventos, R.M. 2010. Phenolic compounds: Chemistry and occurrence in fruits and vegetables. In *Fruit and Vegetable Phytochemicals,* ed. L.A. De la Rosa, E. Alvarez-Parrilla, and G.A. González-Aguilar, pp. 53–88. Ames, IA: Wiley-Blackwell.

Ani, V., Varadaraj, M.C., and Naidu, K.A. 2006. Antioxidant and antibacterial activities of polyphenolic compounds from bitter cumin (*Cuminum nigrum* L.). *European Food Research and Technology* 224:109–115.

Anwar, F., Ali, M., Hussain, A.I., and Shahid, M. 2009. Antioxidant and antimicrobial activities of essential oil and extracts of fennel (*Foeniculum vulgare* Mill.) seeds from Pakistan. *Flavour and Fragrance Journal* 24:170–176.

Aranha, C.P.M., and Jorge, N. 2012. Antioxidant potential of oregano extract (*Origanum vulgare* L.). *British Food Journal* 114:954–965.

Bae, H., Jayaprakasha, G.K., Jifon, J., Patil, B.S. 2012. Variation of antioxidant activity and the levels of bioactive compounds in lipophilic and hydrophilic extracts from hot pepper (*Capsicum* spp.) cultivars. *Food Chemistry* 134:1912–1918.

Bae, H., Jayaprakasha, G.K., Crosby, K. et al. 2014. Ascorbic acid, capsaicinoid, and flavonoid aglycone concentrations as a function of fruit maturity stage in greenhouse-grown peppers. *Journal of Food Composition and Analysis* 33:195–202.

Barros, B., Pereira, E., Calhelha, R.C. et al. 2013. Bioactivity and chemical characterization in hydrophilic and lipophilic compounds of *Chenopodium ambrosioides* L. *Journal of Functional Foods* 5:1732–1740.

Beato, V.M., Orgaz, F., Mansilla, F., and Montaño, A. 2011. Changes in phenolic compounds in garlic (*Allium sativum* L.) owing to the cultivar and location of growth. *Plant Foods for Human Nutrition* 66:218–223.

Bhagwat, S., Haytowitz, D.B., and Holden, J.M. (Ret.). 2014. USDA Database for the flavonoid content of selected foods, Release 3.1. U.S. Department of Agriculture, Agricultural Research Service. Accessed from the Nutrient Data Laboratory home page, http://www.ars.usda.gov/nutrientdata/flav.

Cameroni, G. 2012. Historia de las hierbas aromáticas, especias y aceites esenciales. 1–6. Accessed from http://www.alimentosargentinos.gob.ar/contenido/sectores/sectores.php?secc=aromaticas.

Capecka, E., Mareczek, A., and Leja, M. 2005. Antioxidant activity of fresh and dry herbs of some Lamiaceae species. *Food Chemistry* 93:223–226.

Cardarelli, C.R., Benassi, M.T., and Mercadante, A.Z. 2008. Characterization of different annatto extracts based on antioxidant and colour properties. *LWT—Food Science and Technology* 41:1689–1693.

Chen, S., Shen, X., Cheng, S. et al. 2013. Evaluation of garlic cultivars for polyphenolic content and antioxidant properties. *PLoS ONE* 8:e79730. Accessed from http://doi.org/10.1371/journal.pone.0079730.

Cheng, A., Chen, X., Jin, Q., Wang, W., and Shi, J. 2013. Comparison of phenolic content and antioxidant capacity of red and yellow onions. *Czech Journal of Food Sciences* 31:501–508.

Cheng, X.L., Liu, Q., Peng, Y.B., Qi, L.W., and Li, P. 2011. Steamed ginger (*Zingiber officinale*): Changed chemical profile and increased anticancer potential. *Food Chemistry* 129:1785–1792.

Christova-Bagdassarian, V.L., Bagdassarian, K.S., Atanassova, M.S., and Ahmad, M.A. 2014. Comparative analysis of total phenolic and total flavonoid contents, rutin, tannins and antioxidant capacity in Apiaceae and Lamiaceae families. *Indian Horticulture Journal* 4:131–140.

Chung, H.J. 2009. Evaluation of the biological activity of extracts from star-anise (*Illicium verum*). *Journal of Food Science and Nutrition* 14:195–200.

Civitello, L. 2008. *Cuisine and Culture. A history of food and people.* Hoboken, NJ: John Wiley.

Danila, A.O., Gatea, F., and Radu, G.L. 2011. Polyphenol composition and antioxidant activity of selected medicinal herbs. *Chemistry of Natural Compounds* 47:22–25.

Dinesha, R., Thammannagowda, S.S., Prabhu, M.S.L., Madhu, C.S., and Srinivas, L. 2014. The antioxidant and DNA protectant activities of Star Anise (*Illicium verum*) aqueous extracts. *Journal of Pharmacology and Phytochemistry* 2:98–103.

Dörrie, J., Sapala, K., and Zunino, S.J. 2001. Carnosol-induced apoptosis and downregulation of Bcl-2 in B-lineage leukemia cells. *Cancer Letters* 170:33–39.

Dubie, J., Stancik, A., Morra, M., and Nindo, C. 2013. Antioxidant extraction from mustard (*Brassica juncea*) seed meal using high-intensity ultrasound. *Journal of Food Science* 78:542–548.

Engels, C., Schieber, A., and Gänzle, M.G. 2012. Sinapic acid derivatives in defatted Oriental mustard (*Brassica juncea* L.) seed meal extracts using UHPLC-DAD-ESI-MS n and identification of compounds with antibacterial activity. *European Food Research and Technology* 234:535–542.

Fálder-Rivero, A. 2005. Enciclopedia de los alimentos: Especias, hierbas aromáticas y condimentos. *Distribución y Consumo* 15:133–147.

Guo, Y., Sakulnarmrat, K., and Konczak, I. 2014. Anti-inflammatory potential of native Australian herbs polyphenols. *Toxicology Reports* 1:385–390.

Hoque, M.S., Benjakul, S., and Prodpran, T. 2011. Properties of film from cuttle fish (Sepiapharaonis) skin gelatin incorporated with cinnamon, clove and star anise extracts. *Food Hydrocolloids* 25.1085–1097.

Jayaprakasha, G.K., Jena, B.S., Negi, P.S., and Sakariah, K.K. 2002. Evaluation of antioxidant activities and antimutagenicity of turmeric oil: A byproduct from curcumin production. *Zeitschrift für Naturforschung C* 57:828–835.

Kang, J.H., Kim, C.S., Han, I.S., Kawada, T., and Yu, R. 2007. Capsaicin, a spicy component of hot peppers, modulates adipokine gene expression and protein release from obese-mouse adipose tissues and isolated adipocytes, and suppresses the inflammatory responses of adipose tissue macrophages. *FEBS Letters* 581:4389–4396.

Kannatt, S.R., Chawla, S.P., and Sharma, A. 2014. Antioxidant and radio-protective activities of lemon grass and star anise extracts. *Food Bioscience* 6:24–30.

Kim, J.S., Kang, O.J., and Gweon, O.C. 2013. Comparison of phenolic acids and flavonoids in black garlic at different thermal processing steps. *Journal of Functional Foods* 5:80–86.

Lee, J.J., Jin, Y.R., Lim, Y. et al. 2006. Antiplatelet activity of carnosol is mediated by the inhibition of TXA2 receptor and cytosolic calcium mobilization. *Vascular Pharmacology* 45:148–153.

Licina, B.Z., Stefanovi, O.D., Radojevi, I.D., Deki, M.S., and Comi, L.R. 2013. Biological activities of the extracts from wild growing *Origanum vulgare* L. *Food Control* 33:498–504.

Lin, L.Z., Lu, S.M., and Harnly, J.M. 2007. Detection and quantification of glycosylated flavonoid malonates in celery, Chinese celery, and celery seed by LC-DAD-ESI/MS. *Journal of Agricultural and Food Chemistry* 55:321–1326.

Liu, H.Y., Qiu, N.X., Ding, H.H., and Yao, R.Q. 2008. Polyphenols contents and antioxidant capacity of 68 Chinese herbals suitable for medical or food uses. *Food Research International* 41:363–370.

Lu, M., Yuan, B., Zeng, M., and Chen, J. 2011. Antioxidant capacity and major phenolic compounds of spices commonly consumed in China. *Food Research International* 44:530–536.

Lv, J., Huang, H., Yua, L. et al. 2012. Phenolic composition and nutraceutical properties of organic and conventional cinnamon and peppermint. *Food Chemistry* 132:1442–1450.

Manderfeld, M.M., Schafer, H.W., Davidson, P.M., and Zottola, E.A. 1997. Isolation and identification of antimicrobial furanocoumarins from parsley. *Journal of Food Protection* 60:72–77.

Mariod, A.A., Ibrahim, R.M., Ismail, M., Ismail, N. 2009. Antioxidant activity and phenolic content of phenolic rich fractions obtained from black cumin (*Nigella sativa*) seedcake. *Food Chemistry* 116:306–312.

Martins, N., Barros, L., Santos-Buelga, C., Henriques, M., Silva, S., and Ferreira, I.C.F.R. 2014. Decoction, infusion and hydroalcoholic extract of *Origanum vulgare* L.: Different performances regarding bioactivity and phenolic compounds. *Food Chemistry* 158:73–80.

Maruenda, H., Vico, M.L., Householder, J.E. et al. 2013. Exploration of *Vanilla pompona* from the Peruvian Amazon as a potential source of vanilla essence: Quantification of phenolics by HPLC-DAD. *Food Chemistry* 138:161–167.

Mercado-Mercado, G., Carrillo, L.D.L.R., Wall-Medrano, A., Díaz, J.A.L., and Alvarez-Parrilla, E. 2013. Compuestos polifenólicos y capacidad antioxidante de especias típicas consumidas en México. *Nutrición Hospitalaria* 28:36–46.

Mukherjee, A., Mukherjee, S., Biswas, J., and Roy, M. 2015. Phytochemicals in obesity control. *International Journal of Current Microbiology and Applied Sciences* 4:558–567.

Naczk, M., and Shahidi, F. 2004. Review: Extraction and analysis of phenolics in food. *Journal of Chromatography A* 1054:95–111.

Oboh, G., Akinyemi, A.J., and Ademiluyi, A.O. 2012. Antioxidant and inhibitory effect of red ginger (*Zingiber officinale* var. Rubra) and white ginger (*Zingiber officinale* Roscoe) on Fe2+ induced lipid peroxidation in rat brain *in vitro*. *Experimental and Toxicologic Pathology* 64:31–36.

Ohira, H., Torii, N., Aida, T.M., Watanabe, M., and Smith Jr., R.L. 2009. Rapid separation of shikimic acid from Chinese star anise (*Illicium verum* Hook. f.) with hot water extraction. *Separation and Purification Technology* 69:102–108.

Oktay, M., Gülçin, I., and Küfrevioğlu, Ö.I. 2003. Determination of *in vitro* antioxidant activity of fennel (*Foeniculum vulgare*) seed extracts. *LWT—Food Science and Technology* 36:263–271.

Orellana-Escobedo, L., Garcia-Amezquita, L.E., Olivas, G.I., Ornelas-Paz, J.J., and Sepulveda, D.R. 2013. Capsaicinoids content and proximate composition of Mexican chili peppers (*Capsicum* spp.) cultivated in the state of Chihuahua: Contenido de capsaicinoides y composicion proximal de chiles mexicanos (*Capsicum* spp.) cultivados en el estado de Chihuahua. *CYTA-Journal of Food* 11:179–184.

Padmashree, A., Roopa, N., Semwal, A.D., Sharma, G.K., Agathian, G., and Bawa, A.S. 2007. Star-anise (*Illicium verum*) and black caraway (*Carum nigrum*) as natural antioxidants. *Food Chemistry* 104:59–66.

Park, J., Kim, S., and Moon, B. 2011. Changes in carotenoids, ascorbic acids, and quality characteristics by the pickling of paprika (*Capsicum annuum* L.) cultivated in Korea. *Journal of Food Science* 76:C1075–C1080.

Pawar, N., Pai, S., Nimbalkar, M., and Dixit, G. 2011. RP-HPLC analysis of phenolic antioxidant compound 6-gingerol from different ginger cultivars. *Food Chemistry* 126:1330–1336.

Pérez, M.G.F., Rocha-Guzmán, N.E., Mercado-Silva, E., Loarca-Piña, G., and Reynoso-Camacho, R. 2014. Effect of chemical elicitors on peppermint (*Mentha piperita*) plants and their impact on the metabolite profile and antioxidant capacity of resulting infusions. *Food Chemistry* 156:273–278.

Pérez-Jiménez, J., Neveu, V., Vos, F., and Scalbert, A. 2010. Identification of the 100 richest dietary sources of polyphenols: An application of the Phenol-Explorer database. *European Journal of Clinical Nutrition* 64:S112–S120.

Pérez-Silva, A., Gunata, Z., Lepoutre, J.P., and Odoux, E. 2011. New insight on the genesis and fate of odor-active compounds in vanilla beans (*Vanilla planifolia* G. Jackson) during traditional curing. *Food Research International* 44:2930–2937.

Prakash, D., Singh, B.N., and Upadhyay, G. 2007. Antioxidant and free radical scavenging activities of phenols from onion (*Allium cepa*). *Food Chemistry* 102:1389–1393.

Prasad, S., Gupta, S.C., Tyagi, A.K., and Aggarwal, B.B. 2014. Curcumin, a component of golden spice: From bedside to bench and back. *Biotechnology Advances* 32:1053–1064

Przygodzka, M., Zielińska, D., Ciesarová, Z., Kukurová, K., and Zieliński, H. 2014. Comparison of methods for evaluation of the antioxidant capacity and phenolic compounds in common spices. *LWT—Food Science and Technology* 58:321–326.

Rababah, T.M., Banat, F., Rababah, A., Ereifej, K., and Yang, W. 2010. Optimization of extraction conditions of total phenolics, antioxidant activities, and anthocyanin of oregano, thyme, terebinth, and pomegranate. *Journal of Food Science* 75:626–632.

Ranilla, L.G., Kwon, Y.I., Apostolidis, E., and Shetty, K. 2010. Phenolic compounds, antioxidant activity and *in vitro* inhibitory potential against key enzymes relevant for hyperglycemia and hypertension of commonly used medicinal plants, herbs and spices in Latin America. *Bioresource Technology* 101:4676–4689.

Rebey, I.B., Bourgou, S., Debez, I.B. et al. 2012. Effects of extraction solvents and provenances on phenolic contents and antioxidant activities of cumin (*Cuminum cyminum* L.) seeds. *Food and Bioprocess Technology* 5:2827–2836.

Riachi, L.G., and De Maria, C.A.B. 2015. Peppermint antioxidants revisited. *Food Chemistry* 176:72–81.

Rothwell, J.A., Pérez-Jiménez, J., Neveu, V. et al. 2013. Phenol-Explorer 3.0: A major update of the Phenol-Explorer database to incorporate data on the effects of food processing on polyphenol content. *Database*, bat070.

Salazar-Rojas, V.M., Herrera-Cabrera, B.E., Delgado-Alvarado, A., Soto-Hernández, M., Castillo-González, F., and Cobos-Peralta, M. 2012. Chemotypical variation in *Vanilla planifolia* Jack. (Orchidaceae) from the Puebla-Veracruz Totonacapan region *Genetic Resources Crop Evolution* 59:875–887.

Sánchez, A.M., Martínez-Botas, J., Malagarie-Cazenave, S. et al. 2008. Induction of the endoplasmic reticulum stress protein GADD153/CHOP by capsaicin in prostate PC-3 cells: A microarray study. *Biochemical Biophysical Research Communications* 372:785–791.

Scalbert, A., and Williamson, G. 2000. Chocolate: Modern science investigates an ancient medicine. *Journal of Medicinal Food* 3:121–125.

Schwertner, H.A., and Rios, D.C. 2007. High-performance liquid chromatographic analysis of 6-gingerol, 8-gingerol, 10-gingerol, and 6-shogaol in ginger-containing dietary supplements, spices, teas, and beverages. *Journal of Chromatography B: Analytical Technologies in the Biomedical and Life Sciences* 856:41–47.

Shan, B., Cai, Y.Z., Sun, M., and Corke, H. 2005. Antioxidant capacity of 26 spice extracts and characterization of their phenolic constituents. *Journal of Agricultural and Food Chemistry* 53:7749–7759.

Shyu, Y.S., Lin, J.T., Chang, Y.T., Chiang, C.J., and Yang, D.J. 2009. Evaluation of antioxidant ability of ethanolic extract from dill (*Anethum graveolens* L.) flower. *Food Chemistry* 115:515–521.

Singleton, V.L., and Rossi, J.A. 1965. Colorimetry of total phenolics with phosphomolybdic-phosphotungstic acid reagent. *American Journal of Enology and Viticulture* 16:144–158.

Sinha, A.K., Verma, S.C., and Sharma, U.K. 2007. Development and validation of an RP-HPLC method for quantitative determination of vanillin and related phenolic compounds in *Vanilla planifolia*. *Journal of Separation Science* 30:15–20.

Srinivasan K. 2005. Spices as influencers of body metabolism: An overview of three decades of research. *Food Research International* 38:77–86.

Stankevičius, M., Akuåeca, I., Jákobsone, I., and Maruška, A. 2011. Comparative analysis of radical scavenging and antioxidant activity of phenolic compounds present in everyday use spice plants by means of spectrophotometric and chromatographic methods. *Journal of Separation Science* 34:1261–1267.

Su, L., Yin, J.J., Charles, D., Zhou, K., Moore, J., and Yu, L.L. 2007. Total phenolic contents, chelating capacities, and radical-scavenging properties of black peppercorn, nutmeg, rosehip, cinnamon and oregano leaf. *Food Chemistry* 100:990–997.

Suhaj, M. 2006. Spice antioxidants isolation and their antiradical activity: A review. *Journal of Food Composition and Analysis* 19:531–537.

Teixeira, B., Marques, A., Ramos, C. et al. 2013. Chemical composition and bioactivity of different oregano (*Origanum vulgare*) extracts and essential oil. *Journal of the Science of Food and Agriculture* 93:2707–2714.

U.S. Department of Agriculture, A.R.S. 2014. USDA National Nutrient Database for Standard Reference, Release 27. Accessed from the Nutrient Data Laboratory home page: http://www.ars.usda.gov/Services/docs.htm?docid=8964.

Vallverdú-Queralt, A., Regueiro, J., Martínez-Huélamo, M., Rinaldi Alvarenga, J.F., Leal, L.N., and Lamuela-Raventos, R.M. 2014. A comprehensive study on the phenolic profile of widely used culinary herbs and spices: Rosemary, thyme, oregano, cinnamon, cumin and bay. *Food Chemistry* 154:299–307.

Vanisha, S. Nambiar, M.D., and Guin, P. 2010. Characterization of polyphenols from coriander leaves (*Coriandrum sativum*), red amaranthus (*A. paniculatus*) and green amaranthus (*A. frumentaceus*) using paper chromatography: And their health implications. *Journal of Herbal Medicine and Toxicology* 4:173–177.

Vega-Gálvez, A., Di Scala, K., Rodríguez, K. et al. 2009. Effect of air-drying temperature on physico-chemical properties, antioxidant capacity, colour and total phenolic content of red pepper (*Capsicum annuum*, L. var. Hungarian). *Food Chemistry* 117:647–653.

Viuda-Martos, M., Ciro-Gómez, G.L., Ruiz-Navajas, Y. et al. 2012. *In vitro* antioxidant and antibacterial activities of extracts from annatto (*Bixa Orellana* L.) leaves and seeds. *Journal of Food Safety* 32:399–406.

Viuda-Martos, M., Ruiz-Navajas, Y., Fernández-López, J., and Pérez-Álvarez, J.A. 2011. Spices as functional foods. *Critical Reviews in Food Science and Nutrition* 51:13–28.

Wangensteen, H., Samuelsen, A.B., and Malterud, K.E. 2004. Antioxidant activity in extracts from coriander. *Food Chemistry* 88:293–297.

Wargovich, M.J., Woods, C., Hollis, D.M., and Zander, M.E. 2001. Herbals, cancer prevention and health. *Journal of Nutrition* 131:30345–30365.

WHO (World Health Organization). 2013. Obesity and overweight. Accessed from http://www.who.int/about/contacthq/es/index.html.

Wongsa, P., Chaiwarit, J., and Zamaludien, A. 2012. *In vitro* screening of phenolic compounds, potential inhibition against α-amylase and α-glucosidase of culinary herbs in Thailand. *Food Chemistry* 131:964–971.

Wu, X., Wang, W.P., Zhu, T. et al. 2013. Phenylpropanoid glycoside inhibition of pepsin, trypsin and α-chymotrypsin enzyme activity in Kudingcha leaves from *Ligustrum purpurascens*. *Food Research International* 54:1376–1382.

Zhishen, J., Mengcheng, T., and Jianming, W. 1999. The determination of flavonoid contents in mulberry and their scavenging effects of superoxide radical. *Food Chemistry* 64:555–559.

CHAPTER 18

Phenolic Compounds in Fruits

J. Abraham Domínguez-Avila, Gustavo R. Velderrain-Rodríguez, Maribel Ovando-Martínez, Ana Elena Quirós-Sauceda, Mónica A. Villegas-Ochoa, and Gustavo A. González Aguilar

CONTENTS

18.1 INTRODUCTION

Diverse scientific studies support the fact that consumption of fruits and vegetables supplies bioactive compounds with health-related effects to organisms. Characterization of the phenolic compound (PC) profile in fruits has shown that they exert effects such as antibacterial, antiviral, antioxidant, and others. Publications related to fruit PCs have increased approximately fortyfold from 1991 to 2011 (Haminiuk et al. 2012), partly due to their high bioactivity and ubiquity in fruits (Guimarães et al. 2013). PCs are secondary metabolites produced by plants as a defense mechanism against predators to aid in reproduction and plant-to-plant communication (Velderrain-Rodríguez et al. 2014). PCs can be divided into sixteen classes according to their molecular structure, but the most common and abundant in fruits are flavonoids and phenolic acids (Manganaris et al. 2014). PCs can form diverse interactions with the food matrix, and some of them depend on the molecular identity of the PC and the matrix. In that sense, authors such as Arranz et al. (2009) argue that the fruit PC content may be underestimated when extracted by conventional methods because PCs that are strongly bound to the food matrix resist extraction and are not accounted for. This chapter will summarize the most representative PCs from fruits and discuss current methods to extract, quantify, and evaluate the PC profile and antioxidant activity of fruits.

18.2 TROPICAL, SUBTROPICAL, AND TRADITIONAL FRUITS AS SOURCES OF PCs

Tropical fruits include mango (*Mangifera indica* L.), pineapple (*Ananas comosus* L.), papaya (*Carica papaya* L.), banana (*Musa* sp.), guava (*Psidium guajava* L.), jackfruit (*Artocarpus heterophyllus* Lam.), carambola (*Averrhoa carambola* L.), and others; subtropical fruits include avocado (*Persea americana* Mill.), carob (*Ceratonia siliqua* L.), citrus fruits, figs (*Ficus carica* L.), kiwifruit (*Actinidia chinensis* Planch.), pomegranate (*Punica granatum* L.), lychee (*Litchi chinensis* Sonn.), and others (Kader and Yahia 2011, FAO 2002). Tropical and subtropical fruits differ in their morphology, composition, and physiology, but both are good sources of PCs. Among the previously mentioned ones, there are traditional fruits that are commonly consumed around the world, such as bananas, apples (*Malus sylvestris*), green grapes (*Vitis riparia*), and strawberries (*Fragaria* × *ananassa* Duschesne) (Park et al. 2015a). Flavonoids are the most representative compounds found in these types of fruits, and they occur in esterified and glycosylated forms with a wide variety of structures such as flavones, flavonols, anthocyanins, isoflavones, and flavan-3-ols (catechins and tannins) (Lim et al. 2007, Alothman et al. 2009), followed by phenolic acids that have simpler molecular structures (Balasundram et al. 2006). Their redox properties make them primary and secondary antioxidants that can act as reducing agents, singlet oxygen quenchers, hydrogen donators, and chelating agents of metal ions (Mustafa et al. 2010).

The total phenolic content (TPC), total flavonoid content (TFC), and antioxidant capacity of fruits can vary widely. Table 18.1 lists the TPCs, TFCs, and antioxidant capacities of different tropical and subtropical fruits discussed in the main text.

Different mango cultivars have particular PC profiles that can result in synergistic or antagonistic effects due to PC–PC or PC–matrix interactions, ultimately resulting in the wide range of antioxidant capacities reported by several authors. For example, the PC profiles of Haden, Tommy Atkins, and Ubá mangoes include mangiferin and quercetin-3-O-glucoside as the main components, while Ataulfo mangoes contain gallic acid, protocatechuic acid, chlorogenic acid, and vanillic acid (Balasundram et al. 2006, Heo et al. 2007, Palafox-Carlos et al. 2012b). Molecular features of individual PCs, such as the number and position of aromatic and hydroxyl groups, are commonly named factors that have a significant influence on the antioxidant capacity (Balasundram et al. 2006). Papaya and pineapple also show differences in their PC profile and antioxidant capacity due to similar factors as described for mangoes, where the cultivar is a key factor in determining the antioxidant capacity. According to Kelebek et al. (2015), papaya cultivar Sel 42 has higher contents of protocatechuic acid, caffeoyl hexose-deoxyhexoside, ferulic acid, and *p*-coumaric acid than the Tainung cultivar, whereas the latter has higher concentrations of rutin and kaempferol-3-O-glucoside, all of which contribute to the antioxidant capacity of papaya. Analysis of TPC and PC profiles of six avocado cultivars shows significant variations in PC content, with the Orotawa cultivar having the highest TPC and the Bacon and Hass cultivars having the lowest TPC (Di Stefano et al. 2017).

Avocado by-products such as peel and seeds are also sources of PCs, such as flavanols, proanthocyanidins, hidroxycinnamic acids, and flavonol glycosides. Kosinska et al. (2012) suggested that the bioactive compounds present therein can be extracted and used as functional ingredients for various applications.

Reports on the PC profile of kiwifruits show that hydroxycinnamic acids, procyanidins, and quercetin glycosides are the main compounds (Pinelli et al. 2013). Among different kiwifruit cultivars, SKK12, Bidan and Hwamei have the highest antioxidant capacity values (Park et al. 2014). In addition, the standard kiwifruit has a higher TPC than red plums (*Prunus*

TABLE 18.1 Total Phenolic Content (TPC), Total Flavonoid Content (TFC), and Antioxidant Activity of PCs in Fruits

Fruit	TPC (mg/100 g)	TFC (mg/100 g)	Antioxidant Activity	PC	Reference
mango cv. Ataulfo (whole fruit)	140 GAE FW	7.5 CE FW	240 mg TE/100 g FW (DPPH), 170 mg TE/100 g FW (FRAP), 210 mg TE/100 g FW (relative TPAOC)	gallic acid 95, protocatechuic acid 1, chlorogenic acid 300, and vanillic acid 25 mg/100 g DW by HPLC-MS	Palafox-Carlos et al. 2012b
mango cv. Kent (fresh-cut)	50 GAE FW	20 QE FW	280 μmol TE/100 g FW (TEAC), 40 percent (DPPH)	ND	Robles-Sánchez et al. 2013
mango cv. Ataulfo (fresh-cut)	116 GAE FW	11.2 QE FW	57 percent (DPPH, % RSA), 790 μmol TE/100 g FW (ORAC)	ND	Robles-Sánchez et al. 2011
mango cv. Ataulfo (whole fruit)	110 GAE FW	16.9 QE FW	60 percent (DPPH, % RSA), 850 μmol TE/100 g FW (ORAC)	ND	
mango cv. Haden	55 GAE FW	ND	ND	mangiferin 2.9 and quercetin 3-O-glc 0.6 mg/kg DW by HPLC-MS	Rocha Ribeiro et al. 2007
mango cv. Tommy Atkins	50 GAE FW	ND	ND	mangiferin 2.2 and isomangiferin 0.5 mg/kg, DW by HPLC-MS	
mango cv. Palmer	140 GAE FW	ND	ND	ND	
mango cv. Ubá	210 GAE FW	ND	ND	mangiferin 12.4, isomangiferin 1.1, mangiferin gallate 1.3, isomangiferin gallate 4.5, quercetin 3-O-gal 2.5, quercetin 3-O-glc 6.3, quercetin 3-O-xyl 1.7, quercetin 3-O-arap 1.2, quercetin 3-O-araf 1.2, quercetin 3-O-rha 0.5, kaempferol 3-O-glc 0.6, and quercetin 0.6 mg/kg DW by HPLC-MS	

(Continued)

TABLE 18.1 (CONTINUED) Total Phenolic Content (TPC), Total Flavonoid Content (TFC), and Antioxidant Activity of PCs in Fruits

Fruit	TPC (mg/100 g)	TFC (mg/100 g)	Antioxidant Activity	PC	Reference
mango cv. Ataulfo	274.30 GAE FW	ND	ND	ND	Velderrain-Rodríguez et al. 2016
mango cv. Ataulfo	56.7 GAE FW	ND	326.6 µmol TE/100 g FW (ORAC)	ND	Noratto et al. 2010
mango cv. Haden	19.1 GAE FW	ND	225.8 µmol TE/100 g FW (ORAC)	ND	
mango cv. Francis	17.8 GAE FW	ND	219 µmol TE/100 g FW (ORAC)	ND	
mango cv. Kent	16.4 GAE FW	ND	150 µmol TE/100 g FW (ORAC)	ND	
mango cv. Tommy Atkins	15.3 GAE FW	ND	156.6 µmol TE/100 g FW (ORAC)	ND	
papaya cv Sel-42	171.68 FW	ND	330.12 µmol TE/100 g, FW (DPPH)	*p*-hydroxybenzoic acid 8.66, protocatechuic acid-hexoside 38.89, gallic acid-deoxyhexoside 25.20, protocatechuic acid 10.57, chlorogenic acid 14.89, caffeic acid 8.78, caffeoyl hexose-deoxyhexoside 11.15, rutin 1.71, ferulic acid 35.18, *p*-coumaric acid 13.27, kaempferol-3-O-glc 1.79 and myricetin 1.57 mg/100g, FW by HPLC-MS	Kelebek et al. 2015

(*Continued*)

TABLE 18.1 (CONTINUED) Total Phenolic Content (TPC), Total Flavonoid Content (TFC), and Antioxidant Activity of PCs in Fruits

Fruit	TPC (mg/100 g)	TFC (mg/100 g)	Antioxidant Activity	PC	Reference
papaya cv Tainung	119.79 FW	ND	263 63 μmol TE/100 g, FW (DPPH)	*p*-hydroxybenzoic acid 8.52, protocatechuic acid-hexoside 27.92, gallic acid-deoxyhexoside 22.26, protocatechuic acid 9.11, chlorogenic acid 12.36, caffeic acid 7.99, caffeoyl hexose-deoxyhexoside 3.22, rutin 5.07, ferulic acid 11.14, *p*-coumaric acid 9.05, kaempferol-3-O-glc 2.35 and myricetin 0.81 mg/100g, FW by HPLC-MS	
Papaya	272.66 RE DW	92.95 GAE DW	ND	ND	Maisarah et al. 2013
avocado cv Bacon	6.35 mg/kg FW	ND	ND	4-hydroxybenzoic acid, benzoic acid, caffeic acid, chlorogenic acid, epicatechin, ferulic acid, gallic acid, gentisic acid, isorhamnetin, *p*-coumaric acid, poncirin, protocatechuic acid, rutin, sinapic acid, taxifolin, *trans*-cinnamic acid and vanillin by UHPLC-HESI-MS	Di Stefano et al. 2017
avocado cv Fuerte	2.32 FW	ND	ND		
avocado cv Hass	1.46 FW	ND	ND		
avocado cv Orotawa	2.83 FW	ND	ND		
avocado cv Pinkerton	2.12 FW	ND	ND		

(*Continued*)

TABLE 18.1 (CONTINUED) Total Phenolic Content (TPC), Total Flavonoid Content (TFC), and Antioxidant Activity of PCs in Fruits

Fruit	TPC (mg/100 g)	TFC (mg/100 g)	Antioxidant Activity	PC	Reference
Avocado	297.72 GAE DW	49.58 QE DW	32.05 μmol $FeSO_4$/g DW (FRAP), 593.98 μg/mL (DPPH IC_{50}), 0.07 μg/mL/IC_{50} μg/mL (DPPH AAI)	ND	Morais et al. 2015
Pineapple	197.87 GAE DW	28.31 QE DW	44.77 μmol $FeSO_4$/g DW (FRAP), 666.07 μg/mL, (DPPH IC_{50}), 0.06 μg/mL/IC_{50} μg/mL (DPPH AAI)	ND	
Banana	264.08 GAE DW	28.92 QE DW	16.20 μmol $FeSO_4$/g DW (FRAP), 558.59 μg/mL, (DPPH IC_{50}), 0.07 μg/mL/IC_{50} μg/mL (DPPH AAI)	ND	
Papaya	325.97 GAE DW	65.37 QE DW	32.16 μmol $FeSO_4$/g DW (FRAP), 748.47 μg/mL (DPPH IC_{50}), 0.05 μg/mL/IC_{50} μg/mL (DPPH AAI)	ND	
Passionfruit	378.13 GAE DW	144.42 QE DW	27.50 μmol $FeSO_4$/g DW (FRAP), 869.05 μg/mL (DPPH IC_{50}), 0.04 μg/mL/IC_{50} μg/mL (DPPH AAI)	ND	
Watermelon	241.99 GAE DW	36.81 QE DW	16.37 μmol $FeSO_4$/g DW (FRAP), 902.15 μg/mL (DPPH IC_{50}), 0.04 μg/mL/IC_{50} μg/mL (DPPH AAI)	ND	
Melon	313.45 GAE DW	72.62 QE DW	28.57 μmol $FeSO_4$/g DW (FRAP), 814.24 μg/mL (DPPH IC_{50}), 0.05 μg/mL/IC_{50} μg/mL (DPPH AAI)	ND	

Abbreviations: AAI: antioxidant activity index: final concentration of DPPH; DW: dry weight; FW: fresh weight; GAE: gallic acid equivalents; IC_{50}: extract concentration in μg/mL necessary for 50 percent inhibition of DPPH radical; ND: not determined; QE: quercetin equivalents; RE: rutin equivalents; TE: trolox equivalents.

domestica L.), mangoes, white grapes (*Vitis vinifera* L.), persimmons (*Diospyros* sp.), apples (*Malus domestica* L.), pears (*Pyrus* sp.), red grapefruits (*Citrus* × *paradisi* Macfad.), lemons (*Citrus* × *limon* L.), oranges (*Citrus sinensis* L.), pomelos (*Citrus maxima* Merr.), bananas, and peaches (*Prunus persica* L.), but it is similar to the reported TPC of strawberries.

Differences in PC profile, TPC and antioxidant capacity of tropical, subtropical, and traditional fruits are significant. Some authors have proposed that this information can be used to determine maximum and minimum doses of PCs from fruits (Park et al. 2015b).

18.3 CONVENTIONAL AND NONCONVENTIONAL QUANTIFICATION OF PCs

Several methodologies have been developed to extract, isolate, and identify PCs in fruits, which is due to the differences between food matrices. Some fruits contain free PCs that are easily extractable, while others interact strongly with cell wall polysaccharides, lipids, or proteins, and resist conventional extraction techniques (Acosta-Estrada et al. 2014). Figure 18.1 shows the general workflow of conventional extraction, quantification, and identification of fruit PCs. Because fruits are alive and respond to stress signals, it is necessary to find alternative non-conventional methods for fast PC analysis in the fresh fruit to avoid any alteration to their structures and content (Macheix and Fleuriet 1990, Liu et al. 2015).

According to conventional methods, PCs are classified into two groups depending on their interaction with the matrix, and some of the names used are soluble and insoluble; extractable (EPCs) and nonextractable PCs (NEPCs); or free and bound (Acosta-Estrada et al. 2014, Saura-Calixto 2012). The most common methods include liquid solvent extraction, supercritical fluid extraction (SFE), and chemical or enzymatic hydrolysis. Table 18.2 lists common solvents used to extract EPCs, while Table 18.3 shows different hydrolysis methods used to extract NEPCs from several fruits.

Extractions are commonly performed on ground, dried, or freeze-dried samples, or they can simply be performed by soaking plant material in the solvent. PC extraction should ideally be performed on fresh fruits to avoid the loss or chemical modification of PCs, but harvest season and shelf life often require the use of previously dried or frozen samples (Haminiuk et al. 2012). Even so, nonconventional methods such as near (NIR) and mid-infrared (MIR) reflectance spectroscopy and multispectral imaging has been used for PC determination in blueberries (Sinelli et al. 2008) and grapes (Ferrer-Gallego et al. 2011), respectively. The importance and use of conventional and nonconventional methods in PC quantification will vary according to the area of interest and required knowledge. Nonconventional methods are good for fast labeling of fruit PCs to the food industry. However, when a deep knowledge of the PC profile and prediction of their functional or biological properties are required, conventional methods are mostly used.

18.3.1 Quantification of PCs

Quantification of PCs is the next step after extraction from the fruit matrix in conventional methods. A preliminary step consists of determining all of the PCs or one category of these compounds from the more or less purified extract. Conventional methods of PC quantification include direct measurement of UV-Vis absorbance or colorimetric methods (Macheix and Fleuriet 1990). For example, anthocyanins and yellow flavonoids can

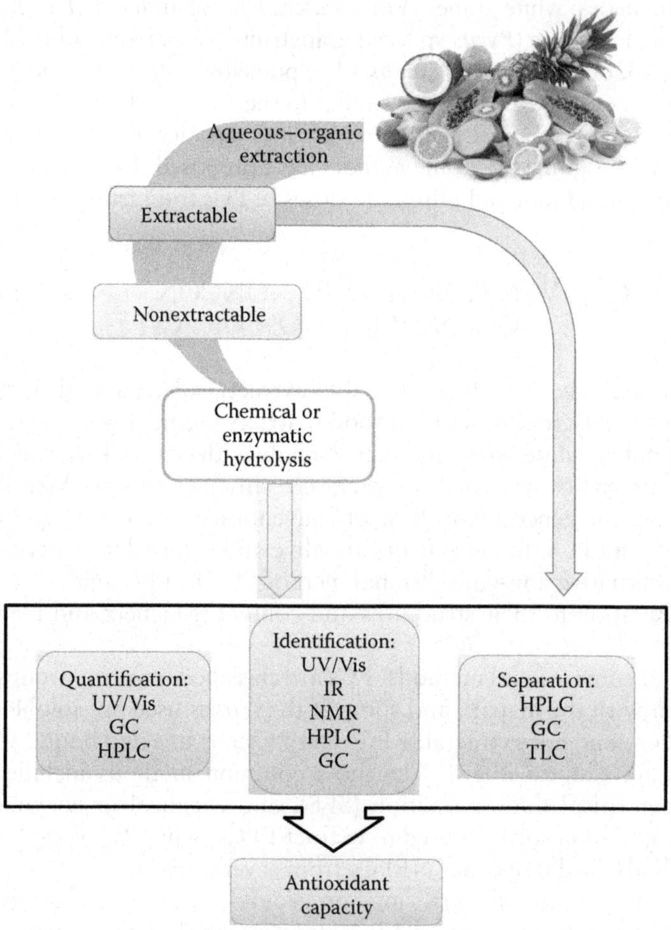

FIGURE 18.1 Extractable PCs are recovered after an aqueous–organic extraction, and can be immediately analyzed by various methods. Nonextractable PCs must be chemically- or enzymatically-treated before being analyzed.

be quantified using absorption coefficients of 982 and 766 $(g/100\ mL)^{-1}\ cm^{-1}$, respectively (Silva et al. 2014). Total anthocyanins in red grapes and red-fleshed apples can be determined by a pH-differential method, where the absorbance of the sample is read at 520 and 700 nm (Rockenbach et al. 2011b, Wang et al. 2014).

The Folin–Ciocalteu reagent has been used for decades to quantify TPC due to its simple procedure (Singleton and Rossi 1965), but lack of specificity can overestimate TPC, because other nonphenolic compounds can react (George et al. 2005). Some of the interfering compounds in TPC determination in fruits are ascorbic acid, sugars (fructose and sucrose), and organic acids (Medina 2011a, Lester et al. 2012). According to Lester et al. (2012), the Folin–Ciocalteu assay had a significant correlation with strawberry fruit total ascorbic acid, indicating an overestimation of TPC in this fruit. Despite interfering substances, the Folin–Ciocalteu assay remains a routine analysis in most laboratories. But a more specific method that has a direct reaction with PCs in fruits, such as the Fast Blue BB assay, can also be used (Medina 2011b). Another common method

TABLE 18.2 Solvents Used to Extract EPCs

Fruit	Solvent	EPC	Reference
Red grape pomace: skin and seeds	Methanol/water/acetic acid (80:20:5)	Total phenolics and total flavonoids	Rockenbach et al. 2011a
Litchi pericarp	Ethanol–water (7:3, v/v) with calcium carbonate	Total phenolics, flavonoids, proanthocyanidins	Wang et al. 2011
Olive	100 percent methanol with agitation and ultrasonic homogenizer	Hydroxytyrosol glucoside, dimethyloleuropein, oleuropein, verbascoside, caffeoyl-6'-secologanoside, quercitrin, comselogoside, luteolin-7-*O*-glc, rutin	Jerman et al. 2010
Pomegranate	50 percent methanol	Total phenolics, total flavonoids, total tannins, total monomeric anthocyanins	Mphahlele et al. 2014
Mango, red mombin, cherimoya, zapote, guava, passionfruit, granadilla, banana passionfruit, strawberry, capulí cherry, plum, Andean blackberry, naranjilla, sweet pepino, physalis, purple/red tree tomato, golden/yellow tree tomato, and tomato	Methanol:water (50:50 v/v) and acetone:water (70/30 v/v)	Total soluble phenolics	Vasco et al. 2008
Spine grape	Acidified methanol (1 M HCl in 80 percent methanol)	Total phenolics, total flavonoids, total flavonols, total monomeric anthocyanins	Meng et al. 2012
Red-fleshed apple	70 percent methanol containing 2 percent formic acid	Total phenolics, total flavonoids, total flavonols, total monomeric anthocyanins	Wang et al. 2014
Jujube	70 percent ethanol	Total phenolics and total flavonoids	Koley et al. 2011
Chinese waxberry	Ethyl acetate and sodium carbonate	Ferulic acid, caffeic acid, sinapic acid, and salicylic acid	Wang et al. 2012

(*Continued*)

TABLE 18.2 (CONTINUED) Solvents Used to Extract EPCs

Fruit	Solvent	EPC	Reference
Apricot	Phosphate buffer and ethyl acetate	Total phenolics	Melgarejo et al. 2014
Mango	80 percent methanol	Total phenolics	Sellamuthu et al. 2013
Strawberry	Acetone	Flavonoids, ellagitannins, ellagic acid conjugates, anthocyanins	Aaby et al. 2012

TABLE 18.3 Type of Hydrolysis Used for NEPCs Extractions

Fruit	Hydrolysis	NEPC type	Reference
Chinese waxberry	6 N HCl under nitrogen, 35°C, 90 min	Bound ferulic acid, caffeic acid, sinapic acid, and salicylic acid	Wang et al. 2012
Apple, orange, pear, banana, melon, grape, mandarin, peach, watermelon	Methanol–sulfuric acid, 85°C, 20 h, pH 5.5	Hydrolyzable tannins, hydroxybenzoic acids, hydroxycinnamic acids, flavonols, flavanones	Pérez-Jiménez and Saura-Calixto 2015
Apple, orange, pear, banana, melon, grape, mandarin, peach, watermelon	Butanol/HCl/$FeCl_3$, 100°C, 1 h	Condensed tannins	Pérez-Jiménez and Saura-Calixto 2015
Unripe apple	Viscozyme l (from *Aspergillus aculeatus*, fungal β-glucanase) Celluclast (from *Trichoderma reesei*, endoglucanase) Pectinex 5xl (from *Aspergillus niger*, depectination)	Bound *p*-coumaric acid, ferulic acid, caffeic acid	Zheng et al. 2009
Black currant pomace	Protease and pectinase	Total phenolics and anthocyanins	Landbo and Meyer 2001
Cranberry, apple, red grape, strawberry, pineapple, banana, peach, lemon, orange, pear, grapefruit	4 N NaOH, room temperature, 1 h, shaking under nitrogen	Bound-E and bound-W phenolic compounds	Sun et al. 2002

uses aluminum chloride to quantify flavonoids, but it has also received criticism for lack of reaction consistency between different molecules (Pekal and Pyrzynska 2014). Ferric chloride along with potassium ferricyanide as the Prussian blue developing agent and the vanillin method have been used to quantify phenolic acids and proanthocyanidins in lychee pericarp, respectively (Wang et al. 2011).

Other methods available to quantify or identify PCs are thin-layer chromatography (TLC), Fourier transform–infrared spectroscopy (FT-IR), high-pressure liquid chromatography–mass spectrometry (HPLC-MS), gas chromatography–mass spectrometry (GC-MS), and nuclear magnetic resonance (NMR) (Macheix and Fleuriet 1990, Oroian and Escriche 2015, Antolovich et al. 2000, Robbins 2003, Haminiuk et al. 2012, Khoddami et al. 2013, Cozzolino 2015). When PC composition is totally unknown, the TLC method is a simple option for a screening analysis and can aid in identifying the compounds present in fruit extracts (Oliveira et al. 2014, Oroian and Escriche 2015). For example, TLC chromatography has been used to detect the presence of naringin, meranzin hydrate, 5,7-dimethoxycoumarin and scoparone in fruit extracts of *Citrus wilsonii* Takana, and *Citrus medica* L. (Zhao et al. 2015).

FT-IR has been used to determine the presence of PCs in apples, bananas, peaches, pears, grapefruits, pomelos, oranges, lemons, red plums, white grapes, mango, and strawberries, using kiwifruits as standard. The spectra of all samples showed bands between 1800 and 600 cm^{-1}, which are typical for the hydroxyl–aromatic molecules of hydrolyzable tannins, while condensed tannins presented bands between 1400 and 1100 cm^{-1}, confirming the presence of catechin, gallic acid, and quercetin when the sample spectra were compared to the respective standards (Park et al. 2015b). Notable disadvantages of TLC and FT-IR are that they only give an idea about the presence and separation but do not quantify individual PCs (Ignat et al. 2011). In some cases, the fractions analyzed by TLC may be further analyzed by HPLC-MS or GC-MS, which allows identification and quantification. Recently, a GC-MS method has been used to separate and identify PCs in samples of olives (*Olea europaea* L.), jujubes (*Ziziphus jujube* Mill.) and figs, where the optimization of the temperature and carrier gas flow rate allowed good resolution of up to 14 compounds and a short elution time of 25 minutes (Ahmad et al. 2016).

Fractioning involves the use of a chromatography column packed with Sephadex LH 20 or a reversed-phase (C_{18}) column, which separates PCs according to polarity, solubility, and molecular weight. This allows easy identification and quantification of PCs using different methods such as those mentioned above (Oroian and Escriche 2015, Falleh et al. 2013). Using small self-made polyamide and Sephadex LH 20 columns along with a Sep-Pak C18 plus cartridge allows fractionation of PCs from Hawthorn (*Crataegus laevigata*) extracts, yielding polymeric procyanidins, phenolic carboxylic acids, and flavonoid glycosides. HPLC-diode array detector (HPLC-DAD) analyses were subsequently used to further separate and identify the PC profile of the Hawthorn extracts (Svedström et al. 2006).

HPLC is used to separate and quantify fruit PCs. This typically involves a C18 column, UV-Vis diode array detector (HPLC-DAD), and binary solvent system containing an acidified aqueous solvent (solvent A) and a polar organic solvent (solvent B) (Ignat et al. 2011, Haminiuk et al. 2012). HPLC-DAD coupled to electrospray ionization mass spectrometry (HPLC-DAD-ESI/MS) was used to analyze the PC profiles of passion fruits (*Passiflora edulis* Sims.), cherimoyas (*Annona cherimola*), lemons, papayas, and strawberries (Spínola et al. 2015). This method allowed the identification of flavonoids, phenolic acids, tannins, and anthocyanins by comparing their UV-Vis spectra and mass spectrophotometric data obtained under both negative and positive electron spray ionization (ESI^-/ESI^+) conditions, and they were all quantified using representative standards for each PC group. Additionally, quantification of catechins in apples, black grapes, and other fruits was performed with reversed-phase HPLC by UV (270 nm) or fluorescence (280/310 nm excitation/emission) detection (Arts and Hollman 1998, Arts et al. 2000, Tsanova-Savova et al. 2005). It has been demonstrated that chromatographic determination is accurate and precise, but it can be difficult to identify the entire PC profile because of the complexity of the mixture due to the complexity of the NEPCs that have not been yet determined (Stratil et al. 2006, Pérez-Jiménez and Torres 2011).

18.4 ANTIOXIDANT CAPACITIES OF FRUIT EXTRACTS

Once fruit PCs have been extracted, quantified, and characterized, their antioxidant capacity is quantified by any of several methodologies currently in use (see Chapter 14 for further details on the mechanisms of action of the assays described). Several of the health benefits associated with PC consumption are directly related to their high antioxidant capacity because they can act as reducing agents, hydrogen-donating antioxidants, and singlet oxygen quenchers (Chun et al. 2003, Chamorro et al. 2013).

The antioxidant capacity of a fruit extract can be quantified by several methodologies and can use Trolox, vitamin C, gallic acid, catechin, quercetin, or several other compounds as standards. TPC and antioxidant capacity can sometimes correlate with each other. For example, TPC and antioxidant capacity by the Trolox equivalent antioxidant capacity (TEAC) and ferric reducing antioxidant power (FRAP) assays were evaluated in 104 apple cultivars and presented a correlation of 0.6. The authors note that the choice of standard will change absolute antioxidant capacity values, but the correlations were still acceptable (Ceymann et al. 2012). Similarly, a weak correlation between TPC and TEAC values of tamarinds (*Tamarindus indica* L.) has also been reported (Romaric et al. 2011). Such trends indicate that some interfering compounds also contribute to the antioxidant capacity in fruits, as was reported by Babbar et al. (2011).

Because these assays are colorimetric, colored compounds can lead to overestimation of the results of some assays. For example, samples that contain anthocyanins can interfere with the DPPH (2,2-diphenyl-1-picrylhydrazyl) assay but not with the TEAC assay. TEAC and oxygen radical absorbance capacity (ORAC) assays can be used to quantify the antioxidant capacity of soluble and insoluble PCs, while DPPH is better suited to work in more polar environments.

Unfortunately, in vitro values may not accurately reflect in vivo antioxidant capacity, because no method takes into account bioaccessibility, bioavailability, metabolism, excretion, and so on of PCs that happen in living organisms. Some in vivo methods have been developed to overcome these limitations, such as the cellular antioxidant activity (CAA) (Haminiuk et al. 2012). CAA is based on the peroxyl radical quenching ability of $HepG_2$ cells as well as inhibition of the generation of dichlorodihydrofluorescein (DCF). When CAA was compared to in vitro results [ORAC and rapid peroxyl radical scavenging capacity (PSC)] of Chinese hawthorn extracts, the authors concluded that CAA may be a better predictor of antioxidant behavior in biological systems because CAA simulates bioaccessibility, uptake, distribution, and metabolism of the antioxidants (Wen et al. 2015).

The antioxidant capacity of PCs can be predicted by spectroscopic methods. NIR and MIR spectroscopy are considered powerful, fast, accurate, and nondestructive analytical tools to quantify antioxidant compounds such as PCs and carotenoids. Identification of IR bands can be done by specialized software, which then uses partial least squares (PLS) and principal component regression (PCR) to predict antioxidant activity in food matrices (Cozzolino 2015).

FT-IR was compared to ORAC for flavonoid-rich extracts from blueberries (*Vaccinium* sp.), grapes, and blackberries (*Rubus* sp.). The spectra of all samples together with their ORAC values were used to perform PLS regression analysis to obtain a calibration model for predicting the antioxidant capacity of the extracts, which showed satisfactory results using the 2000 cm^{-1} to 900 cm^{-1} region. In general, FT-IR can be suitable for rapid and accurate measurement of the antioxidant capacity of different fruit matrices (Lam et al. 2005).

18.5 ANTIOXIDANT CONTRIBUTION AND HEALTH BENEFITS

Data gathered regarding PC profiles and antioxidant capacities of various fruits have been used to estimate average intake, and PCs were found to be the most abundant antioxidants in the diet. PC intake may reach 1 g/day; in contrast, vitamin C intake is reported as 0.3 g/day (Padayatty et al. 2004). Fruits can significantly maintain the antioxidant balance in cells due to their high PC content. Although PCs are not the only antioxidants found in plant tissues, they can be responsible for most of the antioxidant contribution. In this sense, some studies have established a linear correlation between the PC content in plant tissues and its antioxidant contribution (Alonso et al. 2002, Dubost et al. 2007, Gramza et al. 2006, Kiselova et al. 2006). Few studies have evaluated the individual contributions of PCs identified in plant tissue extracts; for example, Dudonné et al. (2011) characterized the PC profile in their poplar bud (*Populus nigra*) extract by HPLC-MS and determined its antioxidant contribution by ORAC, and even when caffeic and *p*-coumaric acids represent only 3.5 percent of the poplar extract dry weight, they exert approximately 50 percent of total antioxidant activity.

Individual evaluation of major PCs present in plant tissue extracts facilitates the evaluation of molecular interactions between antioxidants to determine whether synergistic or antagonistic effects occur. For example, Palafox-Carlos et al. (2012a) evaluated molecular interactions between major phenolic acids (chlorogenic, gallic, protocatechuic, and vanillic acids) found in mango cv. Ataulfo. They found that more than 80% of PC combinations showed a synergistic molecular interaction against DPPH radicals, while only vanillic acid exerted antagonistic effects. Antioxidant synergism between PCs can be enhanced by adding other antioxidant molecules such as α-tocopherol or vitamin C. Jia et al. (1998) reported that the major PCs found in green tea have a synergistic antioxidant effect in combination with α-tocopherol due to PCs recycling tocopherol molecules.

The antioxidant contribution of PCs in fruit extracts can be evaluated by fractioning the individual components. Del Carlo et al. (2004) evaluated the contributions of PC fractions to the antioxidant activity and oxidative stability of olive oil, attributing most of the radical scavenging activity to syringic acid, *p*-coumaric acid, and 3'4'-DHPEA-EA, an isomer of oleuropein aglycone. Likewise, Saini et al. (2014) reported that among six PCs (catechin, caffeic acid, ellagic acid, gallic acid, tannic acid, and *trans*-cinnamic acid) in Himalayan yellow raspberry, the major contributors to the antioxidant activity are ellagic acid and gallic acid. Both gallic and ellagic acids are related with an antiproliferative activity of the fruit extract against cervical cancer cells. According to the authors, consumption of this fruit may play an important role against oxidative stress and prevention of degenerative diseases and cancer.

Regular intake of fruits and vegetables is associated with a reduced risk of cardiovascular diseases and to an overall improvement in health, and recent studies suggest that health benefits of PC consumption extend beyond their antioxidant activity to effects such as weight loss, reduced blood pressure, favored insulin sensitivity, and reduced plasma glucose (Andújar et al. 2012, Bräunlich et al. 2013, Chiva-Blanch and Visioli

2012, Oboh et al. 2012). For example, edible berries have demonstrated a wide spectrum of biomedical functions because their anthocyanin content triggers genetic signaling that promotes human health and prevents disease (Zafra-Stone et al. 2007). It has also been reported that fruits rich in antioxidant dietary fiber can create an antioxidant environment in the large intestine during fermentation by the microbiota (Pérez-Jiménez and Saura-Calixto 2015). In addition, PCs metabolized by the intestinal microbiota, can generate metabolites with more biological properties than the parent compound. This indicates an interaction between PCs and microbiota, where gut microorganisms' impact is on PC absorption, while PC metabolites affect the growth of certain bacteria (Valdés et al. 2015). These studies suggest that the benefits of PC intake are exerted through a series of biological events caused in parallel by their antioxidant activity, such as modulation of enzyme activity, nuclear receptors, gene expression, and multiple signaling pathways.

18.6 SUMMARY POINTS

- PC extraction from fruits requires methods that isolate them without altering their chemical structures.
- PCs may be considered extractable or nonextractable according to their association with the food matrix. NEPCs resist common solvents and require chemical or enzymatic hydrolysis to separate them from the food matrix.
- Identification and quantification of fruit PCs can be achieved by spectrophotometric methods such as UV-Vis and FT-IR, but chromatographic methods such as HPLC and GC are more robust.
- Standardization of extraction, identification, and quantification methods allows comparable and reproducible results.
- Antioxidant capacities of fruits are commonly high, but the particular PC profile will ultimately determine the antioxidant capacity value of each fruit through interactions between PCs.
- Although antioxidant capacity is the most known and described effect of PCs, several other mechanisms can modulate cellular processes that contribute to overall health.

ACKNOWLEDGMENTS

This work is part of CONACYT Fronteras de la Ciencia project 563: "Un enfoque multidisciplinario de la farmacocinética de polifenoles de mango Ataulfo. Interacciones moleculares, estudios preclínicos y clínicos."

REFERENCES

Aaby, K., Mazur, S., Nes, A., and Skrede, G. 2012. Phenolic compounds in strawberry (*Fragaria* x *ananassa* Duch.) fruits: Composition in 27 cultivars and changes during ripening. *Food Chem* 132 (1):86–97.

Acosta-Estrada, B. A., Gutiérrez-Uribe, J. A., and Serna-Saldívar, S. O. 2014. Bound phenolics in foods, a review. *Food Chem* 152:46–55.

Ahmad, N., Zuo, Y., Lu, X., Anwar, F., and Hameed, S. 2016. Characterization of free and conjugated phenolic compounds in fruits of selected wild plants. *Food Chem* 190:80–89.

Alonso, Á. M., Guillén, D. A., Barroso, C. G., Puertas, B., and García, A. 2002. Determination of antioxidant activity of wine byproducts and its correlation with polyphenolic content. *J Agric Food Chem* 50 (21):5832–5836.

Alothman, M., Bhat, R., and Karim, A. 2009. Antioxidant capacity and phenolic content of selected tropical fruits from Malaysia, extracted with different solvents. *Food Chem* 115 (3):785–788.

Andújar, I., Recio, M., Giner, R., and Ríos, J. 2012. Cocoa polyphenols and their potential benefits for human health. *Oxid Med Cell Longev* Vol 2012, Article ID 906252, pp. 1–23.

Antolovich, M., Prenzler, P., Robards, K., and Ryan, D. 2000. Sample preparation in the determination of phenolic compounds in fruits. *Analyst* 125 (5):989–1009.

Arranz, S., Saura-Calixto, F., Shaha, S., and Kroon, P. A. 2009. High contents of non extractable polyphenols in fruits suggest that polyphenol contents of plant foods have been underestimated. *J Agric Food Chem* 57 (16):7298–7303.

Arts, I. C. W., and Hollman, P. C. H. 1998. Optimization of a quantitative method for the determination of catechins in fruits and legumes. *J Agric Food Chem* 46 (12):5156–5162.

Arts, I. C. W., van de Putte, B., and Hollman, P. C. H. 2000. Catechin contents of foods commonly consumed in the Netherlands. 1. Fruits, vegetables, staple foods, and processed foods. *J Agric Food Chem* 48 (5):1746–1751.

Babbar, N., Oberoi, H. S., Uppal, D. S., and Patil, R. T. 2011. Total phenolic content and antioxidant capacity of extracts obtained from six important fruit residues. *Food Res Int* 44 (1):391–396.

Balasundram, N., Sundram, K., and Samman, S. 2006. Phenolic compounds in plants and agri-industrial by-products: Antioxidant activity, occurrence, and potential uses. *Food Chem* 99 (1):191–203.

Bräunlich, M., Slimestad, R., Wangensteen, H. et al. 2013. Extracts, anthocyanins and procyanidins from Aronia melanocarpa as radical scavengers and enzyme inhibitors. *Nutrients* 5 (3):663–678.

Ceymann, M., Arrigoni, E., Schärer, H., Nising, A. B., and Hurrell, R. F. 2012. Identification of apples rich in health-promoting flavan-3-ols and phenolic acids by measuring the polyphenol profile. *J Food Compos Anal* 26 (1–2):128–135.

Chamorro, S., Viveros, A., Arija, I. et al. 2013. Iron chelating property of dietary grape polyphenols and its role in the prevention of lipid oxidation. *Am J Hematol* 88 (5):E135–E136.

Chiva-Blanch, G., and Visioli, F. 2012. Polyphenols and health: Moving beyond antioxidants. *J Berry Res* 2 (2):63–71.

Chun, O. K., Kim, D.-O., Moon, H. Y., Kang, H. G., and Lee, C. Y. 2003. Contribution of individual polyphenolics to total antioxidant capacity of plums. *J Agric Food Chem* 51 (25):7240–7245.

Cozzolino, D. 2015. Infrared spectroscopy as a versatile analytical tool for the quantitative determination of antioxidants in agricultural products, foods and plants. *Antioxidants* 4 (3):482–497.

Del Carlo, M., Sacchetti, G., Di Mattia, C. et al. 2004. Contribution of the phenolic fraction to the antioxidant activity and oxidative stability of olive oil. *J Agric Food Chem* 52 (13):4072–4079.

Di Stefano, V., Avellone, G., Bongiorno, D. et al. 2017. Quantitative evaluation of the phenolic profile in fruits of six avocado (*Persea americana*) cultivars by UHPLC-HESI-MS. *Int J Food Prop* 20 (6):1302–1312.

Dubost, N. J., Ou, B., and Beelman, R. B. 2007. Quantification of polyphenols and ergothioneine in cultivated mushrooms and correlation to total antioxidant capacity. *Food Chem* 105 (2):727–735.

Dudonné, S., Poupard, P., Coutiere, P. et al. 2011. Phenolic composition and antioxidant properties of poplar bud (*Populus nigra*) extract: Individual antioxidant contribution of phenolics and transcriptional effect on skin aging. *J Agric Food Chem* 59 (9):4527–4536.

Falleh, H., Trabelsi, N., Bonenfant-Magné, M. et al. 2013. Polyphenol content and biological activities of Mesembryanthemum edule organs after fractionation. *Ind Crop Prod* 42:145–152.

FAO. 2002. Agricultural commodities: Profiles and relevant wto negotiating issues. FAO, accessed August 2017 from http://www.fao.org/docrep/006/y4343e/y4343e00.htm#Contents.

Ferrer-Gallego, R., Hernández-Hierro, J. M., Rivas-Gonzalo, J. C., and Escribano-Bailón, M. T. 2011. Determination of phenolic compounds of grape skins during ripening by NIR spectroscopy. *LWT-Food Sci Technol* 44 (4):847–853.

George, S., Brat, P., Alter, P., and Amiot, M. J. 2005. Rapid determination of polyphenols and vitamin C in plant-derived products. *J Agric Food Chem* 53 (5):1370–1373.

Gramza, A., Khokhar, S., Yoko, S. et al. 2006. Antioxidant activity of tea extracts in lipids and correlation with polyphenol content. *Eur J Lipid Sci Technol* 108 (4):351–362.

Guimarães, R., Barros, L., Dueñas, M. et al. 2013. Characterisation of phenolic compounds in wild fruits from Northeastern Portugal. *Food Chem* 141 (4):3721–3730.

Haminiuk, C. W., Maciel, G. M., Plata-Oviedo, M. S., and Peralta, R. M. 2012. Phenolic compounds in fruits: An overview. *Int J Food Sci Technol* 47 (10):2023–2044.

Heo, H. J., Kim, Y. J., Chung, D., and Kim, D.-O. 2007. Antioxidant capacities of individual and combined phenolics in a model system. *Food Chem* 104 (1):87–92.

Ignat, I., Volf, I., and Popa, V. I. 2011. A critical review of methods for characterisation of polyphenolic compounds in fruits and vegetables. *Food Chem* 126 (4):1821–1835.

Jerman, T., Trebše, P., and Vodopivec, B. M. 2010. Ultrasound-assisted solid liquid extraction (USLE) of olive fruit (*Olea europaea*) phenolic compounds. *Food Chem* 123 (1):175–182.

Jia, Z.-S., Zhou, B., Yang, L., Wu, L.-M., and Liu, Z.-L. 1998. Antioxidant synergism of tea polyphenols and α-tocopherol against free radical induced peroxidation of linoleic acid in solution. *J Chem Soc Perk T* 2 (4):911–916.

Kader, A., and Yahia, E. 2011. *Postharvest biology of tropical and subtropical fruits*. Vol. 1, *Postharvest biology and technology of tropical and subtropical fruits*. Cambridge, UK: Woodhead Publishing.

Kelebek, H., Selli, S., Gubbuk, H., and Gunes, E. 2015. Comparative evaluation of volatiles, phenolics, sugars, organic acids and antioxidant properties of Sel-42 and Tainung papaya varieties. *Food Chem* 173:912–919.

Khoddami, A., Wilkes, M., and Roberts, T. 2013. Techniques for analysis of plant phenolic compounds. *Molecules* 18 (2):2328.

Kiselova, Y., Ivanova, D., Chervenkov, T. et al. 2006. Correlation between the in vitro antioxidant activity and polyphenol content of aqueous extracts from Bulgarian herbs. *Phytother Res* 20 (11):961–965.

Koley, T. K., Kaur, C., Nagal, S., Walia, S., and Jaggi, S. 2011. Antioxidant activity and phenolic content in genotypes of Indian jujube (*Zizyphus mauritiana* Lamk.). *Arab J Chem* 9, Supplement 2:S1044–S1052.

Kosinska, A., Karamac, M., Estrella, I. et al. 2012. Phenolic compound profiles and antioxidant capacity of *Persea americana* Mill. peels and seeds of two varieties. *J Agric Food Chem* 60 (18):4613–4619.

Lam, H. S., Proctor, A., Howard, L., and Cho, M. J. 2005. Rapid fruit extracts antioxidant capacity determination by Fourier transform infrared spectroscopy. *J Food Sci* 70 (9):C545–C549.

Landbo, A.-K., and Meyer, A. S. 2001. Enzyme-assisted extraction of antioxidative phenols from black currant juice press residues (*Ribes nigrum*). *J Agric Food Chem* 49 (7):3169–3177.

Lester, G. E., Lewers, K. S., Medina, M. B., and Saftner, R. A. 2012. Comparative analysis of strawberry total phenolics via Fast Blue BB vs. Folin–Ciocalteu: Assay interference by ascorbic acid. *J Food Comp Anal* 27 (1):102–107.

Lim, Y., Lim, T., and Tee, J. 2007. Antioxidant properties of several tropical fruits: A comparative study. *Food Chem* 103 (3):1003–1008.

Liu, C., Liu, W., Chen, W., Yang, J., and Zheng, L. 2015. Feasibility in multispectral imaging for predicting the content of bioactive compounds in intact tomato fruit. *Food Chem* 173:482–488.

Macheix, J.-J., and Fleuriet, A. 1990. *Fruit Phenolics*. Boca Raton, FL: CRC Press.

Maisarah, A., Nurul Amira, B., Asmah, R., and Fauziah, O. 2013. Antioxidant analysis of different parts of *Carica papaya*. *Int Food Res J* 20 (3):1043–1048.

Manganaris, G. A., Goulas, V., Vicente, A. R., and Terry, L. A. 2014. Berry antioxidants: Small fruits providing large benefits. *J Sci Food Agric* 94 (5):825–833.

Medina, M. B. 2011a. Determination of the total phenolics in juices and superfruits by a novel chemical method. *J Funct Foods* 3 (2):79–87.

Medina, M. B. 2011b. Simple and rapid method for the analysis of phenolic compounds in beverages and grains. *J Agric Food Chem* 59 (5):1565–1571.

Melgarejo, P., Calín-Sánchez, Á., Carbonell-Barrachina, Á. A. et al. 2014. Antioxidant activity, volatile composition and sensory profile of four new very-early apricots (*Prunus armeniaca* L.). *J Sci Food Agric* 94 (1):85–94.

Meng, J.-F., Fang, Y.-L., Qin, M.-Y., Zhuang, X.-F., and Zhang, Z.-W. 2012. Varietal differences among the phenolic profiles and antioxidant properties of four cultivars of spine grape (*Vitis davidii* Foex) in Chongyi County (China). *Food Chem* 134 (4):2049–2056.

Morais, D. R., Rotta, E. M., Sargi, S. C. et al. 2015. Antioxidant activity, phenolics and UPLC–ESI (–)–MS of extracts from different tropical fruits parts and processed peels. *Food Res Int* 77:392–399.

Mphahlele, R. R., Stander, M. A., Fawole, O. A., and Opara, U. L. 2014. Effect of fruit maturity and growing location on the postharvest contents of flavonoids, phenolic acids, vitamin C and antioxidant activity of pomegranate juice (cv. Wonderful). *Sci Hortic* 179:36–45.

Mustafa, R., Hamid, A. A., Mohamed, S., and Bakar, F. A. 2010. Total phenolic compounds, flavonoids, and radical scavenging activity of 21 selected tropical plants. *J Food Sci* 75 (1):C28–C35.

Noratto, G. D., Bertoldi, M. C., Krenek, K. et al. 2010. Anticarcinogenic effects of polyphenolics from mango (*Mangifera indica*) varieties. *J Agric Food Chem* 58 (7):4104–4112.

Oboh, G., Ademiluyi, A. O., Akinyemi, A. J. et al. 2012. Inhibitory effect of polyphenol-rich extracts of jute leaf (*Corchorus olitorius*) on key enzyme linked to type 2 diabetes (α-amylase and α-glucosidase) and hypertension (angiotensin I converting) in vitro. *J Funct Food* 4 (2):450–458.

Oliveira, A. L., Destandau, E., Fougère, L., and Lafosse, M. 2014. Isolation by pressurised fluid extraction (PFE) and identification using CPC and HPLC/ESI/MS of phenolic compounds from Brazilian cherry seeds (*Eugenia uniflora* L.). *Food Chem* 145:522–529.

Oroian, M., and Escriche, I. 2015. Antioxidants: Characterization, natural sources, extraction and analysis. *Food Res Int* 74:10–36.

Padayatty, S. J., Sun, H., Wang, Y. et al. 2004. Vitamin C pharmacokinetics: Implications for oral and intravenous use. *Ann Intern Med* 140 (7):533–537.

Palafox-Carlos, H., Gil-Chávez, J., Sotelo-Mundo, R. R. et al. 2012. Antioxidant interactions between major phenolic compounds found in 'Ataulfo'mango pulp: Chlorogenic, gallic, protocatechuic and vanillic acids. *Molecules* 17 (11):12657–12664.

Palafox-Carlos, H., Yahia, E. M., and Gonzalez-Aguilar, G. A. 2012. Identification and quantification of major phenolic compounds from mango (*Mangifera indica*, cv. Ataulfo) fruit by HPLC-DAD-MS/MS-ESI and their individual contribution to the antioxidant activity during ripening. *Food Chem* 135 (1):105–111.

Park, Y.-S., Cvikrová, M., Martincová, O. et al. 2015. In vitro antioxidative and binding properties of phenolics in traditional, citrus and exotic fruits. *Food Res Int* 74:37–47.

Park, Y. S., Im, M. H., Choi, J. H. et al. 2014. Effect of long- term cold storage on physicochemical attributes and bioactive components of kiwi fruit cultivars. *CYTA J Food* 12 (4):360–368.

Park, Y.-S., Im, M. H., Ham, K.-S. et al. 2015. Quantitative assessment of the main antioxidant compounds, antioxidant activities and FTIR spectra from commonly consumed fruits, compared to standard kiwi fruit. *LWT-Food Sci Technol* 63 (1):346–352.

Pekal, A., and Pyrzynska, K. 2014. Evaluation of aluminium complexation reaction for flavonoid content assay. *Food Anal Method* 7 (9):1776–1782.

Pérez-Jiménez, J., and Saura-Calixto, F. 2015. Macromolecular antioxidants or non-extractable polyphenols in fruit and vegetables: Intake in four European countries. *Food Res Int* 74:315–323.

Pérez-Jiménez, J., and Torres, J. L. 2011. Analysis of nonextractable phenolic compounds in foods: The current state of the art. *J Agric Food Chem* 59 (24):12713–12724.

Pinelli, P., Romani, A., Fierini, E., Remorini, D., and Agati, G. 2013. Characterisation of the polyphenol content in the kiwifruit (*Actinidia deliciosa*) exocarp for the calibration of a fruit-sorting optical sensor. *Phytochem Anal* 24 (5):460–466.

Robbins, R. J. 2003. Phenolic acids in foods: An overview of analytical methodology. *J Agric Food Chem* 51 (10):2866–2887.

Robles-Sánchez, M., Astiazarán-García, H., Martín-Belloso, O. et al. 2011. Influence of whole and fresh-cut mango intake on plasma lipids and antioxidant Capacity of healthy adults. *Food Res Int* 44 (5):1386–1391.

Robles-Sánchez, R. M., Rojas-Graü, M. A., Odriozola-Serrano, I., González-Aguilar, G., and Martin-Belloso, O. 2013. Influence of alginate-based edible coating as carrier of antibrowning agents on bioactive compounds and antioxidant activity in fresh-cut Kent mangoes. *LWT-Food Sci Technol* 50 (1):240–246.

Rocha Ribeiro, S. M., Queiroz, J. H., Lopes Ribeiro de Queiroz, M. E., Campos, F. M., and Pinheiro Sant'Ana, H. M. 2007. Antioxidant in mango (*Mangifera indica* L.) pulp. *Plant Food Hum Nutr* 62 (1):13–17.

Rockenbach, I. I., Gonzaga, L. V., Rizelio, V. M. et al. 2011a. Phenolic compounds and antioxidant activity of seed and skin extracts of red grape (*Vitis vinifera* and *Vitis labrusca*) pomace from Brazilian winemaking. *Food Res Int* 44 (4):897–901.

Rockenbach, I. I., Rodrigues, E., Gonzaga, L. V. et al. 2011b. Phenolic compounds content and antioxidant activity in pomace from selected red grapes (*Vitis vinifera* L. and *Vitis labrusca* L.) widely produced in Brazil. *Food Chem* 127 (1):174–179.

Romaric, G., Oumou, H., and Mamoudou, H. 2011. Phenolic compounds and antioxidant activities in some fruits and vegetables from Burkina Faso. *Afr J Biotechnol* 10 (62):13548–13547.

Saini, R., Dangwal, K., Singh, H., and Garg, V. 2014. Antioxidant and antiproliferative activities of phenolics isolated from fruits of Himalayan yellow raspberry (*Rubus ellipticus*). *J Food Sci Technol* 51 (11):3369–3375.

Saura-Calixto, F. 2012. Concept and health-related properties of nonextractable polyphenols: The missing dietary polyphenols. *J Agric Food Chem* 60 (45):11195–11200.

Sellamuthu, P. S., Denoya, G. I., Sivakumar, D., Polenta, G. A., and Soundy, P. 2013. Comparison of the contents of bioactive compounds and quality parameters in selected mango cultivars. *J Food Qual* 36 (6):394–402.

Silva, L. M. R. d., Figueiredo, E. A. T. d., Ricardo, N. M. P. S. et al. 2014. Quantification of bioactive compounds in pulps and by-products of tropical fruits from Brazil. *Food Chem* 143:398–404.

Sinelli, N., Spinardi, A., Di Egidio, V., Mignani, I., and Casiraghi, E. 2008. Evaluation of quality and nutraceutical content of blueberries (*Vaccinium corymbosum* L.) by near and mid-infrared spectroscopy. *Postharvest Biol Technol* 50 (1):31–36.

Singleton, V. L., and Rossi, J. A. 1965. Colorimetry of total phenolics with phosphomolybdic-phosphotungstic acid reagents. *Am J Enol Viticult* 16 (3):144.

Spínola, V., Pinto, J., and Castilho, P. C. 2015. Identification and quantification of phenolic compounds of selected fruits from Madeira Island by HPLC-DAD-ESI-MSn and screening for their antioxidant activity. *Food Chem* 173:14–30.

Stratil, P., Klejdus, B., and Kuban, V. 2006. Determination of total content of phenolic compounds and their antioxidant activity in vegetables evaluation of spectrophotometric methods. *J Agric Food Chem* 54 (3):607–616.

Sun, J., Chu, Y.-F., Wu, X., and Liu, R. H. 2002. Antioxidant and antiproliferative activities of common fruits. *J Agric Food Chem* 50 (25):7449–7454.

Svedström, U., Vuorela, H., Kostiainen, R., Laakso, I., and Hiltunen, R. 2006. Fractionation of polyphenols in hawthorn into polymeric procyanidins, phenolic acids and flavonoids prior to high-performance liquid chromatographic analysis. *J Chromatogr A* 1112 (1–2):103–111.

Tsanova-Savova, S., Ribarova, F., and Gerova, M. 2005. (+)-Catechin and (–)-epicatechin in Bulgarian fruits. *J Food Compos Anal* 18 (7):691–698.

Valdés, L., Cuervo, A., Salazar, N. et al. 2015. The relationship between phenolic compounds from diet and microbiota: Impact on human health. *Food Funct* 6 (8):2424–2439.

Vasco, C., Ruales, J., and Kamal-Eldin, A. 2008. Total phenolic compounds and antioxidant capacities of major fruits from Ecuador. *Food Chem* 111 (4):816–823.

Velderrain-Rodríguez, G., Palafox-Carlos, H., Wall-Medrano, A. et al. 2014. Phenolic compounds: Their journey after intake. *Food Funct* 5 (2):189–197.

Velderrain-Rodríguez, G., Quirós-Sauceda, A., Mercado-Mercado, G. et al. 2016. Effect of dietary fiber on the bioaccessibility of phenolic compounds of mango, papaya and pineapple fruits by an in vitro digestion model. *Food Sci Technol (Campinas)* 36 (2):188–194.

Wang, C., Zhao, J., Chen, F., Cheng, Y., and Guo, A. 2012. Separation, identification, and quantitation of phenolic acids in Chinese waxberry (*Myrica rubra*) juice by HPLC-PDA-ESI-MS. *J Food Sci* 77 (2):C272–C277.

Wang, H.-C., Hu, Z.-Q., Wang, Y., Chen, H.-B., and Huang, X.-M. 2011. Phenolic compounds and the antioxidant activities in litchi pericarp: Difference among cultivars. *Sci Hortic* 129 (4):784–789.

Wang, X., Li, C., Liang, D. et al. 2014. Phenolic compounds and antioxidant activity in red-fleshed apples. *J Funct Food* 18:1086–1094.

Wen, L., Guo, X., Liu, R. H. et al. 2015. Phenolic contents and cellular antioxidant activity of Chinese hawthorn "*Crataegus pinnatifida.*" *Food Chem* 186:54–62.

Zafra-Stone, S., Yasmin, T., Bagchi, M. et al. 2007. Berry anthocyanins as novel antioxidants in human health and disease prevention. *Mol Nutr Food Res* 51 (6):675–683.

Zhao, P., Duan, L., Guo, L. et al. 2015. Chemical and biological comparison of the fruit extracts of *Citrus wilsonii* Tanaka and *Citrus medica* L. *Food Chem* 173:54–60.

Zheng, H.-Z., Hwang, I.-W., and Chung, S.-K. 2009. Enhancing polyphenol extraction from unripe apples by carbohydrate-hydrolyzing enzymes. *J Zhejiang Univ Sci B* 10 (12):912–919.

CHAPTER 19

Phenolic Compounds in Cocoa and Chocolate

Griselda Rabadan-Chavez and Eugenia Lugo-Cervantes

CONTENTS

19.1 CACAO ORIGINS AND VARIETIES

Cacao beans are derived from the wide-branched evergreen tree *Theobroma cacao L* (*Sterculiaceae* family), which currently grows throughout the rainforests of the Western hemisphere between latitudes 18°N and 15°S. The species demand a warm temperature (18/21–30/32°C) with small oscillations, a constant high humidity (70–100 percent), and shade (protection against direct radiation). Several species are often found at the edges of rivers or marshes in temporarily flooded areas, while others always grow on elevated (up to 600 m) drained places (Young 2007; Colombo et al. 2012).

The standard height of the cocoa tree is kept at 2–5 meters, though it can be as tall as 8–12 meters. Their seedpods, named *caboose*, are up to 40 centimeters long and 10 centimeters wide and show different colors (green, yellow, red purple) and shapes (e.g., even, warty, round, oval). They contain 20–50 almond-sized seeds that are surrounded by mucilaginous pulp that arises from the bean teguments (Young 2007; Colombo et al. 2012).

Although some controversy still exists about the origin of the cocoa tree, several authors support a neotropical origin with natural dispersion in tropical lowland rainforests extending from the Amazon basin through Southern Mexico. Nonetheless, the upper

Amazon basin is considered as the center of origin of *Theobroma cacao L*, since the highest genetic diversity has been found in this region (Brooks and Guard 1952; Cuatrecasas 1964; Motamayor et al. 2002, 2008).

Based on the geographical location and morphological diversity observed in Central America as well as in South America, two main subspecies were originally defined: *T. cacao ssp. cacao* (cacao Criollo) and *T. cacao ssp. sphaerocarpum* (cacao Forastero) (Cuatrecasas 1964; Laurent et al. 1994; Motamayor et al. 2002, 2008).

The Criollo variety is composed of trees with thick white or rosy beans yielding the most flavored and finest chocolate. It is highly aromatic and it develops mild, nutty, earthy, flowery, or tea-like flavors. They were the first cocoa trees to be cultivated and domesticated in Mesoamerica (southern Mexico and Central America) during the pre-Columbian and colonial period. Currently, this variety is cultivated mainly in Mexico, Venezuela, Dominican Republic, Peru, Colombia, New Guinea, Java, and Madagascar. Their ripe pods are yellow or red, and the beans are large and round with white-colored cotyledons. At present, Criollo trees are infrequently grown because of their low resistance to climatic changes, their susceptibility to disease and pests, and their low yields. They have been steadily replaced by more disease-resistant and productive Forastero and Trinitario clones (Laurent et al. 1994; Motamayor and Lanaud 2002; Young 2007; Colombo et al. 2012).

The Forastero variety is a productive and vigorous type cultivated since historic times. The beans are small and flat with violet cotyledons. It is subdivided into diverse populations with different geographic origins: Upper Amazon, lower Amazon, Orinoco, and the Guianas. Lower Amazon Forastero trees were initially cultivated in the Amazon basin and were the first to be introduced into Africa. On the other hand, upper Amazon Forastero cultivars are often used in breeding programs due to their greater resistance to disease and higher yield. Although the flavor and taste of these beans are much less aromatic than the fine varieties (classified as bulk, basic, or ordinary cocoa grade), Nacional cocoa seeds are characterized by excellent aroma and taste similar to Criollo and Trinitario. Today, Forastero trees are mainly grown in Africa, Ecuador, and Brazil and account for 80 percent of the world's cocoa supply (Motamayor and Lanaud 2002; Young 2007; Colombo et al. 2012).

The Trinitario cacao is a third hybrid group originating from crosses between Criollo and Forastero, developed to save the Criollo tree from becoming extinct. It has higher yields and it is less susceptible to diseases than the others. Trinitario cacao trees are grown mainly in Colombia and Central America, but also in Cameroon, Samoa, Sri Lanka, Java, and Papua New Guinea. The features of Trinitario beans vary depending on tree growth habit and the genetic information of the parent trees. However, they are often described as flavorful cocoa beans, perceived as aromatic or smooth with fruity, raisin, floral, spicy, nutty, molasses, and caramel notes coming from vigorous and productive trees (Motamayor et al. 2003; Young 2007; Colombo et al. 2012).

19.2 CACAO BEAN COMPOSITION

The physics and chemistry of cocoa beans are very complex and change throughout the life of the bean. Additionally, variations in bean composition may arise from botanical origin, location of growth, and agronomic and processing conditions (Young 2007; Colombo et al. 2012).

Fresh cocoa beans consist essentially of two cotyledons or kernels (86–90 percent dry weight of the bean) enclosed in a shell or testa (10–14 percent dry weight of the bean). The testa acts as a semipermeable barrier to the flow of substances between the seed and pulp. It is freely permeable to water, ethanol, acetic, and lactic acids and some volatile organic compounds. The cotyledons act as storage organs containing two types of parenchyma storage cells. Polyphenolic cells make up about 14 to 20 percent of the cocoa seed (dry basis) and consist of one single large vacuole filled with polyphenols and alkaloids (caffeine, theobromine, and theophylline). On the other hand, lipid–protein cells have cytoplasms tightly packed with multiple small protein and lipid vacuoles and other components, all of which play roles in defining cocoa flavor and aroma characteristics. Cotyledons are approximately composed of 32–39 percent water, 30–32 percent fat, 8–10 percent protein, 4–6 percent starch, 2–3 percent cellulose, 2–3 percent sucrose, 5–6 percent polyphenols, 1–2 percent theobromine, 1 percent organic acids, and 1 percent caffeine (Wollgast and Anklam 2000; Elwers et al. 2009; Oracz, Zyzelewicz, and Nebesny 2015).

19.2.1 Nutrients

19.2.1.1 Lipid Content

Total fat content accounts for 50 to 57 percent of the dry weight of cocoa beans. It has been reported that cocoa butter contains a higher proportion of neutral lipids (98 percent) compared to polar lipids (1–2 percent). Neutral lipids are mainly composed by triglycerides (75 percent) containing mostly stearic acid (18:0; 35 percent), palmitic acid (16:0; 25 percent), oleic acid (18:1; 35 percent), and linoleic acid (18:2; 3 percent). Among the polar lipids, phospholipids and glycolipids account for approximately 30 percent and 70 percent, respectively. The phospholipid fraction is mainly composed of phosphatidylcholine, phosphatidylinositol, phosphatidylethanolamine, and lyso-phosphatidylcholine (Parsons et al. 1969; Hernandez et al. 1991; Gould et al. 2016).

19.2.1.2 Carbohydrate Content

Total polysaccharides have been reported to represent approximately 12 percent of cocoa bean composition after drying. Starch is the major digestible polysaccharide in cotyledons, ranging from 3 to 7 percent on a dry basis (36 percent amylose and 64 percent amylopectin), whereas cellulose is one of the predominant components of the cell wall polysaccharides (~35 percent), together with pectic polysaccharides (~45 percent; heterogeneous mixture of rhamnogalacturonans with variable degrees of branching) and hemicelluloses (~20 percent; mixture of a fucosylated xyloglucan, galactoglucomannans, and glucuronoarabinoxylan). The predominant sugar in cocoa beans is sucrose (~90 percent of total sugars), followed by fructose and glucose (~6 percent of total sugars), and trace amounts of sorbose, mannitol, and inositol (Redgwell and Hansen 2000; Redgwell et al. 2003).

19.2.1.3 Protein Content

Protein is the second-most abundant constituent of cocoa beans (10–15 percent on dry basis). The total protein content can be classified into four predominant fractions: Albumin (water-soluble), globulin (salt-soluble), prolamin (alcohol-soluble), and glutelin (soluble in dilute acids or alkali). Albumin and globulin are the two major fractions, accounting for 52 percent and 43 percent, respectively, of total protein in cocoa beans; glutelins and prolamins are present in a lower concentrations (~5 percent and 1 pecent, respectively). Fermentation, roasting, and drying conditions greatly influence microbiological and enzymatic reactions that lead to an extensive breakdown of cocoa proteins into peptides and free amino acids that, together with reducing sugars, lead to the flavor formation in cocoa-derived products (Voigt and Biehl 1993; Bertazzo et al. 2011).

19.2.2 Nonnutrients Content

Cocoa is rich in phenolic compounds contained in the polyphenol storage cells of cocoa cotyledons. Depending on the content of anthocyanins, these pigment cells can vary significantly in color, from white or pale pink to deep purple (Elwers et al. 2009; Hii et al. 2009; Oracz, Zyzelewicz, and Nebesny 2015).

Polyphenols are biologically active secondary metabolites in plants, acting as cell wall support materials, colorful attractants for birds and insects, and defensive protections under different environmental stress conditions (wounding, infection, excessive light, or UV irradiation). Moreover, they are also responsible for the bitter and astringent taste developed during the processing steps of cocoa beans (Elwers et al. 2009; Hii et al. 2009; Oracz, Zyzelewicz, and Nebesny 2015).

The three main groups of polyphenols that have been reported in cocoa beans are flavan-3-ols (~37 percent), procyanidins (~58 percent), and anthocyanins (~4 percent). Among flavan-3-ols, (−)-epicatechin is the predominant monomeric form (up to 35 percent of total polyphenols) in defatted freshly harvested cocoa beans, followed by smaller quantities of (+)-catechin, and trace amounts of (+)-gallocatechin, (−)-epigallocatechin, and (−)-epicatechin-3-O-gallate. Procyanidins are composed of flavan-3-ol monomers units linked through an interflavanoid linkage. Those with 2–10 units are defined as oligomeric and those with >10 units as polymeric procyanidins. The anthocyanin fraction is dominated by two cyanidin glycosides: Cyanidin-3-α-L-arabinoside and cyanidin-3-ß-D-galactoside (Wollgast and Anklam 2000; Elwers et al. 2009; Oracz, Zyzelewicz, and Nebesny 2015).

Other phenolic compounds, such as flavonols (quercetin aglycone, quercetin-3-O-glucoside, quercetin-3-O-arabinoside, quercetin-3-O-galactoside, and quercetin-3-O-glucuronide), flavones (apigenin, luteolin and some of their glycosides), and flavanones (naringenin and naringenin-7-O-glucoside) have been identified in cocoa beans (Wollgast and Anklam 2000; Oracz, Zyzelewicz, and Nebesny 2015).

Additionally, nonflavonoid polyphenols such as hydroxybenzoic and hydroxycinnamic acid derivatives have also been found in cocoa beans. Gallic, p-hydroxybenzoic, protocatechuic, vanillic, and syringic acids are the main benzoic acid derivatives detected in cocoa beans, whereas caffeic, ferulic, and p-coumaric acids are the predominant hydroxycinnamic acids (Oracz, Zyzelewicz, and Nebesny 2015).

Total polyphenols in cocoa beans have been reported in the range of 40 to 84.2 milligrams GAE/gram, depending on geographical origin and variety. Among all the tested varieties, the lowest total polyphenol content was found in the Criollo beans. Their polyphenol content is only two-thirds of the amount of these compounds in the Forastero variety. However, it has been shown that high levels of procyanidins in Criollo seeds contribute to the astringency and bitter taste of the chocolate produced with this cultivar. Additionally, Criollo beans resulted a variety with very high amounts of pyrazines, which comprise over 40 percent of the cocoa powder essence and can be used as tracers for the cocoa flavor. Compared to Criollo, Forastero beans have shown lower procyanidin levels and almost similar amounts of pyrazines. As the concentration of aromatic compounds in Forastero beans is lower than in Criollo, the flavor precursors before roasting are also slightly lower (Wollgast and Anklam 2000; Counet et al. 2004; Elwers et al. 2009; Saltini et al. 2013; Oracz, Zyzelewicz, and Nebesny 2015).

Chemical differences between Criollo and Forastero beans are reflected in the sensorial profile of the final product, since chocolate produced from the Forastero beans has shown to be less bitter, less astringent, and less acidic than chocolate produced with Criollo or Trinitario beans. Even though Trinitario cultivar has shown similar levels of procyanidins as Forastero beans, they have strong basic chocolate characters that are not found in other varieties (Wollgast and Anklam 2000; Elwers et al. 2009; Oracz, Zyzelewicz, and Nebesny 2015).

The effect of genotype on the variability of polyphenolic substances in cocoa seeds of the Forastero, Criollo, and Trinitario varieties from different origins is shown in Figure 19.1. Significant quantitative differences in anthocyanins content between Criollo and other cocoa types and subgroups have been reported. Criollo cocoa beans contain few or no anthocyanins in its composition. However, these seeds are characterized by a significantly higher concentration of (+)-N-(E)-caffeoyl-L-aspartic acid (caffeic acid aspartate) among the other cocoa samples (Elwers et al. 2009; Oracz, Zyzelewicz, and Nebesny 2015).

Flavanols are the most abundant flavonoids in cocoa beans (~60 percent of total phenolic compounds). They comprise the monomeric flavanols, (+)-catechin, and (−)- epicatechin, and their oligomeric and polymeric forms (procyanidins) (Beecher 2003; Fraga and Oteiza 2011; Oracz, Zyzelewicz, and Nebesny 2015).

Monomeric flavanols are characterized for having a C6–C3–C6 skeleton with a hydroxyl group in position three of the C ring. Monomeric catechin and epicatechin are stereoisomers at position 3 of the C ring, but have the same configuration at position 2 (Figure 19.2). Their respective enantiomers, namely (−)-catechin and (+)-epicatechin, are not commonly found in nature (Wollgast and Anklam 2000; Fraga and Oteiza 2011; Blumberg et al. 2014).

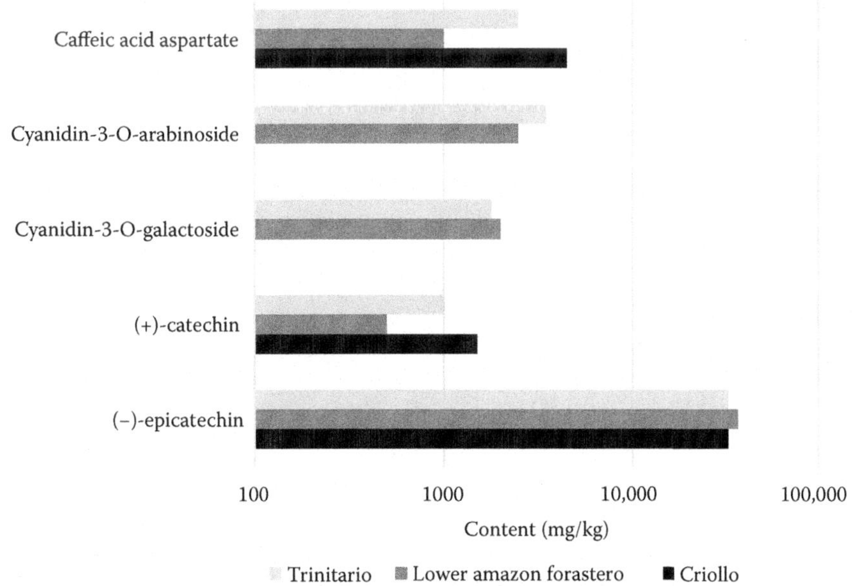

FIGURE 19.1 Content of main phenolic compounds in fresh and unfermented cocoa beans of different varieties (mg/kg fat-free dry mass). (Adapted from Oracz, J., Zyzelewicz, D., and Nebesny, E., *Critical Reviews in Food Science and Nutrition*, 55, 1176–1192, 2015.)

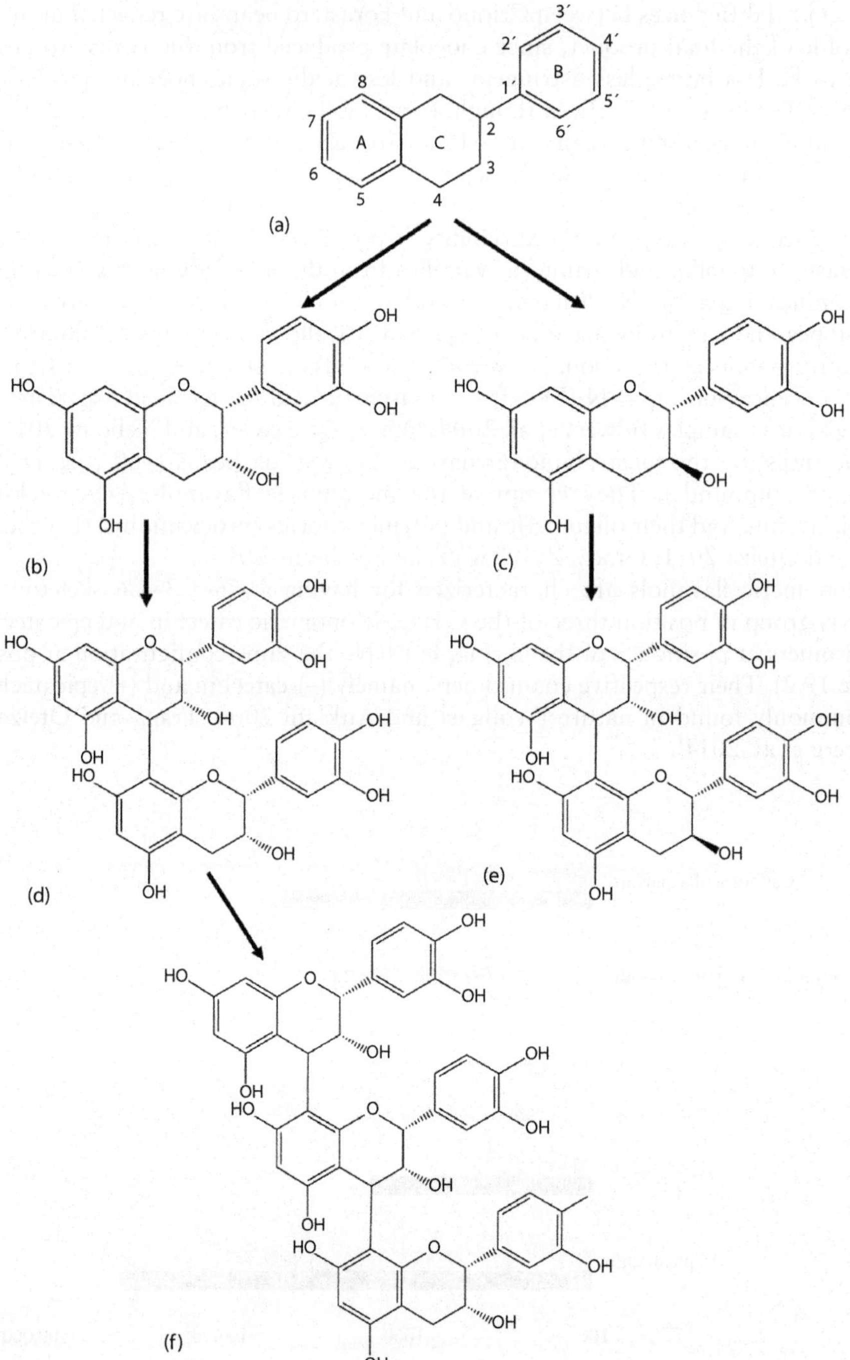

FIGURE 19.2 Basic structure and numbering system of flavonoids (a). Structure of the main monomeric flavan-3-ols: (–)-epicatechin (b) and (+)-catechin (c); and oligomeric flavanols: procyanidin dimer B2 (d), procyanidin dimer B1 (e), and procyanidin trimer C1 (f) from *Theobroma Cacao*. (Adapted from Oracz, J., Zyzelewicz, D., and Nebesny, E., *Critical Reviews in Food Science and Nutrition*, 55, 1176–1192, 2015.)

Flavan-3-ols are commonly found in their polymerized forms as oligomers (dimers to pentamers) or polymers (six or more units). The most common oligomers are the B series (B1 to B8), formed by two flavanol units (either catechin or epicatechin) and joined by a C4–C8 linkage (B1 to B4) or a C4–C6 linkage (B5 to B8). The least frequent dimers are the A series, characterized by the presence of double linkages between two monomeric flavanol units, one C4–C8 or C4–C6 and an additional one between C2 and C5 or C7 (Fraga and Oteiza 2011; Blumberg et al. 2014).

Apart from polyphenols, cocoa beans also contain methylxanthines, such as caffeine, theobromine, and theophylline. Theobromine is the major purine alkaloid present in cocoa beans (up to 4 percent), followed by caffeine (< 1 percent) and very low amounts of theophylline. In general, these compounds are widely known for their effects on the nervous system (increased concentration and attention), and vasodilatation. Organic acids are also present in cocoa beans, and their kind and amount will depend on the maturation and fermentation of the seeds. The most common organic acids are citric, oxalic, malic, acetic, and formic. Though, the most important is acetic acid because of its influence on the taste of cocoa-derived products (Wollgast and Anklam 2000; Oracz, Zyzelewicz, and Nebesny 2015).

Volatile organic compounds (such as alcohols, organic acids, aldehydes, ketones, esters, carboxylic acids, and pyrazines), mainly generated during fermentation and roasting of cocoa beans, are of great importance due to their influence in chocolate flavor formation. Thus, they have been widely studied to evaluate and improve the aroma quality of cocoa and chocolate. However, given the influence of genotype and origin on volatile compounds composition, it has been suggested that they could be useful markers for chocolate authentication (Acierno et al. 2016).

19.3 COCOA FLAVANOLS DURING BEAN PROCESSING

After harvesting, cocoa beans undergo complex processing that modifies their original chemical and physical properties in order to increase the palatability of final products. The phenolic content of cocoa beans has been shown to decrease up to 90 percent in the final product (cocoa powder or chocolate) throughout the different manufacturing processes. Particularly, primary processing, which includes the fermentation and drying stages, leads to considerable losses of cocoa flavanols and procyanidins (Figure 19.3). Nonetheless, secondary processing (roasting, alkalization, and conching) highly contributes to decrease in the concentration of cocoa polyphenols in the finished products (Afoakwa et al. 2008; Elwers et al. 2009; Saltini et al. 2013; Oracz, Zyzelewicz, and Nebesny 2015).

19.3.1 Fermentation

After the pods are cut from the trees, the beans with the adhering pulp are removed and transferred to heaps, boxes, or baskets for fermentation to take place. Fermentation is the first step in cocoa bean processing and it is crucial to generate aroma precursors in the final product. It is characterized by a well-known systematic microbial succession that starts with a yeast colonization that degrades pectin and produces ethanol by the secretion of pectinolytic enzymes and anaerobic fermentation of pulp sugars (glucose and fructose). Afterward, the growth of lactic acid bacteria (LAB) and acetic acid bacteria (AAB) leads (1) to microaerophilic fermentation of sugars and citric acid

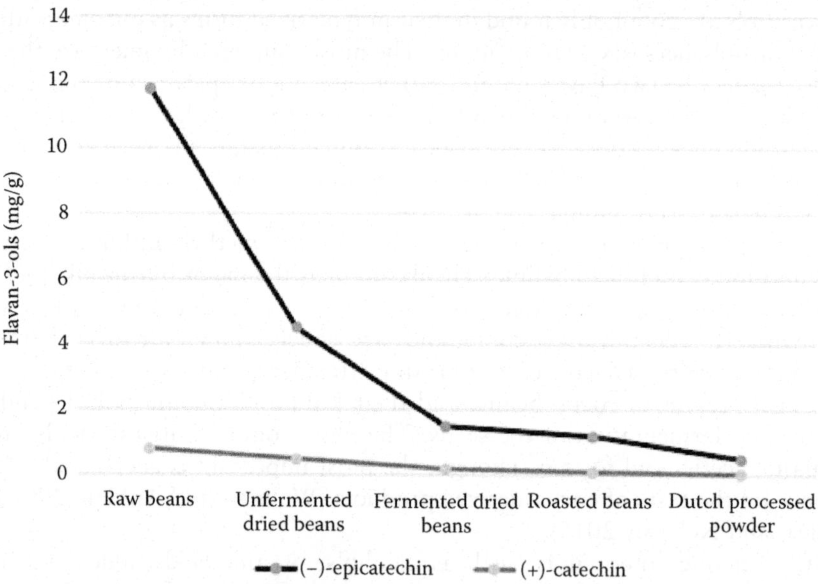

FIGURE 19.3 Effect of processing on flavan-3-ol content of cocoa beans (mg/g). (Adapted from Oracz, J., Zyzelewicz, D., and Nebesny, E., *Critical Reviews in Food Science and Nutrition,* 55, 1176–1192, 2015.)

into lactic acid, acetic acid and mannitol, and (2) to aerobic exothermic (increase the temperature up to 50°C) bioconversion of ethanol into acetic acid (Afoakwa et al. 2008, 2013, 2014).

These microbial activities result in (1) the death of the embryo within the seed due to penetration of mainly ethanol and acetic acid through the husk into the cotyledons; (2) cell membrane disruption, followed by the release of the cell storage components; and (3) the creation of an appropriate environment for development of flavor precursors and pigment degradation by endogenous enzymes, such as invertases, glycosidases, proteases (endoproteinases, carboxypeptidases, aminopeptidases), and polyphenol oxidases (Afoakwa et al. 2008, 2014; Camu et al. 2008; Beckett 2009).

Enzymatic hydrolysis of vacuolar storage proteins (mainly vicilin-class globulins) and polysaccharides from the cell wall results in peptides of various chain length, free amino acids, sucrose, and reducing sugars that are required for the Maillard reaction during the roasting process (Afoakwa et al. 2008; Camu et al. 2008; Hurst et al. 2011; Saltini et al. 2013).

Moreover, during the fermentation of cocoa beans, polyphenolic compounds diffuse with cell liquids from their storage cells (lixiviation) and undergo enzymatic oxidation by polyphenol oxidase, which is a copper-dependent enzyme that in the presence of oxygen catalyses two different reactions: The hydroxylation of monophenols to o-diphenols (monophenolase activity) and the oxidation of o-diphenols to o-quinones. The latter can form complexes with amino acids, proteins and other polyphenols to form complex insoluble tannins (Camu et al. 2008; Afoakwa et al. 2013; Esatbeyoglu et al. 2015).

Since the very beginning of fermentation, a significant decrease (up to 20 percent) in epicatechin and catechin content has been reported, whereas dimeric and trimeric procyanidins have been shown to decrease drastically only after the fourth day of fermentation (Mayorga-Gross et al. 2016). Since polyphenol oxidase activity is strongly reduced during the first days of fermentation, it has been suggested that nonenzymatic oxidation reactions, which include quinones formation and subsequent polyphenol complexing with oligopeptides, amino acids, and cell wall material, contribute to the significant decrease of polyphenols content (Kim and Keeney 1984; Andres-Lacueva et al. 2008; Payne et al. 2010).

Additionally, during the fermentation process, anthocyanins content decreases rapidly (up to 93 percent after four days) as they are hydrolyzed to anthocyanidins and sugars (galactose and arabinose) by glycosidases. Since anthocyanins (3-β-D-galactosyl- and 3-α-L-arabinosylcyanidin) are responsible for the purple pigments of fresh cocoa beans; their hydrolyzation causes the bleaching of cotyledons. Free anthocyanidins are oxidized to quinones that subsequently can polymerize with other flavonoids to form tannins. High molecular weight tannins complex with peptides and proteins through hydrogen bonding, which decreases their solubility and astringency and gives rise to the characteristic brown coloration of the final product (Wollgast and Anklam 2000; Saltini et al. 2013; Oracz, Zyzelewicz, and Nebesny 2015).

In quality-control applications, anthocyanin content has been considered a good marker for the fermentation of cocoa beans, along with the formation of a brown color. The latter has been widely used to predict flavor potential of cocoa beans and hence their suitability for chocolate manufacture (Saltini et al. 2013).

Given that polyphenols (mainly epicatechin, catechin, procyanidin B2, procyanidin B5, and procyanidin C1) along with methylxanthines (caffeine and theobromine) are the major compounds responsible for the bitterness and astringency of cocoa beans and that a negative correlation between their content and aroma acceptability has been suggested, the loss of polyphenols (up to 95 percent) during fermentation is considered a sign of well-fermented beans and thus, a quality control of the manufacturing process (Kim and Keeney 1984; Andres-Lacueva et al. 2008; Payne et al. 2010; Saltini et al. 2013; Mayorga-Gross et al. 2016).

19.3.2 Drying and Roasting

After fermentation, cocoa beans must be dried immediately to avoid overfermentation and product deterioration, as excessive degradation of polyphenolic compounds leads to a high production of nitrogen ammonium and undesirable flavors. Whereas undesirable bitter and astringent tastes decrease, desirable fruity, floral, and cocoa flavors develop during fermentation and drying (Rohan and Connell 1964; Paul et al. 1981; Afoakwa et al. 2008; Kongor et al. 2016).

Moisture content of fermented cocoa beans is between 55 and 60 percent. Drying of the fermented cocoa beans is a process of heating that reduces the moisture content to less than 7.5 percent (w/w). Moreover, nonenzymatic oxidation and migration of polyphenols with vaporizing water during the drying process contribute to astringency, bitterness, and acidity reduction in fermented cocoa beans.

Full enzymatic (invertase activity) and nonenzymatic (acidic environment) sucrose hydrolysis and decreased concentrations of nonvolatile (citric and lactic acids) and volatile (acetic acid) compounds have been reported in dried fermented cocoa beans (Robinson et al. 1961; Paul et al. 1981; Saltini et al. 2013).

Drying rate is of crucial importance for the cocoa bean's final quality. If the drying rate is too fast, the beans will tend to retain an excessive amount of acids (including acetic acid), which are deleterious to the flavor and cause high rejection rates of cocoa. On the other hand, too slow of a drying rate results in low acidity, poor color, and high presence of molds (Paul and Jeanne 1981; Robinson et al. 1961; Saltini et al. 2013).

Once fermented and dried, cacao nibs must be roasted to produce typical aromatic compounds of chocolate (alcohols, ethers, esters, furans, thiazoles, oxazoles, pyrazines, and pyrroles) through nonenzymatic browning Maillard reactions that take place between the reactive carbonyl groups of reducing sugars and the nucleophilic amino groups of amino acid residues (Wollgast and Anklam 2000; Noor-Soffalina et al. 2009; Payne et al. 2010).

Pyrazines are recognized as key heterocyclic aromatic organic compounds that significantly contribute to nutty, earthy, roasty, and green cocoa flavors. About 95 pyrazines have been identified in cocoa aroma, though tetramethylpyrazine has been reported to constitute about 90 percent of the total pyrazines (Frauendorfer and Schieberle 2008; Aprotosoaie et al. 2016).

Although temperature and duration of the roasting treatment influences the final concentration of pyrazines, it has been demonstrated that the fermentation process of cocoa beans is a more critical factor. Several studies have indicated that, during roasting, high polyphenol content in nonfermented cocoa beans can react and form insoluble complexes with amino acids and reducing sugars, thus reducing their availability for pyrazine formation. The latter promotes the development of astringency and bitterness properties over chocolate flavor in the final products (Misnawi et al. 2004; Oracz, Zyzelewicz, and Nebesny 2015; Aprotosoaie et al. 2016).

On the other hand, the roasting of well-fermented cocoa beans significantly affects the levels of phenolic compounds. Procyanidins (mainly procyanidin dimers B1, B2, B5, and procyanidin trimer C1) and anthocyanins show extensive degradation due to hydrolysis, oxidation, or condensation reactions under the effect of high temperatures, elevated humidity, and enhanced exposure to oxygen (Counet et al. 2004; Oracz, Zyzelewicz, and Nebesny 2015; Gültekin-Özgüven et al. 2016a; Żyżelewicz et al. 2016).

A similar trend has been observed for (–)-epicatechin, which shows intensive oxidation or degradation during the roasting process. Moreover, oxidation reactions of flavan-3-ol monomers may be followed by polymerization and condensation with other phenolic compounds leading to the formation of high molecular weight structures (condensed tannins).

Interestingly, it has been observed that the content of catechin increases during the roasting process. The latter has been attributed to the degradation of procyanidins into (+)-catechin and (–)-epicatechin, combined with the epimerization of (–)-epicatechin into (–)-catechin, due to high temperatures of roasting (Oracz, Zyzelewicz, and Nebesny 2015; Oracz, Nebesny, and Żyżelewicz 2015; Gültekin-Özgüven et al. 2016a; Żyżelewicz et al. 2016).

19.3.3 Grinding and Alkalization

Followed by roasting, nib grinding is typically performed in two stages: An initial stage to convert the solid nibs into a fluid paste and a final stage to achieve the desired particle size. Cacao nibs are ground into a dark brown fluid mass known as cocoa liquor, which contains cocoa butter and nonfat particles. The fluidity is due to the breakdown of the cell walls and the release of the cocoa butter during the processing (Paul et al. 1981; Afoakwa et al. 2007; Kongor et al. 2016).

Afterward, cocoa butter is separated from the liquor by either hydraulic presses or screw presses. After pressing the liquor to remove most of the cocoa butter, the remaining solid cake is ground into cocoa powder, which is typically 88 to 90 percent nonfat solids and 10 to 12 percent residual cocoa butter. Most of the chocolate flavor and the polyphenol content remain in the nonfat cocoa solids (Paul et al. 1981; Wollgast and Anklam 2000; Afoakwa et al. 2007; Oracz, Zyzelewicz, and Nebesny 2015).

Alkalization, also known as Dutching, is a treatment frequently used before or after grinding to modify the color (from red-brown to dark mahogany-red to extremely dark) and flavor of cocoa-derived products. The process involves soaking the nib or the cocoa mass in alkalies (such as bicarbonate, carbonate, sodium, ammonium or potassium hydroxide, and magnesium oxide or carbonate) to raise the pH level to 8 (Miller et al. 2008; Sulistyowati and Misnawi 2008).

Concentration and type of alkali, reaction time, temperature, and moisture content influence the color and flavor properties as well as the polyphenol content in cocoa products (liquor or powder). It has been reported that alkalization leads to a remarkable decrease in the content of flavanols and procyanidins (up to 60 percent loss of the mean total flavonoid content). Among monomeric flavanols, (–)-epicatechin has shown a larger decline than (+)-catechin (~67 percent vs. 38 percent, respectively). Moreover, procyanidin dimer B2 and trimer C1 have shown great losses (~69 percent and 67 percent, respectively) due to the alkalinization process (Andres-Lacueva et al. 2008).

It has been suggested that polyphenol changes during alkalinization treatment could be attributed to oxidation reactions under basic pH conditions that lead to brown pigments that are polymerized to different degrees. O-quinones, previously formed during fermentation of cocoa beans, have been suggested to be responsible for the different color formations found in alkalized products (Andres-Lacueva et al. 2008; Miller et al. 2008; Sulistyowati and Misnawi 2008; Payne et al. 2010; Gültekin-Özgüven et al. 2016a; Żyżelewicz et al. 2016).

19.3.4 Cocoa Powder and Chocolate

Cocoa powder is traditionally produced by mechanically pressing cocoa liquor to remove part or most of the cocoa butter (depending on the manufacturer specifications), leaving a solid mass (cocoa cake) that is then pulverized to form cocoa powder (Wollgast and Anklam 2000; Gültekin-Özgüven et al. 2016a).

Grinding of the cakes into powder has been reported to account for losses from 10 to 30 percent of flavanols due to heat (friction), oxygen, and low amounts of fat that cause their oxidation. Moreover, if alkalizing treatment exists, an additional loss of flavonoids is observed. It has been reported that total flavanol content of cocoa powder is inversely proportional to the degree of alkalinization (light, medium, or heavily alkali-processed cocoa powder) (Andres-Lacueva et al. 2008; Payne et al. 2010; Hurst et al. 2011; Gültekin-Özgüven et al. 2016a).

Given that polyphenols reside in the nonfat cocoa solids portion of the cocoa bean, natural cocoa powder contains higher levels of flavanols and procyanidins compared to other cocoa-derived products, such as dark chocolate (up to threefold higher content on an equal weight basis) (Andres-Lacueva et al. 2008; Miller et al. 2008; Stahl et al. 2009).

Cocoa liquor or cocoa powder can be the base from which chocolate is made. Though other ingredients such as additional cocoa butter, sugar, milk, and emulsifying agents can be added, depending on the type of chocolate being made (Paul et al. 1981; Wollgast and Anklam 2000b; Sulistyowati and Misnawi 2008).

After mixing all ingredients, the resulting mixture undergoes a multistep refining process to ensure the homogeneity of particle sizes and enhance their dispersion in a continuous fat phase. Subsequently, the refined paste is subjected to a multiday heat treatment, known as *conching*, to obtain optimal flavor and smoothness within the final product. Although conching is typically applied to reduce moisture and the concentration of acetic acid that remains from the upstream fermentation, a decrease in volatile small molecules that positively influence flavor may also occur (Paul et al. 1981; Wollgast and Anklam 2000; Sulistyowati and Misnawi 2008; Afoakwa et al. 2007; Gültekin-Özgüven et al. 2016b).

Remarkably, it has been reported that the conching process does not contribute to a significant degradation of flavanols. The latter has been attributed to the conduction of the process at mild temperatures (50–70°C) and the presence of large amounts of cocoa butter with the ability to protect flavanols (McShea et al. 2008). The cocoa mass derived from conching needs to be tempered to obtain stable cocoa butter crystals, which are responsible for the good melting properties and glossy surface of good-quality chocolate (Gültekin-Özgüven et al. 2016b).

Given the wide variations in polyphenol concentration among different types of commercial chocolates, the content of nonfat cocoa solids (NFCS) is used as an excellent marker to determine the total phenolic content in the final product. Several studies have reported a strong linear correlation between NFCS in different cocoa products (including cocoa liquor and dark chocolates) and total polyphenols and antioxidant capacity (Belščak et al. 2009; Meng et al. 2009; Laličić-Petronijević et al. 2016). According to these findings, dark chocolates containing the highest content of NFCS (as much as 60 percent) will have a higher content of total and individual polyphenols than milk and white chocolates (Miller et al. 2006; Belščak et al. 2009; Meng et al. 2009).

To date, most existing studies have only focused on the individual effects of fermentation, drying, roasting, or alkalization on cocoa flavonoids content, and limited information on the changes of total or individual flavonoids during chocolate manufacturing has been published. Thus, further studies evaluating phenolic content during the refining, conching, and tempering steps are still needed.

19.4 POTENTIAL HEALTH IMPLICATIONS OF COCOA FLAVANOLS

Few natural products have been claimed to effectively treat such a wide variety of medical disorders as have cocoa powder and dark chocolate. Despite the losses caused by cacao bean processing, they have high flavanol monomer (epicatechin and catechin) and oligomer (procyanidins) content and therefore can be considered significant contributors to the total dietary intake of flavonoids (Rusconi and Conti 2010; Lippi 2013).

The earliest evidence of cocoa consumption and medicinal use comes from Mesoamerican civilizations, such as Olmec, Maya, Zapotec, Mixtex, and Aztec. Aztecs and Mayas made a dark unsweetened drink, called *xocoatl*, that was used to treat a wide variety of ailments, including angina, constipation, dental problems, dysentery, dyspepsia, indigestion, fatigue, gout, and hemorrhoids. Likewise, when the Spaniards brought cacao beans to Europe, they were initially used as medicine rather than as foodstuff (Lippi 2013).

Over the past decade, cacao and its derived products, mainly chocolate and cocoa powder, have attracted the attention of many scientific investigations because of their potential nutritional and medicinal properties (Hii et al. 2009; Rusconi and Conti 2010).

Cocoa flavanols have promising potential for reducing cardiovascular disease risk that has been extensively researched during the last few years, as they have chemical structures that support antioxidant activity against cell free-radicals and chelation properties of redox-active metals (Fe^{2+}, Fe^{3+}, and Cu^{2+}) that react with compounds that originate highly reactive species. It is well-known that the latter depends on their aromatic rings with hydroxyl groups, which give them an adequate configuration to reduce free radicals (through one electron (e-) donation) and to stabilize the new radical derived from the flavanol (Fraga et al. 2010; Fraga and Oteiza 2011; Verstraeten et al. 2015).

Besides their capacity to act as redox regulators, mechanistic aspects defining the beneficial effect of cocoa flavanols have been shown to involve interaction with signaling proteins, enzymes, DNA, and cell membranes (Fraga et al. 2010; Fraga and Oteiza 2011).

In the last decade, scientific evidence has shown a protective role of cocoa flavanols in lowering the risk of cardiovascular disease (CVD) mainly through maintaining adequate nitric oxide (NO) bioavailability, as it is considered as a major determinant of endothelial dysfunction and hypertension (Figure 19.4) (Wollgast and Anklam 2000; Jalil and Ismail 2008; Ackar et al. 2013).

Several studies have proposed that cocoa and its main flavanols regulate several processes involved in the maintenance of NO bioavailability: (1) Free radical scavenging action (superoxide anion and related oxidants); (2) decreasing the activity/expression of NADPH oxidase; (3) increasing the activity/expression of endothelial nitric oxide synthase (eNOS); (4) protection of oxidative loss of tetrahydrobiopterin to avoid eNOS uncoupling (a condition which produces large quantities of superoxide anion rather than NO); and (5) decreasing the activity of inducible nitric oxide synthase (iNOS) (Fraga et al. 2011; Osakabe 2013; Rabadan-Chavez et al. 2016a).

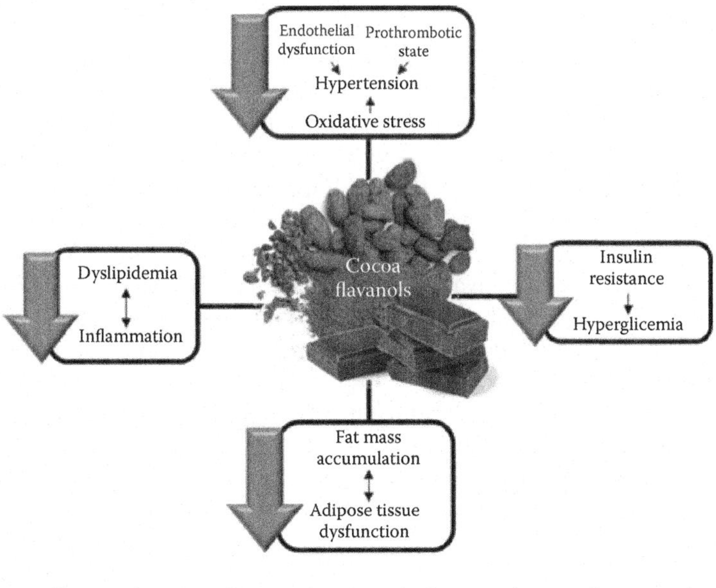

FIGURE 19.4 Effects of cocoa flavanols on risk factors for cardiovascular disease and metabolic disorders.

Current investigations on cocoa flavanols and procyanidins are focused on the mechanisms underlying their beneficial effects in reducing the risk or delaying the development of obesity-related metabolic disorders (Figure 19.4). Experimental evidence has demonstrated the positive effects of cocoa flavanols' body weight, body fat content, serum lipid profile, and glucose homeostasis (Jalil and Ismail 2008; Rabadan-Chavez et al. 2016b).

Cocoa and its main flavanols have been shown to improve hyperglycemia and glucose intolerance by antagonizing digestive enzymes, such as α-amylase (breaks down starch into glucose oligomers) and α-glucosidase (cleaves small oligosaccharides, facilitating absorption of monomeric sugars). Moreover, it has been demonstrated that flavanols exert an inhibitory effect on intestinal glucose transporter 2 (GLUT2) and sodium/glucose cotransporter 1 (SGLT1), resulting in an attenuated elevation of blood glucose after a meal (Ruzaidi et al. 2005; Tomaru et al. 2007; Ali et al. 2014).

Epicatechin and cocoa extract have also been shown to improve insulin signaling and translocation of GLUT4 to the plasma membrane by enhancing the activity of insulin receptor, insulin receptor substrate (IRS-1, IRS-2), and master regulator AMPK (Vinayagam and Xu 2015).

Data from numerous studies have shown that cocoa flavanols can effectively reduce molecules involved in the inflammatory cascade, such as nuclear factor κB (NF-κB), which regulates the expression of a large family of genes including those encoding proteins involved in inflammation. Flavanols and procyanidins can interfere with NF-κB activation by decreasing cell oxidants, or by bonding specific proteins involved in the NF-κB pathway (RelA or p50 subunits). Further in vitro studies have shown that cocoa flavanols can reduce the production of proinflammatory cytokines and chemokines produced by stimulated macrophages, such as TNF-α, IL-1, IL-6, and monocyte chemoattractant protein-1 (MCP-1) (Fraga et al. 2010; Andujar et al. 2012; Osakabe 2013).

Given that dyslipidemia is an important criterion of metabolic syndrome, some studies have addressed that, along with the digestive enzymes already mentioned, flavanols also reduce digestive lipases activity by modifying the size of the intestinal fat droplets that form the lipid emulsion. The latter results in an increased lipid content in fecal matter and reduced lipid absorption (Yasuda et al. 2008; Fraga et al. 2010). Moreover, many studies have demonstrated that cocoa flavanols decrease fasting plasma triglycerides (TG), total cholesterol (TC), and low density lipoproteins (LDL) levels, as well as they increase plasmatic levels of high density lipoproteins (HDL) (Baba et al. 2007).

Finally, a group of nuclear hormone receptors (peroxisome proliferator-activated receptors [PPARs]) and uncoupling chain proteins (UCP-1) have also been implicated as molecular targets of cocoa flavanols, as they regulate numerous biological processes including glucose uptake, lipolysis, fatty acid oxidation, and energy expenditure, in which these compounds have been shown to be involved (Matsui et al. 2005; Ali et al. 2014; Rabadan-Chávez et al. 2016b).

Cocoa powder, dark chocolate, and pure flavanol monomers (mainly epicatechin) have been used to demonstrate that cocoa flavanols are likely to alleviate several obesity-related metabolic disorders through distinct mechanisms due to differences in structure as well as bioavailability. However, it remains unknown which of these mechanisms are primarily responsible for observed effects in vivo. Therefore, additional in vivo mechanistic studies are needed to isolate and assess individual primary and intermediate mechanisms of action. Moreover, future investigations should utilize acute and chronic study designs, as well as a full characterization of the cocoa-derived products utilized to obtain better insights into the mechanisms by which cocoa flavanols act and how they can be included in the diet and lifestyle of populations with high metabolic risk.

REFERENCES

Acierno, V., Sine, Y., Martin, A. et al. 2016. Factors contributing to the variation in the volatile composition of chocolate: Botanical and geographical origins of the cocoa beans, and brand-related formulation and processing. *Food Research International.* 84:86–95.

Ackar, D., Valek, K., Valek, M. et al. 2013. Cocoa Polyphenols: Can we consider cocoa and chocolate as potential functional food? *Journal of Chemistry.* 2013:7.

Afoakwa, E., Budu, A., Mensah-Brown, H. et al. 2014. Changes in biochemical and physico-chemical qualities during drying of pulp preconditioned and fermented cocoa (*Theobroma cacao*) beans. *Journal of Nutritional Health & Food Science.* 2:1–8.

Afoakwa, E., Paterson, A. and Fowler, M. 2007. Factors influencing rheological and textural qualities in chocolate—A review. *Trends in Food Science & Technology.* 18:290–298.

Afoakwa, E., Paterson, A., Fowler, M. et al. 2008. Flavor formation and character in cocoa and chocolate: A critical review. *Critical Reviews in Food Science and Nutrition.* 48:840–857.

Afoakwa, E., Quao, J., Takrama, J. et al. 2013. Chemical composition and physical quality characteristics of Ghanaian cocoa beans as affected by pulp pre-conditioning and fermentation. *Journal of Food Science and Technology.* 50:1097–105.

Ali, F., Ismail, A. and Kersten, S. 2014. Molecular mechanisms underlying the potential antiobesity-related diseases effect of cocoa polyphenols. *Mol Nutr Food Res.* 58:33–48.

Andres-Lacueva, C., Monagas, M., Khan, N. et al. 2008. Flavanol and flavonol contents of cocoa powder products: Influence of the manufacturing process. *Journal of Agricultural and Food Chemistry.* 56:3111–3117.

Andujar, I., Recio, M. and Giner, R. 2012. Cocoa polyphenols and their potential benefits for human health. *Oxidative Medicine and Cellular Longevity.* 2012:23.

Aprotosoaie, A.C., Luca, S.V. and Miron, A. 2016. Flavor chemistry of cocoa and cocoa products-an overview. *Comprehensive Reviews in Food Science and Food Safety.* 15:73–91.

Baba, S., Osakabe, N., Kato, Y. et al. 2007. Continuous intake of polyphenolic compounds containing cocoa powder reduces LDL oxidative susceptibility and has beneficial effects on plasma HDL-cholesterol concentrations in humans. *American Journal of Clinical Nutrition* 85:709–717.

Beckett, S. 2009. Chocolate manufacture. In *Science and Technology of Enrobed and Filled Chocolate, Confectionery and Bakery Products,* pp. 11–28. Cambridge: Woodhead Publishing.

Beecher, G. 2003. Overview of dietary flavonoids: Nomenclature, occurrence and intake. *The Journal of Nutrition.* 133:3248s–3254s.

Belščak, A., Komes, D., Horžić, D. et al. 2009. Comparative study of commercially available cocoa products in terms of their bioactive composition. *Food Research International.* 42:707–716.

Bertazzo, A., Comai, S., Brunato, I. et al. 2011. The content of protein and non-protein (free and protein-bound) tryptophan in *Theobroma cacao* beans. *Food Chemistry.* 124:93–96.

Blumberg, J., Ding, E., Dixon, R. et al. 2014. The science of cocoa flavanols: Bioavailability, emerging evidence, and proposed mechanisms. *Advances in Nutrition.* 5:547–549.

Brooks, E. and Guard, A. 1952. Vegetative anatomy of *Theobroma cacao*. *Botanical Gazette*. 113:444–454.

Camu, N., De Winter, T., Addo, S. et al. 2008. Fermentation of cocoa beans: Influence of microbial activities and polyphenol concentrations on the flavour of chocolate. *Journal of Science of Food and Agriculture*. 88:2288–2297.

Colombo, M., Pinorini-Godly, M., and Conti A., 2012. Botany and pharmacognosy of the cacao tree. In *Chocolate and Health*, pp. 41–62. Milano: Springer Milan.

Counet, C., Ouwerx, C., Rosoux, D. et al. 2004. Relationship between procyanidin and flavor contents of cocoa liquors from different origins. *Journal of Agricultural and Food Chemistry*. 52:6243–6249.

Cuatrecasas, J. 1964. Cacao and its allies; a taxonomic revision of the genus *Theobroma*. Washington: Smithsonian Institution.

Elwers, S., Zambrano, A., Rohsius, C. et al. 2009. Differences between the content of phenolic compounds in Criollo, Forastero and Trinitario cocoa seed (*Theobroma cacao L.*). *European Food Research and Technology*. 229:937–948.

Esatbeyoglu, T., Wray, V., and Winterhalter, P. 2015. Isolation of dimeric, trimeric, tetrameric and pentameric procyanidins from unroasted cocoa beans (*Theobroma cacao L.*) using countercurrent chromatography. *Food Chemistry*. 179:278–289.

Fraga, C. and Oteiza, P. 2011. Dietary flavonoids: Role of (-)-epicatechin and related procyanidins in cell signaling. *Free Radical Biology & Medicine*. 51:813–23.

Fraga, C., Galleano, M., Verstraeten, V. et al. 2010. Basic biochemical mechanisms behind the health benefits of polyphenols. *Molecular Aspects of Medicine*. 31:435–445.

Fraga, C., Litterio, M., Prince, P.D. et al. 2011. Cocoa flavanols: Effects on vascular nitric oxide and blood pressure. *Journal of Clinical Biochemistry and Nutrition*. 48:63–67.

Frauendorfer, F., and Schieberle, P. 2008. Changes in key aroma compounds of Criollo cocoa beans during roasting. *Journal of Agricultural and Food Chemistry*. 56:10244–10251.

Gould, J., Furse, S., and Wolf, B. 2016. The role of endogenous lipids in the emulsifying properties of cocoa. *Frontiers in Chemistry*. 4:11.

Gültekin-Özgüven, M., Berktaş, I., and Özçelik, B. 2016a. Change in stability of procyanidins, antioxidant capacity and in-vitro bioaccessibility during processing of cocoa powder from cocoa beans. *LWT-Food Science and Technology*. 72:559–565.

Gültekin-Özgüven, M., Berktaş, I., and Özçelik, B. 2016b. Influence of processing conditions on procyanidin profiles and antioxidant capacity of chocolates: Optimization of dark chocolate manufacturing by response surface methodology. *LWT—Food Science and Technology*. 66:252–259.

Hernandez, B., Castellote, A., and Permanyer, J. 1991. Triglyceride analysis of cocoa beans from different geographical origins. *Food Chemistry*. 41:269–276.

Hii, C., Law, C., Suzannah, C. et al. 2009. Polyphenols in cocoa (*Theobroma cacao L.*). *Asian Journal of Food and Agro-Industry*. 2:702–722.

Hurst, W., Krake, S., Bergmeier, S. et al. 2011. Impact of fermentation, drying, roasting and Dutch processing on flavan-3-ol stereochemistry in cacao beans and cocoa ingredients. *Chemistry Central Journal*. 5:53.

Jalil, A., and Ismail, A. 2008. Polyphenols in cocoa and cocoa products: Is there a link between antioxidant properties and health? *Molecules* 13:2190–219.

Kim, H. and Keeney, P. 1984. (-)-Epicatechin content in fermented and unfermented cocoa beans. *Journal of Food Science*. 49:1090–1092.

Kongor, J., Hinneh, M., Van de Walle, D. et al. 2016. Factors influencing quality variation in cocoa (*Theobroma cacao*) bean flavour profile—A review. *Food Research International.* 82:44–52.

Laličić-Petronijević, J., Komes, D., Gorjanović, S. et al. 2016. Content of total phenolics, flavan-3-ols and proanthocyanidins, oxidative stability and antioxidant capacity of chocolate during storage. *Food Technoogy and Biotechnology* 54:13–20.

Laurent, V., Risterucci, A., and Lanaud, C. 1994. Genetic diversity in cocoa revealed by cDNA probes. *Theoretical and Applied Genetics.* 88:193–198.

Lippi, D. 2013. Chocolate in history: Food and medicine. *Nutrients* 5:1573–1584.

Matsui, N., Ito, R., Nishimura, E. et al. 2005. Ingested cocoa can prevent high-fat diet-induced obesity by regulating the expression of genes for fatty acid metabolism. *Nutrition* 21:594–601.

Mayorga-Gross, A.L., Quirós-Guerrero, L.M., Fourny, G. et al. 2016. An untargeted metabolomic assessment of cocoa beans during fermentation. *Food Research International.* 89:901–909.

McShea, A., Ramiro-Puig, E., Munro, S.B. et al. 2008. Clinical benefit and preservation of flavonols in dark chocolate manufacturing. *Nutrition Reviews.* 66:630–641.

Meng, C.C., Jalil, A.M., and Ismail, A. 2009. Phenolic and theobromine contents of commercial dark, milk and white chocolates on the Malaysian market. *Molecules.* 14:200–209.

Miller, K., Hurst, W., Payne, M. et al. 2008. Impact of alkalization on the antioxidant and flavanol content of commercial cocoa powders. *Journal of Agricultural and Food Chemistry.* 56:8527–8533.

Miller, K.B., David, A. et al. 2006. Antioxidant activity and polyphenol and procyanidin contents of selected commercially available cocoa-containing and chocolate products in the United States. *Journal of Agricultural and Food Chemistry.* 54:4062–4068.

Misnawi, S., Jinap, B., Jamilah, B. et al. 2004. Effect of polyphenol concentration on pyrazine formation during cocoa liquor roasting. *Food Chemistry.* 85:73–80.

Motamayor, J., and Lanaud, C. 2002. Molecular analysis of the origin and domestication of *Theobroma cacao L.* Paper read at International Conference on Science and Technology for Managing Plant Genetic Diversity in the 21st Century.

Motamayor, J., Lachenaud, P., Wallace, J. et al. 2008. Geographic and genetic population differentiation of the Amazonian chocolate tree. *PLoS ONE* 3:e3311.

Motamayor, J., Risterucci, A., Heath, M. et al. 2003. Cacao domestication II: Progenitor germplasm of the Trinitario cacao cultivar. *Heredity* 91:322–330.

Motamayor, J., Risterucci, A., Lopez, P. et al. 2002. Cacao domestication I: The origin of the cacao cultivated by the Mayas. *Heredity* 89:380–386.

Noor-Soffalina, S., Jinap, S., Nazamid, S. et al. 2009. Effect of polyphenol and pH on cocoa Maillard-related flavour precursors in a lipidic model system. *International Journal of Food Science & Technology.* 44:168–180.

Oracz, J., Nebesny, E., and Żyżelewicz, D. 2015. Changes in the flavan-3-ols, anthocyanins, and flavanols composition of cocoa beans of different *Theobroma cacao* L. groups affected by roasting conditions. *European Food Research and Technology.* 241:663–681.

Oracz, J., Zyzelewicz, D., and Nebesny, E. 2015. The content of polyphenolic compounds in cocoa beans (*Theobroma cacao L.*), depending on variety, growing region, and processing operations: A review. *Critical Reviews in Food Science and Nutrition.* 55:1176–1192.

Osakabe, N. 2013. Flavan 3-ols improve metabolic syndrome risk factors: Evidence and mechanisms. *Journal of Clinical Biochemistry and Nutrition*. 52:186–192.

Parsons, J., Keeney, P., and Patton, S. 1969. Identification and quantitative analysis of phospholipids in cocoa beans. *Journal of Food Science*. 34:497–499.

Paul, D., Jeanne, D., and Hoskin, M. 1981. Chemico-physical aspects of chocolate processing—A review. *Canadian Institute of Food Science and Technology Journal*. 14:269–282.

Payne, M., Hurst, W., Miller, K. et al. 2010. Impact of fermentation, drying, roasting, and Dutch processing on epicatechin and catechin content of cacao beans and cocoa ingredients. *Journal of Agricultural and Food Chemistry*. 58:10518–10527.

Rabadan-Chávez, G., Quevedo-Corona, L., Miliar, A. et al. 2016. Cocoa powder, cocoa extract and epicatechin attenuate hypercaloric diet-induced obesity through enhanced β-oxidation and energy expenditure in white adipose tissue. *Journal of Functional Foods*. 20:54–67.

Rabadan-Chavez, G., Reyes-Maldonado, E., Quevedo-Corona, L. et al. 2016. The prothrombotic state associated with obesity-induced hypertension is reduced by cocoa and its main flavanols. *Food and Function*. 7:4880–4888.

Redgwell, R., and Hansen, C. 2000. Isolation and characterisation of cell wall polysaccharides from cocoa (*Theobroma cacao L.*) beans. *Planta*. 210:823–830.

Redgwell, R., Trovato, V., Merinat, S. et al. 2003. Dietary fibre in cocoa shell: Characterisation of component polysaccharides. *Food Chemistry*. 81:103–112.

Robinson, T., Ranalli, A., and Phillips, A. 1961. Cocoa polyphenols, changes in cocoa tannins during processing. *Journal of Agricultural and Food Chemistry*. 9:295–298.

Rohan, T., and Connell, M. 1964. The precursors of chocolate aroma: A study of the flavonoids and phenolic acids. *Journal of Food Science*. 29:460–463.

Rusconi, M., and Conti, A. 2010. *Theobroma cacao L.*, the food of the gods: A scientific approach beyond myths and claims. *Pharmacological Research*. 61:5–13.

Ruzaidi, A., Amin, I., Nawalyah, A. et al. 2005. The effect of Malaysian cocoa extract on glucose levels and lipid profiles in diabetic rats. *Journal of Ethnopharmacology*. 98:55–60.

Saltini, R., Akkerman, R., and Frosch, S. 2013. Optimizing chocolate production through traceability: A review of the influence of farming practices on cocoa bean quality. *Food Control*. 29:167–187.

Stahl, L., Miller, K., Apgar, J. et al. 2009. Preservation of cocoa antioxidant activity, total polyphenols, flavan-3-ols, and procyanidin content in foods prepared with cocoa powder. *Journal of Food Science*. 74:C456–461.

Sulistyowati, E., and Misnawi, T. 2008. Effects of alkali concentration and conching temperature on antioxidant activity and physical properties of chocolate. *International Food Research Journal*. 15:297–304.

Tomaru, M., Takano, N., Osakabe, N. et al. 2007. Dietary supplementation with cacao liquor proanthocyanidins prevents elevation of blood glucose levels in diabetic obese mice. *Nutrition*. 23:351–355.

Verstraeten, S., Fraga, C., and Oteiza, P. 2015. Interactions of flavan-3-ols and procyanidins with membranes: Mechanisms and the physiological relevance. *Food and Function*. 6:32–41.

Vinayagam, R., and Xu, B. 2015. Antidiabetic properties of dietary flavonoids: A cellular mechanism review. *Nutrition and Metabolism*. 12:60.

Voigt, J., and Biehl, B. 1993. The major seed proteins of *Theobroma cacao* L. *Food Chemistry*. 47:145–151.

Wollgast, J., and Anklam, E. 2000. Review on polyphenols in *Theobroma cacao*: Changes in composition during the manufacture of chocolate and methodology for identification and quantification. *Food Research International*. 33:423–447.

Yasuda, A., Natsume, M., Sasaki, K. et al. (2008). Cacao procyanidins reduce plasma cholesterol and increase fecal steroid excretion in rats fed a high-cholesterol diet. *BioFactors*. 33:211–223.

Young, A.M. 2007. *The Chocolate Tree: A Natural History of Cacao*. Gainesville, FL: The University Press of Florida.

Żyżelewicz, D., Krysiak, W., Oracz, J. et al. 2016. The influence of the roasting process conditions on the polyphenol content in cocoa beans, nibs and chocolates. *Food Research International*. 89:918–929.

CHAPTER 20

Phenolic Compounds in Processed Foods

Claudia Ivette Gamboa-Gómez, Nuria Elizabeth Rocha-Guzmán, José Alberto Gallegos-Infante, Rubén Francisco González-Laredo, and Martha Rocío Moreno-Jiménez

CONTENTS

20.1 INTRODUCTION

Polyphenol compounds are synthesized in plants as secondary metabolites in response to ecological and physiological stresses such as adverse climate, insect attack, UV radiation, and wounding (Kennedy et al., 2011). The basic structural feature of phenolic compounds is an aromatic ring bearing one or more hydroxyl groups (Chirinos et al., 2009). Plant polyphenols are classified as simple phenols or polyphenols based on the number of phenol units in the molecule. Thus, plant polyphenols comprise simple phenols, coumarins, lignins, lignans, condensed and hydrolyzable tannins, phenolic acids, and flavonoids (Soto-Vaca et al., 2012).

Plant foods (including fruits, cereal grains, legumes, and vegetables) and beverages (including tea, coffee, fruit juices, and cocoa) are major sources of polyphenols in the human diet (Khoddami et al., 2013). However, processing of foods and food ingredients often exerts a major effect on their constituents (Shahidi, 2009). For instance, most of the vegetables that we consume are cooked by being boiled in water or microwaved. These processes bring about a number of changes in the physical characteristics and chemical composition of vegetables (Rehman et al., 2003; Zhang and Hamauzu, 2004). Sahlin et al. (2004) showed that boiling and baking had a small effect on the ascorbic acid, total phenolics, lycopene, and antioxidant activity of tomatoes, while frying significantly reduced the ascorbic, total phenolic, and lycopene contents of tomatoes. Additionally, Rocha-Guzmán et al. (2007) reported that processes such as pressure-cooking affected not only the total polyphenols concentration, but also their distribution in the different effluents of the process (e.g., bean seeds and cooking water).

For this reason, we summarize the information on some common processes used for food products and their effect on polyphenols based in the current scientific literature.

20.2 PROCESSED FOOD

Food processing is defined as all methods and techniques used by the industry to turn whole fresh foods into food products (Monteiro et al., 2010a). This classification divides all foodstuffs into three groups: (1) Unprocessed or minimally processed foods, (2) processed culinary ingredients, and (3) ultraprocessed products. These distinctions address social, economic, cultural, and other aspects of public health nutrition, as well as biological issues (Monteiro et al., 2010b).

Unprocessed foods are parts of animals immediately after they have been slaughtered and parts of plants after harvesting or collection. Minimally processed foods are foods subjected to changes, mostly physical, that do not substantially change the nutritional properties and uses of the original foods. These processes are used to extend the duration and storage of unprocessed foods, and often to reduce the time and effort involved in their preparation. Such processes include cleaning and removal of inedible fractions, portioning, grating, flaking, drying, chilling, freezing, pasteurization, fermentation, fat reduction, vacuum and gas packing, squeezing, and simple wrapping. This group includes fresh or frozen meat; fresh or pasteurized milk and plain yogurt; whole or polished grains; fresh, frozen, or dried fruits and unsweetened fruit juices; fresh and frozen vegetables; whole or peeled roots and tubers; unsalted nuts and seeds; and tea and coffee (Moubarac et al., 2013).

Group 2 is made of processed culinary ingredients. These are inexpensive substances extracted from Group 1 foods through physical and chemical transformations, such as refining, milling, and hydrolysis. They have nutritional properties and entirely different uses from the original whole foods. Group 2 ingredients include vegetable oils, animal fats, sucrose, and flours and pastas (when made of flour and water). Most are depleted of nutrients and essentially provide energy. Furthermore, they are typically inedible in present form but rather cooked in households and also in restaurants to prepare and enhance the flavor of meals and dishes fixed with unprocessed or minimally processed foods (Moubarac et al., 2013).

Finally, Group 3 is made of ultraprocessed food and drink products. These are ready-to-consume industry formulations that are manufactured from cheap ingredients directly extracted from whole foods (oils, fats, sucrose, and flours) or processed from components extracted from whole foods (high-fructose corn syrup, hydrogenated oils, a variety of starches) and the cheap parts or remnants of meat. Several preservatives and cosmetic additives are typically added to these products, with little or no content of whole foods. Some ultraprocessed products, such as breads and sausages, have been part of dietary patterns in many countries since before industrialization. Others, such as burgers, chips, cookies, cakes, sweets, pizzas, chicken nuggets, energy bars, soft drinks, and other sugared drinks, are more recent, at least in the quantity now manufactured. Because of the nature of their formulation, which includes packaging, these products have a long shelf-life and dispense with culinary preparation and the need for dishes and cutlery because they are intensely palatable and appealing to the senses. They are typically energy-dense, with a high content in total of saturated and trans-fats, free sugars, and sodium, and little or no water, fiber, micronutrients and other protective bioactive compounds existing in whole foods (Moubarac et al., 2013).

20.3 POLYPHENOLS IN PROCESSED FOOD

Phenolic compounds are highly unstable and may undergo numerous reactions in the course of food processing and storage. However, in several researches an improvement was observed of antioxidant properties in polyphenol rich foods by processing; for instance, Maillard reaction products such as melanoidins.

It has been reported that melanoidins exhibit antioxidant properties in vitro due to their protective effect against reactive oxygen species (Valls-Bellés et al., 2004). The Maillard reaction involves condensation reactions between sugars and amino acids, leading to the formation of melanoidins. These reactions are considered an important pathway in natural humification processes, in which polyphenol polymers are regarded as important precursors in the formation of humic substances. Therefore, in nature, it is most likely that these two processes do not occur separately but rather interact with each other (Jokic et al., 2004). Therefore, different processing of food will have different effects in relation to polyphenol compound concentration and its activity.

20.4 ENZYMATIC AND CHEMICAL OXIDATION

One of the most important biochemical processes is enzymatic oxidation. This starts as soon as the integrity of the cell is broken, in association with enzymes such as esterases, glycosidases, and decarboxylases that produce transformations and degradations of polyphenolic compounds (Cheynier, 2005). Enzymatic oxidation is ubiquitous in plant foods. The resulting browning is usually detrimental to quality, particularly in post-harvested, storage of fresh fruits or juice and puree technology, but may be desirable for some products such as black tea, cocoa, raisins, and coffee (Cheynier, 2005). For instance, del Castillo et al. (2002) reported an increase in antioxidant capacity in brews from medium roasted coffee to green coffee. This is attributed to the formation of melanoidin-bound phenolic compounds. Perrone et al. (2012) observed that phenolic compounds were incorporated into melanoidins mainly at early stages of the process, being thereafter partly oxidized to dihydrocaffeic acid. The relative content of melanoidin-bound phenolic acids increases significantly during roasting, reaching up to 29 percent of total phenolic compounds in brews from dark roasted coffees. These compounds contribute 25–47 percent of the antioxidant capacity of roasted coffees. In addition, Ludwig et al. (2014) reported that the formation of melanoidin-bound phenolic acids contributes to the antiglycative, anti-inflammatory, and antioxidant properties of coffee.

For black tea production, fresh leaves of *Camellia sinensis* L., better known as green tea, are fermented, biotransforming the native phenolic compounds. The main polyphenols in *Camellia sinensis* fresh leaves are flavanol monomers, such as epicatechin, epigallocatechin, and their gallic acid esters (Figure 20.1). The fermentation process of these leaves consists of enzymatic oxidation of native green tea polyphenols, catalyzed by polyphenoloxidase, followed by chemical reactions where quinones are primary produced (Balentine et al., 1997). Flavonols from the leaves are transformed in several compounds such as thearubigins and theaflavins, which are responsible for the dark brown color

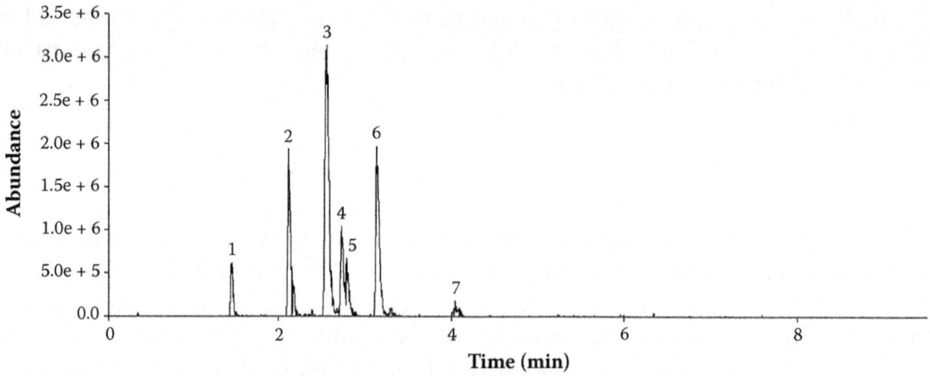

FIGURE 20.1 Main polyphenols in *Camellia sinensis* fresh leaves: (1) gallocatechin, (2) catechin, (3) gallocatechin gallate, (4) epicatechin, (5) epigallocatechin gallate, (6) epicatechin gallate, and (7) rutin.

of black tea. Additionally, other compounds in lower quantities are produced such as theaflavic acids and bis-flavanols, which are also called theasinensins (Cheynier, 2005).

Although it has been reported that chemical or enzymatic oxidations lead to a progressive decrease in polyphenol concentration and then in antioxidant properties, polyphenols with an intermediate oxidation state can exhibit higher radical scavenging efficiency than the nonoxidized ones. For example, Nicoli et al. (2000) reported that after enzymatic (mushroom tyrosinase) and chemical oxidation (butyl septa and metallic caps) of aqueous solution of catechin (1×10^{-4} M), an initial increase and a following decrease in the chain-breaking activity was observed.

Fermentation has also been applied to increase the content of bioactive phenolic compounds in legumes, thus enhancing their antioxidant activity (Lee et al., 2008; Torino et al., 2013). Authors such Torino et al. (2013) reported that the bioconversion of the conjugated forms of phenolic compounds into their free forms during fermentation improves their bioavailability and antioxidant potential. The latter is due to several ligninolytic and carbohydrate enzymes that hydrolyze phenolic glycosides and release free aglycones, which have the potential for high antioxidative activity (Vattem and Shetty, 2003).

Several studies have reported that solid-state fermentation increased phenolic compounds. For instance, Guzmán-Uriarte et al. (2013) reported that solid-state fermentation for producing common bean functional flour substantially increased the concentration of total polyphenols, total hydrophilic antioxidant capacity, and antihypertensive potential of final products. Similarly, Sanjukta et al. (2015) reported that proteolitic *Bacillus subtillis* fermentation on soybean enhanced the concentration of polyphenols and increased scavenging activity (3.1–24 folds) of DPPH and superoxide radicals than the unfermented counterparts.

On the other hand, Watawana et al. (2015) also reported an enhancement of polyphenols such chlorogenic and caffeic acids ($P < 0.05$) from coffee through fermentation by "tea fungus" (*Kombucha*).

20.5 THERMAL PROCESSING

Thermal process is applied in food with the purpose of extending its shelf life. However, during this process, natural nutrients and other compounds such antioxidants are affected significantly due to the fact that these compounds are relatively unstable to heat. For example, Vallverdú-Queralt et al. (2014) reported that concentration of polyphenols such as quercetin in tomato sauces decreased during the cooking process (15, 30, 45, and 60 minutes), whereas caffeic acid and tyrosol concentration were not affected.

Blanching is a processing step that inactivates enzymes, maintaining the color and nutritional aspects of food products. For this process many methods can be used, including water, steam, vacuum-steam, in-can, and hot-air. The first method (e.g., 75–95°C for 1–10 minutes) is the most commonly applied in industry for fruits and vegetables because costs are relatively low (Rawson et al., 2011). Several studies reported that blanching affected polyphenol compounds. For example, Turkmen et al. (2005) reported that treatment for one minute in boiling water reduced (12–26 percent) total polyphenol content in spinach, swamp cabbage, kale, shallots, and cabbage due to breakdown of these compounds during cooking. Similarly, Pacheco-Palencia et al. (2009) reported a degradation of anthocyanins in fruit puree of two acai fruit species (80°C for 60 minutes).

Other common thermal processes are pasteurization and sterilization. Heat pasteurization is carried out to eliminate microorganisms and inactivate enzymes (e.g., pectin methylesterase in fruit juices). Similar to blanching, changes in the bioactive content of the foodstuff were observed. For instance, pasteurization of mango puree generally leads to a decrease in the levels of pholyphenols, although it depends on the severity of the process (Kim et al., 2009). In addition, Hoffmann-Ribani et al. (2009) reported that levels of quercetin, kaempferol, and myricetin in industrially processed acerola, cashew apple, and pitanga juice and pulp decreased significantly by the pasteurization process. Azofeifa et al. (2015) reported that pasteurized blackberry juices prepared at 75°C for 15 seconds and 92°C for 10 seconds decreased anthocyanin concentrations, compared with nonpasteurized juices, whereas no changes were observed on ellagitannins. Therefore, the degree of degradation of polyphenols compounds is variable.

Thermal drying is another important process used in the food industry. Drying of foods allows the extension of shelf life and the reduction in volume of several products such as fruits (Prakash et al., 2004). Several methods for dehydration are commonly used for fruits and vegetables, such as sun drying, oven drying, fluidized bed, spray, microwave drying, osmotic air-drying, puff, cross-flow, drum, and freeze-drying. Even though thermal drying increases the shelf life of food products, this process affects the concentration and stability of bioactive compounds such as polyphenols. For instance, during the manufacture of cocoa beans a drying process is performed to retain the "chocolate" flavor that develops during fermentation, in addition to reducing the moisture content of the cocoa beans (approximately 7 percent). Teh et al. (2015) investigated the effect of hot air drying and a constant humidity-controlled (50 percent relative humidity) oven at temperature range of 60–80°C on cocoa beans, reporting that the concentration of polyphenol declined rapidly. Additionally, polyphenol oxidation (specifically epicatechin and procyanidins) resulted from the production of polymeric brown pigments (Dimick, 1993). On other hand, Vega-Gálvez et al. (2012) determined the total phenolic content for dehydrated apple samples, and observed that an increase in drying temperature caused a degradation of total polyphenols with respect to the corresponding content in fresh sample (158.28 ± 0.65 mg acid gallic equivalents/100 g) ($p < 0.05$). However, prolonged drying times did not necessarily produce the higher degradation of total polyphenols. Gümüşay et al. (2015) studied the effects of three different drying processes (sun drying, oven drying, and vacuum oven drying) for tomatoes (*Solanum lycopersicum*) and ginger (*Zingiber officinale*) and reported that phenolic contents decreased compared to fresh samples (1351.10 ± 62.16 for ginger and 792.22 ± 43.35 for tomato) in 76 and 60 percent for sun, 73 and 56 percent for oven, and 78 and 55 percent for vacuum oven for ginger and tomato, respectively.

20.6 FREEZING PROCESS

Freezing is recognized worldwide as one of the best methods available in the food industry for preserving food products such as fruits and vegetables. Lower temperatures inhibit metabolic processes that occur in fruits and vegetables after harvesting; additionally, it slows down the rate of microbiological growth (Jaiswal et al., 2012). During the freezing process the transformation of liquid water into ice occurs, and leads to a variety of potential stress mechanisms for vegetable tissues, deteriorating these products (Van Buggenhout et al., 2006).

Previous research has reported that in several vegetables the freezing process effects the final concentration of polyphenol compounds. For example, Ninfali and Bacchiocca (2003) studied six fresh and frozen vegetables (beet green, spinach, broccoli, carrot, onion, and celery) from the same cultivar, and analyzed the whole juice and the acetone extract of the squeezed pulp. The results showed that four of the six frozen vegetables had lower polyphenol concentration than the fresh counterparts, whereas in the other two cases the concentration increased. The authors attributed these differences in concentration of polyphenols to the genotype of plants and differences between native compounds in each vegetable.

Shofian et al. (2011) studied the effect of freeze-drying on bioactive compounds and antioxidant activity of five tropical fruits: mango (*Mangifera indica* L.), starfruit (*Averrhoa carambola* L.), muskmelon (*Cucumis melo* L.), papaya (*Carica papaya* L.), and watermelon (*Citruluss lanatus* Thunb.). Except for muskmelon, total phenolic concentration between fresh and freeze-dried fruit samples was reduced by 24 percent for starfruit, 23 percent for mango, 39 percent for papaya, 10 percent for muskmelon, and 48 percent for watermelon. On antioxidant activity their results showed that only fresh samples of starfruit and mango had relatively higher antioxidant activities by DPPH (2,2-diphenyl-1-picrylhydrazyl) radical scavenging and reducing power assays than the freeze-dried fruits. These results suggest that changes depend on sample characteristics and on the native compounds found in its fresh form. Michalczyk et al. (2009) reported that freeze-drying in raspberry (*Rubus ideaus* L.), strawberry (*Fragaria ananassa* Duch), and bilberry (*Vaccinum myrtillus*) retained the raw material properties in storage better than the air-dried products. This was especially true for total phenolics and anthocyanin contents as well as for antioxidant properties, despite the great losses of these compounds by processing. Therefore, this technology could be considered a good option for food processing in relation with polyphenol compounds and their biological effects.

20.7 EMERGENT TECHNOLOGIES FOR PROCESSED FOODS

In recent years, several technologies have been proposed for processing foods to extend their shelf life and decrease the physical and chemical changes that may affect the beneficial properties of food products. Pulsed electric fields (PEF) is a valid technology for the production of safe beverage products that have shown a positive influence in the texture of solid plant foods, leading to enhanced yields of metabolites extraction as well to increased juice yields. PEF consists in higher intensity fields: 15–40 kV/cm, 5–100 pulses, 40 to 700 μs, 1.1 to 100 Hz (Zulueta et al., 2010). Several studies on exotic fruits demonstrated the effect of pulsed electric field on bioactive compounds. For example, Oms-Oliu et al. (2009) reported that PEF treatments reduced vitamin C content in watermelon juice; however, a lycopene retention was observed fluctuating from 87.6 percent to 121.2 percent on the range of processing parameters (field strength 25–35 kV/cm, frequency 50–250 Hz, pulse width 1–7 μs, and treatment time 50–2050 μs). They attributed this lycopene enhancement to the fact that PEF induces cell permeabilization and releases intracellular pigments (lycopene) from watermelon.

Other relevant technology used to process food is ozone processing. The main interest in the application of ozone is the high efficacy and wide antimicrobial spectrum. In the food industry, ozone has been used for washing and storing fruits and vegetables. Different effects on total polyphenol concentration by application of ozone in fruits were reported. Alothman et al. (2009) reported that ozone treatment (8 ± 0.2 mL/s for 0, 10, 20, and 30 minutes) on fresh-cut honey pineapple, banana, and guava resulted in an increase of total phenolic and flavonoid contents of pineapple and banana when exposed to ozone for up to 20 minutes, whereas the opposite effect was observed for guava. These results have shown a positive correlation with the antioxidant capacity measured by FRAP and DPPH assays. However, using ozone at enough effective doses for decontamination could change sensory qualities of food. Therefore, ozone is not beneficial in all cases since it promotes oxidative spoilage in several foods (Khadre et al., 2001).

One important emerging technology used in the food industry is ultrasound as an alternative for conventional pasteurization and sterilization. Ultrasound is an effective process for microbial inactivation and phytochemical retention that can be used on its own or in combination with heat or pressure (Zenker et al., 2003). Previous studies reported that ultrasound processing increased extraction yield of bioactive compounds by about 6 and 35 percent, depending on the processing conditions (Vilkhu et al., 2008). Rawson et al. (2011) studied the effect of temperature on bioactive compounds and reported a reduced concentration of polyphenol concentration in watermelon juice sonicated at 25 to 45°C. This effect was more pronounced at higher processing times (10 minutes).

One nonthermal process used in the food industry is high hydrostatic pressure, which is a preservation technique that reduces affectations on nutritional and quality parameters better than the thermal process. On high hydrostatic pressure treatments, the pressure is given in position and time, and at all directions, transmitted uniformly through the transferring medium and independently of geometry (Oey et al., 2008). Previous studies have reported that high hydrostatic pressure treatments minimize the degradation of bioactive compounds in several fruits and vegetables (Oey et al., 2008). However, authors such as Ferrari et al. (2010) reported that high pressures (400–600 MPa) at 25, 45, and 50°C for 5 or 10 minutes on pomegranate juice influenced anthocyanin and polyphenol contents. Their results showed that at room temperature the concentration of polyphenols decreased with the intensity of the treatment (pressure level and processing time). Therefore, care on high pressures at lower times and pressures of processing are more adequate for polyphenol conservation.

Finally, irradiation treatment involves the exposure of food products to ionizing or nonionizing radiation for food preservation. Ionizing radiation sources could be high-energy electrons, x-rays, or gamma rays (from cobalt-60 or cesium-137), while nonionizing radiation is electromagnetic radiation that does not carry enough energy/quanta to ionize atoms or molecules, represented mainly by ultraviolet rays (UV-A, UV-B, and UV-C), visible light, microwaves, and infrared (Rawson et al., 2011). Authors such as Wood and Bruhn (2000) reported that irradiation induces negligible or subtle losses of bioactive compounds compared with thermal processes. However, Alighourchi et al. (2008) demonstrated significant reduction in the total and individual anthocyanin content in pomegranate juice after irradiation at higher doses (3.5–10 kGy). The latter effect depends upon the nature of the polyphenols. For instance, diglycosides are relatively stable toward irradiation doses when compared with monoglycosides (Reyes and Cisneros-Zevallos, 2007).

On the other hand, Alothman et al. (2009) investigated the effect of ultraviolet treatment (2.158 J/m^2) on total phenol and flavonoid on fresh-cut honey pineapple, banana, and guava. Their results showed an increased concentration of flavonoids content after 10 minutes of treatment. In this regard, Fan et al. (2003) explained that free radicals generated during irradiation might act as stress signals and may trigger stress responses, resulting in an increased antioxidant synthesis.

20.8 EFFECT OF STORAGE ON POLYPHENOLS IN PROCESSED FOOD

One of the more important parameters for consideration with processed food is the storage. Several processed food products are saved for long time, and although the technology applied was not as aggressive for polyphenol degradation, the storage alone affects such compounds. Previous studies have agreed with the above mentioned. For instance, Klimczak et al. (2007) analyzed fresh orange juice after storage at 18, 28, and 38°C for two, four, and six months, and results showed a decrease in the content of polyphenols and vitamin C, which was reflected by the decrease in the antioxidant capacity of orange juices. However, small changes in flavanone content were observed, indicating high stability of these compounds upon storage.

As stated, the range of degradation of polyphenols depends on the process applied to the food product. For example, anthocyanins are the most sensible compounds for thermal processing. These compounds decompose upon heating into a chalcone structure, which is being further transformed into a coumarin glucoside derivative with a loss of the B ring. Degradation is caused primarily by oxidation, cleavage of covalent bonds, or enhanced oxidation reactions due to thermal processing. Thermal degradation of anthocyanins can result in a variety of species depending upon the severity and nature of heating (Rawson et al., 2011).

In the case of nonthermal treatments, the degradation of polyphenols compounds follow another route, for instance, when ozone treatments cause the loss of phytochemicals because of its strong oxidizing activity (Rawson et al., 2011).

20.9 CONCLUSION

According to changes in lifestyle, the consumption of processed foods is part of everyday human life. Ensuring food safety and at the same time meeting the demand for foods with more than nutrients (bioactive compound such as polyphenols) has been a challenge for the food industry. The mechanism by which bioactive compounds degrade are numerous, complex, and perplexing, and sometimes unknown, but in general, high-temperature treatments accelerate these reactions. Furthermore, novel nonthermal processing is promising for bioactive conservation as little information on several food products is available and sometimes the information is contradictory since mostly it depends on the nature of food.

It is relevant to understand the mechanism of action in the different processes and technologies as well their effects on the bioactive compounds that influence the functional properties of food products.

REFERENCES

Alighourchi, H., Barzegar, M., and Abbasi, S. 2008. Effect of gamma irradiation on the stability of anthocyanins and shelf-life of various pomegranate juices. *Food Chemistry*. 110(4): 1036–1040.

Alothman, M., Bhat, R., and Karim, A.A. 2009. Effects of radiation processing on phytochemicals and antioxidants in plant produce. *Trends in Food Science & Technology*. 20(5): 201–212.

Azofeifa, G., Quesada, S., Pérez, A.M. et al. 2015. Pasteurization of blackberry juice preserves polyphenol-dependent inhibition for lipid peroxidation and intracellular radicals. *Journal of Food Composition and Analysis*. 42: 56–62.

Balentine, D., Wiserman, S.A., and Bouwens, L.C.M. 1997. The chemistry of tea flavonoids. *Critical Reviews in Food Science and Nutrition*. 37: 693–704.

Cheynier, V. 2005. Polyphenols in foods are more complex than often thought. *The American Journal of Clinical Nutrition*. 81(1): 223S–229S.

Chirinos, R., Betalleluz-Pallardel, I., Huamán, A. et al. 2009. HPLC-DAD characterization of phenolic compounds from Andean oca (*Oxalis tuberosa* Mol.) tubers and their contribution to the antioxidant capacity. *Food Chemistry*. 113: 1243–1251.

Del Castillo, M.D., Ames, J.M., and Gordon, M.H. 2002. Effect of roasting on the antioxidant activity of coffee brews. *Journal of Agricultural and Food Chemistry*. 50(13): 3698–3703.

Dimick, P.S. 1993. Penn State Chocolate Manufacture Short Course. 29–465. University Park, PA: Penn State University.

Fan, X., Toivonen, P.M.A., Rajkowski, K.T. et al. 2003. Warm water treatment in combination with modified atmosphere packaging reduces undesirable effects of irradiation on the quality of fresh-cut iceberg lettuce. *Journal of Agricultural and Food Chemistry*. 51: 1231–1236.

Ferrari, G., Maresca, P., and Ciccarone, R. 2010. The application of high hydrostatic pressure for the stabilization of functional foods: Pomegranate juice. *Journal of Food Engineering*. 100: 245–253.

Gümüşay, Ö.A., Borazan, A.A., Ercal, N. et al. 2015. Drying effects on the antioxidant properties of tomatoes and ginger. *Food Chemistry*. 173: 156–162.

Guzmán-Uriarte, M.L., Sánchez-Maga, L.M., Angulo-Meza, G.Y. et al. 2013. Solid-state bioconversion for producing common bean (*Phaseolus vulgaris* L.) functional flour with high antioxidant activity and antihypertensive potential. *Food and Nutrition Sciences*. 4: 480–490.

Hoffmann-Ribani, R., Huber, L.S., and Rodriguez-Amaya, D.B. 2009. Flavonols in fresh and processed Brazilian fruits. *Journal of Food Composition and Analysis*. 22(4): 263–268.

Jaiswal, A.K., Gupta, S., and Abu-Ghannam, N. 2012. Kinetic evaluation of color, texture, polyphenols and antioxidant capacity of Irish York cabbage after blanching treatment. *Food Chemistry*. 131: 63–72.

Jokic, A., Wang, M.C., Liu, C. et al. 2004. Integration of the polyphenol and Maillard reactions into a unified abiotic pathway for humification in nature: The role of δ-MnO2. *Organic Geochemistry*. 35(6): 747–762.

Kennedy, D.O., and Wightman, E.L. 2011. Herbal extracts and phytochemicals: Plant secondary metabolites and the enhancement of human brain function. *Advances in Nutrition*. 2: 32–50.

Khadre, M.A., Yousef, A.E., and Kim, J.G. 2001. Microbiological aspects of ozone applications in food: A Review. *Journal of Food Science*. 66(9): 1242–1251.

Khoddami, A., Wilkes, M.A., and Roberts, T.H. 2013. Techniques for analysis of plant phenolic compounds. *Molecules*. 18(2): 2328–2375.

Kim, Y., Lounds-Singleton, A.J., and Talcott, S.T. 2009. Antioxidant phytochemical and quality changes associated with hot water immersion treatment of mangoes (*Mangifera indica* L.). *Food Chemistry*. 115(3): 989–993.

Klimczak, I., Małecka, M., Szlachta, M. et al. 2007. Effect of storage on the content of polyphenols, vitamin C and the antioxidant activity of orange juices. *Journal of Food Composition and Analysis*. 20(3): 313–322.

Lee, S.H., Lee, W.S., Lee, C.H. et al. 2008. Degradation of phenanthrene and pyrene in rhizosphere of grasses and legumes. *Journal of Hazardous Materials*. 153(1): 892–898.

Ludwig, I.A., Clifford, M.N., Lean, M.E. et al. 2014. Coffee: Biochemistry and potential impact on health. *Food and Function*. Aug;5(8): 1695–717. Review.

Michalczyk, M., Macura, R., and Matuszak, I. 2009. The effect of air-drying, freeze-drying and storage on the quality and antioxidant activity of some selected berries. *Journal of Food Processing and Preservation*. 33(1): 11–21.

Monteiro, C.A., Levy, R.B., Claro, R.M. et al. 2010a. A new classification of foods based on the extent and purpose of their processing. *Cadernos de Saúde Pública*. 26: 2039–2049.

Monteiro, C.A., Levy, R.B., Claro, R.M. et al. 2010b. Increasing consumption of ultra-processed foods and likely impact on human health: Evidence from Brazil. *Public Health Nutrition*. 14(1):5–13.

Moubarac, J.C., Martins, A.P.B., Claro, R.M. et al. 2013. Consumption of ultra-processed foods and likely impact on human health. Evidence from Canada. *Public Health Nutrition*. 16(12): 2240–2248.

Nicoli, M.C., Calligaris, S., and Manzocco, L. 2000. Effect of enzymatic and chemical oxidation on the antioxidant capacity of catechin model systems and apple derivatives. *Journal of Agricultural and Food Chemistry*. 48(10): 4576–4580.

Ninfali, P., and Bacchiocca, M. 2003. Polyphenols and antioxidant capacity of vegetables under fresh and frozen conditions. *Journal of Agricultural and Food Chemistry*. 51(8): 2222–2226.

Oey, I., Plancken, I.V., Loey, A.V. et al. 2008. Does high pressure processing influence nutritional aspects of plant based food systems? *Trends in Food Science & Technology*. 19(6): 300–308.

Oms-Oliu, G., Odriozola-Serrano, I., Soliva-Fortuny, R. et al. 2009. Effects of high-intensity pulsed electric field processing conditions on lycopene, vitamin C and antioxidant capacity of watermelon juice. *Food Chemistry*. 115: 1312–1319.

Pacheco-Palencia, L.A., Duncan, C.E., and Talcott, S.T. 2009. Phytochemical composition and thermal stability of two commercial açai species, *Euterpe oleracea* and *Euterpe precatoria*. *Food Chemistry*. 115: 1199–1205.

Perrone, D., Farah, A., and Donangelo, C.M. 2012. Influence of coffee roasting on the incorporation of phenolic compounds into melanoidins and their relationship with antioxidant activity of the brew. *Journal of Agricultural and Food Chemistry*. 60(17): 4265–4275.

Prakash, S., Jha, S.K., and Datta, N. 2004. Performance evaluation of blanched carrots dried by three different driers. *Journal of Food Engineering*. 62(3): 305–313.
Rawson, A., Patras, A., Tiwari, B.K. et al. 2011. Effect of thermal and non-thermal processing technologies on the bioactive content of exotic fruits and their products: Review of recent advances. *Food Research International*. 44(7): 1875–1887.
Rehman, Z.U., Islam, M., and Shah, W.H. 2003. Effect of microwave and conventional cooking on insoluble dietary fibre components of vegetables. *Food Chemistry*. 80: 237–240.
Reyes, L.F., and Cisneros-Zevallos, L. 2007. Electron-beam ionizing radiation stress effects on mango fruit (*Mangifera indica* L.) antioxidant constituents before and during postharvest storage. *Journal of Agricultural and Food Chemistry*. 55: 6132–6139.
Rocha-Guzmán, N.E., González-Laredo, R.F., Ibarra-Pérez, F.J. et al. 2007. Effect of pressure cooking on the antioxidant activity of extracts from three common bean (*Phaseolus vulgaris* L.) cultivars. *Food Chemistry*. 100(1): 31–35.
Sahlin, E., Savage, G.P., and Lister, C.E. 2004. Investigation of the antioxidant properties of tomatoes after processing. *Journal of Food Composition and Analysis*. 17: 635–647.
Sanjukta, S., Rai, A.K., Muhammed, A. et al. 2015. Enhancement of antioxidant properties of two soybean varieties of Sikkim Himalayan region by proteolytic *Bacillus subtilis* fermentation. *Journal of Functional Foods*. 14: 650–658.
Shahidi, F. 2009. Nutraceuticals and functional foods: Whole versus processed foods. *Trends in Food Science & Technology*. 20(9): 376–387.
Shofian, N.M., Hamid, A.A., Osman, A. et al. 2011. Effect of freeze-drying on the antioxidant compounds and antioxidant activity of selected tropical fruits. *International Journal of Molecular Sciences*. 12(7): 4678–4692.
Soto-Vaca, A., Gutierrez, A., Losso, J.N. et al. 2012. Review: Evolution of phenolic compounds from color and flavor problems to health benefits. *Journal of Agricultural and Food Chemistry*. 60(27): 6658–6677.
Teh, Q.T.M., Tan, G.L.Y., Loo, S.M. et al. 2015. The drying kinetics and polyphenol degradation of cocoa beans. *Journal of Food Process Engineering*. 40: 323–331.
Torino, M.I., Limón, R.I., Martínez-Villaluenga, C. et al. 2013. Antioxidant and antihypertensive properties of liquid and solid state fermented lentils. *Food Chemistry*. 136(2): 1030–1037.
Turkmen, N., Sari, F., and Velioglu, Y.S. 2005. The effect of cooking methods on total phenolics and antioxidant activity of selected green vegetables. *Food Chemistry*. 93(4): 713–718.
Valls-Bellés, V., Torres, M.C., Muñiz, P. et al. 2004. The protective effects of melanoidins in adriamycin-induced oxidative stress in isolated rat hepatocytes. *Journal of the Science of Food and Agriculture*. 84(13): 1701–1707.
Vallverdú-Queralt, A., Regueiro, J., Rinaldi de Alvarenga, J.F. et al. 2014. Home cooking and phenolics: Effect of thermal treatment and addition of extra virgin olive oil on the phenolic profile of tomato sauces. *Journal of Agricultural and Food Chemistry*. 62(14): 3314–3320.
Van Buggenhout, S., Lille, M., Messagie, I. et al. 2006. Impact of pretreatment and freezing conditions on the microstructure of frozen carrots: Quantification and relation to texture loss. *European Food Research and Technology*. 222(5–6): 543–553.

Vattem, D.A., and Shetty, K. 2003. Ellagic acid production and phenolic antioxidant activity in cranberry pomace (*Vaccinium macrocarpon*) mediated by *Lentinus edodes* using a solid-state system. *Process Biochemistry*. 39(3): 367–379.
Vega-Gálvez, A., Ah-Hen, K., Chacana, M. et al. 2012. Effect of temperature and air velocity on drying kinetics, antioxidant capacity, total phenolic content, colour, texture and microstructure of apple (var. Granny Smith) slices. *Food Chemistry*. 132(1): 51–59.
Vilkhu, K., Mawson, R., Simons, L. et al. 2008. Applications and opportunities for ultrasound assisted extraction in the food industry—A review. *Innovative Food Science and Emerging Technologies*. 9(2): 161–169.
Watawana, M.I., Jayawardena, N., and Waisundara, V.Y. 2015. Enhancement of the functional properties of coffee through fermentation by "Tea Fungus"(Kombucha). *Journal of Food Processing and Preservation*. 39(6): 2596–2603.
Wood, O.B., and Bruhn, C.M. 2000. Position of the American Dietetic Association: Food irradiation. *Journal of the American Dietetic Association*. 100: 246–253.
Zenker, M., Heinz, V., and Knorr, D. 2003. Application of ultrasound assisted thermal processing for preservation and quality retention of liquid foods. *Journal of Food Protection*. 66: 1642–1649.
Zhang, D., and Hamauzu, Y. 2004. Phenolics, ascorbic acid, carotenoids and antioxidant activity of broccoli and their changes during conventional and microwave cooking. *Food Chemistry*. 88: 503–509.
Zulueta, A., Esteve, M.J., and Frígola, A. 2010. Ascorbic acid in orange juice–milk beverage treated by high intensity pulsed electric fields and its stability during storage. *Innovative Food Science and Emerging Technologies*. 11(1): 84–90.

CHAPTER 21

Phenolics in Vegetable Oils

Janu Chandran, Nayana N, and P Nisha

CONTENTS

21.1 INTRODUCTION

Oils and fats, along with carbohydrates and proteins, are major components of the human diet. Oils provide energy, fat soluble vitamins (vitamins A, D, and E), and essential fatty acids that are required for proper growth and development. Beneficial micronutrients that function as natural antioxidants such as phospholipids, ascorbic acids, α-tocopherol, β-carotene, and phenolic compounds are also naturally present in vegetable oils. But most of these bioactive and nutritionally significant compounds are removed during oil processing. There are great possibilities for using compounds in oil seeds that are known to have nutritional and health benefits to develop new functional vegetable oils. As most people use a vegetable oil during food preparation, the health benefits to a population, for example, in reducing heart disease, could be substantial.

Any oil is adjudged as "healthy" based on the ratio of saturated/mono unsaturated/polyunsaturated fatty acid, ratio of essential fatty acids (omega 6/omega 3) and presence of natural antioxidants. Plant oils contain small amounts of natural antioxidants such as phenolic compounds, tocopherols, sterols, stanols, phospholipids, waxes, squalene, and other hydrocarbons (Lecker and Rodriguez-Estrada 2000).

Oil-bearing plants have been reported as a good source of phenolic compounds and numerous phenolic compounds have been reported (Alu'datt et al. 2017). However, a major part of the phenolic compounds are lost during various processing of commercial manufacture of oils. Phenolic compounds are very important for the oxidative stability of oils, especially those with polyunsaturated fatty acids. As these compounds are gaining lot of importance owing to their health significances, the phenolic content of edible vegetables oils have been investigated recently. Janu et al. (2014) investigated the phenolic content and antioxidant activity of common edible vegetable oils such as coconut, sunflower, rice bran, groundnut, sesame, and mustard oil. Siger et al. (2008) reported the phenolic profile and antioxidant activities of methanolic extracts of cold-pressed oils such as soybean, sunflower, rapeseed, corn, grapeseed, hemp, flax, rice bran, and pumpkin. However, the phenolic composition of oils/fats except olive oil is not studied in depth.

Extraction of phenolics from the oil is one important stage in the analysis of phenolics in oils. Different methods have been adopted for the extraction of phenolics from oils and fat. Liquid–liquid extraction, using methanol and methanol/water, is the most reported method (Brenes et al. 2000; Siger et al. 2008; Janu et al. 2014). A known quantity of oil is dissolved in a solvent and extracted three times. Solvent extracts may be pooled together and evaporated under vacuum. Siger et al. (2008) used SPE column for the separation of phenolics from oil and the phenolics were eluted from the column with methanol. Extraction of phenolics from olive oil using deep eutectic solvents (DESs) as green extraction method has been reported (García et al. 2016). Several analytical methods such as TLC, HPLC, GC-MS, and LC-MS/MS were employed to identify and quantify the phenolic acids in oils and fats.

The phenolic composition and antioxidant activities of major vegetable oils and some of the minor oils are discussed in the following sections.

21.2 OLIVE OIL

Olive oil, extracted from the fruit of *Olea europea*, is an unsaturated oil. It can be consumed in the crude form as an unrefined (extra virgin/virgin oil) and refined oil. It has a pleasant flavor with a pale green color and high nutritive characteristics. Olive oil is reported to be the richest source of various phenolic compounds and its health benefits have been established by various researchers (Tuck and Hayball 2002). Phenolic content in olive oil depends on the fruit variety, region where it is grown, agro climatic conditions, agricultural techniques used, maturity at which fruit is harvested, the type of oil extraction, processing, storage methods, and time since harvest (Pérez-Rodrigo and Aranceta 2016).

Unlike other vegetable oils, olive oil contains both lipophilic and hydrophilic phenols (Boskou 1996). The lipophilic phenols include tocopherols and tocotrienols, which are commonly found in other vegetables oils. The hydrophilic phenols are less abundant in other oils. The predominant hydrophilic phenols in olive oil are oleuropein, hydroxytyrosol (Hy), tyrosol (Ty), and ligstroside (Tuck and Hayball 2002; Shahidi 1997). Olive oil contain four groups of phenolic compounds with positive bioactive properties: Simple phenols (phenolic acids and phenolic alcohols, the latter from hydrolyzed secoiridoids such as Hy and Ty), secoiridoid derivatives (the aglycone of oleuropein and ligstroside, elenolic acid linked to Hy and Ty [Hy-EA and Ty-EA, respectively], and their respective decarboxylated dialdehydic derivatives), lignans (pino- and acetoxypinoresinol), and flavones (luteolin and apigenin) (Garcia et al. 2016). Phenolic acids such as caffeic, vanillic, syringic, p-coumaric, o-coumaric, protocatechuic, sinapic, and p-hydroxybenzoic acid were found in olive oil, the richest source of phenolic compounds among the vegetable oils (Carrasco-Pancorbo et al. 2005). The total phenolic content in olive oil is reported to be 2 mg/kg in refined oil, 100–800 mg/kg in virgin oil, and more than 1500 mg/kg in extra virgin oil (Tuck and Hayball 2002; Oliveras-Lopez et al. 2008).

β-glucosidase is the key enzyme responsible for the determination of the virgin olive oil phenolic profile as the decrease in its activity after three weeks of storage at 20°C was parallel to a dramatic decrease in the phenolic content of the oils (Hbaieb et al. 2015). Olive phenolics have been reported to contribute to the health benefits of Mediterranean diets. It is reported that phenol-rich virgin olive oil reduces the post-prandial inflammatory response (Mena et al. 2009; Camargo et al. 2014). High phenolic olive oil has been reported to provide beneficial effects on systolic blood pressure and serum oxidative status (oxLDL) (Hohmann et al. 2015). Olive oil phenolics are reported to reduce the risk of cardiovascular diseases by preventing lipo-protein oxidation; however, further cohort studies are warranted to affirm activity (Ruiz-Canela and Martínez-González 2011).

21.3 SESAME OIL

Sesame oils are unsaturated oils that belong to the oleic–linoleic acid group extracted from the seeds of *Sesamum indicum*. Sesame oil contains substantial amounts of unique lignans components, namely, sesamin and sesamolin (Figure 21.1) and minor components sesaminol and sesamolinol. These phytochemicals are reported to contribute to the inherent higher stability of sesame oil when compared to the other unsaturated oils (Budowski 1950; Nuchanart et al. 2010; Moazzami et al. 2006). Due to the high virtual appearance of sesame oil it is called "queen of the oilseed crops." Various other bioactive phytochemicals such as sterols and tocopherols are also present in sesame oil. Studies indicate the beneficial health effects of sesame lignans in lowering of plasma cholesterol levels (Hirata et al. 1996; Chen et al. 2005) and elevation of g-tocopherol levels (Cooney et al. 2001). The seeds are also a rich source of lignan glucosides, mainly sesaminol triglucoside (Moazzami et al. 2006). The range of lignan content of sesame seeds in general is between 224–1148 mg/100 g seed (Moazzami and Kamal-Eldin, 2006).

Coumaric acid, 2-hydroxycinnamic acid, vanillic acid, gallic acid, caffeic acid, ferulic acid, syringic acid, sinapyl alcohol, sinapic acid, *trans* resveratrol, apigenin, luteolin, catechin, epicatechin, quercetin, daidzein, genistein, daidzin, genistin, and sesamin were detected in sesame oil using LC–MS/MS combined with magnetic carboxylated

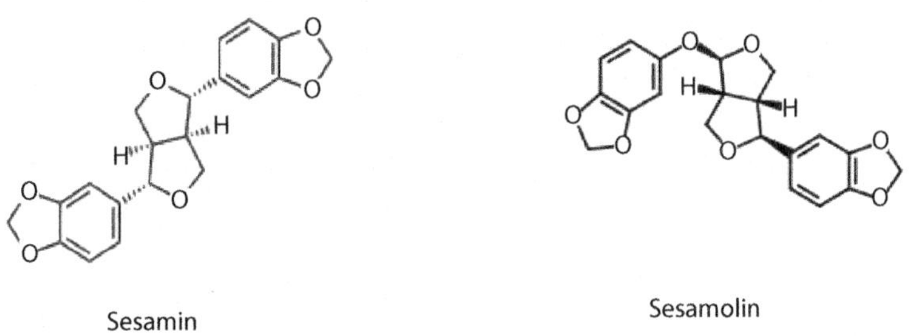

FIGURE 21.1 Chemical structures of sesamin and sesamolin.

multiwalled carbon nanotubes (Wu et al. 2016). The free-radical scavenging activities of sesame oil have been studied and compared with other edible oils (Janu et al. 2014).

21.4 RICE BRAN OIL

Rice bran oil is obtained as a by-product in the rice industry. It is popular in Asian countries including Japan, India, Korea, China, and Indonesia as a cooking oil. Phytochemicals such as tocotrienols, tocopherols, phytosterols, polyphenols, squalene, and gamma-oryzanol, which are reported to have beneficial effects to human health, are present in rice bran oil. The unsaturated portion mainly consists of unsaponifiable matters, which include bioactive phytonutrients such as α, β, γ, and δ-tocopherols, tocotrienols, and γ-oryzanol (Rogers et al. 1993; Seetharamaiah and Chandrasekhara 1990). The strong antioxidant property of c-oryzanols has been widely recognized. Studies have shown several physiological effects of oryzanols, such as reducing plasma cholesterol (Rogers et al. 1993), reducing cholesterol absorption, suppressing early atherosclerosis (Perez-Ternero et al. 2017; Seetharamaiah and Chandrasekhara 1990), inhibiting platelet aggregation (Seetharamaiah and Chandrasekhara 1989), and increasing fecal bile acid excretion (Seetharamaiah and Prabhakar 1986).

Crude rice bran oil also consists of some derivatives of phytosterols and ferulic acid such as cloartenyl ferulate, 24-methylenecycloar-tanyl ferulate, campesteryl ferulate, and b-sitosteryl ferulate (Xu and Godber 1999).

21.5 SOYBEAN OIL

Soybean oil extracted from the seeds of soybeans (*Glycine max*) is the second-most widely consumed cooking oil. The presence of tocopherols of about 930.5 µg/g oil was identified by the HPLC profiling of soybean oil. The total phenolic content of soybean oil is reported to be 1.48 ± 0.05 mg caffeic acid equivalent (CAE)/100 g oil and the major phenolic compounds reported were *p*-hydroxybenzoic, caffeic, vanillic, *p*-coumaric, ferulic, and sinapic acid (Siger et al. 2008).

21.6 COCONUT OIL

The highly saturated coconut oil is obtained from the extraction of *Cocos nucifera* seeds. Indonesia, Srilanka, Philippines, Brazil, Thailand, Malaysia, Vietnam, and some regions of Europe and North America use coconut oil for cooking. The saturated nature of the oil, along with the tocopherols and unsaponifiable matters, strongly correlate with its high oxidative stability. A total phenolic content of 1.8 mg/100 g of oil was reported (Janu et al. 2014). Some of the phenolic compounds, such as ferulic acid, vanillic acid, syringic acid, gallocatechin gallate, *p*-coumaric acid, sinapic acid, and cinnamic acid were identified from the methanolic extract of coconut oil (Janu et al. 2014; Marina et al. 2009). The nonsaponifiable fraction of coconut oil was reported to contain phenolic compounds such as catechin, ferulic acid, and *p*-coumaric acid (Seneviratne et al. 2008). The antioxidant activity of coconut oil has been reported in terms of DPPH, ABTS, superoxide, and nitric oxide radical scavenging activities with IC_{50} values of 229.76 mg/mL, 51.75 mg/mL, 2.76 mg/mL (TEAC value) and 11.14 mg/mL, respectively.

21.7 RAPESEED OIL/MUSTARD OIL

Rapeseed/mustard oil ranks third in levels of oil consumption. It is obtained from the *Brassicaceae* family—rapeseed oil from *Brassica rapa* and mustard oil from *Brassica juncea*. The presence of tocopherols, chlorophylls, β-carotene, and squalene were identified in rapeseed oil (Tuberoso et al. 2007). The total phenolic content of rapeseed oil was reported to be 1.31 ± 0.04 mg caffeic acid equivalent/100 g of oil and its phenolic acids content was 256.6 μg/100 g (Siger et al. 2007). The major phenolic compound in the rapeseed oil was syringic acid (6.8 mg/kg). The total phenolic content of mustard oil is reported to be 0.56 mg/100 g of oil and vanillic acid, *p*-coumaric acid, sinapic acid, and apigenin were the major compounds (Janu et al. 2014). Both rapeseed and mustard oil exhibited antioxidant properties in terms of various radical scavenging assays performed (Siger et al. 2008; Janu et al. 2014).

21.8 PALM OIL

The oil from palm, or *Elaeis guineensis jacquin,* a native of South Africa, is the source of the world's most-consumed oil. The crude oil is a bright orange-red viscous liquid. The typical content of carotenoids in palm oil gives it this unique color. Palm oil is one of the largest sources of vitamin E, which includes tocotrienols (70 percent) and tocopherols (30 percent). The presence of the antioxidant Vitamin E protects in stabilizing the oil towards oxidative rancidity. Palm oil is reported to contain several phenolic acids namely syringic acid, ferulic acid, vanillic acid, *p*-hydroxybenzoic acid, and 3,4-dihydroxybenzaldehyde (Atawodi et al. 2011). The authors also reported that that the oil exhibits promising antioxidant and radical scavenging activities in terms of hypoxanthine/xanthine oxidase and 2-deoxyguanosine assays.

21.9 GRAPE SEED OIL

Grape seed oil is a by-product of wine pressing that is used in human nutrition, cosmetics, the soap industry, and lipochemistry. Oil content is about 6 percent in the dark varieties and about 14 percent in white varieties. The composition of grape seed oil is very similar to that of sunflower oil, with 63 percent linoleic, 22 percent oleic, and 9.5 percent palmitic acids. The phenolic content of grape seed oil is reported to be lower, 0.51 mg CAE/100 g, when compared to other minor oils (Siger et al. 2008). Grape seeds are reported to be a rich source of monomeric phenols such as catechin; epicatechin; and di-, tri, and tetrameric procyanidin (Kim et al. 2006). Numerous in vitro and in vivo evidence suggests antioxidant, anti-inflammatory, cardioprotective, and anticancer effects of grape seed oil (Garavaglia et al. 2016).

21.10 SAFFLOWER OIL

Safflower (*Carthamustintorius, Composae*) has one of the highest linoleic acid contents (75–80 percent) and is utilized as a cooking oil, salad oil, mayonnaise, margarine. and in processed foods. The study conducted by Yu et al. (2013) showed that the total phenolic and flavonoid contents of safflower oil were 126.0 ± 2.4 mg GAE/g and 62.2 ± 1.9 mg QE/g, respectively. The major phenolic compounds reported were (–) epigallocatechin and gallocatechin. Safflower oil exhibited radical scavenging activities, ferric reducing antioxidant power (FRAP), and reducing power in a dose-dependent manner. Moreover, the oxygen radical absorbance capacity (ORAC) value of safflower (0.1 mg/mL) was 62.9 ± 4.7 μM TE (trolox equivalent)/g (Yu et al. 2013).

21.11 FLAX SEED OIL

Flaxseed oil (*Linum usitatissimum* L.) is gaining a lot of recognition in food applications as a functional food ingredient as it is a rich source of omega-3 fatty acids and fiber with beneficial health effects. Choo et al. (2007) showed a total phenolic acid content of 76.8–307.3 mg/100 g in cold-pressed flaxseed oils. The total flavonoid contents in seven samples of cold-pressed flaxseed oils as reported by Choo et al. (2007) ranged from 12.7 to 25.6 mg/100 g, as luteolin equivalents. Vanillin was identified as the most abundant phenolic compound in the flaxseed oil. The other components are nonhydrolyzable (proanthocyanidins) and hydrolyzable tannins, *p*-coumaric acid, ferulic acid, caffeic acid, coniferyl, and syringic aldehyde and small amounts of flavonoids, probably luteolin and kaempferol derivatives (Hasiewicz-Derkacz 2015). Flax seed oil was also reported to contain the lignans matairesinol, pinoresinol, and isolariciresinol, in relatively low levels (Herchi et al. 2011).

REFERENCES

Alu'datt, M.H., Rababah, T., Mohammad N. et al. 2017. A review of phenolic compounds in oil-bearing plants: Distribution, identification and occurrence of phenolic compounds. *Food Chem.* 218: 99–106.

Atawodi, S.E., Yusufu, M.D.L., Atawodi, C.J. et al. 2011. Phenolic compounds and antioxidant potential of Nigerian red palm oil. *International Journal of Biology.* 3: 153–161.

Boskou, D. 1996. *Olive Oil: Chemistry and Technology.* Champaign, IL: AOCS Press.

Brenes, M., Hidalgo, F., Garcia, A. et al. 2000. Pinoresinol and 1-acetoxypinoresinol, Two New Phenolic Compounds Identified in Olive Oil. *JAOCS.* 77: 715–720.

Budowski, P. 1950. Sesame oil. III. Antioxidant properties of sesamol. *Journal of American Oil Chemists Society.* 27: 264–267.

Camargo, A., Rangel-Zuñiga, O.A., Haro, C. et al. 2014. Virgin olive oil phenolic compounds reduce the postprandial inflammatory response in association with postprandial plasma lipopolysaccharide levels. *Food Chem.* 162: 161–171.

Carrasco-Pancorbo, A., Cerretani, L., Bendini, L. et al. 2005. Analytical determination of polyphenols in olive oils. *J. Sep. Sci.* 28: 837–858.

Chen, P.R., Chien, K.L., Su, T.C., Chang, C.J., Liu, T., Cheng, H., and Tsai, C. 2005. Dietary sesame reduces serum cholesteroland enhances antioxidant capacity in hypercholesterolemia. *Nutr Res.* 25: 559–567.

Choo, W.S., Birch, J., and Dufour, J.P. 2007. Physicochemical and quality characteristics of cold-pressed flaxseed oils. *J. Food Compost. Anal.* 20: 202–211.

Cooney, R.V., Custer, L.J., Okinaka, L. et al. 2001. Effects of dietary sesame seeds on plasma tocopherol levels. *Nutrition and Cancer.* 39: 66–71.

Garavaglia, J., Markoski, M.M., Oliveira, A., and Marcadenti, A. 2016. Grape seed oil compounds: Biological and chemical actions for health. *Nutr Metab Insights.* 9: 59–64.

García, A., Rodríguez-Juan, E., Rodríguez-Gutiérrez, G., Rios, J.J., and Fernández-Bolaños, J. 2016. Extraction of phenolic compounds from virgin olive oil by deep eutectic solvents (DESs). *Food Chemistry*. 197: 554–561.

Hasiewicz-Derkacz, K., Kulma, A., Czuj, T. et al. 2015. Natural phenolics greatly increase flax (*Linum usitatissimum*) oil stability. *BMC Biotechnology*. 15: 62.

Hbaieb, R.H., Kotti, F., García-Rodríguez, R. et al. 2015. Monitoring endogenous enzymes during olive fruit ripening and storage: Correlation with virgin olive oil phenolic profiles. *Food Chemistry*. 174: 240–247.

Herchi, W., Sawalha, S., Arráez-Román, D. et al. 2011. Determination of phenolic and other polar compounds in flaxseed oil using liquid chromatography coupled with time-of-flight mass spectrometry. *Food Chemistry*. 126: 332–338.

Hirata, F., Fujita, K., Ishikura, Y. et al. 1996. Hypocholesterolemic effect of sesame lignan inhumans. *Artheriosclerosis*. 122: 135–136.

Hohmann, C.D., Cramer, H., Michalsen, A., and Steckhan, N. 2015. Effects of high phenolic olive oil on cardiovascular risk factors: A systematic review and meta-analysis. *Phytomedicine*. 22: 631–640.

Janu, C., Kumar, D.R., Reshma, M.V., Jayamurthy, P., Sundaresan, A., and Nisha, P. 2014. Comparative study on the total phenolic content and radical scavenging activity of common edible vegetable oils. *J Food Biochem*. 38: 38–49.

Kim, S.J., Jeong, S.M., Park, W.P., Nam, K.C., Ahn, D.U., and Lee, S.C. 2006. Effect of heating conditions of grape seeds on the antioxidant activity of grape seed extracts. *Food Chemistry*. 97: 472–479.

Lecker, G. and Rodriguez-Estrada, M.T. 2000. Chromatographic analysis of unsaponifiable compounds of olive oils and fat-containing foods. *J. Chromatogr. A*. 88: 105–129.

Marina, A.M., Che Man, Y.B., Nazimah, S.A.H., and Amin, I. 2009. Antioxidant capacity and phenolic acids of virgin coconut oil. *International Journal of Food Sciences and Nutrition*. 60(sup2): 114–123.

Mena, M.P., Sacanella, E., Vazquez-Agell, M. et al. 2009. Inhibit ion of circulating immune cell activation: A molecular antiinflammatory effect of the Mediterranean diet. *Am J Clin Nutr*. 89: 248–256.

Moazzami, A.A., Andersson, R.E., and Kamal-Eldin, A. 2006. HPLC analysis of sesaminol glucosides in sesame seeds. *J AgricFood Chem*. 54: 633–638.

Moazzami, A.A., and Kamal-Eldin, A. 2006. Sesame seed is a rich source of dietary lignans. *Journal of American Oil Chemists Society*. 83: 719–723.

Nuchanart, R., Nanthanit, P., and Chulabhorn, M. 2010. Variation of sesamin, sesamolin and tocopherols in sesame (*Sesamum indicum* L.) seeds and oil products in Thailand. *Food Chemistry*. 122: 724–730.

Oliveras-López, M.J., Berná, G., Carneiro, E.M., de la Serrana, H.L.G., Martín, F., and López, M.C. 2008. An extra-virgin olive oil rich in polyphenolic compounds has antioxidant effects in OF1 mice. *The Journal of Nutrition*. 138: 1074–1078.

Pérez-Rodrigo, C., and Aranceta, J. 2016. Olive oil: Its role in the diet. In *Encyclopedia of Food and Health*, pp. 158–166. Oxford: Academic Press.

Perez-Ternero, C., Sotomayor, M.A., and Herrera, M.D. 2017. Contribution of ferulic acid, γ-oryzanol and tocotrienols to the cardiometabolic protective effects of rice bran. *Journal of Functional Food*. 32: 58–71.

Rogers, E.J., Rice, S.M., Nicolosi, R.J. et al. 1993. Identification and quantitation of γ-oryzanol components and simultaneous assessment of tocols in rice bran oil. *Journal of the American Oil Chemists' Society*. 70: 301–307.

Ruiz-Canela, A., and Martínez-González, M.A. 2011. Olive oil in the primary prevention of cardiovascular disease. *Maturitas*. 68: 245–250.

Seetharamaiah, G.S., and Chandrasekhara, N. 1989. Studies on hypocholesterolemic activity of rice bran oil. *Atherosclerosis*. 78: 219–223.

Seetharamaiah, G.S., and Chandrasekhara, N. 1990. Effect of oryzanol on cholesterol absorption and biliary and fecal bile acids in rats. *The Indian Journal of Medical Research*. 92: 471–475.

Seetharamaiah, G.S., and Prabhakar, J.V. 1986. Oryzanol content of Indian rice bran oil and its extraction from soap stock. *Journal of Food Science and Technology*. 23: 270–274.

Seneviratne, K.N., and Sudarshana Dissanayake, D.M. 2008. Variation of phenolic content in coconut oil extracted by two conventional methods. *International Journal of Food Science and Technology*. 43: 597–602.

Shahidi, F. 1997. *Natural Antioxidants: Chemistry, Health Effects and Applications*. Champaign, IL: AOCS Press.

Siger, A., Nogala-Kalucka, M., and Lampart-Szczapa, E. 2008. The content and antioxidant activity of phenolic compounds in cold-pressed plant oils. *J Food Lipids*. 15: 137–149.

Tuberoso, C.I., Kowalczyk, T.A., Sarritzu, E., and Cabras, P. 2007. Determination of antioxidant compounds and antioxidant activity in commercial oilseeds for food use. *Food Chemistry*. 103: 1494–1501.

Tuck, K.L., and Hayball, P.J. 2002. Major phenolic compounds in olive oil: Metabolism and health effects. *J Nutr Biochem*. 13: 636–644.

Wu, R., Ma, F., Zhang, L. et al. 2016. Simultaneous determination of phenolic compounds in sesame oil using LC–MS/MS combined with magnetic carboxylated multi-walled carbon nanotubes. *Food Chemistry*. 204: 334–342.

Xu, Z., and Godber, J.S. 1999. Purification and identification of components of gamma-oryzanol in rice bran oil. *J. Agric. Food Chem*. 47: 2724–2728.

Yu, S.Y., Lee, Y.J., Kim, J.D. et al. 2013. Phenolic composition, antioxidant activity and anti-adipogenic effect of hot water extract from safflower (*Carthamus tinctorius* L.) seed. *Nutrients*. 5: 4894–4907.

Index

Page numbers followed by f and t indicate figures and tables, respectively.